'The classical mechanics of systems of finitely many point particles belongs to the bedrock of theoretical physics every physicist has to be familiar with. Giorgilli's book on Hamiltonian mechanics is a treasure chest. It conveys the author's profound knowledge of the history of the subject and provides a pedagogical exposition of the basic mathematical techniques. Very many important examples of systems are discussed, and the reader is guided towards fairly recent and new developments in this important subject. This book will become a classic.'

— Jürg Fröhlich, *ETH Zürich*

'This is an impressive book by one of the protagonists of the modern theory of dynamical systems. It contains the basic steps of the Hamiltonian theory, but it emphasizes the modern developments of the theory that started with Poincaré and Birkhoff, reaching the study of chaos. The book has two important advantages. It gives the successive steps needed by a beginner who enters into this field, up to its most recent developments. At the same time it provides a remarkable historical account of the developments of the theory. For example, the author uses the Lie series instead of the previous cumbersome canonical transformations, he gives a new classical proof of the Kolmogorov–Arnold–Moser theorem, he considers exponential and superexponential stability and so on. Giorgilli introduced many new ideas in the theory of dynamical systems, but he presents the new developments in a systematic way without emphasizing his own contributions. Finally, the book contains an extensive list of useful references. This book will be of great value for anyone interested in dynamics, and it is absolutely necessary for any library in physics, astronomy and related fields.'

— George Contopoulos, *Academy of Athens*

'This is a book that the reader will refer to over and over again: it provides a theoretical and practical framework for understanding Hamiltonian formalism and classical perturbation theory. It contains a readable complete proof of the most important results in the field (Kolmogorov–Arnold–Moser theorem and Nekhoroshev's theorem) as well as their applications to the fundamental problems of celestial mechanics. It also explains how small divisors are at the origin of the divergence of perturbation series and Poincaré's discovery of homoclinic intersections and of chaotic behaviour in near-to-integrable systems. What a remarkably useful and exciting book!'

— Stefano Marmi, *Scuola Normale Superiore, Pisa*

'This amazing book, written by a prominent master of the theory of Hamiltonian systems, is a wonderful gift for anyone interested in classical dynamics, from a novice student to a sophisticated expert. The author, together with the reader, goes from the definition of canonical equations to such shining peaks as the Kolmogorov theorem on invariant tori and the Nekhoroshev theorem on exponential stability (with complete proofs, and for each of these fundamental theorems two entirely different proofs are presented!). Carefully selected examples and exercises and historical digressions greatly facilitate learning the material and turn reading the book into an intellectual festivity.'

— Mikhail Sevryuk, *Semenov Federal Research Center of Chemical Physics, Moscow*

LONDON MATHEMATICAL SOCIETY STUDENT TEXTS

Managing Editor: Ian J. Leary,
Mathematical Sciences, University of Southampton, UK

64 A short course on Banach space theory, N. L. CAROTHERS
65 Elements of the representation theory of associative algebras I, IBRAHIM ASSEM, DANIEL SIMSON & ANDRZEJ SKOWROŃSKI
66 An introduction to sieve methods and their applications, ALINA CARMEN COJOCARU & M. RAM MURTY
67 Elliptic functions, J. V. ARMITAGE & W. F. EBERLEIN
68 Hyperbolic geometry from a local viewpoint, LINDA KEEN & NIKOLA LAKIC
69 Lectures on Kähler geometry, ANDREI MOROIANU
70 Dependence logic, JOUKU VÄÄNÄNEN
71 Elements of the representation theory of associative algebras II, DANIEL SIMSON & ANDRZEJ SKOWROŃSKI
72 Elements of the representation theory of associative algebras III, DANIEL SIMSON & ANDRZEJ SKOWROŃSKI
73 Groups, graphs and trees, JOHN MEIER
74 Representation theorems in Hardy spaces, JAVAD MASHREGHI
75 An introduction to the theory of graph spectra, DRAGOŠ CVETKOVIĆ, PETER ROWLINSON & SLOBODAN SIMIĆ
76 Number theory in the spirit of Liouville, KENNETH S. WILLIAMS
77 Lectures on profinite topics in group theory, BENJAMIN KLOPSCH, NIKOLAY NIKOLOV & CHRISTOPHER VOLL
78 Clifford algebras: An introduction, D. J. H. GARLING
79 Introduction to compact Riemann surfaces and dessins d'enfants, ERNESTO GIRONDO & GABINO GONZÁLEZ-DIEZ
80 The Riemann hypothesis for function fields, MACHIEL VAN FRANKENHUIJSEN
81 Number theory, Fourier analysis and geometric discrepancy, GIANCARLO TRAVAGLINI
82 Finite geometry and combinatorial applications, SIMEON BALL
83 The geometry of celestial mechanics, HANSJÖRG GEIGES
84 Random graphs, geometry and asymptotic structure, MICHAEL KRIVELEVICH *et al.*
85 Fourier analysis: Part I - Theory, ADRIAN CONSTANTIN
86 Dispersive partial differential equations, M. BURAK ERDOĞAN & NIKOLAOS TZIRAKIS
87 Riemann surfaces and algebraic curves, R. CAVALIERI & E. MILES
88 Groups, languages and automata, DEREK F. HOLT, SARAH REES & CLAAS E. RÖVER
89 Analysis on Polish spaces and an introduction to optimal transportation, D. J. H. GARLING
90 The homotopy theory of $(\infty, 1)$-categories, JULIA E. BERGNER
91 The block theory of finite group algebras I, M. LINCKELMANN
92 The block theory of finite group algebras II, M. LINCKELMANN
93 Semigroups of linear operators, D. APPLEBAUM
94 Introduction to approximate groups, M. C. H. TOINTON
95 Representations of finite groups of Lie type (2nd Edition), F. DIGNE & J. MICHEL
96 Tensor products of C*-algebras and operator spaces, G. PISIER
97 Topics in cyclic theory, D. G. QUILLEN & G. BLOWER
98 Fast track to forcing, M. DŽAMONJA
99 A gentle introduction to homological mirror symmetry, R. BOCKLANDT
100 The calculus of braids, P. DEHORNOY
101 Classical and discrete functional analysis with measure theory, M. BUNTINAS
102 Notes on Hamiltonian dynamical systems, ANTONIO GIORGILLI
103 A course in stochastic game theory, EILON SOLAN

London Mathematical Society Student Texts 102

Notes on Hamiltonian Dynamical Systems

ANTONIO GIORGILLI
University of Milan

CAMBRIDGE
UNIVERSITY PRESS

University Printing House, Cambridge CB2 8BS, United Kingdom

One Liberty Plaza, 20th Floor, New York, NY 10006, USA

477 Williamstown Road, Port Melbourne, VIC 3207, Australia

314–321, 3rd Floor, Plot 3, Splendor Forum, Jasola District Centre,
New Delhi – 110025, India

103 Penang Road, #05–06/07, Visioncrest Commercial, Singapore 238467

Cambridge University Press is part of the University of Cambridge.

It furthers the University's mission by disseminating knowledge in the pursuit of education, learning, and research at the highest international levels of excellence.

www.cambridge.org
Information on this title: www.cambridge.org/9781009151146
DOI: 10.1017/9781009151122

First published 2022

A catalogue record for this publication is available from the British Library.

ISBN 978-1-009-15114-6 Hardback
ISBN 978-1-009-15113-9 Paperback

Contents

Apology xiii

Plan of the Book xv

Expressions of Gratitude xviii

1 Hamiltonian Formalism 1

1.1 Phase Space and Hamilton's Equations 1

Autonomous versus Non-autonomous Systems (2) – Connection with the Lagrangian Formalism (3) – Compact Notation for the Canonical Equations (10)

1.2 Dynamical Variables and First Integrals 10

The Algebra of Poisson Brackets (11) – First Integrals (14)

1.3 Use of First Integrals 21

Motion on a One-Dimensional Manifold (21) – Systems with One Degree of Freedom (25) – The Period of Oscillation (26) – Higher-Dimensional Models (28)

2 Canonical Transformations 31

2.1 Preserving the Hamiltonian Form of the Equations 32

Conditions for Canonicity (32) – Symplecticity of the Jacobian Matrix (35) – Preservation of Poisson Brackets (36) – The Hamiltonian Flow as a Canonical Transformation (39) – Preservation of Lagrange Brackets (40)

2.2 Differential Forms and Integral Invariants 42

Preservation of a 2-Form (42) – Action Integral (43)

2.3 Generating Functions 45

A First Form of Generating Function (46) – A Second Form of the Generating Function (47) – The General Class of Generating Functions (50)

2.4 Time-Dependent Canonical Transformations 52

Transformations Which Leave Time Unchanged (53) – Using the Fundamental Poisson Brackets (54) – Time-Depending

Generating Functions (55) – Changing the Time Variable (55)

2.5 *The Hamilton–Jacobi Equation* 57

3 Integrable Systems 63

3.1 *Involution Systems* 65

Some Useful Lemmas (65) – Variational Equations and First Integrals (67) – Commutation of Canonical Flows (69) – Coordinates on Invariant Manifolds (72) – Local Coordinates in Phase Space (73) – Liouville's Canonical Coordinates (75) – Changing the Involution System (77)

3.2 *Liouville's Theorem* 78

Integration Procedure (79) – Some Comments on Liouville's Theorem (83) – Action-Angle Variables for Systems with One Degree of Freedom (84)

3.3 *On Manifolds with Non-Singular Vector Fields* 88

The Flow in the Large (89) – The Stationary Group (91) – Angular Coordinates (91)

3.4 *Action-Angle Variables* 92

Periods in a Neighbourhood of the Torus (93) – Global Coordinates and Action-Angle Variables (98) – Non-uniqueness of the Action-Angle Variables (99) – Explicit Construction of Action-Angle Variables (100)

3.5 *The Arnold–Jost Theorem* 102

3.6 *Delaunay Variables for the Keplerian Problem* 102

Determination of Cycles (103) – Construction of the Action Variables (104) – Delaunay variables (105) – Construction of the Angle Variables (106)

3.7 *The Linear Chain* 107

A Complete Involution System (108) – The Frequencies of the Linear Chain (109)

3.8 *The Toda Lattice* 112

Lax Pairs (112) – First Integrals for the Toda Lattice (113)

4 First Integrals 117

4.1 *Periodic and Quasi-Periodic Motion on a Torus* 118

Motion on a Two-Dimensional Torus (119) – The Many-Dimensional Case (121) – Changing the Frequencies on a Torus (122) – Dynamics of the Kronecker Flow (123)

4.2 The Kronecker Map 124
General Properties (124) – Resonances (125) – Dynamics of the Kronecker Map (127)

4.3 Ergodic Properties of the Kronecker Flow 128
Time Average and Phase Average (129) – Quasiperiodic Motions (131)

4.4 Isochronous and Anisochronous Systems 132
Anisochronous Non-degenerate Systems (132) – Isochronous Systems (133)

4.5 The Theorem of Poincaré 134
Equations for a First Integral (135) – Nonexistence of First Integrals (137)

4.6 Some Remarks on the Theorem of Poincaré 140
On the Genericness Condition (140) – A Puzzling Example (142) – Truncated First Integrals (144)

5 Nonlinear Oscillations 147

5.1 Normal Form for Linear Systems 148
The Classification of Poincaré (149) – Normal Form of a Quadratic Hamiltonian (153) – The Linear Canonical Transformation (154) – Complex Normal Form of the Hamiltonian (157) – First Integrals (158) – The Case of Real Eigenvalues (158) – The Case of Pure Imaginary Eigenvalues (159) – The Case of Complex Conjugate Eigenvalues (160)

5.2 Non-linear Elliptic Equilibrium 163
Use of Action-Angle Variables (163) – A Formally Integrable Case (165)

5.3 Old-Fashioned Numerical Exploration 168
The Galactic Models of Contopoulos and of Hénon and Heiles (169) – Poincaré Sections (171) – The Onset of Chaos (172) – Stability of a Non-linear Equilibrium (178) – Exploiting the First Integrals (183)

5.4 Quantitative Estimates 187
Algebraic and Analytic Setting (187) – Analytical Estimates (188) – Truncated Integrals (189) – Exponential Estimates (192)

6 The Method of Lie Series and of Lie Transforms 195

6.1 Formal Expansions 196

6.2 Lie Series 198

The Lie Series Operator (199) – The Triangle of Lie Series (201) – Composition of Lie Series (202)

6.3 Lie Transforms 206

The Triangular Diagram for the Lie Transform (207)

6.4 Analytic Framework 212

Cauchy estimates (212) – Complexification of Domains (213) – Generalized Cauchy Estimates (215)

6.5 Analyticity of Lie Series 218

Convergence of Lie Series (218) – Analyticity of the Composition of Lie Series (221)

6.6 Analyticity of the Lie Transforms 222

Convergence of the Lie Transforms (222)

6.7 Weighted Fourier Norms 226

Analytic Fourier Series (226) – Generalized Cauchy Estimates (227) – Quantitative Estimates for Lie Series (231)

7 The Normal Form of Poincaré and Birkhoff 233

7.1 The Case of an Elliptic Equilibrium 233

The Formal Algorithm (234) – The Solution of Poincaré and Birkhoff (237) – The Canonical Transformation (237) – First Integrals (238) – Back to the Direct Construction of First Integrals (240)

7.2 Action-Angle Variables for the Elliptic Equilibrium 243

The Normal Form (244) – The Non-resonant Case (244) – The Resonant Case (245) – An Example of Resonant Hamiltonian (246)

7.3 The General Problem 251

The Naive But Inconclusive Approach (251) – The Case of Constant Frequencies (252) – Expansion in Trigonometric Polynomials (254) – Some Analytical Estimates (257) – Truncated Normal Form (258) – Using Composition of Lie Series (259)

7.4 The Dark Side of Small Divisors 261

The Accumulation of Small Divisors (261)

8 Persistence of Invariant Tori 267

8.1 The Work of Kolmogorov 268

Statement of the Theorem (268) – The Normal Form of Kolmogorov (269) – Sketch of the Formal Scheme (269) – The Method of Fast Convergence (270)

8.2 The Proof According to the Scheme of Kolmogorov 272

Analytic Setting (272) – Formal Algorithm (275) – Iterative Lemma (278) – Lemma on Small Divisors (279) – Proof of the Iterative Lemma (281) – Conclusion of the Proof of Proposition 8.3 (285)

8.3 A Proof in Classical Style 289

The Formal Constructive Algorithm (290) – Quantitative Estimates (298) – Estimates for the Generating Functions (299) – The General Scheme of Estimates (301) – The Accumulation of Small Divisors (304) – The Kindness of Small Divisors (305) – Small Divisors in the Algorithm of Kolmogorov (308) – A Detailed Statement (311) – Recurrent Estimates (312) – Completion of the Proof (314) – Estimate of the Sequence ν (316)

8.4 Concluding Remarks 318

A General Scheme (319) – A Comment on Lindstedt's Series (321)

9 Long Time Stability 323

9.1 Overview on the Concept of Stability 323

A Short Historical Note (326)

9.2 The Theorem of Nekhoroshev 327

Statement of the Theorem (328) – An Overview of the Method (329)

9.3 Analytic Part 331

Formal Scheme (331) – An Explicit Expression for the Remainder (333) – Truncated First Integrals (333) – Quantitative Estimates on the Generating Sequence (334) – Estimates for the Normal Form (338) – Local Stability Lemma (339)

9.4 Geometric Part 341

Geography of Resonances (342) – Three Technical Lemmas (349) – Geometric Properties of the Geography of Resonances (352)

9.5 The Exponential Estimates 355
Global Estimates Depending on Parameters (356) – Choice of the Parameters and Exponential Estimate (358)

9.6 An Alternative Proof by Lochak 360

10 Stability and Chaos 363

10.1 The Neighbourhood of an Invariant Torus 364
Poincaré–Birkhoff Normal Form (365) – Exponential Stability (367) – Superexponential Stability (369)

10.2 The Roots of Chaos and Diffusion 370
Asymptotic Orbits (371) – Stable and Unstable Manifolds (374) – Existence of a Homoclinic Orbit (375) – Existence of a Second Homoclinic Orbit (379) – Existence of Infinitely Many Homoclinic Orbits (380) – Heteroclinic Orbits and Diffusion (383)

10.3 An Example in Dimension 2: the Standard Map 386
Boxes into Boxes (386) – Homoclinic and Heteroclinic Tangles (390) – The Case of Higher Dimension (397)

10.4 Stability in the Large 399
Statement of the Theorem (400) – Sketch of the Proof (401)

10.5 Some Final Considerations 403

A The Geometry of Resonances 407

A.1 Discrete Subgroups and Resonance Moduli 407
Construction of a Basis of a Discrete Subgroup (409) – Unimodular Matrices (411) – Resonance Moduli (412) – Basis of a Resonance Module (412)

A.2 Strong Non-resonance 413
A Result from Diophantine Theory (415) – The Condition of Bruno (416) – Some Examples (417)

B A Quick Introduction to Symplectic Geometry 421

B.1 Basic Elements of Symplectic Geometry 421
The Symplectic Group (422) – Symplectic Spaces and Symplectic Orthogonality (422) – Canonical Basis of a Symplectic Space (425) – Properties of the Subspaces of a Symplectic Space (426)

References 429

Index 445

Apology

Durissima est hodie conditio scribendi libros Mathematicos, præcipue Astronomicos. Nisi enim servaveris genuinam subtilitatem propositionum, instructionum, demonstrationum, conclusionum, liber non erit Mathematicus: sin autem servaveris, lectio efficitur morosissima, præsertim in Latina lingua, quæ caret articulis, & illa gratia quam habet Græca, cum per signa literaria loquitur.

Johannes Kepler, *Astronomia Nova*

This book is a somewhat reordered collection of notes accumulated over the years while I was preparing my lectures at different levels, ranging from second-year faculty courses on Mechanics, to more advanced courses on Celestial Mechanics and Dynamical Systems, to series of lectures at the post-graduate and PhD level on perturbation methods. I should stress that it is neither a treatise nor an encyclopedic collection: the objective is mainly didactical. The guiding line is to make a journey from a basic knowledge of the Hamiltonian formalism, without assuming it to be known in advance to the reader, up to some most recent results in the field of Dynamical Systems, in particular in the Hamiltonian framework.

The choice of the arguments reflects my personal experience, both in research and in teaching; therefore it is unavoidably incomplete. The best way to teach, in my modest opinion, is to respect the historical development, whenever possible – which I did only partially. I tried to follow the development of our knowledge starting with the dream of the nineteenth century – to find an effective way to solve the problems of Mechanics and in particular of the dynamics of the Planetary System – and ending with the amazing coexistence of order and chaos that has been discovered by Poincaré, but has progressively become known only in the second half of the past century and nowadays is an inexhaustible argument of research and of many books and conferences. The style of exposition is that of a lecture or a talk; I attempt to place the mathematical results in the context (in a broad sense, not merely mathematical) of the problems that they answer.

All this may explain the perhaps surprising epigraph at the beginning of this apology. The idea, to quote the sentence of Kepler – with all due respect to that great mathematician and astronomer, if he cares to accept my apologies – is closely linked to the style of the presentation. The text mixes technical sections, including formal statements and detailed proofs of the main theorems, with more discursive parts including the results of numerical experiments and, here and there, some historical digression; in this sense it is not written as a typical mathematical book. So to speak, I tried to use theorems in order to explain the outcome of numerical experiments, and numerical experiments in order to suggest how to interpret the theorems. At the same time, I used history in order to instill a feeling that our current knowledge is only a provisional step on a long way paved with problems and questions, successes and failures, moments of enthusiasm or frustration, small steps forward and, from time to time, some profound intuition or brilliant discovery that we owe to great scientists and which indicates the way for the future.

A last comment on the epigraph. A reader who considers the reference to Latin and Greek languages as outdated is definitely right; but let me suggest to replace *Latina lingua* with *'CTL'* (Common Technical Language, that funny communication tool, resembling English, which is popular in the scientific community), and to replace *Græca* with the reader's own native or preferred language.

My hope is that the reader – a student, say – will find this book a companion and a useful guide in the journey that I propose. But it is not my intention to invite the reader – in particular a reader wanting to spend part of his or her life in scientific research – to stop at the end of the book. I will consider my objectives as fully reached when the reader concludes that my notes are too limited and casts this book into some trunk in the attic, looking forward to something better or perhaps to adding a next step.

Locutus sum in corde meo dicens:
ecce magnus effectus sum
et præcessi sapientia omnes
qui fuerunt ante me in Hierusalem,
et mens mea contemplata est multa sapienter et didicit,
dedique cor meum ut scirem prudentiam
atque doctrinam erroresque et stultitiam,
et agnovi quod in his quoque esset labor et adflictio spiritus,
eo quod in multa sapientia multa sit indignatio,
et qui addit scientiam addit et laborem.

Qohelet 1, 16–18

Plan of the Book

The present notes may be considered as structured in three parts with an *intermezzo* – like an *opera*.

The first part is composed of Chapters 1 and 2. The arguments covered are essentially the ones developed in the second half of the nineteenth century, up to Poincaré.

Chapter 1 is like an *overture*; the Hamiltonian formalism is introduced from scratch. It is addressed mainly to undergraduate faculty students, assuming that they have a basic knowledge of Analysis and of the Lagrangian methods of Mechanics, but not necessarily of the Hamiltonian formalism. Readers who are already familiar with the canonical formalism are encouraged to skip this chapter.

Chapter 2 deals with canonical transformations. The aim is to provide the practical tools for the rest of the book; therefore it intends to go a little beyond a short exposition such as can be usually found in many first-level textbooks on Mechanics; at the same time there is no pretension to completeness. The purpose is to present the argument in a self-consistent form and in traditional terms, so that the exposition is reasonably simple – at least according to the author's experience.

The second part consists of three chapters. It is concerned with the transition from the dream of solving the equations of any mechanical system to the discovery that this was only an unachievable dream.

Chapter 3 is devoted to classical integration methods based on use of first integrals. Its historical location begins with the middle of the nineteenth century, with the theorem of Liouville, and extends till a century later, jumping to the theorem of Arnold and Jost on existence of invariant tori which characterizes most integrable Hamiltonian systems – the modern version of the epicycles of Greek Astronomy.

Chapter 4 is a complement and a counterpart of the previous one. On the one hand it illustrates the Kronecker flow on tori that characterizes integrable systems; on the other hand it discusses in detail the theorem of Poincaré on nonexistence of first integrals which represents a first breach on the dream of integrability.

Chapter 5 deals with the dynamics in a neighbourhood of an equilibrium, paying attention to the problem of stability. The purpose is to create a bridge between the first half and the second half of the twentieth century. In view of the didactical purposes the first section is devoted to a discussion of the methods of solution of linear systems of differential equations, revisiting the classification of Poincaré. A detailed discussion of the Hamiltonian case is included. The rest of the chapter is concerned with the dynamics of a nonlinear system around an elliptic equilibrium. Its content is both theoretical, with a discussion of the existence of formal first integrals, and numerical, including both the methods based on the Poincaré section and the methods of algebraic manipulation. The last section introduces some analytical estimates, which in this case are not particularly difficult but open the way towards the strong results of the second half of the twentieth century.

Chapter 6 is a sort of technical *intermezzo*. It provides an introduction to the methods of Lie series and Lie transform that will be used in the next chapters. In particular it provides the basic analytical tools which allow us to prove the two main theorems of the next chapter. The formal methods discussed in this chapter have been developed starting around 1950, mainly in the milieu of Celestial Mechanics.

The third part begins with Chapter 7 and extends to the end of the notes. Most of its content is concerned with research developed in the second half of the twentieth century, with a remarkable exception: the homoclinic orbits discovered by Poincaré.

Chapter 7 introduces the methods of normal form for perturbed, close-to-integrable systems – the *general problem of dynamics*, as it has been qualified by Poincaré. The purpose is to provide an exposition of the methods developed by Poincaré and Birkhoff, and to discuss the long-standing problem of resonances and of small divisors. It also includes some comments concerning the way out from the difficulties raised by the theorem of Poincaré which will be fully developed in the next chapters.

Chapter 8 is devoted to the proof of the celebrated theorem of Kolmogorov on persistence of invariant tori. The first part of the chapter illustrates the concept of the Kolmogorov normal form, together with the fast convergence method which allowed him to solve a two-century-old problem. The proof of Kolmogorov is reproduced in complete form, only replacing the traditional method of generating functions in mixed variables with the method of Lie series. The second part provides a proof in a classical style, avoiding the fast convergence method, which hopefully sheds light on the behaviour of small divisors.

Chapter 9 deals with a further celebrated result: the exponential stability of Nekhoroshev. The theorem may be seen as a complement of Kolmogorov's one: it assures long-time stability for orbits in an open set, not just on strongly nonresonant invariant tori. The price to be paid is that stability is assured only for finite times, but hopefully is very long. The exposition is very technical but provides a complete proof.

Chapter 10, the final one, represents an attempt – a partial one – to illustrate how chaos and order may coexist in a perturbed Hamiltonian system.

The exposition is less technical with respect to the two previous chapters. It contains two basic pieces of information. The first one is an enhancement of the exponential stability of Nekhoroshev, showing that Kolmogorov's invariant tori are superexponentially stable. The second one is a reproposition of the last chapter of *Méthodes Nouvelles* of Poincaré, with the discovery of the phenomenon of homoclinic intersection and of chaotic behaviour.

The last chapter, Chapter 10, is intended as an encouragement to the reader who wants to keep exploring the Dynamics, thus filling the many holes that are left in the present notes and making a step forward. Therefore, it is not merely the end of the book: my hope is it will represent the beginning of a new journey. I will consider this as the best success of my work.

Note: Proofs of theorems, propositions, lemmas and corollaries are terminated by *'Q.E.D.'*. Examples are terminated by *'E.D.'*. Exercises are terminated by *'A.E.L.'*.

Expressions of Gratitude

Some years ago – it was between 1972 and 1973 – I was a student of Physics at the University of Milan. There I happened to meet my mentor Luigi Galgani – since then my good friend, whom I consider my scientific father. He was interested in the dynamics of the FPU system and in a work by P. Bocchieri, A. Scotti, B. Bearzi and A. Loinger on that subject, published a couple of years before. Coming back from a meeting in Cagliari, where Luigi Galgani and Antonio Scotti had met G. Contopoulos and learned from him about his methods of construction of a third integral, Luigi told a few people about Contopoulos's ideas. I was among these people, due to my dawning interest in the FPU problem; I went away so impressed that eventually I decided to begin my thesis work by studying the method of Contopoulos. A few months later, still during my thesis work, I was lucky to meet Contopoulos personally. My encounter with Luigi Galgani and George Contopoulos marked the beginning of my career; they are the first people to whom I must express my deep gratitude. This book represents the evolution of that lucky event, a bifurcation point in my life.

During the first years of my career I had a long and fruitful collaboration with Giancarlo Benettin and Jean Marie Strelcyn, together with Luigi. Our common work on Kolmogorov's and Nekhoroshev's theorems lies at the very basis of a significant part of the present notes. In more recent years I cultivated my interest in Celestial Mechanics. In that field I met three excellent students (among some others): Alessandro Morbidelli, Ugo Locatelli and Marco Sansottera. The present notes include also work that was developed thanks to their valuable collaboration.

I could add a long list of people I've met over the years, but it would be too long. I say that I met many people in a broad sense: an encounter may be direct – seminars, meetings and reciprocal visits are excellent opportunities to discuss with many people and to create a collaboration – or indirect – old books may establish a connection with scientists of the past and are a valuable source of knowledge. Some of them are mentioned in the references at the end of the book.

Now a few words concerning the preparation of this book. A few years ago Jürg Fröhlich told me, 'You should write a book.' I had written a considerable part of my notes at that time, making them available to students, but I constantly refrained from publishing them as a book despite the advice of friends and colleagues. Eventually I took Fröhlich's advice seriously: it was a good reason for undertaking such a job.

As I said at the beginning of my apology, the book is a reordered and partially rewritten collection of notes accumulated over many years. I have profited from the remarks, sometimes the criticism, of many students. I can not remember all their names, hence I do not name them, not to be unfair to anyone; but if my notes have been improved over time, this is due also to all of them.

Ugo Locatelli, Marco Sansottera, Simone Paleari, Giancarlo Benettin, Giuseppe Pucacco and Giuseppe Gaeta have taken the burden of reading at least part of the notes; their remarks and suggestions have been extremely useful.

Finally, I am grateful to my family: Anna, my wife; Cristina and Elena, my daughters; Elisa, Sergio and Pietro, my grandchildren. Perhaps they had no direct influence on the contents of this book, but they have helped create the family atmosphere and have endured my frequent moments of inattention when I was focused on writing. Without their patience I would never have finished this job.

In the run-up to the conclusion of the editorial process for this book, I would like to thank the editor Kaitlin Leach, at Cambridge University Press, for her courteous and valuable assistance.

1
Hamiltonian Formalism

A major role in the development of the theory of classical dynamical systems is played by the Hamiltonian formulation of the equations of dynamics. This chapter is intended to provide a basic knowledge of the Hamiltonian formalism, assuming that the Lagrangian formalism is known. A reader already familiar with the canonical formalism may want to skip the present chapter.

The canonical equations were first written by Giuseppe Luigi Lagrangia (best known by the French version of his name, Joseph Louis Lagrange) as the last improvement of his theory of secular motions of the planets [140]. The complete form, later developed in what we now call Hamiltonian formalism, is due to William Rowan Hamilton [106][107][108]. A short sketch concerning the anticipations of Hamilton's work can be found in the treatise of Edmund Taylor Whittaker [209], §109.

In view of the didactical purpose of the present notes, the exposition in this chapter follows the traditional lines. The chapter includes some basic tools: the algebra of Poisson brackets and the elementary integration methods. Many examples are also included in order to illustrate how to write the Hamiltonian function for some models, often investigated using Newton's or Lagrange's equations.

1.1 Phase Space and Hamilton's Equations

The dynamical state of a system with n degrees of freedom is identified with a point on a $2n$-dimensional differentiable manifold, denoted by $\mathscr{F}$, endowed with canonically conjugated coordinates $(q, p) \equiv (q_1, \ldots, q_n, p_1, \ldots, p_n)$. The object of investigation is the evolution of the state of the system. The manifold $\mathscr{F}$ was named a *phase space* by Josiah Willard Gibbs [75].

The evolution of the system is determined by a real-valued Hamiltonian function $H\colon \mathscr{F} \to \mathbb{R}$ through the vector field defined by *Hamilton's equations* (also called *canonical equations*),

$$\dot{q}_j = \frac{\partial H}{\partial p_j}\ ,\quad \dot{p}_j = -\frac{\partial H}{\partial q_j}\ ,\quad j = 1, \ldots, n\ , \tag{1.1}$$

where $H = H(q,p)$. The Hamiltonian will be assumed to be a smooth (differentiable) function. In most examples the cases of $C^\infty(\mathscr{F}, \mathbb{R})$ or even $C^\omega(\mathscr{F}, \mathbb{R})$ Hamiltonians will be considered.

An *orbit* of the system is a smooth curve $\bigl(q(t), p(t)\bigr)$, for t in some (possibly infinite) interval, which is a solution of the canonical equations (1.1); the initial condition is written $q(0) = q_0$, $p(0) = p_0$. As a general fact, the solutions of a system of differential equations may be interpreted as a flow in phase space which transports every point (q_0, p_0) to another point $\bigl(q(t), p(t)\bigr) = \phi^t(q_0, p_0)$ after a time interval t, thus representing in some sense the dynamics in phase space as the motion of a fluid; the symbol ϕ^t represents the action of the flow. The corresponding orbit is thus represented as $\Omega(q_0, p_0) = \bigcup_t \phi^t(q_0, p_0)$, the union being made over the (possibly infinite) time interval of existence of the solution.

1.1.1 Autonomous versus Non-autonomous Systems

In most treatises the general case of a time-dependent Hamiltonian $H(q, p, t)$ is considered. It is customary to call such a system *non-autonomous*, in contrast with the time-independent case that is called *autonomous*.[1]

As a matter of fact, the non-autonomous case can be reduced to the autonomous one by a standard technique, already used by Henri Poincaré ([190], vol. I, § 12). The suggestion is to introduce the *extended phase space* by adding one more pair of canonically conjugated coordinates q_+, p_+, thus increasing the dimension of the phase space by 2; the special role played by the new variables will be emphasized by the special notation (q, q_+, p, p_+), although all canonical pairs should be considered on the same footing. Having given the Hamiltonian $H(q, p, t)$, let us introduce a new Hamiltonian

$$\tilde{H}\bigl(q, q_+, p, p_+\bigr) = H(q, p, q_+) + p_+\ . \tag{1.2}$$

---

[1] The non-autonomous case typically arises when one considers a small system subjected to the time-dependent action of some external device, the state of which is not affected by the small part one is interested in. Examples are the restricted problem of three bodies, where the motion of a small body (e.g., an asteroid or a spacecraft) under the action of two big bodies moving on Keplerian orbits (e.g., Sun–Jupiter or Earth–Moon) is investigated; an electric charge acted on by an electromagnetic wave; a particle in an accelerator; &c. The autonomous case arises when an isolated system is considered.

Thus, one has to consider the extended set of canonical equations

$$
\begin{aligned}
\dot{q}_j &= \frac{\partial H}{\partial p_j}\ , \quad \dot{p}_j = -\frac{\partial H}{\partial q_j}\ , \quad j = 1, \ldots, n\ , \\
\dot{q}_+ &= 1 \quad\ , \quad \dot{p}_+ = -\frac{\partial H}{\partial q_+}\ .
\end{aligned}
\tag{1.3}
$$

The third equation has the trivial solution $q_+(t) = t - t_0$, where t_0 is the initial time. In view of this, and having fixed the initial time t_0, the first two equations actually coincide with (1.1). Suppose now that we know a solution $q(t), p(t)$ of (1.1); if we replace it in the equation for $\dot{p}_+$, the r.h.s. turns out to be a known function of t only, so that the equation can be solved by a quadrature.[2] Therefore, the autonomous system (1.1) and the non-autonomous one (1.3) are fully equivalent.

Extending the phase space in order to transform a non-autonomous system into an autonomous one may appear to be an unnecessary complication: we have one more variable to take care of, although in a straightforward manner. However, this will offer us the opportunity of developing most of the theory in the framework of the autonomous systems, which makes the exposition definitely simpler.

1.1.2 Connection with the Lagrangian Formalism

It is traditional to introduce the Hamiltonian formalism starting from the equations of Lagrange.[3] The Hamiltonian function is introduced as the Legendre transform of the Lagrangian. One considers an n-dimensional differentiable manifold (the *configuration space*) endowed with (local) coordinates $q_1, \ldots, q_n$, and its tangent space described by the generalized components $\dot{q}_1, \ldots, \dot{q}_n$ of the velocities. The dynamical state of the system

---

[2] Actually no quadrature is needed, for an autonomous Hamiltonian $H(q,p)$ is a first integral, i.e., along the orbit $q(t), p(t)$ one has $H\bigl(q(t), p(t)\bigr) = E$, where $E = H\bigl(q(0), p(0)\bigr)$ is the initial value – in most mechanical models it is the energy of the system. Since the Hamiltonian $\tilde{H}(q, q_+, p, p_+)$ in (1.2) is seen as an autonomous one we get the first integral

$$\tilde{H}(q(t), t - t_0, p(t), p_+(t)) = H(q(t), p(t), t) + p_+(t) = C\ .$$

The dynamics does not depend on the value of the constant C; hence we may well restrict the study to the manifold $\tilde{H} = 0$; this gives $p_+(t) = -H(q(t), p(t), t)$. Thus, the canonical momentum p_+ conjugated to the time t is identified with the energy E, which depends on time.

[3] For a deduction of the Lagrangian form of the equations of a mechanical system from Newton's equations see, for instance, [209], § 26. The original formulation can be found in Lagrange's treatise [139] and [143]. For a deduction of Hamilton's equations see [209], § 109 or [144], ch. VI.

at a given time t is completely determined by the knowledge of $q(t), \dot{q}(t)$. The dynamics is determined by the Lagrangian function $L(q, \dot{q}, t)$ through the n differential equations of second order

$$\frac{d}{dt}\frac{\partial L}{\partial \dot{q}_j} - \frac{\partial L}{\partial q_j} = 0 , \quad 1 \le j \le n . \tag{1.4}$$

In order to give the equations the Hamiltonian form, one introduces the momenta $p_1, \ldots, p_n$ conjugated to $q_1, \ldots, q_n$ defined as

$$p_j = \frac{\partial L}{\partial \dot{q}_j} , \quad 1 \le j \le n ; \tag{1.5}$$

thus, the momenta are given as functions of q, $\dot{q}$ and t. The latter equations together with

$$\dot{p}_j = \frac{\partial L}{\partial q_j} , \quad 1 \le j \le n \tag{1.6}$$

are equivalent to the equations of Lagrange (1.4). If the condition

$$\det\left(\frac{\partial^2 L}{\partial \dot{q}_j \partial \dot{q}_k}\right) \ne 0$$

is fulfilled, then (1.5) can be solved with respect to $\dot{q}_1, \ldots, \dot{q}_n$, thus giving $\dot{q}_j = \dot{q}_j(q, p, t)$, and the momenta can be used in place of the velocities $\dot{q}$ in order to determine the dynamical state.

The Hamiltonian function is defined as

$$H(q, p, t) = \sum_{j=1}^{n} p_j \dot{q}_j - L(q, \dot{q}, t)\Big|_{\dot{q}=\dot{q}(q,p,t)} , \tag{1.7}$$

where $\dot{q}$ must be replaced everywhere with its expression as a function of q, p, t. With the latter function the dynamics is determined by the canonical equations (1.1). The latter claim is proved as follows. Differentiate the function H as depending on the $3n+1$ variables q, $\dot{q}$, p and t, and then substitute the functions $\dot{q}(q, p, t)$, thus getting

$$\begin{aligned} dH &= \sum_{j=1}^{n} \left(\dot{q}_j \, dp_j + p_j \, d\dot{q}_j - \frac{\partial L}{\partial \dot{q}_j} \, d\dot{q}_j - \frac{\partial L}{\partial q_j} \, dq_j \right) - \frac{\partial L}{\partial t} \, dt \\ &= \sum_{j=1}^{n} \left(\dot{q}_j \, dp_j - \frac{\partial L}{\partial q_j} \, dq_j \right) - \frac{\partial L}{\partial t} \, dt . \end{aligned}$$

Here the definition, (1.5), of the momenta p has been used. The latter expression must be compared with the differential of H as a function of q, p, t, namely

$$dH = \sum_{j=1}^{n} \left(\frac{\partial H}{\partial p_j} dp_j + \frac{\partial H}{\partial q_j} dq_j \right) + \frac{\partial H}{\partial t} dt \ .$$

By comparison of coefficients, also recalling (1.6), we get

$$\frac{\partial H}{\partial p_j} = \dot{q}_j \ , \quad \frac{\partial H}{\partial q_j} = -\frac{\partial L}{\partial q_j} = -\dot{p}_j \ , \quad \frac{\partial H}{\partial t} = -\frac{\partial L}{\partial t} \ ,$$

namely the equations of Hamilton plus a relation between the time derivatives of H and L.

Example 1.1: *Free particle.* Let a point of mass m move in space under no forces. Using rectangular coordinates, the configuration space is identified with $\mathbb{R}^3$. The Lagrangian function coincides with the kinetic energy and reads

$$L = \frac{1}{2} m(\dot{x}^2 + \dot{y}^2 + \dot{z}^2) \ .$$

The momenta conjugated to the coordinates are $p_x = m\dot{x}, p_y = m\dot{y}, p_z = m\dot{z}$; they coincide with the components of the momentum of the particle. The phase space in this case is naturally identified with $\mathbb{R}^6$, and the Hamiltonian coincides with the kinetic energy, being

$$H = \frac{1}{2m}(p_x^2 + p_y^2 + p_z^2) \ . \tag{1.8}$$

If the particle is acted on by a force depending on a potential $V(x, y, z)$, then the Lagrangian and the Hamiltonian functions are, respectively,

$$\begin{aligned} L &= \frac{1}{2} m(\dot{x}^2 + \dot{y}^2 + \dot{z}^2) - V(x, y, z) \ , \\ H &= \frac{1}{2m}(p_x^2 + p_y^2 + p_z^2) + V(x, y, z) \ . \end{aligned} \tag{1.9}$$

In particular, the Hamiltonian represents the energy of the particle, the kinetic energy having been expressed in terms of the momenta. *E.D.*

Example 1.2: *One-dimensional harmonic oscillator.* Let now a point of mass m move on a straight line under the action of a perfectly elastic spring. The configuration space is $\mathbb{R}$, and the Lagrangian function is

$$L = \frac{1}{2} m\dot{x}^2 - \frac{1}{2} kx^2 \ .$$

Introducing the *angular frequency* $\omega = \sqrt{k/m}$, the Lagrangian changes to the following:[4]

$$L = \frac{1}{2}\dot{x}^2 - \frac{1}{2}\omega^2 x^2 \ .$$

[4] Recall that if the Lagrangian function is multiplied by an arbitrary non-zero factor, the equations of Lagrange do not change.

The momentum conjugated to x is $p = \dot{x}$, so that the phase space is naturally identified with $\mathbb{R}^2$. The Hamiltonian is again the energy and is written as

$$H = \frac{1}{2}p^2 + \frac{1}{2}\omega^2 x^2 . \tag{1.10}$$

E.D.

Example 1.3: *The pendulum.* A mass point is moving without friction on a vertical circle, so that it is subjected to gravity. The configuration space can be chosen to be a one-dimensional torus $\mathbb{T} = \mathbb{R}/(2\pi\,\mathbb{Z})$, namely the real line where the points x and $x + 2k\pi$ ($k \in \mathbb{Z}$) are identified. The resulting angular coordinate will be denoted by ϑ. The Lagrangian function may be written as

$$L = \frac{1}{2}\dot{\vartheta}^2 + \frac{g}{l}\cos\vartheta ,$$

where l is the length of the pendulum and g the constant acceleration due to local gravity. The momentum conjugated to ϑ is $p = \dot{\vartheta} \in \mathbb{R}$, and the phase space is naturally identified with $\mathbb{T} \times \mathbb{R}$. The Hamiltonian is

$$H = \frac{1}{2}p^2 - \frac{g}{l}\cos\vartheta . \tag{1.11}$$

E.D.

Example 1.4: *Motion under central forces.* The problem is to investigate the motion of a particle P of mass m acted on by a force in the direction PO, where O is a fixed point. Let us assume that the force field is spherically symmetric, so that there exists a potential $V(r)$, where r is the distance of P from the fixed point O. In view of the symmetry of the problem it is convenient to use spherical coordinates r, ϑ, φ, related to the rectangular coordinates x, y, z by

$$x = r\sin\vartheta\cos\varphi , \quad y = r\sin\vartheta\sin\varphi , \quad z = r\cos\vartheta .$$

Thus, the configuration space is $\mathbb{R}^+ \times (0,\pi) \times \mathbb{T}$ and the Lagrangian function reads

$$L = \frac{1}{2}m\left(\dot{r}^2 + r^2\dot{\vartheta}^2 + r^2\dot{\varphi}^2\sin^2\vartheta\right) - V(r) .$$

The momenta conjugated to the coordinates are $p_r = m\dot{r}$, $p_\vartheta = mr^2\dot{\vartheta}$ and $p_\varphi = mr^2\dot{\varphi}\sin^2\vartheta$. The Hamiltonian is

$$H = \frac{1}{2m}\left(p_r^2 + \frac{p_\vartheta^2}{r^2} + \frac{p_\varphi^2}{r^2\sin^2\vartheta}\right) + V(r) , \tag{1.12}$$

on the phase space $\mathbb{R}^+ \times (0,\pi) \times \mathbb{T} \times \mathbb{R}^3$. E.D.

Exercise 1.5: In galactic dynamics one often assumes the galaxy to be composed of a central nucleus with a large number of stars, and many other stars distributed in some way close to a plane orthogonal to the axis of rotation of the galaxy. The simplest approach consists in assuming that the distribution of the stars possesses an axial symmetry. Thus, one is led to study the motion of a single star close to the galactic plane, far from the nucleus, as subjected to an average potential $V(r, z)$, independent of ϑ in view of the axial symmetry. Show that in cylindrical coordinates r, ϑ, z with

$$x = r\cos\vartheta\ , \quad y = r\sin\vartheta$$

the Hamiltonian reads

$$H = \frac{1}{2}\left(p_r^2 + \frac{p_\vartheta^2}{r^2} + p_z^2\right) + V(r, z)\ .$$

A.E.L.

Example 1.6: *The problem of n bodies.* The problem is to investigate the motion of a system of n particles in space, with a two-body interaction due to some potential. Let $\mathbf{x}_1, \ldots, \mathbf{x}_n$ be the positions of the n bodies in a rectangular frame, so that the configuration space can be identified with $\mathbb{R}^{3n}$, and let $m_1, \ldots, m_n$ be the masses of the bodies. Then the Lagrangian function reads

$$L = \frac{1}{2}\sum_{j=1}^{n} m_j \dot{\mathbf{x}}_j^2 - V(\mathbf{x}_1, \ldots, \mathbf{x}_n)\ ,$$

where $V(\mathbf{x}_1, \ldots, \mathbf{x}_n)$ is the potential energy. The momenta conjugated to the coordinates are $\mathbf{p}_j = m_j \dot{\mathbf{x}}_j\ , \ 1 \le j \le n$. Therefore, the phase space can be identified with $\mathbb{R}^{6n}$, and the Hamiltonian reads

$$H = \sum_{j=1}^{n} \frac{\mathbf{p}_j^2}{2m_j} + V(\mathbf{x}_1, \ldots, \mathbf{x}_n)\ . \tag{1.13}$$

The cases $n = 2$ and $n = 3$ are particularly interesting. The former one, the *problem of two bodies*, is the most natural approximation for the motion of a planet revolving around the Sun when the law of gravitation is applied in its correct form (i.e., the gravitational action of the planet on the Sun is taken into account). The latter one is the celebrated *problem of three bodies*, which was at the origin of the discovery of chaos by Poincaré. *E.D.*

Example 1.7: *Particle in a rotating frame.* We consider again the problem of a particle of mass m moving in space, acted on by some potential, and investigate its motion with respect to a frame uniformly rotating around a fixed axis. Let ξ, η, ζ be the coordinates of the particle with respect to a fixed rectangular frame, with the ζ axis coinciding with the fixed axis of the rotating frame. Let now x, y, z be rectangular coordinates in the rotating frame, chosen so that the two frames have a common origin, and the axes ζ and z coincide. Therefore, the two sets of coordinates are related via

$$\xi = x\cos\omega t - y\sin\omega t\ , \quad \eta = x\sin\omega t + y\cos\omega t\ , \quad \zeta = z\ ,$$

ω being the angular velocity of the rotating frame with respect to the fixed one. By substitution of the latter transformation in the Lagrangian function for a point in a fixed frame (see Example 1.1) one has

$$L = \frac{1}{2}m\left[(\dot{x} - \omega y)^2 + (\dot{y} + \omega x)^2 + \dot{z}^2\right] - V(x, y, z, t) ,$$

$V(x, y, z, t)$ being the potential energy, which may depend on time due to the change of coordinates. The momenta conjugated to the coordinates are

$$p_x = m(\dot{x} - \omega y) , \quad p_y = m(\dot{y} + \omega x) , \quad p_z = m\dot{z} .$$

The Hamiltonian reads

$$H = \frac{1}{2m}\left(p_x^2 + p_y^2 + p_z^2\right) + \omega\left(yp_x - xp_y\right) + V(x, y, z, t) . \tag{1.14}$$

E.D.

Example 1.8: *The circular restricted problem of three bodies.* A remarkable example, which plays a fundamental role in Celestial Mechanics, is the so-called *restricted problem of three bodies*: the particle (named the *planetoid*) is attracted by two much bigger masses (the *primaries*, e.g., Sun–Jupiter or Earth–Moon) moving on Keplerian orbits without being affected by the planetoid. This model is a fundamental one when dealing with the dynamics of small bodies (e.g., inner planets, asteroids or spacecraft). The simplest approach is to assume that the primaries revolve uniformly on a circular orbit around their common centre of mass. It is also common to choose the units as follows: (i) the unit of length is the distance between the primaries; (ii) the period of the primaries is $T = 2\pi$; (iii) the masses of the primaries are set to be μ and $1 - \mu$ with $0 < \mu < 1$, so that the total mass coincides with the mass unit. By the way, in these units the gravitational constant G takes the value $G = 1$. The rotating system is chosen so that the origin coincides with the barycentre and the primaries are in fixed positions $1 - \mu$ and μ on the x axis. Therefore the potential turns out to be independent of t, since

$$V(x, y, z) = -\frac{m(1-\mu)}{\sqrt{(x-\mu)^2 + y^2 + z^2}} - \frac{m\mu}{\sqrt{(x+1-\mu)^2 + y^2 + z^2}} .$$

We are also allowed to remove the factor m from the Lagrangian, which is tantamount to substituting $m = 1$. Therefore, the Hamiltonian reads

$$\begin{aligned} H(x, y, z, p_x, p_y, p_z) = &\frac{1}{2}(p_x^2 + p_y^2 + p_z^2) - \omega x p_y + \omega y p_x \\ &- \frac{1-\mu}{\sqrt{(x-\mu)^2 + y^2 + z^2}} - \frac{\mu}{\sqrt{(x+1-\mu)^2 + y^2 + z^2}} . \end{aligned}$$

E.D.

Example 1.9: *Forced oscillations.* A simple but remarkable example in Physics is a harmonic oscillator acted on by a (small) periodic forcing. It is usual to write immediately the equation as, for example,

$$\ddot{x} + \omega^2 x = \varepsilon \cos \nu t , \tag{1.15}$$

where ω is the proper frequency of the oscillator and ν is the frequency of the forcing term, while ε is a real parameter. The equation may be derived from the Lagrangian function

$$L = \frac{1}{2}\dot{x}^2 - \frac{\omega^2}{2}x^2 + \varepsilon x \cos \nu t \ .$$

The Hamiltonian is easily found to be

$$H(x, p, t) = \frac{1}{2}(p^2 + \omega^2 x^2) - \varepsilon x \cos \nu t \ ,$$

where $p = \dot{x}$ is the momentum. The reader will notice that the Hamiltonian has the generic form $H = T + V(x, t)$, where $T = \dot{x}^2/2$ is the kinetic energy and $V(x, t)$ is the potential energy. An immediate generalization is to replace the forcing term with a more general function $xf(t)$, where $f(t)$ is time-periodic (i.e., $f(t + T) = f(t)$ for all t and some fixed T, e.g., $T = 2\pi/\nu$ in the preceding case). One may also add multiple periods or even include a non-periodic dependence on time, when useful. *E.D.*

Example 1.10: *The forced pendulum.* A definitely more difficult problem arises when one considers a forced *nonlinear* oscillator. A widely studied model is the forced pendulum, as described by the equation (denoting again by ϑ the coordinate)

$$\ddot{\vartheta} + \sin \vartheta = \varepsilon \cos \nu t \ , \tag{1.16}$$

where the forcing term may be replaced by a generic periodic function. The equation may be derived from the Lagrangian function

$$L = \frac{1}{2}\dot{\vartheta}^2 + \cos \vartheta + \varepsilon \vartheta \cos \nu t \ .$$

The Hamiltonian is easily found to be

$$H(\vartheta, p, t) = \frac{1}{2}p^2 - \cos \vartheta - \varepsilon \vartheta \cos \nu t \ .$$

A slightly different form of the Hamiltonian that is often used is

$$H_\varepsilon(\vartheta, p) = \frac{p^2}{2} - \cos \vartheta - \varepsilon \cos(\vartheta - t) \ , \quad \vartheta \in \mathbb{T} \ , \ p \in \mathbb{R} \ .$$

E.D.

Example 1.11: *Autonomous mechanical system.* The general form of the Lagrangian function for an autonomous mechanical system may be written as

$$L(q, \dot{q}) = T(q, \dot{q}) - V(q) \ , \quad T(q, \dot{q}) = \frac{1}{2}\sum_{j,k=1}^{n} g_{j,k}(q)\dot{q}_j\dot{q}_k \ ,$$

where $T(q, \dot{q})$ is the kinetic energy in a general coordinate system, defined through a symmetric non-degenerate matrix $g(q)$ with elements $g_{j,k}(q)$, and

$V(q)$ is the potential energy, independent of the velocities. The relation between generalized velocities and momenta is

$$p_j = \sum_{k=1}^{n} g_{jk}(q)\,\dot q_k\ ,\qquad \dot q_k = \sum_{j=1}^{n} [g^{-1}]_{kj}(q) p_j\ ,$$

where we have denoted by $[g^{-1}]_{j,k}$ the elements of the inverse matrix $g^{-1}(q)$. Thus we get

$$H = \frac{1}{2}\sum_{j,k=1}^{n} [g^{-1}]_{j,k} p_j p_k + V(q)\ . \tag{1.17}$$

E.D.

1.1.3 Compact Notation for the Canonical Equations

The canonical equations may be given a more compact form by introducing the $2n$ column vector $\mathbf{z}$ and the $2n \times 2n$ skew symmetric matrix J as

$$\mathbf{z} = (q_1, \ldots, q_n, p_1, \ldots, p_n)^\top\ ,\qquad \mathsf{J} = \begin{pmatrix} 0 & \mathsf{I}_n \\ -\mathsf{I}_n & 0 \end{pmatrix}\ , \tag{1.18}$$

where I_n is the $n \times n$ identity matrix. Remark that $\mathsf{J}^2 = -\mathsf{I}_{2n}$. We shall also introduce the operator

$$\partial_z = \left(\frac{\partial}{\partial z_1}, \ldots, \frac{\partial}{\partial z_{2n}}\right)^\top\ . \tag{1.19}$$

With these notations the canonical equations take the compact form

$$\dot{\mathbf{z}} = \mathsf{J}\partial_z H\ . \tag{1.20}$$

The compact notation used here turns out to be useful in a few cases, since it allows us to write shorter expressions. Although most of the present notes are written in the traditional language, in some cases the compact notation will be adopted in order to simplify some calculations.

1.2 Dynamical Variables and First Integrals

A *dynamical variable* is a differentiable real function $f : \mathscr{F} \to \mathbb{R}$ with domain on the phase space (e.g., the kinetic energy, the potential energy and the Hamiltonian itself are dynamical variables, and so are the coordinates themselves). If the canonical coordinates evolve in time as $\bigl(q(t), p(t)\bigr)$, so does the dynamical variable $f\bigl(q(t), p(t)\bigr)$. In particular, if the flow $\bigl(q(t), p(t)\bigr)$ is determined by the canonical equations, (1.1), then the time evolution of $f(p, q)$ obeys

$$\dot{f} = \sum_{j=1}^{n} \left(\frac{\partial f}{\partial q_j} \frac{\partial H}{\partial p_j} - \frac{\partial f}{\partial p_j} \frac{\partial H}{\partial q_j} \right) .$$

This is the time derivative of f along the Hamiltonian flow induced by H, called the *Lie derivative* of f and denoted by $L_H f$. In canonical coordinates the operator L_H representing the Lie derivative takes the form

$$L_H := \sum_{j=1}^{n} \left(\frac{\partial H}{\partial p_j} \frac{\partial}{\partial q_j} - \frac{\partial H}{\partial q_j} \frac{\partial}{\partial p_j} \right) . \tag{1.21}$$

1.2.1 The Algebra of Poisson Brackets

Let $f(q,p)$ and $g(q,p)$ be differentiable dynamical variables; then the *Poisson bracket* is defined as

$$\{g,f\} = \sum_{j=1}^{n} \left(\frac{\partial g}{\partial q_j} \frac{\partial f}{\partial p_j} - \frac{\partial g}{\partial p_j} \frac{\partial f}{\partial q_j} \right) . \tag{1.22}$$

It is immediately seen that the latter expression is the Lie derivative of the function g along the Hamiltonian field generated by f. Thus, it is natural to associate to any function f the *Lie derivative* $L_f \cdot = \{\cdot, f\}$, defined in coordinates as in (1.21).

The operation of the Poisson brackets satisfies some relevant properties which may be stated either in traditional notations or in terms of the Lie derivative. It is also useful to introduce the notation $[\cdot,\cdot]$ for the commutator between two operators; for example, $[L_f, L_g] = L_f L_g - L_g L_f$.

Proposition 1.12: *If f, g and h are differentiable dynamical variables and α is a real constant, we have:*

(i) linearity, $\{f, g+h\} = \{f,g\} + \{f,h\}\,,\ \{f,\alpha g\} = \alpha\{f,g\}\,;$

(ii) anticommutativity, $\{f,g\} = -\{g,f\}\,;$

(iii) Jacobi's identity,

$$\{f,\{g,h\}\} + \{g,\{h,f\}\} + \{h,\{f,g\}\} = 0\ .$$

In terms of Lie derivatives the same properties are written as

$$L_f(g+h) = L_f g + L_f h\ ,\quad L_f(\alpha g) = \alpha L_f g\ , \tag{1.23}$$

$$L_f g = -L_g f\ , \tag{1.24}$$

$$[L_f, L_g] = L_{L_f g} = L_{\{g,f\}}\ . \tag{1.25}$$

Jacobi's identity (1.25) deserves particular attention. The left member is a composition of two differential operators; hence it is natural to imagine that it is a second-order operator. This is untrue, in fact. The second member is a first-order operator, with a remarkable interpretation: the commutation between the Hamiltonian vector fields generated by the functions f and g is the Hamiltonian vector field generated by $\{g, f\}$.

Corollary 1.13: *The following additional properties apply*

(iv) distributivity over the product: $\{f, gh\} = g\{f, h\} + \{f, g\}h$;

(v) Leibniz's rule: for $1 \le j \le n$

$$\frac{\partial}{\partial q_j}\{f, g\} = \left\{\frac{\partial f}{\partial q_j}, g\right\} + \left\{f, \frac{\partial g}{\partial q_j}\right\} , \quad \frac{\partial}{\partial p_j}\{f, g\} = \left\{\frac{\partial f}{\partial p_j}, g\right\} + \left\{f, \frac{\partial g}{\partial p_j}\right\} .$$

In terms of Lie derivatives the same properties are written as

$$L_f(gh) = gL_f h + hL_f g \ , \tag{1.26}$$

$$\left[\frac{\partial}{\partial q_j}, L_f\right] = L_{\frac{\partial f}{\partial q_j}} \ , \quad \left[\frac{\partial}{\partial p_j}, L_f\right] = L_{\frac{\partial f}{\partial p_j}} \ , \quad 1 \le j \le n \ . \tag{1.27}$$

Proof of Proposition 1.12 and of Corollary 1.13. We leave to the reader the easy check that the properties (1.23) to (1.27) are just a rewriting of the corresponding properties (i)–(v). Also, with the exception of Jacobi's identity (iii) or equivalently (1.25), the proof of the other identities is an easy matter and is left to the reader. So let us prove (1.25). Begin by calculating

$$\begin{aligned}
L_f L_g &= \sum_{j,k} \left(\frac{\partial f}{\partial p_j}\frac{\partial}{\partial q_j} - \frac{\partial f}{\partial q_j}\frac{\partial}{\partial p_j}\right)\left(\frac{\partial g}{\partial p_k}\frac{\partial}{\partial q_k} - \frac{\partial g}{\partial q_k}\frac{\partial}{\partial p_k}\right) \\
&= \sum_{j,k} \left[\frac{\partial f}{\partial p_j}\left(\frac{\partial^2 g}{\partial q_j \partial p_k}\frac{\partial}{\partial q_k} - \frac{\partial^2 g}{\partial q_j \partial q_k}\frac{\partial}{\partial p_k}\right)\right] \\
&\quad - \sum_{j,k} \left[\frac{\partial f}{\partial q_j}\left(\frac{\partial^2 g}{\partial p_j \partial p_k}\frac{\partial}{\partial q_k} - \frac{\partial^2 g}{\partial p_j \partial q_k}\frac{\partial}{\partial p_k}\right)\right] \\
&\quad + \sum_{j,k} \left[\frac{\partial f}{\partial p_j}\left(\frac{\partial g}{\partial p_k}\frac{\partial^2}{\partial q_j \partial q_k} - \frac{\partial g}{\partial q_k}\frac{\partial^2}{\partial q_j \partial p_k}\right)\right] \\
&\quad - \sum_{j,k} \left[\frac{\partial f}{\partial q_j}\left(\frac{\partial g}{\partial p_k}\frac{\partial^2}{\partial p_j \partial q_k} - \frac{\partial g}{\partial p_k}\frac{\partial^2}{\partial p_j \partial p_k}\right)\right] .
\end{aligned}$$

The similar calculation for $L_g L_f$ consists merely in exchanging the symbols f and g in the previous expression. Hence when subtracting $L_f L_g - L_g L_f$, all terms that generate a second-order differentiation (namely the two last lines in the formula, with an exchange of the summation indices j, k where useful) compensate each other, and we are left with

$$\begin{aligned} L_f L_g - L_g L_f = \sum_{j,k} \left[\frac{\partial f}{\partial p_j} \left(\frac{\partial^2 g}{\partial q_j \partial p_k} \frac{\partial}{\partial q_k} - \frac{\partial^2 g}{\partial q_j \partial q_k} \frac{\partial}{\partial p_k} \right) \right] \\ - \sum_{j,k} \left[\frac{\partial f}{\partial q_j} \left(\frac{\partial^2 g}{\partial p_j \partial p_k} \frac{\partial}{\partial q_k} - \frac{\partial^2 g}{\partial p_j \partial q_k} \frac{\partial}{\partial p_k} \right) \right] \\ - \sum_{j,k} \left[\frac{\partial g}{\partial p_j} \left(\frac{\partial^2 f}{\partial q_j \partial p_k} \frac{\partial}{\partial q_k} + \frac{\partial^2 f}{\partial q_j \partial q_k} \frac{\partial}{\partial p_k} \right) \right] \\ + \sum_{j,k} \left[\frac{\partial g}{\partial q_j} \left(\frac{\partial^2 f}{\partial p_j \partial p_k} \frac{\partial}{\partial q_k} - \frac{\partial^2 f}{\partial p_j \partial q_k} \frac{\partial}{\partial p_k} \right) \right] . \end{aligned}$$

Letting the index k be fixed and collecting the coefficients of the operator $\frac{\partial}{\partial q_k}$, we get

$$\begin{aligned} \sum_j \left(\frac{\partial f}{\partial p_j} \frac{\partial}{\partial q_j} - \frac{\partial f}{\partial q_j} \frac{\partial}{\partial p_j} \right) \frac{\partial g}{\partial p_k} - \left(\frac{\partial g}{\partial p_j} \frac{\partial}{\partial q_j} - \frac{\partial g}{\partial q_j} \frac{\partial}{\partial p_j} \right) \frac{\partial f}{\partial p_k} \\ = \left\{ \frac{\partial g}{\partial p_k}, f \right\} + \left\{ g, \frac{\partial f}{\partial p_k} \right\} = \frac{\partial}{\partial p_k} \{g, f\} . \end{aligned}$$

Here Leibniz's rule (iii) has been used. Similarly, the coefficients of the operator $\frac{\partial}{\partial p_k}$ give $-\frac{\partial}{\partial q_k}\{g, f\}$. Thus, restoring the sum over k, we get

$$L_f L_g - L_g L_f = \frac{\partial}{\partial p_k} \{g, f\} \frac{\partial}{\partial q_k} - \frac{\partial}{\partial q_k} \{g, f\} \frac{\partial}{\partial p_k} = L_{\{g,f\}} ,$$

namely Jacobi's identity in the form (1.25). *Q.E.D.*

As we have seen, in terms of Poisson brackets (or Lie derivatives) the time evolution of a differentiable dynamical variable $f(q, p)$ satisfies the partial differential equation $\dot{f} = \{f, H\}$, or equivalently $\dot{f} = L_H f$. As a consequence, the Hamiltonian dynamics can be expressed in terms of Poisson brackets (see [129]). In particular, the canonical equations may be written in one of the following more symmetric forms:

$$\dot{q}_j = \{q_j, H\} , \quad \dot{p}_j = \{p_j, H\} , \quad 1 \le j \le n , \tag{1.28}$$

or

$$\dot{q}_j = L_H q_j , \quad \dot{p}_j = L_H p_j , \quad 1 \le j \le n . \tag{1.29}$$

This exploits the fact that the coordinates themselves can be considered as dynamical variables.

1.2.2 First Integrals

A dynamical variable $\Phi(p,q)$ is said to be a *first integral* if it keeps its value constant under the canonical flow generated by $H(p,q)$, namely by Eq. (1.1). The name *constant of motion* is often used. If $\Phi(p,q)$ is differentiable, then one has $\dot{\Phi} = 0$ under the flow. Thus, a differentiable first integral $\Phi(q,p)$ must satisfy the partial differential equation

$$L_H \Phi = 0 \ . \tag{1.30}$$

From the properties of the Lie derivative the following results are immediately obtained:

(i) The Hamiltonian of an autonomous system is a first integral;[5] for a typical autonomous mechanical system this is actually the total energy, so that it is usually named the energy integral.

(ii) If $\Phi(q,p)$ and $\Psi(q,p)$ are differentiable first integrals, then $\{\Phi,\Psi\}$ is a first integral.

The first statement follows from anticommutativity of the Poisson bracket; the second one is a straightforward consequence of Jacobi's identity and is often referred to as *Poisson's theorem.*

Everybody knows that the existence of first integrals for a system of differential equations may greatly help in investigating the behaviour of the flow, possibly leading to a complete integration of the equations when a sufficient number of independent first integrals is known.[6] In Chapter 3 the

---

[5] A time-dependent dynamical variable $f(q,p,t)$ satisfies $\dot{f} = \{f,H\} + \frac{\partial f}{\partial t}$. Thus, for a non-autonomous Hamiltonian $H(q,p,t)$ one has $\frac{dH}{dt} = \frac{\partial H}{\partial t}$. This is nothing but the equation $\dot{p}_+ = -\frac{\partial H}{\partial t}$ in (1.3).

[6] For an autonomous system $\dot{x} = f(x)$ of ordinary differential equations on an n-dimensional manifold, a differentiable first integral defines a $(n-1)$-dimensional manifold through the implicit function theorem, which is invariant for the flow. Similarly, if $k < n$ independent first integrals are known, $\varphi_1(x) = c_1, \ldots, \varphi_k(x) = c_k$ say, then the implicit function theorem states the existence of an invariant $(n-k)$-dimensional manifold. If $k = n-1$, then the resulting one-dimensional invariant manifold is a set of orbits. In the latter case a complete integration can be worked out by suitably choosing one of the coordinates, x_j say, and determining the remaining $n-1$ as functions of x_j and of the $n-1$ constants $c_1, \ldots, c_{n-1}$. Thus, one is left with a single one-dimensional differential equation of the form $\dot{x}_j = g(x_j; c_1, \ldots, c_{n-1})$, depending on the constants as parameters, which may be solved by a quadrature. This makes it evident that no more than $n-1$ independent first integrals may be found. By the way, the name 'first integrals' comes from the remark that finding the complete solution of a system of differential equations of order n requires n integrations, thus introducing n integration constants. The knowledge of a function which is constant under the flow corresponds to having performed one of the

particular case of systems possessing a sufficient number of first integrals will be investigated – luckily, an exceptional case.

The argument goes as follows. Consider the subset of $\mathscr{F}$,

$$M_\Phi = \{(q,p) \in \mathscr{F} : \Phi(q,p) = \Phi(q_0,p_0)\} ,$$

where $(q_0,p_0) = (q_{0,1},\ldots,q_{0,n},p_{0,1},\ldots,p_{0,n}) \in \mathscr{F}$ is the initial point. Then the orbit through (q_0,p_0) is entirely contained in M_Φ. In particular, if the canonical vector field $\mathsf{J}\partial_z\Phi$ (with the compact notation of Section 1.1.3) does not vanish on M_Φ, then M_Φ is a differentiable submanifold of $\mathscr{F}$ which is invariant for the flow induced by H.

Let now $\Phi_1(q,p),\ldots,\Phi_r(q,p)$ be independent first integrals, that is, assume that the Jacobian $r\times 2n$ matrix satisfies

$$\operatorname{rank}\begin{pmatrix} \frac{\partial\Phi_1}{\partial q_1} & \cdots & \frac{\partial\Phi_1}{\partial q_n} & \frac{\partial\Phi_1}{\partial p_1} & \cdots & \frac{\partial\Phi_1}{\partial p_n} \\ \vdots & \ddots & \vdots & \vdots & \ddots & \vdots \\ \frac{\partial\Phi_r}{\partial q_1} & \cdots & \frac{\partial\Phi_r}{\partial q_n} & \frac{\partial\Phi_r}{\partial p_1} & \cdots & \frac{\partial\Phi_r}{\partial p_n} \end{pmatrix} = r .$$

Then the equations

$$\Phi_1(q,p) = \Phi_1(q_0,p_0),\ldots,\Phi_r(q,p) = \Phi_r(q_0,p_0)$$

define a $(2n-r)$-dimensional submanifold of $\mathscr{F}$ which is still invariant for the flow induced by H.

As already remarked, an autonomous Hamiltonian system always possesses a first integral – the Hamiltonian itself, also named energy integral. Therefore the orbits of an autonomous Hamiltonian system with one degree of freedom lie on one-dimensional manifolds determined as the level curves of $H(q,p) = E$. This is enough in order to perform a complete integration of the system, as will be discussed in Section 1.3.

For systems with more than one degree of freedom the existence of the energy integral is not enough to perform a complete integration. According to the argument in note 6 one could naively expect that $2n-1$ independent first integrals must be found to integrate the system by quadratures (where n is the number of degrees of freedom). It is a remarkable fact that the Hamiltonian structure requires the existence of only n independent first integrals, provided they satisfy the further condition of being in involution, namely that $\{\Phi_j,\Phi_k\} = 0$ for every pair j,k. This is the contents of Liouville's theorem; it will be discussed in Chapter 3. This interesting fact has a reverse side: as a general fact, no more than n nontrivial first integrals in involution

---

necessary integrations, the constant being the value of the function at the initial point. The discussion in this section is essentially an adaptation of the general argument to the particular case of Hamiltonian systems.

exist in a global sense,[7] due to the peculiar behaviour of the orbits. On the other hand, as discovered by Poincaré, integrability itself turns out to be an exceptional property.

Example 1.14: *Autonomous Hamiltonian system with one degree of freedom.* As already remarked, the Hamiltonian $H(q,p)$ is a first integral and the orbits are subsets of the level sets defined by the equation $H(q,p) = E$, where $E = H\bigl(p(0), q(0)\bigr)$ is a constant determined by the initial point. Let us begin with two examples of linear systems (i.e., with a quadratic Hamiltonian), which describe the typical behaviour in a neighbourhood of an equilibrium point. In the case of an elastic attractive force (the harmonic oscillator) the Hamiltonian is $H(x,p) = p^2/2 + \omega^2 x^2/2$ (see panel (a) of Fig. 1.1). For $E = 0$ the orbit is a point, namely the equilibrium state $x = p = 0$. For $E > 0$ the level curves are ellipses with centre at the origin, representing oscillations around the equilibrium. Thus the equilibrium is stable. The case of an elastic repulsive force (see panel (b) of Fig. 1.1) is described by the Hamiltonian $H(p,x) = p^2/2 - \lambda^2 x^2/2$. For $E = 0$ the level curves are two straight lines intersecting at the origin, named *separatrices.* They actually represent five different orbits. Two of them represent orbits asymptotic to the origin for $t \to -\infty$ (unstable manifolds); two represent orbits asymptotic to the origin for $t \to +\infty$ (stable manifolds); the fifth orbit is the origin, which is an unstable equilibrium. For $E \neq 0$ the level curves are hyperbolæ representing orbits reflected back by the repulsive potential for $E < 0$, and orbits overtaking the equilibrium position for $E > 0$. A typical non-linear example is the pendulum. In this case the phase space is a cylinder, and the Hamiltonian is $H(\vartheta,p) = p^2/2 - \cos\vartheta$ (see panel (c) of Fig. 1.1). For $E = -1$ the orbit is the equilibrium point $\vartheta = p = 0$, similar to the equilibrium of the harmonic oscillator. For $E = 1$ the orbits are the upper unstable equilibrium $\vartheta = \pi$, with stable and unstable manifolds emanating from it (recall that ϑ is an angle, so the points $\vartheta = \pi$ and $\vartheta = -\pi$ are just different representations of the same point). In the neighbourhood of the unstable equilibrium the figure looks similar to that of the repulsive force, but far from the equilibrium one sees that the stable and unstable manifolds coincide: the corresponding orbits are asymptotic to the unstable equilibrium both for $t \to -\infty$ and for $t \to +\infty$. The stable and unstable manifolds separate orbits representing oscillations from orbits representing rotations. This should be enough to explain how to represent the phase portrait for the Hamiltonian of any natural mechanical system with one degree of freedom. *E.D.*

[7] It is relevant here to distinguish between *local* and *global* first integrals. Local first integrals can always be found for a system of differential equations provided that the vector field satisfies suitable regularity constraints (e.g., Lipschitz condition). Global first integrals are a more delicate matter. The examples in the rest of the section should help to clarify this matter.

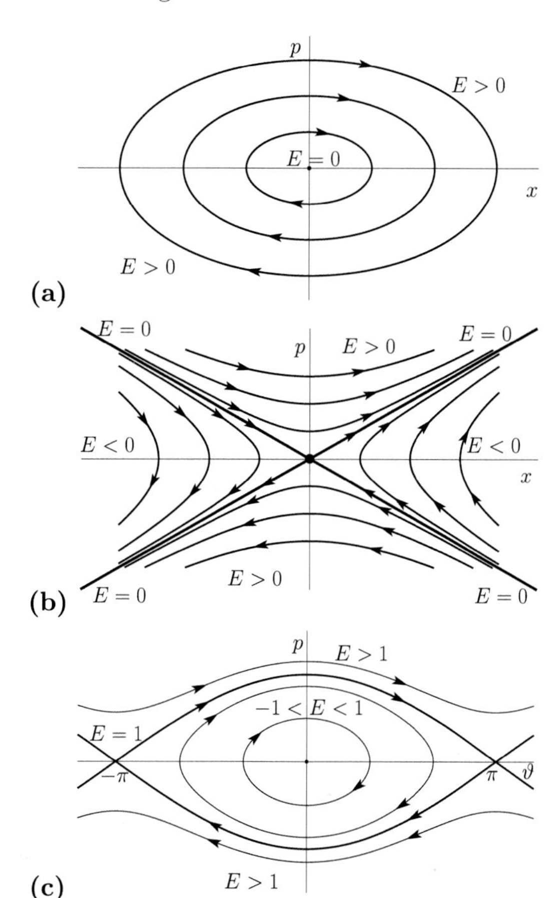

Figure 1.1 Phase portrait for three paradigmatic systems:
(**a**) elastic attractive force,

$$H = p^2/2 + \omega^2 x^2/2 \ ;$$

(**b**) elastic repulsive force,

$$H = p^2/2 - \lambda^2 x^2/2 \ ;$$

(**c**) the pendulum,

$$H = p^2/2 - \cos\vartheta \ .$$

The curves are level lines of energy, $H(x, p) = E$. The arrows indicate the direction of the flow ϕ^t.

Exercise 1.15: Draw the phase portrait for the Hamiltonian

$$H(x, p) = \frac{p^2}{2} + V(x)$$

representing the motion of a particle with unit mass moving on a straight line under the potential $V(x)$. Consider the following cases:

(i)	$V(x) = \frac{1}{2}x^2 + \frac{1}{4}x^4$;	(v)	$V(x) = \frac{1}{3}x^3$;
(ii)	$V(x) = \frac{1}{2}x^2 + \frac{1}{3}x^3$;	(vi)	$V(x) = -\frac{1}{4}x^4$;
(iii)	$V(x) = \frac{1}{2}x^2 - \frac{1}{4}x^4$;	(vii)	$V(x) = -\frac{1}{4}x^4$;
(iv)	$V(x) = -\frac{1}{2}x^2 + \frac{1}{4}x^4$;	(viii)	$V(x) = \frac{\cos x}{1+x^2}$.

A.E.L.

Exercise 1.16: Sometimes it is useful to consider potentials $V(x)$ which are not differentiable or even not continuous at some point. This may be useful for describing, for example, collisions. With a little attention, the reader will realize that a phase portrait can still be traced. Here are some examples.

(ix) $V(x) = \begin{cases} 0 & \text{for } x < 0 \,, \\ x & \text{for } x \geq 0 \,. \end{cases}$

(x) $V(x) = |x| \,.$

(xi) $V(x) = \begin{cases} -1 & \text{for } x < 0 \,, \\ 1 & \text{for } x \geq 0 \,. \end{cases}$

(xii) $V(x) = \begin{cases} -1 & \text{for } |x| < 1 \,, \\ 1 & \text{for } |x| \geq 1 \,. \end{cases}$

(xiii) $V(x) = \begin{cases} 0 & \text{for } x < 0 \,, \\ \infty & \text{for } x \geq 0 \,. \end{cases}$

A.E.L.

Exercise 1.17: Draw the phase portrait for the Hamiltonians

$$H = p(1-p)\sin\vartheta \quad \text{and} \quad H = \cos p \sin\vartheta \,, \qquad \vartheta \in \mathbb{T} \,,\; p \in \mathbb{R} \,.$$

A.E.L.

Example 1.18: *Free particle.* The Hamiltonian is $H = \frac{1}{2m}\left(p_x^2 + p_y^2 + p_z^2\right)$ (see Example 1.1). Besides the Hamiltonian, the following quantities are immediately checked to be first integrals:

$$p_x \,,\; p_y \,,\; p_z \,,\; M_x = yp_z - zp_y \,,\; M_y = zp_x - xp_z \,,\; M_z = xp_y - yp_x$$

(the linear and the angular momentum). Not all of them are independent, of course,[8] but we can select five independent first integrals among them, for instance: p_x , p_y , p_z , M_x and M_y. As a curiosity, one could ask what happens if one tries to construct new first integrals via Poisson brackets (recall that the Poisson bracket between two first integrals is a first integral). The answer is found in Table 1.1. Using the first integral, we can easily construct the orbit. For the intersection in $\mathbb{R}^6$ of the planes $p_x = c_1$, $p_y = c_2$, $p_z = c_3$, $M_x = c_4$, $M_y = c_5$, where $c_1, \ldots, c_5$ are constants determined by the initial point, is a straight line representing the orbit. *E.D.*

Example 1.19: *The problem of n bodies.* The Hamiltonian reads (see Example 1.6)

$$H = \sum_{j=1}^{n} \frac{\mathbf{p}_j^2}{2m_j} + V(\mathbf{x}_1, \ldots, \mathbf{x}_n) \,.$$

The system possesses seven independent first integrals, namely: the total energy H, the three components of the linear momentum

$$P_x = \sum_j p_{j,x} \,, \quad P_y = \sum_j p_{j,y} \,, \quad P_z = \sum_j p_{j,z}$$

(where $p_{j,x}$, $p_{j,y}$ and $p_{j,z}$ denote the x , y and z components, respectively, of the momentum $\mathbf{p}_j$ of the jth particle) and the three components of the angular momentum

[8] Recall that in general a system of differential equations on an n-dimensional manifold cannot possess more than $n-1$ independent first integrals (see note 6).

Table 1.1 Poisson brackets between the components of the momentum and the angular momentum for one particle.

$\{\cdot,\cdot\}$	p_x	p_y	p_z	M_x	M_y	M_z
p_x	0	0	0	0	p_z	$-p_y$
p_y	0	0	0	$-p_z$	0	p_x
p_z	0	0	0	p_y	$-p_x$	0
M_x	0	p_z	$-p_y$	0	M_z	$-M_y$
M_y	$-p_z$	0	p_x	$-M_z$	0	M_x
M_z	p_y	$-p_x$	0	M_y	$-M_x$	0

$$M_x = \sum_j y_j p_{j,z} - z_j p_{j,y}\,,\ M_y = \sum_j z_j p_{j,x} - x_j p_{j,z}\,,\ M_z = \sum_j x_j p_{j,y} - y_j p_{j,x}\,.$$

No new first integrals can be constructed by Poisson brackets between the known ones: the results of Table 1.1 still apply. For $n = 2$ (the problem of two bodies) the existence of such integrals allows us to conclude that the orbit lies on five-dimensional manifolds in the phase space $\mathbb{R}^{12}$. At first sight, this appears insufficient in order to perform a complete integration, seemingly contradicting the well-known fact that the system is actually integrable. However, we have already anticipated that the existence of n (in this case 6) independent first integrals which are in involution is enough. The common choice for the six first integrals is H , P_x , P_y , P_z , M_z and $\Gamma^2 = M_x^2 + M_y^2 + M_z^2$. For $n > 2$ the seven known first integrals are insufficient, and we should look for more. However, the theorems of Heinrich Bruns and Henry Poincaré state that in general there are no more first integrals. *E.D.*

Example 1.20: *Motion under central forces.* In Cartesian coordinates the Hamiltonian reads

$$H = \frac{1}{2m}\left(p_x^2 + p_y^2 + p_z^2\right) + V(r)\ ,$$

where $r = \sqrt{x^2 + y^2 + z^2}$. The phase space is $\mathbb{R}^6$, and the system possesses four independent first integrals, namely the Hamiltonian H and the three components M_x , M_y and M_z of the angular momentum (check it). We conclude that the orbits lie on a two-dimensional manifold in the phase space. *E.D.*

Example 1.21: *The Keplerian case.* According to Kepler's laws, the orbit of a planet in physical space is an ellipse. This does not follow from the general discussion of the previous example, because we could only conclude that the orbit in phase space lies on a two-dimensional manifold.[9] The fact that the orbit in a Keplerian potential is closed is due to the existence of a further first integral known as the *Runge–Lenz vector* (but already known to Laplace). In Cartesian coordinates the vector is

$$\mathbf{A} = \mathbf{p} \wedge \mathbf{M} - \frac{km\mathbf{x}}{r} , \tag{1.31}$$

where k is the constant in the Keplerian potential and $\mathbf{M}$ is the angular momentum. The vector $\mathbf{A}$ points to the pericentre of the orbit. Not all the components of the vector are independent of the four first integrals of the general case. However, one can extract a fifth integral which is independent of them, so that the invariant manifold has dimension 1 and the orbit is closed.[10] *E.D.*

Exercise 1.22: Let the phase space be $\mathscr{F} = \mathbb{T}^2 \times \mathbb{R}^2$, with $q \in \mathbb{T}^2$ and $p \in \mathbb{R}^2$, and consider the Hamiltonian

$$H(q,p) = \omega_1 p_1 + \omega_2 p_2$$

with real ω_1, ω_2. The momenta p_1 and p_2 are obviously first integrals, so that every orbit lies on a torus $\mathbb{T}^2$. Check that the function $\omega_1 q_2 - \omega_2 q_1$ is a third first integral, which, however, is only *local*, in general, in the sense that it does not provide a global one-value function on the torus (it is not periodic in q_1, q_2). Prove that if ω_1/ω_2 is an irrational number, then there are no more global first integrals. If, however, $\omega_1/\omega_2 = r/s$, where r, s are integers, then the function $f(q) = \sin(rq_2 - sq_1)$ is a global first integral.[11] Try to extend the example to a higher-dimensional case $\mathscr{F} = \mathbb{T}^n \times \mathbb{R}^n$ with Hamiltonian $H(q,p) = \omega_1 p_1 + \cdots + \omega_n p_n$. For a detailed discussion see Section 4.1.1. *A.E.L.*

---

[9] A theorem by Newton, later proved in extended form by Joseph Bertrand [25], states that there are only two potentials for which all bounded orbits are closed: (a) the Keplerian potential $V(r) = -k/r$ for negative energy, the orbit being an ellipse with the centre of force being placed in one of the foci, and (b) the harmonic potential $V(r) = kr^2$, the orbit being an ellipse with centre coinciding with the centre of force. For a discussion of Newton's theorem see [101]. For a proof of Bertrand's theorem see also [206].

[10] For a short historical note see [102]. See also [44].

[11] In some classical books, e.g., Wintner's treatise [212], it is usual to distinguish between *isolating* first integrals, which define a global invariant surface, and *non-isolating* ones, which are only local in the language adopted here.

1.3 Use of First Integrals

In the previous section the first integrals have been used in order to get qualitative information on the orbits, that is, without proceeding to integrate the dynamical equations and finding the time dependence of the evolution. Actually we can do better. The classical approach consists in exploiting the first integrals in order to perform the so-called *integration by quadratures:* one looks for a solution in terms of algebraic operations, including inversion of functions, and calculation of integrals of known functions. The general theory in this sense will be developed in Chapter 3, with the theorems of Liouville and of Arnold–Jost. In this section a first simplified approach is made, with the aim of describing in detail the process of integration by quadratures.

1.3.1 Motion on a One-Dimensional Manifold

The simplest case is the motion of a particle with mass m on a straight line (or possibly a circle, or another smooth curve) under the action of a given autonomous potential, which is assumed to be smooth. This is the case considered in Examples 1.7 and Exercises 1.15 and 1.16, where the energy integral has been used in order to draw the phase portrait; we have seen that the energy integral provides a complete qualitative description of the dynamics. Here we perform the next step: to write the solution of the canonical equations in a somehow explicit form. This is indeed the process of integration by quadrature. The discussion here provides the basis for the further extension to the case of higher-dimensional systems which possess a sufficient number of first integrals.

The Hamiltonian has the general form

$$H(x,p) = \frac{p^2}{2m} + V(x) \ , \tag{1.32}$$

where the potential $V(x)$ is assumed to be a smooth function, and the corresponding Hamilton's equations are

$$\dot{x} = \frac{p}{m} \ , \quad \dot{p} = -\frac{dV}{dx} \ . \tag{1.33}$$

The general integration scheme goes as follows. Solve Eq. (1.32) for p, thus getting

$$p = \pm\sqrt{2m\big(E - V(x)\big)} \ , \tag{1.34}$$

and substitute it into Eq. (1.33), thus getting

$$\dot{x} = \pm\sqrt{\frac{2}{m}\big(E - V(x)\big)} \ . \tag{1.35}$$

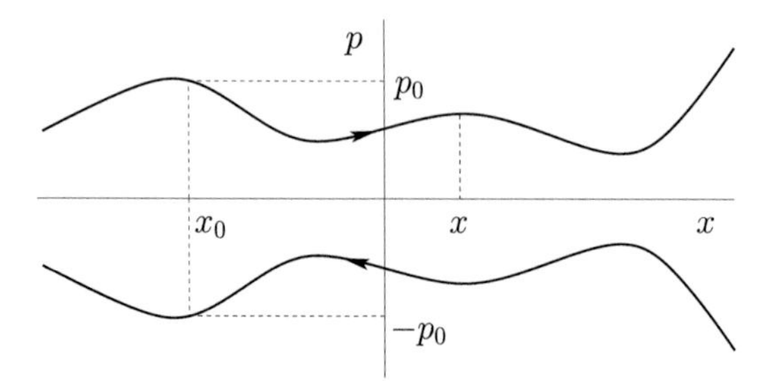

Figure 1.2 Illustrating the case of an orbit in the phase plane without reflection points.

This is a separable differential equation, and the solution satisfying the initial condition $q(t_0) = q_0$, $p(t_0) = p_0$ is written as

$$t - t_0 = \pm\sqrt{\frac{m}{2}} \int_{x_0}^{x} \frac{d\xi}{\sqrt{E - V(\xi)}} \,. \tag{1.36}$$

This is the general solution. However, understanding this formula may be not completely immediate; so let us add a short discussion that an expert reader may skip.

Let us refer to the phase portrait, which is obtained by drawing the level curves implicitly defined by the equation $H(x,p) = E$, the value of the energy. Remark that for systems with Hamiltonian of the form (1.32) all equilibria are located on the axis $p = 0$ and correspond to stationary points of the potential. On the other hand, all curves that intersect the axis $p = 0$ on a point x which is not an equilibrium have a vertical tangent at that point.

We may focus our attention on four examples of curves in the phase plane:

(i) isolated equilibrium points;
(ii) open curves not containing equilibrium points;
(iii) closed curves topologically equivalent to a circle;
(iv) curves which intersect on an equilibrium.

If $\overline{x}$ is an equilibrium for (1.33), that is, if $\frac{dV}{dx}(\overline{x}) = 0$, then $x(t) = \overline{x}$, $p(t) = 0$ is a solution (check it). By the way, from the canonical equation (1.33) it is clear that equilibrium points (if any) are located on the x axis.

Let us consider the simplest case of a value E of energy such that $E > V(x)$ holds true for all x. Assuming also that $V(x)$ is bounded from below, we have two separate curves in the phase plane, symmetric with respect to the x axis, which lie on the upper and lower half plane, respectively, as illustrated in Fig. 1.2. Let us pick an initial point x_0 together with the fixed value $E > V(x_0)$ of the energy. The corresponding initial value p_0 is calculated by (1.34) by choosing the appropriate sign, thus selecting one between the upper or the lower curve (i.e., the direction of the motion). The integral formula (1.36) tells us how much time the point needs in order to move from the initial point x_0 to an arbitrary point x. This gives us a smooth function

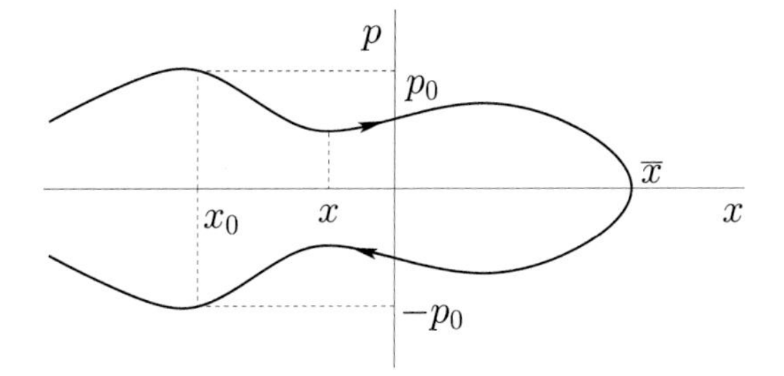

Figure 1.3 Illustrating the case of an orbit in the phase plane with a reflection point.

$t(x)$ which is clearly monotonic (the function under the integral does not change its sign), and so it may be inverted, thus giving the solution $x(t)$. The equation $p(x)$ of the curve is found by replacing $x(t)$ in (1.34), still keeping the chosen sign of the square root. Thus, the problem is completely solved.

The case illustrated in Fig. 1.3 requires some more attention. Here, we should assure that the argument of the square root is always non-negative. Thus, the upper limit x of the integral must satisfy $x \leq \overline{x}$, where $V(\overline{x}) = E$. Choose $p_0 > 0$, on the upper part of the curve. Then we are able to calculate the integral for any $x \leq \overline{x}$, as in the previous case. If we set the upper limit to $\overline{x}$, then we calculate a time

$$t_1 = t_0 + \sqrt{\frac{2}{m}} \int_{x_0}^{\overline{x}} \frac{d\xi}{\sqrt{E - V(\xi)}}$$

at which the point reaches $\overline{x}$. A moment's thought will allow us to realize that (for a smooth potential) such a time t_1 is finite.[12] Since $\overline{x}$ is not an equilibrium, the point on the phase plane must continue its motion along the curve: $\overline{x}$ is a reflection point for the motion. Thus, changing the sign of p, we may continue our integration by calculating

$$t - t_1 = -\sqrt{\frac{2}{m}} \int_{\overline{x}}^{x} \frac{d\xi}{\sqrt{E - V(\xi)}} = \int_{x}^{\overline{x}} \frac{d\xi}{\sqrt{E - V(\xi)}} ,$$

where $x < \overline{x}$ is still arbitrary. Remark that t is still increasing, so that we end up again with a monotonic function $t(x)$, to be inverted in order to obtain the motion $x(t)$. Remark also that the second integral actually coincides with the previous one (just replace the initial point x_0 with an arbitrary point x), so that we don't need to compute it again.[13]

---

[12] Since $V(\overline{x}) = E$ and $V'(\overline{x}) \neq 0$, then we have $E - V(x) = E - V(\overline{x}) + O(x - \overline{x})$, so that the integral has a finite value.

[13] A reader who looks carefully at Eq. (1.35) may remark that we are in a paradoxical situation. If $E = V(x_0)$, then taking x_0 as initial condition and replacing

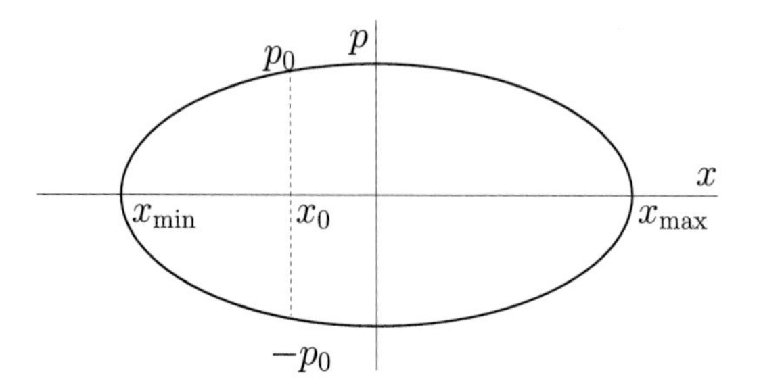

Figure 1.4 Illustrating the case of closed orbits in the phase plane.

A similar situation occurs if we consider a closed orbit, as in fig. 1.4. Here we have two reflection points $x_{\rm min}$ and $x_{\rm max}$, with $V(x_{\rm min}) = V(x_{\rm max}) = E$ and $V(x) > E$ in the interval $(x_{\rm min}, x_{\rm max})$. The point oscillates between the two extrema, and if we calculate

$$\tau = \sqrt{\frac{2}{m}} \int_{x_{\rm min}}^{x} \frac{d\xi}{\sqrt{E - V(\xi)}} \,, \qquad x_{\rm min} \le x \le x_{\rm max},$$

we know the forward motion from $x_{\rm min}$ to $x_{\rm max}$. Then the backward motion from $x_{\rm max}$ to $x_{\rm min}$ is also known, by symmetry, so that we know the motion over one period. Since the motion is periodic, we know it for all times. The period T may be calculated as

---

it in the right member of (1.33), we immediately realize that x_0 is an equilibrium, but at this point the Lipschitz condition is violated. In its simplest form the equation is $\dot{x} = \sqrt{x}$, a classical example of non-uniqueness of the solution: both functions $x(t) = 0$ and $x(t) = (t - t_0)^2/2$ are solutions satisfying the initial condition $x(t_0) = 0$, whatever t_0 is. E.g., if we cut the stalk of an apple it could happen that the apple stays on the tree for an arbitrary time, until it realizes that the stalk has been severed, and eventually it falls down: this is something that we can *imagine*, and we see sometimes in cartoons, but is hardly seen in real life. The reason for this apparent paradox lies precisely in the reduction that we have performed using conservation of energy. We must remember that the orbit in phase plane is defined *implicitly* by the conservation of energy, namely by (1.32). If $E = V(x_0)$, then we have $p = 0$ and so also $\frac{\partial H}{\partial p} = 0$, so that we cannot determine p as a function of x using the implicit function theorem. If we have also $\frac{\partial H}{\partial x}(x_0) = 0$, then x_0 is a true equilibrium, and the solution $x(t) = x_0$ is unique; else we should express x as a function of p and replace it in the second equation of (1.33), thus getting $\dot{p} = -\frac{dV}{dx}\big|_{x=x(p)}$. The latter equation includes the information that the acceleration is not zero at x_0, so that the solution is unique. Therefore, the paradox disappears if we calculate *local* solutions corresponding to different parts of the curve and then glue them together; the resulting solution is regular. In practice we avoid such a procedure: just solve Eq. (1.34) as explained in the text and throw away the spurious equilibrium solution.

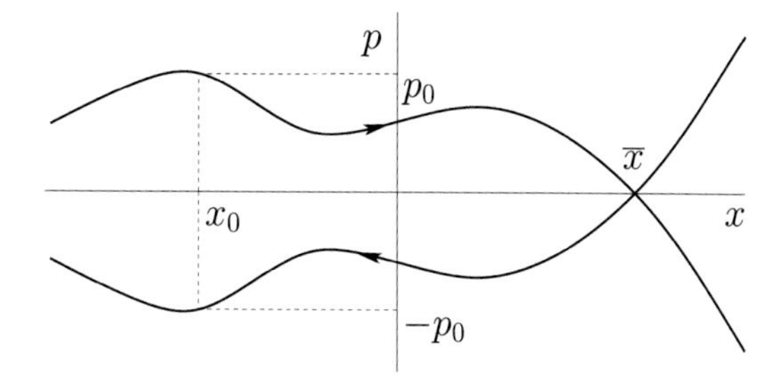

Figure 1.5 Illustrating the case of curves that intersect in an equilibrium point.

$$T = 2\sqrt{\frac{2}{m}} \int_{x_{\min}}^{x_{\max}} \frac{d\xi}{\sqrt{E - V(\xi)}} . \tag{1.37}$$

The last case is concerned with curves that intersect at an equilibrium $\overline{x}$, as illustrated in Fig. 1.5. This happens when the potential has a local maximum in $\overline{x}$, and we set $E = V(\overline{x})$.

1.3.2 Systems with One Degree of Freedom

The discussion of Section 1.3.1 can be repeated, with a few minor changes, for the generic case of a smooth Hamiltonian $H(q, p)$ on a two-dimensional phase space $\mathscr{F}$ (e.g., the Hamiltonian of Exercise 1.17). Write Hamilton's equations

$$\dot{q} = \frac{\partial H}{\partial p} , \quad \dot{p} = -\frac{\partial H}{\partial q}$$

and recall that a point $(\overline{q}, \overline{p}) \in \mathscr{F}$ is an equilibrium if and only if $\frac{\partial H}{\partial q}(\overline{q}, \overline{p}) = \frac{\partial H}{\partial p}(\overline{q}, \overline{p}) = 0$. In this case $q(t) = \overline{q}$, $p(t) = \overline{p}$ is an orbit. If a point (q_0, p_0) is not an equilibrium, then at least one of $\frac{\partial H}{\partial q}(q_0, p_0)$ and $\frac{\partial H}{\partial p}(q_0, p_0)$ does not vanish. Considering the latter case and setting the value of the energy $E = H(q_0, p_0)$, we can solve the equation $H(q, p) = E$ with respect to p, thus getting $p = p(q, E)$ as a smooth function in a neighbourhood of q_0, p_0. Substituting the latter function in the equation for $\dot{q}$, we get

$$\dot{q} = \frac{\partial H}{\partial p}\bigl(q, p(q, E)\bigr) ,$$

which is easily integrated as

$$\int_{q_0}^{q} \frac{d\xi}{\frac{\partial H}{\partial p}\bigl(\xi, p(\xi, E)\bigr)} = t - t_0 , \tag{1.38}$$

t_0 being the initial time. This gives a monotonic function $t(q)$, which can be inverted, thus giving the solution $q(t)$. By substitution in $p(q, E)$, $p(t)$ also is found, so that the motion along the orbit is fully determined. The procedure is, of course, local, but having reached a new point, say $q_1 = q(t_1), p_1 = p(t_1)$ with $t_1 \neq t_0$, the whole process can be repeated so that the solution is continued along the curve. The integrand in (1.38) becomes singular if at some point it happens that $\frac{\partial H}{\partial p}(\xi, p(\xi, E)) = 0$, but at that point one has $\frac{\partial H}{\partial q} \neq 0$, in view of the assumption that there are no equilibria on the phase curve. Thus, by exchanging the roles of q and p, the integration process can be continued again.

If the phase portrait contains a closed curve with no equilibria, then the corresponding orbit is periodic, and the motion over one period can be integrated. The conclusion is: a system with one degree of freedom can be completely solved by quadratures (i.e., calculation of integrals of known functions).

1.3.3 The Period of Oscillation

Let us consider again a system with one degree of freedom, so that the phase space has dimension 2, and assume that $(\overline{q}, \overline{p})$ is a maximum or minimum for the Hamiltonian. Then $(\overline{q}, \overline{p})$ is an equilibrium, and in the phase portrait it is surrounded by closed curves representing oscillations of the system. It is often interesting to know the period of the oscillation, even forgetting the details about the orbit.

Example 1.23: *The period of oscillation of a pendulum.* Let ℓ be the length of the pendulum, and g be the constant acceleration due to gravity. Then, using as coordinate the angle ϑ with respect to the vertical, the Hamiltonian writes

$$H = \frac{p^2}{2} - \frac{g}{\ell} \cos \vartheta \,. \tag{1.39}$$

Let the initial energy E be fixed, with $-g/l < E < g/l$, so that the pendulum librates around the lower equilibrium point. According to (1.37) the period is given by

$$T = 2 \int_{-\vartheta_0}^{\vartheta_0} \frac{d\vartheta}{\sqrt{E + \frac{g}{\ell} \cos \vartheta}} \,,$$

where ϑ_0 (the amplitude of the oscillation) is the solution of $E = \frac{g}{\ell} \cos \vartheta_0$. By replacing this value of E in the integral and exploiting the symmetry $V(-\vartheta) = V(\vartheta)$ of the potential, we get

$$T = \sqrt{\frac{8\ell}{g}} \int_0^{\vartheta_0} \frac{d\vartheta}{\sqrt{\cos \vartheta - \cos \vartheta_0}} \,.$$

Although apparently simple, this integral cannot be evaluated with elementary methods. However, it can be reduced to an elliptic integral as follows. Using the trigonometric formula $\cos \alpha = 1 - 2 \sin^2 \frac{\alpha}{2}$, we get

$$T = 2\sqrt{\frac{\ell}{g}} \int_0^{\vartheta_0} \frac{d\vartheta}{\sqrt{\sin^2\frac{\vartheta_0}{2} - \sin^2\frac{\vartheta}{2}}} \ .$$

Next, we change the variable by setting

$$\sin\frac{\vartheta}{2} = \sin\frac{\vartheta_0}{2}\sin\varphi \ , \quad 0 \le \varphi \le \frac{\pi}{2} \ ,$$

which in turn gives

$$T = 4\sqrt{\frac{l}{g}}\, K\left(\sin\frac{\vartheta_0}{2}\right) \ , \tag{1.40}$$

where

$$K(m) = \int_0^{\pi/2} \frac{d\varphi}{\sqrt{1 - m^2\sin^2\varphi}} \ , \quad 0 \le m < 1 \tag{1.41}$$

is named the *complete elliptic integral of the first kind.* The numerical values of the integral depending on the parameter $m = \sin\frac{\vartheta_0}{2}$ (or sometimes of ϑ_0) can be found in numerical tables, for example, [1].

Let us see how the integral may by calculated using a series expansion. Recall the series expansion of the inverse of the square root:[14]

$$\frac{1}{\sqrt{1+x}} = 1 + \frac{1}{2}x + \frac{1\cdot 3}{2\cdot 4}x^2 + \frac{1\cdot 3\cdot 5}{2\cdot 4\cdot 6}x^3 + \cdots = \sum_{k\ge 0} \frac{(2k-1)!!}{(2k)!!}x^k \ .$$

This is an absolutely and uniformly convergent series for $|x| < 1$, and so it can be integrated term by term. Setting $x = m^2\sin^2\varphi$ and replacing the series expansion in (1.41), we get

$$K(m) = J_0 + \frac{1}{2}m^2 J_2 + \frac{1\cdot 3}{2\cdot 4}m^4 J_4 + \cdots = \sum_{k\ge 0} \frac{(2k-1)!!}{(2k)!!}m^{2k} J_{2k} \ ,$$

where

$$J_{2k} = \int_0^{\pi/2} \sin^{2k}\varphi \, d\varphi \ .$$

The latter integral may be computed by the recurrent formula

$$J_0 = \frac{\pi}{2} \ , \quad J_{2k} = \frac{2k-1}{2k} J_{2k-2} \ .$$

[14] Here the symbol $n!!$ denotes the *semifactorial* (sometimes inappropriately called *double factorial*), namely $0!! = 1!! = 1$ and, in recurrent form, $n!! = n\cdot(n-2)!!$.

We conclude:

$$K(m) = \frac{\pi}{2}\left(1 + \frac{1}{4}m^2 + \frac{9}{64}m^4 + \cdots\right) = \sum_{k\geq 0}\left(\frac{(2k-1)!!}{(2k!!)}\right)^2 m^{2k} .$$

A rough indication of the change of the period due to the nonlinearity is found by writing the first terms of the expansion of the period T as a function of the amplitude ϑ_0. To this end use the expansion of the sine function

$$m = \sin\frac{\vartheta_0}{2} = \frac{\vartheta_0}{2} + \frac{\vartheta_0^3}{2^3\cdot 3!} + \cdots$$

and replace it in the expression of the period, keeping just the first two terms. This gives the approximate value, for small values of ϑ_0,

$$T = 2\pi\sqrt{\frac{\ell}{g}}\left(1 + \frac{1}{16}\vartheta_0^2 + \cdots\right) ,$$

where the first forgotten term is of order ϑ_0^4. It is straightforward to check that if the amplitude ϑ_0 does not exceed 4×10^{-2}, which is about 2°, then the relative change of the period is less than 10^{-4}. For higher values of the amplitude a longer calculation is needed. *E.D.*

Exercise 1.24: Find an approximate expression of the period for the potential $V(x) = x^2/2 - x^3/3$. *A.E.L.*

Exercise 1.25: Show how to calculate the period of rotation of the pendulum, that is, for $E > g/\ell$ in the Hamiltonian (1.39). *A.E.L.*

Exercise 1.26: Calculate the motion on the separatrix, that is, for $E = g/\ell$, for the Hamiltonian (1.39) of the pendulum. *A.E.L.*

Exercise 1.27: Calculate the period of oscillation for the potentials (x) and (xii) of Exercise 1.16. *A.E.L.*

1.3.4 Higher-Dimensional Models

Some interesting Hamiltonian systems with more than one degree of freedom can be integrated by separating the variables, provided that enough first integrals are known. Here are a few examples.

Example 1.28: *Central forces.* The most classical example is the motion under a central force (see Example 1.20).[15] Using the conservation of the

[15] The procedure illustrated here is the common one which can be found in most books on Rational Mechanics or Classical Mechanics where the problem is discussed starting from Newton's or Lagrange equations. The reader may notice that the Hamiltonian approach is definitely more direct: the relevant facts, such as the conservation of angular momentum and the corresponding reduction to the radial problem with an effective potential, emerge rather spontaneously.

angular momentum one proves that the motion is a planar one. Introducing polar coordinates in the plane orthogonal to the direction of the angular momentum, we write the Lagrangian as

$$L = \frac{1}{2}m(\dot{r}^2 + r^2\dot{\vartheta}^2) - V(r) \, .$$

Calculating the momenta as

$$p_r = m\dot{r} \, , \quad p_\vartheta = mr^2\dot{\vartheta},$$

the Hamiltonian is found to be

$$H = \frac{1}{2m}\left(p_r^2 + \frac{p_\vartheta^2}{r^2}\right) + V(r) \tag{1.42}$$

(this is just a straightforward simplification of Example 1.4). The Hamiltonian possesses two first integrals, namely the Hamiltonian itself and the angular momentum $\ell = p_\vartheta$. Hamilton's equations are

$$\begin{aligned} \dot{r} &= \frac{p_r}{m} \, , & \dot{p}_r &= \frac{p_\vartheta^2}{mr^3} - V'(r), \\ \dot{\vartheta} &= \frac{p_\vartheta}{mr^2} \, , & \dot{p}_\vartheta &= 0 \, . \end{aligned} \tag{1.43}$$

Let $\ell = p_\vartheta(0)$ be the constant value of the angular momentum. It is an easy matter to check that a substitution of this value in the first two equations produces a one-degree-of-freedom system with Hamiltonian

$$H = \frac{p_r^2}{2m} + \frac{\ell^2}{2mr^2} + V(r) \tag{1.44}$$

on the phase space $\mathscr{F} = (0, +\infty) \times \mathbb{R}$. Introducing the *effective potential*

$$V^*(r) = \frac{\ell^2}{2mr^2} + V(r),$$

the Hamiltonian is written as

$$H = \frac{p_r^2}{2m} + V^*(r) \, ,$$

belonging to the class of systems considered in Section 1.3.1. We may perform a complete integration, thus finding the solution $r(t)$. Replacing this function in the equation (1.43) for $\dot{\vartheta}$, the r.h.s. turns out to be a function of time only, so that the equation can be integrated. This shows that the problem can be completely solved by quadratures. *E.D.*

Exercise 1.29: The phase space for the Hamiltonian (1.42) is identified as $\mathscr{F} = (0, +\infty) \times \mathbb{T} \times \mathbb{R}^2$. Determine the form of the invariant surfaces defined by the two first integrals H and M_z. *A.E.L.*

Exercise 1.30: Consider the Hamiltonian of example 1.4, that is, the central motion in the spatial case. Show that the problem can be completely integrated. *A.E.L.*

Exercise 1.31: Consider the Hamiltonian of the *spherical pendulum*

$$H = \frac{1}{2}\left(p_\vartheta^2 + \frac{p_\varphi^2}{\sin^2\vartheta}\right) - \cos\vartheta,$$

with $0 < \vartheta < \pi$ and $\varphi \in \mathbb{T}$. Show that the problem can be completely integrated by quadratures. *A.E.L.*

2
Canonical Transformations

As for a generic system of differential equations, coordinate transformations can be used to reduce the system to a simpler form. If the system is Hamiltonian, it is desirable to preserve the canonical form of the equations. Transformations satisfying the latter requirement are called *canonical*.

It should be said that the argument of canonical transformations is a rather complex and elaborate one: the scheme of exposition in existing treatises is far from being uniform. Hence a choice on how to present the matter should be made. The present chapter is a proposal.

The aim is to go a little beyond a synthetic exposition as can be usually found in many elementary textbooks on Mechanics; at the same time, there is no pretension of completeness. The purpose is to present the argument in a self-consistent form and in traditional terms, so that the exposition is reasonably simple – according to the author's experience. Most of the chapter is devoted to answering two questions:

(i) to check whether a given transformation is canonical or not; different criteria are proposed in Section 2.1.

(ii) to find an effective method for constructing canonical transformations; the answer rests on the method of *generating functions*.

The first part of the chapter will be concerned with autonomous systems, and time-independent transformations. The case of time dependence will be discussed in Section 2.4, where it is shown how it can be reduced to the time-independent one.

The Hamilton–Jacobi equation, discussed in Section 2.5, makes a clever use of generating functions in order to integrate the canonical equations. In this respect it represents one of the highest points in the development of Mechanics in the nineteenth century, before the work of Poincaré.

A final remark is in order. Nowadays it is common to place the canonical formalism in the framework of *symplectic geometry*. It is also common to call *symplectic transformations* the class which is traditionally called canonical (with some caveats concerning the local and global aspects). Actually, a moderate use of the language of symplectic geometry helps to simplify some particular discussions. For this reason a short introduction to symplectic geometry is reported in Appendix B.

2.1 Preserving the Hamiltonian Form of the Equations

It may be useful to outline again a connection with the Lagrangian formalism. It is well known that Lagrange's equations have the nice property of being invariant with respect to point transformation (i.e., changes of the coordinates in configuration space, which by differentiation generate the corresponding transformations on the generalized velocities). The Hamiltonian formalism removes the tie between generalized coordinates and velocities, so that arbitrary transformations involving all the canonical coordinates may be devised. However, an arbitrary transformation will likely produce equations which are not in Hamiltonian form. The problem then is to characterize a restricted class of transformation which preserves the form of Hamilton's equations.

2.1.1 *Conditions for Canonicity*

The first approach consists in looking for a class of transformations $(q,p) = \mathcal{C}(Q,P)$ satisfying the following

Condition 1: *To every Hamiltonian function $H(q,p)$ one can associate another function $K(Q,P)$ such that the canonical system of equations*

$$\dot{q}_j = \frac{\partial H}{\partial p_j}\ , \quad \dot{p}_j = -\frac{\partial H}{\partial q_j}\ , \quad j = 1,\ldots,n$$

is changed into the system

$$\dot{Q}_j = \frac{\partial K}{\partial P_j}\ , \quad \dot{P}_j = -\frac{\partial K}{\partial Q_j}\ , \quad j = 1,\ldots,n\ ,$$

which is still canonical.

We shall say that such a transformation *preserves the canonical form of the equations.* Condition 1 applies to both cases of time-independent and time-dependent transformations. A more restrictive request is expressed by

Condition 2: *The transformation preserves the canonical form of the equations with the new Hamiltonian*

$$K(Q,P) = H(q,p)\big|_{q=q(Q,P)\,,\,p=p(Q,P)}\ .$$

Here, the difference with respect to condition 1 is that the new Hamiltonian is constructed by a direct substitution of the transformation in the old one. Transformations satisfying condition 2 are often called *completely canonical* (see, e.g., Tullio Levi Civita and Ugo Amaldi [150], ch. X, § 12).[1] This is the class of transformations which we will be particularly concerned with in the present notes; we shall name them *canonical*, with no further specification.

The weaker condition 1 turns out to be useful in two cases. The first one concerns time-dependent transformations; the second case will be illustrated in Example 2.3.

Exercise 2.1: Prove that the composition of transformations satisfying either condition 1 or 2 satisfies at least one of them. *A.E.L.*

The following examples show that there are transformations satisfying the preceding conditions.

Example 2.2: *Translation.* The transformation

$$q_j = Q_j + a_j \ , \quad p_j = P_j + b_j \ , \quad 1 \le j \le n \ ,$$

where $a = (a_1, \ldots, a_n)$ and $b = (b_1, \ldots, b_n)$ are constants, preserves the canonical form of the equations with the new Hamiltonian

$$K(Q,P) = H(q,p)\Big|_{q=Q+a\,,\,p=P+b} \ .$$

This is easy to show, for we calculate

$$\begin{aligned} \dot{Q}_j = \dot{q}_j &= \left.\frac{\partial H}{\partial p_j}\right|_{q=Q+a,\,p=P=b} = \frac{\partial K}{\partial P_j} \ , \\ \dot{P}_j = \dot{p}_j &= -\left.\frac{\partial H}{\partial q_j}\right|_{q=Q+a,\,p=P=b} = -\frac{\partial K}{\partial Q_j} \ . \end{aligned}$$

Thus, the transformation is canonical. *E.D.*

Example 2.3: *Scaling transformation.* The transformation

$$q_j = \alpha Q_j \ , \quad p_j = \beta P_j \ , \quad 1 \le j \le n \ ,$$

with real constants α and β, preserves the canonical form of the equations with the new Hamiltonian

$$K(Q,P) = \frac{1}{\alpha\beta} H(q,p)\Big|_{q=\alpha Q,\,p=\beta P} \ .$$

---

[1] Some authors call canonical any transformation satisfying the requisite of preserving the canonical form of the equations, as stated by condition 1 in the text. For example, this is the attitude of Aurel F. Wintner [212] and of Felix R. Gantmacher [74]. Others, e.g., Jurgen K. Moser and Eduard J. Zehnder [177], call transformations satisfying condition 1 *generalized canonical.*

Here too, the proof is not difficult. Setting $H'(Q,P) = H(q,p)\big|_{q=\alpha Q,\, p=\beta P}$, we calculate

$$\frac{\partial H'}{\partial Q_j} = \alpha \frac{\partial H}{\partial q_j}\bigg|_{q=\alpha Q,\, p=\beta P} , \qquad \frac{\partial H'}{\partial P_j} = \beta \frac{\partial H}{\partial q_j}\bigg|_{q=\alpha Q,\, p=\beta P} .$$

Therefore we have

$$\begin{aligned} \dot{Q}_j &= \frac{1}{\alpha}\dot{q}_j = \frac{1}{\alpha}\frac{\partial H}{\partial p_j}\bigg|_{q=\alpha Q,\, p=\beta P} = \frac{1}{\alpha\beta}\frac{\partial H'}{\partial P_j} = \frac{\partial K}{\partial P_j} , \\ \dot{P}_j &= \frac{1}{\beta}\dot{p}_j = -\frac{1}{\alpha}\frac{\partial H}{\partial p_j}\bigg|_{q=\alpha Q,\, p=\beta P} = -\frac{1}{\alpha\beta}\frac{\partial H'}{\partial Q_j} = -\frac{\partial K}{\partial Q_j} . \end{aligned}$$

Therefore, the transformation always satisfies condition 1. However, condition 2 is fulfilled only in case $\alpha\beta = 1$, and so we shall call it canonical only in the latter case.[2] E.D.

Exercise 2.4: Show that the transformation

$$q_j = \alpha_j Q_j , \quad p_j = \beta_j P_j , \quad j = 1, \ldots, n ,$$

with α, $\beta \neq 0$, satisfies condition 1 only in case $\alpha_1\beta_1 = \cdots = \alpha_n\beta_n$. A.E.L.

Example 2.5: *Exchange of conjugated coordinates.* The transformation

$$q_j = P_j , \quad p_j = -Q_j , \quad 1 \leq j \leq n$$

preserves the canonical form of the equations with the new Hamiltonian

$$K(Q,P) = H(q,p)|_{q=P\,,\, p=-Q} .$$

Thus, it is a canonical transformation. E.D.

Exercise 2.6: Let $J \subset \{1,\ldots,n\}$ and $K \subset \{1,\ldots,n\}$ be a partition of $\{1,\ldots,n\}$, that is, $J \cap K = \emptyset$ and $J \cup K = \{1,\ldots,n\}$. Show that the following transformation is canonical:

$$\begin{aligned} q_j &= Q_j , \quad p_j = P_j \quad \text{for} \quad j \in J , \\ q_k &= P_k , \quad p_k = -Q_k \quad \text{for} \quad k \in K . \end{aligned}$$

A.E.L.

[2] This example shows that restricting the use of the adjective *canonical* only to time-independent transformations satisfying condition 2 essentially means that we are excluding the class of scaling transformations for which $\alpha\beta \neq 1$. Including all such transformations makes the definition more general, of course. However, it also makes the exposition unnecessarily complicated, which we want to avoid. The more general framework is recovered by keeping in mind that a canonical transformation in the sense intended here can always be composed with a scaling transformation, which sometimes is a useful device.

The exercise shows that the distinction between coordinates and momenta is natural in the framework of Mechanics but becomes rather fuzzy in the general framework of Hamiltonian dynamics. Conversely, the conjugacy between pairs of canonical variables is a distinctive character of the Hamiltonian formalism.

2.1.2 Symplecticity of the Jacobian Matrix

Here it is convenient to use the compact notation of Section 1.1.3. Write the transformation as $\mathbf{x} = u(\mathbf{y})$, and let $K(\mathbf{y}) = H \circ u(\mathbf{y})$ be the Hamiltonian expressed in the new coordinates $\mathbf{y}$.

Proposition 2.7: *The transformation* $\mathbf{x} = u(\mathbf{y})$ *is canonical if and only if the Jacobian matrix* $u_{\mathbf{y}}$ *of the transformation satisfies*

$$u_{\mathbf{y}}^{\top} \mathsf{J} u_{\mathbf{y}} = \mathsf{J} \ . \tag{2.1}$$

Here, J is the matrix defined by (1.18). In the language of symplectic geometry, as in Section B.1, we say that the Jacobian matrix of the transformation must be symplectic.

Proof. Using coordinates, calculate

$$\frac{\partial K}{\partial y_k} = \sum_{j=1}^{2n} \frac{\partial x_j}{\partial y_k} \frac{\partial H}{\partial x_j}\bigg|_{\mathbf{x}=u(\mathbf{y})} \ ,$$

or, in compact notation,

$$\partial_{\mathbf{y}} K = u_{\mathbf{y}}^{\top} \partial_{\mathbf{x}} H \circ u \ .$$

Using $\mathsf{J}^2 = -\mathsf{I}$, write the canonical equations for H as $-\mathsf{J}\dot{\mathbf{x}} = \partial_{\mathbf{x}} H$, and using also $\dot{\mathbf{x}} = u_{\mathbf{y}}\dot{\mathbf{y}}$, calculate $-\mathsf{J}u_{\mathbf{y}}\dot{\mathbf{y}} = \partial_{\mathbf{x}} H \circ u$. Finally, multiply both sides of the latter relation by $u_{\mathbf{y}}^{\top}$ and get $-u_{\mathbf{y}}^{\top}\mathsf{J}u_{\mathbf{y}}\dot{\mathbf{y}} = u_{\mathbf{y}}^{\top}\partial_{\mathbf{x}} H \circ u = \partial_{\mathbf{y}} K$. Therefore, the equations for $\mathbf{y}$ are written

$$-u_{\mathbf{y}}^{\top} \mathsf{J} u_y \dot{\mathbf{y}} = \partial_{\mathbf{y}} K \ .$$

They are canonical with the Hamiltonian $K(\mathbf{y})$ provided that $u_{\mathbf{y}}^{\top}\mathsf{J}u_{\mathbf{y}} = \mathsf{J}$. *Q.E.D.*

Exercise 2.8: Prove that the transformation

$$q_j = Q_j \cos\alpha_j - P_j \sin\alpha_j \ , \quad p_j = Q_j \sin\alpha_j + P_j \cos\alpha_j$$

is canonical. *A.E.L.*

Exercise 2.9: Let R be an orthogonal $2n \times 2n$ real matrix. Show that the transformation $\mathbf{x} = \mathsf{R}\mathbf{y}$ is generally not canonical. *A.E.L.*

2.1.3 Preservation of Poisson Brackets

We revert now to the traditional notation. As we remarked in Section 1.2.2, the Hamiltonian formalism can be expressed in terms of Poisson brackets, saying that *the time evolution of any dynamical variable f is given by equation $\dot{f} = \{f, H\}$* . This leads to a characterization of canonical transformations as possessing the property of leaving invariant the form of the Poisson brackets (see, e.g., [129]).

Let $(q,p) = \mathscr{C}(Q,P)$ be a coordinate transformation, and denote by $\mathscr{C}f$ the transformed function

$$(\mathscr{C}f)(Q,P) = f(q,p)\Big|_{(q,p)=\mathscr{C}(Q,P)} .$$

Also, denote by $\{\cdot,\cdot\}_{q,p}$ and by $\{\cdot,\cdot\}_{Q,P}$ the Poisson brackets with respect to the conjugate variables q,p and Q,P, respectively. Consider now the class of transformations satisfying the condition that Diagram (2.2) is commutative for every pair of functions f and g:

$$\begin{array}{ccc} f, g & \xrightarrow{\mathscr{C}} & \mathscr{C}f, \mathscr{C}g \\ {\scriptstyle \{\cdot,\cdot\}_{q,p}}\big\downarrow & & \big\downarrow{\scriptstyle \{\cdot,\cdot\}_{Q,P}} \\ \{f,g\}_{q,p} & \underset{\mathscr{C}}{\longrightarrow} & \mathscr{C}(\{f,g\}_{q,p}) = \{\mathscr{C}f, \mathscr{C}g\}_{Q,P} . \end{array} \tag{2.2}$$

In words, one obtains the same result by both (a) computing the Poisson brackets with respect to the variables q,p and then changing the variables in the result, and (b) changing the variables and then computing the Poisson brackets with respect to the new variables Q,P. If this happens to be true, we shall say that the transformation *preserves the Poisson brackets.*

Proposition 2.10: *A transformation $(q,p) = \mathscr{C}(Q,P)$ is canonical if and only if it preserves the Poisson brackets, that is, the diagram (2.2) is commutative for every pair of functions f, g.*

The difficult part of the proof is the "only if," that is, that the condition is necessary. To see it, we need to investigate to what extent we can turn the criterion expressed by the latter proposition into a practically applicable one.

Let us consider the coordinates q,p as functions on the phase space; it is an easy matter to check that the relations

$$\begin{aligned} &\{q_j, q_k\} = \{p_j, p_k\} = 0 , \\ &\{q_j, p_k\} = \delta_{jk} , \qquad\qquad 1 \le j \le n , \ 1 \le k \le n \end{aligned} \tag{2.3}$$

hold true, where δ_{jk} is the Kronecker symbol. These expressions are sometimes called the *fundamental Poisson brackets.*

We prove the following.

Lemma 2.11: *A transformation preserves the Poisson brackets between any two functions if and only if it preserves the fundamental Poisson brackets.*

As a direct consequence, Proposition 2.10 can be reformulated in a more useful manner as

Proposition 2.12: *A transformation $(q,p) = \mathscr{C}(Q,P)$, is canonical if and only if it preserves the fundamental Poisson brackets, that is,*

$$(2.4)\qquad \begin{aligned} &\{q_j, q_k\}_{Q,P} = \{p_j, p_k\}_{Q,P} = 0\,, \\ &\{q_j, p_k\}_{Q,P} = \delta_{jk}\,, \qquad\qquad 1 \le j \le n\,,\quad 1 \le k \le n\,. \end{aligned}$$

The latter property is the same thing as the property of symplecticity of the Jacobian matrix of the transformation. However, it is not immediately clear that the stronger condition of Proposition 2.10 applies, namely that the Poisson brackets between *any two functions* must be preserved.

Proof of Lemma 2.11. If the Poisson brackets between any two functions is preserved, then the fundamental Poisson brackets are preserved, too; so, we must prove only the converse. To this end, first check that if we are given a function $f(\varphi_1, \ldots, \varphi_r)$, where, in turn, $\varphi_1, \ldots, \varphi_r$ are functions of the canonical variables q, p, then

$$\{f, g\} = \sum_{l=1}^{r} \frac{\partial f}{\partial \varphi_l} \{\varphi_l, g\}\,,$$

where $g(q,p)$ is any function. This is just a matter of straightforward differentiation. Put now $r = 2n$ and $\varphi_1(Q,P) = q_1(Q,P), \ldots, \varphi_{2n}(Q,P) = p_n(Q,P)$, and consider also $g(q,p)$ as a function of the new variables Q, P through $\varphi_1, \ldots, \varphi_{2n}$. Then, using the identity just presented, calculate

$$\begin{aligned} \{f, g\}_{Q,P} = \sum_{j,k} \biggl(&\frac{\partial f}{\partial q_j}\frac{\partial g}{\partial q_k}\{q_j, q_k\}_{Q,P} + \frac{\partial f}{\partial q_j}\frac{\partial g}{\partial p_k}\{q_j, p_k\}_{Q,P} \\ &+ \frac{\partial f}{\partial p_j}\frac{\partial g}{\partial q_k}\{p_j, q_k\}_{Q,P} + \frac{\partial f}{\partial p_j}\frac{\partial g}{\partial p_k}\{p_j, p_k\}_{Q,P} \biggr)\,. \end{aligned}$$

In view of the preservation of the fundamental Poisson brackets we immediately get $\{f, g\}_{Q,P} = \{f, g\}_{q,p}$, namely that the Poisson brackets between f and g is preserved. *Q.E.D.*

Proof of Proposition 2.10. Let the transformation preserve the Poisson brackets. Denoting $Q = Q(q,p)\,,\ P = P(q,p)$ the inverse transformation, let $f(q,p)$ be any of the new coordinates, namely $f(q,p) = Q_j(q,p)$ or $f(q,p) = P_j(q,p)$ for some j. Then $\dot{f} = \{f, H\}_{q,p}$. On the other hand, by preservation of the Poisson brackets we also have, after changing the variables, $\dot{f} = \{f, H\}_{Q,P}$, that is, $\dot{Q}_j = \frac{\partial H}{\partial P_j}$ or $\dot{P}_j = -\frac{\partial H}{\partial Q_j}$, where $H(Q,P) = H(q,p)\bigr|_{(q,p)=\mathscr{C}(Q,P)}$. Therefore, the transformed equations keep the canonical form by just transforming the Hamiltonian, as required by condition 2. Conversely, let the transformation be canonical, and let, for example, $f = q_j$ and $H = p_k$ for some j and k. Thus, $\dot{f} = \{q_j, p_k\} = \delta_{j,k}$. On the other hand, after transforming to new variables we have $\dot{f} = \{f, H\}_{Q,P}$, because in the new variables the equations are still in canonical form. Since the time derivative of f must be the same after the transformation, we conclude $\{q_j, p_k\}_{Q,P} = \delta_{j,k}$. The argument applies to any pair of canonical coordinates q, p, and this means that the fundamental Poisson brackets are preserved. By Lemma 2.11 this implies that the Poisson brackets are preserved. *Q.E.D.*

Example 2.13: *The case of one degree of freedom.* In the case $n = 1$ the canonicity condition to be met for the transformation to be canonical may be written as

$$\{q, p\}_{Q,P} = \det \begin{pmatrix} \frac{\partial q}{\partial Q} & \frac{\partial q}{\partial P} \\ \frac{\partial p}{\partial Q} & \frac{\partial p}{\partial P} \end{pmatrix} = 1 ,$$

which means that the transformation *must preserve both the area and the orientation.*[3] For instance, the scaling transformation $q = \alpha Q\ ,\quad p = P/\alpha$ is canonical. Conversely, the transformation to polar coordinates in the phase plane is not area preserving. A similar area-preserving transformation is

$$q = \sqrt{2I}\cos\varphi\ ,\quad p = \sqrt{2I}\sin\varphi\ . \tag{2.5}$$

The variables I, φ thus defined are called *action-angle variables for the harmonic oscillator.* *E.D.*

Example 2.14: *Harmonic oscillators.* The Hamiltonian of a system of harmonic oscillators is

$$H(q, p) = \sum_{j=1}^{n} \frac{1}{2}\left(p_j^2 + \omega_j^2 q_j^2\right)\ ,\quad \omega = (\omega_1, \ldots, \omega_n) \in \mathbb{R}^n\ .$$

The scaling transformation $q_j = Q_j/\sqrt{\omega_j}\ , p_j = P_j\sqrt{\omega_j}$ for $1 \le j \le n$ gives the Hamiltonian the more symmetric form

$$K(Q, P) = \sum_j \frac{\omega_j}{2}\left(P_j^2 + Q_j^2\right)\ .$$

[3] If $\{q, p\}_{Q,P} = -1$, then the transformation preserves the area but not the orientation. Strictly, this is the case for the transformation (2.5) if one considers φ (the angle) as the new coordinate and I (the action) as the new momentum; this is indeed the usual choice, and the transformation is widely used in this form. However, it is usually harmless, in particular if the system is a reversible one, as happens for most physical models.

The Hamiltonian can be further simplified by transforming to action-angle variables (2.5), namely $Q_j = \sqrt{2I_j}\cos\varphi_j$, $P_j = \sqrt{2I_j}\sin\varphi_j$ for $1 \le j \le n$, thus getting

$$H(I,\varphi) = \sum_j \omega_j I_j \ .$$

E.D.

2.1.4 The Hamiltonian Flow as a Canonical Transformation

Consider a point (q,p) of the phase space, and let $(q_t,p_t) = \phi^t(q,p)$ be the evolved point under the flow ϕ^t generated by a canonical system with Hamiltonian $H(q,p)$. Let t be fixed. The symbol φ^t for the flow may be interpreted either as an action of the flow that transports a point, or as a transformation that assigns (q_t,p_t) as new coordinates of the point (q,p). In the latter case we are given a one-parameter family of coordinate transformations on the phase space which reduces to identity for $t = 0$.

Proposition 2.15: *Let ϕ^t denote the flow of the Hamiltonian $H(q,p)$. Then for every t the transformation $\bigl(q_t(q,p),p_t(q,p)\bigr) = \phi^t(q,p)$ is canonical.*

Proof. We prove that for every t the fundamental Poisson brackets are preserved, namely that

$$\begin{aligned}
&\{q_{j,t},q_{k,t}\}_{q,p} = \{p_{j,t},p_{k,t}\}_{q,p} = 0\ ,\\
&\{q_{j,t},p_{k,t}\}_{q,p} = \delta_{j,k}\ .
\end{aligned}$$

This is true for $t = 0$, because $\phi^0(q,p) = (q,p)$ is the identity. Let us prove that at every point (q,p) of the phase space one has

$$\frac{d}{dt}\{q_{j,t},q_{k,t}\} = \frac{d}{dt}\{p_{j,t},p_{k,t}\} = \frac{d}{dt}\{q_{j,t},p_{k,t}\} = 0.$$

To this end, recalling again that $q_{j,0} = q_j$, $p_{j,0} = p_j$,

$$q_{j,t} = q_j + t\frac{\partial H}{\partial p_j} + \ldots\ ,\qquad p_{j,t} = p_j - t\frac{\partial H}{\partial q_j} + \ldots\ ,$$

where the dots stand for terms of higher order in t. Thus we have

$$\{q_{j,t},q_{k,t}\} = \{q_j,q_k\} + t\left[\left\{\frac{\partial H}{\partial p_j},q_k\right\} + \left\{q_j,\frac{\partial H}{\partial p_k}\right\}\right] + \ldots\ ,$$

which in turn means that one has

$$\frac{d}{dt}\{q_{j,t},q_{k,t}\} = \left\{\frac{\partial H}{\partial p_j},q_k\right\} + \left\{q_j,\frac{\partial H}{\partial p_k}\right\} = -\frac{\partial}{\partial p_k}\frac{\partial H}{\partial p_j} + \frac{\partial}{\partial p_j}\frac{\partial H}{\partial p_k} = 0\ .$$

With a similar calculation we get

$$\begin{aligned}\frac{d}{dt}\{p_{j,t},p_{k,t}\} &= -\left\{\frac{\partial H}{\partial q_j},p_k\right\} - \left\{p_j,\frac{\partial H}{\partial q_k}\right\} = 0\ ,\\ \frac{d}{dt}\{q_{j,t},p_{k,t}\} &= \left\{\frac{\partial H}{\partial p_j},p_k\right\} - \left\{q_j,\frac{\partial H}{\partial q_k}\right\} = 0\ .\end{aligned}$$

Since the fundamental Poisson brackets have a zero time derivative at every point, they keep a constant value along every orbit; that is, the value at $t = 0$, for which the flow is the identity. Hence the property of preserving the fundamental Poisson brackets is satisfied for any t, as claimed. *Q.E.D.*

The proposition has a suggestive geometrical interpretation: *the Hamiltonian flow can be seen as the unfolding of a canonical transformation parametrically depending on time.*

The latter remark is more than a curious and beautiful property of canonical flows. In Chapter 3 we shall see that it plays a main role in the proof of the theorems of Liouville and of Arnold Jost which characterize integrable systems. Moreover, canonical transformations defined through a canonical flow are a useful tool which will be exploited through the methods based on Lie series; this will be discussed in detail in Chapter 6.

2.1.5 Preservation of Lagrange Brackets

Suppose that the canonical coordinates q, p are given as functions of two variables u, v. The Lagrange bracket $[u, v]$ is defined as[4]

$$[u,v] = \sum_{j=1}^{n}\left(\frac{\partial q_j}{\partial u}\frac{\partial p_j}{\partial v} - \frac{\partial q_j}{\partial v}\frac{\partial p_j}{\partial u}\right)\ . \tag{2.6}$$

Let the canonical coordinates q, p be expressed as differentiable and invertible independent functions of $2n$ variables $u_1, \ldots, u_{2n}$ as

$$q_j = q_j(u_1,\ldots,u_{2n})\ ,\quad p_j = p_j(u_1,\ldots,u_{2n})\ ,\quad j,k = 1,\ldots,n\ , \tag{2.7}$$

so that the variables u can be expressed as differentiable functions of the canonical coordinates q, p, by inversion.

---

[4] The Lagrange bracket first appears in the papers [140], [141] soon followed by a more synthetic presentation in [142]. The argument was later inserted by Lagrange in the third edition of the *Mécanique Analytique* [143], tome I, seconde partie, sect. V, presented by Lagrange as *“Méthode générale d’approximation pour les problèmes de la Dynamique, fondée sur la variation des constantes arbitraires.”* These writings actually represent a large anticipation of the Hamiltonian formalism and of symplectic geometry (see, e.g., [118] and [168]).

Lemma 2.16: *Let the $2n \times 2n$ matrices $\mathsf{A} = \{A_{j,k}\}$ and $\mathsf{B} = \{B_{j,k}\}$ be defined as $A_{jk} = \{u_j, u_k\}$ and $B_{jk} = [u_j, u_k]$. Then we have*

$$\mathsf{A}\mathsf{B}^\top = \mathsf{I} \,, \tag{2.8}$$

the identity matrix.

Proof. Let us denote for a moment $\mathbf{w} = (q_1, \ldots, q_n, p_1, \ldots, p_n)$, so that the transformation (2.7) and its inverse read

$$w_j = w_j(u_1, \ldots, u_{2n}) \,, \quad u_j = u_j(w_1, \ldots, w_{2n}) \,, \quad j = 1, \ldots, 2n \,.$$

Recall the known functional relation $\mathbf{u_w} \cdot \mathbf{w_u} = \mathsf{I}$, where $\mathbf{u_w}$ and $\mathbf{w_u}$ are the Jacobian matrices of the transformations. In more explicit terms we have

$$\sum_{l=1}^{2n} \frac{\partial w_j}{\partial u_l} \frac{\partial u_l}{\partial w_k} = \delta_{j,k} \,.$$

Restoring the symbols $q_1, \ldots, q_n, p_1, \ldots, p_n$ in place of $w_1, \ldots, w_{2n}$ for the canonical variables the latter relation splits as

$$\begin{aligned}
\sum_{l=1}^{2n} \frac{\partial u_l}{\partial q_j} \frac{\partial p_k}{\partial u_l} &= \sum_{l=1}^{2n} \frac{\partial u_l}{\partial p_j} \frac{\partial q_k}{\partial u_l} = 0 \,, \\
\sum_{l=1}^{2n} \frac{\partial u_l}{\partial q_j} \frac{\partial q_k}{\partial u_l} &= \sum_{l=1}^{2n} \frac{\partial u_l}{\partial p_j} \frac{\partial p_k}{\partial u_l} = \delta_{jk} \,.
\end{aligned}$$

Using the latter identities, for $r, s = 1, \ldots, 2n$, calculate

$$\begin{aligned}
&\sum_{l=1}^{2n} \{u_l, u_r\}[u_l, u_s] \\
&\quad = \sum_{l=1}^{2n} \sum_{j=1}^{n} \sum_{k=1}^{n} \left(\frac{\partial u_l}{\partial q_j} \frac{\partial u_r}{\partial p_j} - \frac{\partial u_l}{\partial p_j} \frac{\partial u_r}{\partial q_j} \right) \left(\frac{\partial q_k}{\partial u_l} \frac{\partial p_k}{\partial u_s} - \frac{\partial q_k}{\partial u_s} \frac{\partial p_k}{\partial u_l} \right) \\
&\quad = \sum_{j=1}^{n} \sum_{k=1}^{n} \left[\left(\sum_{l=1}^{2n} \frac{\partial u_l}{\partial q_j} \frac{\partial q_k}{\partial u_l} \right) \frac{\partial u_r}{\partial p_j} \frac{\partial p_k}{\partial u_s} - \left(\sum_{l=1}^{2n} \frac{\partial u_l}{\partial q_j} \frac{\partial p_k}{\partial u_l} \right) \frac{\partial u_r}{\partial p_j} \frac{\partial q_k}{\partial u_s} \right. \\
&\qquad \left. - \left(\sum_{l=1}^{2n} \frac{\partial u_l}{\partial p_j} \frac{\partial q_k}{\partial u_l} \right) \frac{\partial u_r}{\partial q_j} \frac{\partial p_k}{\partial u_s} + \left(\sum_{l=1}^{2n} \frac{\partial u_l}{\partial p_j} \frac{\partial p_k}{\partial u_l} \right) \frac{\partial u_r}{\partial q_j} \frac{\partial q_k}{\partial u_s} \right] \\
&\quad = \sum_{j=1}^{n} \left(\frac{\partial u_r}{\partial p_j} \frac{\partial p_j}{\partial u_s} + \frac{\partial u_r}{\partial q_j} \frac{\partial q_j}{\partial u_s} \right) = \delta_{rs} \,.
\end{aligned}$$

Q.E.D.

It is now convenient to rename $(u_1, \ldots, u_{2n})$ as $(Q_1, \ldots, Q_n, P_1, \ldots, P_n)$.

Choosing the transformation (2.7) to be the identity, we have

$$(2.9)\qquad \begin{aligned} &[Q_j, Q_k] = [P_j, P_k] = 0\,, \\ &[Q_j, P_k] = \delta_{jk}\,, \quad 1 \le j \le n\,,\ 1 \le k \le n\,. \end{aligned}$$

These expressions are called the *fundamental Lagrange brackets.*

Proposition 2.17: *A transformation $(q,p) = \mathscr{C}(Q,P)$ is canonical if and only if it preserves the fundamental Lagrange brackets.*

Proof. Recall Proposition 2.12 on preservation of the fundamental Poisson brackets, and use Lemma 2.16. If the fundamental Poisson brackets are preserved, then $\mathsf{A} = \mathsf{J}$, the symplectic matrix defined by (1.18); using $\mathsf{J}\mathsf{J}^\top = \mathsf{I}$, by (2.8) we conclude $\mathsf{B} = \mathsf{J}$. Conversely, if the fundamental Lagrange brackets are preserved, then $\mathsf{B} = \mathsf{J}$, and by (2.8) we conclude $\mathsf{A} = \mathsf{J}$. *Q.E.D.*

2.2 Differential Forms and Integral Invariants

In the case of one degree of freedom a canonical transformation is characterized by the property of preserving areas (see Example 2.13); that is, writing $p = p(Q,P)\,,\ q = q(Q,P)$ we have $dq\,dp = dQ\,dP$. The argument may be extended to more than one degree of freedom. The traditional method, going back to Poincaré, consists in projecting a small two-dimensional surface on the planes q_j, p_j and adding up algebraically the areas of the projections.[5] The application of Stokes theorem allows us to reduce the calculation of the area to an integral of a 1-form over a closed curve.

2.2.1 *Preservation of a 2-Form*

Still assigning a privileged role to canonically conjugate variables, let us consider the differential 2-form (using the notation of exterior algebra)

$$(2.10)\qquad \omega^2 = \sum_{j=1}^{n} dq_j \wedge dp_j\,,$$

which has been named the *absolute integral invariant* by Poincaré.

Proposition 2.18: *A transformation $(q,p) = \mathscr{C}(Q,P)$ is canonical if and only if it preserves the 2-form $\omega^2 = \sum_j dq_j \wedge dp_j$, namely*

$$(2.11)\qquad \sum_j dq_j \wedge dp_j = \sum_j dQ_j \wedge dP_j.$$

---

[5] The study of integral invariants under the Hamiltonian flow is developed by Poincaré in [187], ch. II, and reelaborated in [188], tome III, ch. XXII; in the latter chapter reference is made to the general discussion on differential forms in [186] and [189]. See also [209], ch. X, or, more recent, [9] and [177].

Proof. Using the formula for changing variables, calculate

$$\begin{aligned}\sum_j dq_j \wedge dp_j &= \sum_j \sum_{k,l} \left(\frac{\partial q_j}{\partial Q_k} dQ_k + \frac{\partial q_j}{\partial P_k} dP_k \right) \wedge \left(\frac{\partial p_j}{\partial Q_l} dQ_l + \frac{\partial p_j}{\partial P_l} dP_l \right) \\ &= \sum_{k<l} \bigl([Q_k, Q_l] dQ_k \wedge dQ_l + [P_k, P_l] dP_k \wedge dP_l \bigr) \\ &\quad + \sum_{k,l} [Q_k, P_l] dQ_k \wedge dP_l \ ,\end{aligned}$$

which shows that the coefficients of the transformed differential form are expressed in terms of Lagrange brackets. If the transformation is canonical, then by Proposition 2.17 we get $\sum_j dp_j \wedge dq_j = \sum_j dP_j \wedge dQ_j$, as claimed. Conversely, if the latter identity is fulfilled, then the Lagrange brackets are preserved. By Proposition 2.17 the claim follows. *Q.E.D.*

2.2.2 Action Integral

Let now γ be a smooth closed curve in the phase space $\mathscr{F}$, which we describe with the canonical coordinates (q, p). Consider the integral over γ of the differential 1-form $\sum_{j=1}^n p_j dq_j$, namely the quantity, or *action*,[6]

$$I = \oint_\gamma \sum_{j=1}^{n} p_j dq_j \tag{2.12}$$

(possibly with a normalizing factor for the length of γ). Poincaré has given to integrals of a differential form on closed surfaces (here curves) the name *relative integral invariants*. In the case of one degree of freedom the integral represents the area inside the curve. The aim is to extend the same concept to more degrees of freedom.

The characterization of canonical transformations via the action integral is expressed as follows. Let $\mathscr{C}(\gamma)$ be the image of the curve γ under a coordinate transformation $(q, p) = \mathscr{C}(Q, P)$.

Proposition 2.19: *A transformation* $(q, p) = \mathscr{C}(Q, P)$ *is canonical if and only if*

$$\oint_\gamma \sum_j p_j dq_j = \oint_{\mathscr{C}(\gamma)} \sum_j P_j dQ_j \ , \tag{2.13}$$

where γ *is any smooth closed curve.*

---

[6] Incidentally, the action so defined played a major role in the first phase of the development of quantum theory, due to Bohr and Sommerfeld. A thorough discussion concerning the quantization process for action variables may be found in Born's treatise [28].

Before coming to the proof, let us make a little detour (maybe a somewhat pedantic one). This is useful in view of the introduction of generating functions in Section 2.3. We recall once more that all considerations made here have a local character.

It is convenient to consider a $4n$-dimensional domain, an open subset of $\mathbb{R}^{4n}$, with coordinates (q, p, Q, P). In this domain the canonical transformation $(q, p) = \mathscr{C}(Q, P)$ determines a $2n$-dimensional manifold, $\mathscr{W}$ say. This seems to introduce a strange and somewhat unpleasant mixing of variables; however, it opens the possibility of finding an effective method to generate transformations that are surely canonical, as we shall discuss in the next section.

In the domain $\mathscr{W}$ above the curve γ is represented as a smooth function $\bigl(q(s), p(s)\bigr)$ with $s \in [0, 1]$ and with the conditions $q(0) = q(1)$, $p(0) = p(1)$. Thus, $\mathscr{C}(\gamma)$ will be represented, with the same parameter s, as $\bigl(Q(s), P(s)\bigr)$ and the relation $\bigl(q(s), p(s))\bigr) = \mathscr{C}\bigl(Q(s), P(s)\bigr)$ apply, with the inherited conditions $\bigl(Q(0), P(0)\bigr) = \bigl(Q(1), P(1)\bigr)$. What we get, in fact, is a closed smooth curve Γ lying on the manifold $\mathscr{W}$, parameterized as $\bigl(q(s), p(s), Q(s), P(s)\bigr)$, which projects on the curves γ and $\mathscr{C}(\gamma)$ on the q, p and Q, P planes, respectively.

With this setting it is convenient to consider the differential 1-form Ω in $\mathscr{W}$ defined as

$$\Omega = \sum_{j=1}^{n} (p_j \, dq_j - P_j \, dQ_j), \tag{2.14}$$

which will play a relevant role.

Lemma 2.20: *The 1-form Ω on $\mathscr{W}$ defined by (2.14) is closed if and only if the transformation $(q, p) = \mathscr{C}(Q, P)$ is canonical.*

Proof. Recall that the differential 1-form satisfies the local condition for being closed in case its external differential is zero.[7] The external differential of the 1-form (2.14) is

---

[7] Consider the 1-form $w = \sum_j \varphi_j(x) dx_j$. The external differential is (recalling that $dx_k \wedge dx_j = -dx_j \wedge dx_k$)

$$dw = \sum_{j,k} \frac{\partial \varphi_j}{\partial x_k} dx_k \wedge dx_j = \sum_{k<j} \left(\frac{\partial \varphi_j}{\partial x_k} - \frac{\partial \varphi_k}{\partial x_j} \right) dx_k \wedge dx_j .$$

In view of the dx being arbitrary, the latter expression is zero if and only if all coefficients of $dx_k \wedge dx_j$ are zero. This is the known local condition for the differential form to be closed.

$$d\Omega = \sum_{j=1}^{n} (dp_j \wedge dq_j - dP_j \wedge dQ_j).$$

In view of Proposition 2.18 we have $d\Omega = 0$ if and only if the transformation is canonical. *Q.E.D.*

We finally come to the

Proof of Proposition 2.19. By construction of the curve Γ, we have

$$\oint_{\Gamma} \sum_{j=1}^{n} (p_j dq_j - P_j dQ_j) = \oint_{\gamma} \sum_{j} p_j dq_j - \oint_{\mathscr{C}(\gamma)} \sum_{j} P_j dQ_j \ .$$

On the other hand, in view of Proposition 2.18, for a closed smooth surface $\Sigma \subset \mathscr{W}$ with border Γ, we have (Stokes' theorem)

$$\oint_{\Gamma} \sum_{j=1}^{n} (p_j dq_j - P_j dQ_j) = \int_{\Sigma} \sum_{j} (dq_j \wedge dp_j - dQ_j \wedge dP_j) = 0 \ ,$$

so that the claim follows. *Q.E.D.*

2.3 Generating Functions

Let us come back to consider the differential 1-form (2.14), namely

$$\Omega = \sum_{j=1}^{n} (p_j \, dq_j - P_j \, dQ_j) \ ,$$

on the $2n$-dimensional manifold $\mathscr{W}$. By Lemma 2.20 it is a closed form. This allows us to reformulate the canonicity criterion of Proposition 2.19 as follows:

A transformation $(q, p) = \mathscr{C}(Q, P)$ is canonical if and only if the differential 1-form Ω on $\mathscr{W}$ is the differential of a function, that is,

$$\sum_{j=1}^{n} (p_j \, dq_j - P_j \, dQ_j) = dS \ . \tag{2.15}$$

The function S is called a *generating function*; it can be considered as a function $S(q, p, Q, P)$ of $4n$ variables, but we must keep in mind that only $2n$ variables are independent, since the relations $(q, p) = \mathscr{C}(Q, P)$ apply. The interesting fact is that by exploiting condition (2.15), we can devise a method that produces transformations which are certainly canonical. However, there is a price to be paid: we have to mix up the $4n$ variables (q, p, Q, P) and make a suitable (local) choice of $2n$ independent variables among them.

2.3.1 A First Form of Generating Function

Let us write the transformation $(q, p) = \mathscr{C}(Q, P)$ in the slightly more explicit form

$$q = q(Q, P) \;, \quad p = p(Q, P) \;. \tag{2.16}$$

Let us also assume that

$$\det \frac{\partial(q_1, \ldots, q_n)}{\partial(P_1, \ldots, P_n)} \neq 0 \;. \tag{2.17}$$

We shall see later that such a strange condition can always be replaced by a similar one (see Lemma 2.30 in Section 2.3.3).

Under condition (2.17) we are allowed to invert the first of (2.16) with respect to P, thus expressing it as a function of (Q, q). It means that we can introduce the mixed variables (q, Q) as independent (local) coordinates on the manifold $\mathscr{W}$. Next, we replace the result in the second part of (2.16), thus getting the functions

$$p_j = \alpha_j(Q, q) \;, \quad P_j = \beta_j(Q, q) \;. \tag{2.18}$$

Hence we rewrite the differential form Ω as

$$\Omega = \sum_{j=1}^{n} \bigl(\alpha_j(Q, q)_j \, dq_j - \beta_j(Q, q) \, dQ_j\bigr) \;.$$

Since it is a closed form, we have

$$\alpha_j(Q, q) = \frac{\partial S}{\partial q_j} \;, \quad \beta_j(Q, q) = \frac{\partial S}{\partial Q_j}$$

with some function $S(Q, q)$. Reversing the argument, we state the following

Proposition 2.21: *Let the function $S(Q, q)$ satisfy the condition*

$$\det \left(\frac{\partial^2 S}{\partial q_j \partial Q_k} \right) \neq 0 \;. \tag{2.19}$$

Then the transformation implicitly defined by

$$p_j = \frac{\partial S}{\partial q_j}(q, Q) \;, \quad P_j = -\frac{\partial S}{\partial Q_j}(q, Q) \;, \quad 1 \le j \le n \tag{2.20}$$

is canonical.

Proof. The transformation clearly satisfies condition (2.15), hence it is canonical. *Q.E.D.*

It should be remarked that the transformation is given in implicit form; however, by condition (2.19), we are allowed to do the inversions needed in order to express either the old variables q, p as functions of the new ones,

Q, P, or the new variables as functions of the old ones. For, inverting the second part of (2.20) with respect to q, we get $q = q(Q, P)$, and replacing the latter function in the first part of (2.20), we get also $p = p(q, Q)\big|_{q=q(Q,P)}$, as required. Conversely, inverting the first part of (2.20) with respect to Q and replacing the result in the second part of (2.20), we obtain the inverse transformation.

Example 2.22: *Exchange of conjugated coordinates.* The canonical transformation $p_j = Q_j$, $P_j = -q_j$ of Example 2.5 is produced by the generating function $S(q, Q) = \sum_j Q_j q_j$. *E.D.*

The class of canonical transformation which can be found with the generating function $S(Q, q)$ of Proposition 2.21 does not exhaust all possibilities. For instance, one will easily realize that the identity does not belong to it – which is quite disappointing. Actually, there are different forms of the generating function, due to the choice of the variables it depends on; the common characteristic is that it must depend on mixed variables, which must be suitably selected.

2.3.2 A Second Form of the Generating Function

A widely used form is given by the following.

Proposition 2.23: *Let the generating function $S(P, q)$ satisfy*

$$\det\left(\frac{\partial^2 S}{\partial P_j \partial q_k}\right) \neq 0 \ . \tag{2.21}$$

Then the transformation implicitly defined by

$$p_j = \frac{\partial S}{\partial q_j}(P, q) \ , \quad Q_j = \frac{\partial S}{\partial P_j}(P, q) \ , \quad 1 \le j \le n \ . \tag{2.22}$$

Proof. It is enough to show that condition (2.15) is satisfied with some function $\tilde{S}$. Such a function is provided by the Legendre transform of $S(P, q)$, namely $\tilde{S} = S - \sum_j P_j Q_j$. For, using (2.22), calculate

$$\begin{aligned} d\tilde{S} &= \sum_j \left(\frac{\partial S}{\partial P_j} dP_j + \frac{\partial S}{\partial q_j} dq_j - P_j dQ_j - Q_j dP_j\right) \\ &= \sum_j (p_j dq_j - P_j dQ_j) \ , \end{aligned}$$

as required. *Q.E.D.*

A direct proof which does not use Legendre's transform can be obtained by replacing condition (2.17) with

$$\det \frac{\partial(q_1, \ldots, q_n)}{\partial(Q_1, \ldots, Q_n)} \neq 0 \ . \tag{2.23}$$

By inverting $q(Q,P)$ with respect to P, one gets the relations

$$p_j = \alpha_j(P,q)\ , \quad Q_j = \beta_j(P,q)$$

in place of (2.18). Then the argument of Section 2.3.1 can be easily adapted.

The form $S(P,q)$ of the generating function is the most common one; actually, many textbooks report only this form, thus skipping the discussion concerning its generality. This is because most useful transformations are expressed with a generating function of the form (2.22). The following examples illustrate some interesting cases.

Example 2.24: *The identity and the scaling transformations.* The function $S(P,q) = \alpha \sum_j P_j q_j$ with $\alpha > 0$ generates the scaling transformation

$$p_j = \alpha P_j\ , \quad Q_j = \alpha q_j\ , \quad 1 \le j \le n\ .$$

For $\alpha = 1$ this is the identity. More generally, the generating function $S(P,q) = \sum_j \alpha_j P_j q_j$ produces a different scaling for different pairs of canonical coordinates. *E.D.*

The next example is particularly relevant because it shows how to handle the class of transformations which are allowed in the framework of the Lagrangian formalism. We should recall that the equations of Lagrange are invariant with respect to coordinate transformation (i.e., diffeomorphisms $q = q(Q)$ involving only the coordinates, usually named *point transformations*). The corresponding transformation for the velocities is calculated as

$$\dot q_j = \sum_{l=1}^{n} \frac{\partial q_j}{\partial Q_l} \dot Q_l\ ,$$

so that it is completely determined. In principle, the corresponding transformation for the momenta p could be found by going back to the expression of the Lagrangian function, applying the transformation and recalculating the new momenta – a definitely cumbersome procedure. Actually, the example provides a standard way to construct a generating function which extends the point transformation to a canonical one. Moreover, it shows that the extension is not unique, a fact which does not occur in the Lagrangian framework.

Example 2.25: *Extended point transformation.* Suppose that we are given a point transformation $q = q(Q)$ which is a diffeomorphism, so that it admits an inverse $Q = Q(q)$, and, moreover,

$$\det\frac{\partial(Q_1,\ldots,Q_n)}{\partial(q_1,\ldots,q_n)} \ne 0\ , \quad \det\frac{\partial(q_1,\ldots,q_n)}{\partial(Q_1,\ldots,Q_n)} \ne 0\ . \tag{2.24}$$

A corresponding canonical transformation can be constructed using the generating function

$$S(P,q) = \sum_k P_k Q_k \big|_{Q=Q(q)} \ .$$

For the complete transformation is

$$q_j = q_j(Q) \ , \quad p_j = \sum_k P_k \frac{\partial Q_k}{\partial q_j}(q) \ , \quad 1 \le j \le n \ .$$

On the other hand, the invertibility condition (2.21) of Proposition 2.23 is satisfied in view of (2.24), since

$$\det\left(\frac{\partial^2 S}{\partial P_k \partial q_j}\right) = \det\left(\frac{\partial Q_k}{\partial q_j}\right) \neq 0 \ .$$

This extension is not unique. The most general extended point transformation is generated by the function

$$S(P,q) = \sum_k P_k Q_k \big|_{Q=Q(q)} + W(q) \ , \tag{2.25}$$

where $W(q)$ is an arbitrary function. *E.D.*

Example 2.26: *Near the identity canonical transformations.* Consider the generating function

$$S(P,q) = \sum_j P_j q_j + \varepsilon f(P,q) \ , \tag{2.26}$$

where $f(P,q)$ is an arbitrary function and ε a real parameter, which is assumed to be small. The invertibility condition (2.21) of Proposition 2.23 is clearly satisfied for small-enough ε. The corresponding canonical transformation in implicit form is

$$p_j = P_j + \varepsilon \frac{\partial f}{\partial q_j}(P,q) \ , \quad Q_j = q_j + \varepsilon \frac{\partial f}{\partial P_j}(P,q) \ , \quad 1 \le j \le n \ .$$

The explicit form can be found, for example, by inverting the second relation with respect to q and replacing the result in the first one. This gives

$$q_j = Q_j - \varepsilon \frac{\partial f}{\partial P_j}(P,Q) + O(\varepsilon^2) \ ,$$

$$p_j = P_j + \varepsilon \frac{\partial f}{\partial q_j}(P,Q) + O(\varepsilon^2) \ .$$

For $\varepsilon = 0$, the transformation is the identity, while for $\varepsilon \neq 0$ the coordinates are changed by a little amount. Such a kind of transformations is the basic tool for the development of perturbation theory. However, it can be remarked that the inversion required to put the transformation in explicit form is often an unpleasant task, especially if one plans to perform an explicit calculation. A different viewpoint consists in representing a near-the-identity canonical transformation as the flow at time ε of a Hamiltonian system (see Section 2.1.4). This is indeed the underlying idea of the algorithms of Lie series and of the Lie transform discussed later, in Chapter 6. These algorithms have the advantage of removing the need of inversions. *E.D.*

2.3.3 The General Class of Generating Functions

Once more, the generating functions discussed till now do not exhaust the class of canonical transformations. Let us exploit the possibility of exchanging a pair of canonically conjugated coordinates, as in Example 2.5 and Exercise 2.6. The underlying idea is the following. The actual difference between the generating functions of Propositions 2.21 and 2.23 is the replacement of the new coordinates Q in condition (2.17) with the conjugated ones P in condition (2.19). We make a different choice by replacing only *some* of the coordinates Q_j with the corresponding P_j.

Formally, we proceed as follows. Pick a partition of $\{1,\ldots,n\}$ into two disjoint subsets J and K, so that $J\cap K=\emptyset$ and $J\cup K=\{1,\ldots,n\}$; there are 2^n such partitions. In condition (2.17) keep Q_j for $j\in J$ and replace Q_k with P_k for $k\in K$.

Proposition 2.27: *Take any partition of the integers $\{1,\ldots,n\}$ into two disjoint sets J,K. Assume that the generating function $S=S(Q_J,P_K,q)$ satisfies the condition*[8]

$$\det\left(\frac{\partial^2 S}{\partial(Q_J,P_K)\partial q_l}\right)\neq 0\,. \tag{2.27}$$

Then the transformation implicitly defined by

$$\begin{aligned} P_j &= -\frac{\partial S}{\partial Q_j} &&\text{for}\quad j\in J\,,\\ Q_k &= \frac{\partial S}{\partial P_k} &&\text{for}\quad k\in K\,,\\ p_l &= \frac{\partial S}{\partial q_l} &&\text{for}\quad 1\le l\le n \end{aligned} \tag{2.28}$$

is canonical. Conversely, for any canonical transformation one can find a partition J,K and a generating function of the form just presented.

The proof that the transformation is canonical requires checking that condition (2.15) is fulfilled. This is just a minor modification of the proof of

[8] Here the short notation $\partial(Q_J,P_K)$ means that the partial derivative is made with respect to ∂Q_j if the label $j\in J$, and with respect to ∂P_k if $k\in K$.

Proposition 2.23, and is left to the reader: just make an appropriate use of Legendre's transform.

The proposition actually shows that we can give the generating functions at least 2^n different forms. Our aim now is to show that Proposition 2.27 actually covers the whole class of time-independent canonical transformations, and that the number 2^n of different generating functions cannot be reduced.

Example 2.28: 2^n *canonical transformations.* Let J, K be a partition of the set $\{1, \ldots, n\}$ into two disjoint subsets, $J \cup K = \{1, \ldots, n\}$, $J \cap K = \emptyset$, and consider the canonical transformation

$$\begin{array}{llll} p_j = Q_j \,, & P_j = -q_j & \text{for} & j \in J \,, \\ p_k = P_k \,, & Q_k = q_k & \text{for} & k \in K \,. \end{array} \tag{2.29}$$

There are 2^n different such transformations. For example, the exchange of conjugated coordinates of Example 2.22 is found by setting $J = \{1, \ldots, n\}$, $K = \emptyset$, and the identity is found by setting $J = \emptyset$, $K = \{1, \ldots, n\}$; see Propositions 2.21 and 2.23, respectively. All the other cases remain uncovered. For a given partition J, K the transformation (2.29) satisfies the condition

$$\det \left(\frac{\partial(q_1, \ldots, q_n)}{\partial(Q_J, P_K)} \right) \neq 0 \,.$$

The corresponding generating function is

$$S\big((Q_J, P_K), q\big) = \sum_{j \in J} Q_j p_j + \sum_{k \in K} P_k q_k \,, \tag{2.30}$$

depending on the new coordinates Q_J, P_K . That is, we use Q_J, P_K as local coordinates on the manifold $\mathscr{W}$. E.D.

Exercise 2.29: Check that for a given partition J, K the canonical transformation (2.29) is produced only by the generating function (2.30). A.E.L.

The exercise shows that the set of 2^n generating functions of Proposition 2.27 is the minimal required one in order to cover the whole class of time-independent canonical transformations.

We come now to show that it is also a sufficient set, that is, it covers the whole class.[9]

---

[9] Some authors allow also a partition of the old coordinates $q_1, \ldots, q_n, p_1, \ldots, p_n$, with the prescription that only one between the pair q_j, p_j of canonically conjugate coordinates can be selected (see, e.g., [74], § 29). This increases the number of different forms of the generating function to 4^n. Such extra freedom is, of course, legitimate, but it is not necessary: the 2^n cases discussed in this section are enough, and their number cannot be decreased (see, e.g., [9], § 48).

Lemma 2.30: *Let $(q,p) = \mathscr{C}(Q,P)$ be a canonical transformation. Then there exists a partition J, K of $\{1,\ldots,n\}$ such that*

$$\det\left(\frac{\partial(q_1,\ldots,q_n)}{\partial(Q_J,P_K)}\right) \neq 0\,. \tag{2.31}$$

Proof. For a given transformation the n functions $q_1(Q,P),\ldots,q_n(Q,P)$ are independent, which means that the $n \times 2n$ Jacobian matrix

$$\frac{\partial(q_1,\ldots,q_n)}{\partial(Q_1,\ldots,Q_n,P_1,\ldots,P_n)} \tag{2.32}$$

has rank n, that is, we can extract a square $n \times n$ submatrix with non-zero determinant. What we should prove is that we can always find a submatrix of the form (2.31), that is, for every pair of canonically conjugate coordinates Q_j, P_j we select only one of the corresponding columns. To this end we make use of the few notions of symplectic geometry collected in Appendix B. In particular we use Lemma B.9. At every point (Q,P) we define a canonical basis $\{\mathbf{e}_1,\ldots,\mathbf{e}_n,\mathbf{d}_1,\ldots,\mathbf{d}_n\}$ by setting

$$\mathbf{e}_j = \left(\frac{\partial Q_j}{\partial Q_1},\ldots,\frac{\partial Q_j}{\partial Q_n},0,\ldots,0\right)\,,\quad \mathbf{d}_j = \left(0,\ldots,0,\frac{\partial P_j}{\partial P_1},\ldots,\frac{\partial P_j}{\partial P_n}\right)\,.$$

The n vectors $\mathsf{J}\left(\frac{\partial q_j}{\partial Q_1},\ldots,\frac{\partial q_j}{\partial Q_n},\frac{\partial q_j}{\partial P_1},\ldots,\frac{\partial q_j}{\partial P_n}\right)$ span an n-dimensional Lagrangian subspace of the tangent space to $\mathscr{F}$ at every point (Q,P); this follows from $\{q_j,q_k\} = 0$ for $j,k = 1,\ldots,n$. By Lemma B.9, the Lagrangian subspace is complementary to at least one of the Lagrangian arithmetic planes of the canonical basis. This means that there exists a partition $\{J,K\}$ of $\{1,\ldots,n\}$ such that the $2n \times 2n$ matrix obtained by adding to (2.32) the n rows $\{\mathbf{e}_k\}_{k\in K}\cup\{\mathbf{d}_j\}_{j\in J}$ has non-zero determinant. On the other hand, the determinant turns out to be exactly (2.31), so that we conclude that J,K is the desired partition. *Q.E.D.*

We conclude this section with the claim:

> *Any time-independent canonical transformation possesses at least one generating function among the 2^n provided by Proposition 2.27.*

For, in view of Lemma 2.30 it is enough to determine the appropriate partition $\{J,K\}$ for the desired transformation.

2.4 Time-Dependent Canonical Transformations

The theory of canonical transformations developed till now can be generalized so that the cases of non-autonomous Hamiltonians and of time-depending transformations are taken into account.

2.4.1 Transformations Which Leave Time Unchanged

Let us exploit the extension of the phase space introduced in Section 1.1.1, that is, add two further canonical variables q_+, p_+, and for a given Hamiltonian $H(q,p,t)$ we consider a second Hamiltonian in the extended phase space[10]

$$\tilde{H}(q,p,q_+,p_+) = H(q,p,q_+) + p_+ \ . \tag{2.33}$$

The theory developed until now fully applies to canonical transformations

$$q = q\,(Q,P,Q_+,P_+)\ , \quad p = p\,(Q,P,Q_+,P_+)\,,$$
$$q_+ = q_+(Q,P,Q_+,P_+)\ , \quad p_+ = p_+(Q,P,Q_+,P_+)$$

on the extended phase space. However, this means that we may change also the time variable; for example, the new variable Q_+ could be non-uniform in time.

The simplest choice is to consider a restricted class of transformations which keeps the coordinate $q_+ = Q_+$ invariant; in turn, $p_+ = p_+(Q,P,Q_+,P_+)$ will be determined so as to fulfill the canonicity conditions. A first consequence is that the condition $\{q_+,p_+\} = 1$ implies $p_+ = P_+ + f(Q,P,q_+)$, with an arbitrary function f independent of P_+. A second consequence is that the canonicity conditions $\{q_+,q_j\} = \{q_+,p_j\} = 0$ for $1 \le j \le n$ imply that q_j,p_j do not depend on P_+. This also means that the Poisson brackets $\{q_j,q_k\}$, $\{q_j,p_k\}$ and $\{p_j,p_k\}$ are actually calculated by differentiating only with respect to the variables Q,P.

With a transformation satisfying the preceding conditions, the transformed Hamiltonian will take the form

$$\begin{aligned}&\tilde{H}(Q,P,Q_+,P_+)\\ &\quad = H(q,p,q_+)\big|_{q=q(Q,P,Q_+),\,p=p(Q,P,Q_+),q_+=Q_+} + P_+ + f(Q,P,Q_+)\ .\end{aligned}$$

In view of the linear dependence on P_+, we remove the extension of the phase space by setting again $Q_+ = t$ and removing the term P_+, thus obtaining the transformed Hamiltonian

$$K(Q,P,t) = H(q,p,t)\big|_{q=q(Q,P,t),\,p=p(Q,P,t)} + f(Q,P,t)\ .$$

We emphasize that the new Hamiltonian is not merely the transformed function of the old one: there is an extra term which should be calculated. So, let us see how to deal with it.

---

[10] The particular role played by the variables q_+,p_+ is emphasized by denoting the canonical coordinates in phase space as (q,p,q_+,p_+), where $q = (q_1,\ldots,q_n)$ and $p = (p_1,\ldots,p_n)$.

2.4.2 Using the Fundamental Poisson Brackets

The preservation of fundamental Poisson brackets is still a criterion for checking the canonicity of a given transformation and for determining the additional term in the transformed Hamiltonian.

Proposition 2.31: *Let $q = q(Q, P, t)$, $p = p(Q, P, t)$ be a time-dependent transformation which preserves the fundamental Poisson brackets identically in t. Then the transformation is canonical, and there exists a function $F(q, p, t)$ such that the transformed Hamiltonian is*

$$K(Q, P, t) = \big[H(q, p, t) - F(q, p, t)\big]_{q=q(Q,P,t),p=p(Q,P,t)}. \tag{2.34}$$

Proof. We prove that in the extended phase space there is a function $F(q, p, Q_+)$ such that the extended transformation

$$\begin{aligned} &q = q(Q, P, Q_+)\ , \quad p = p(Q, P, Q_+)\ , \\ &q_+ = Q_+ \qquad\quad\ , \quad p_+ = P_+ - F(q, p, Q_+)\big|_{q=q(Q,P,Q_+)\,,\,p=p(Q,P,Q_+)} \end{aligned} \tag{2.35}$$

is canonical. Recalling that $\{q_j, q_k\} = \{p_j, p_k\} = 0\,,\ \{q_j, p_k\} = \delta_{jk}$ (true in view of the assumed preservation of the fundamental Poisson brackets), we differentiate these relations with respect to Q_+; we get

$$\begin{aligned} &\left\{\frac{\partial q_j}{\partial Q_+}, q_k\right\} + \left\{q_j, \frac{\partial q_k}{\partial Q_+}\right\} = 0\ , \\ &\left\{\frac{\partial q_j}{\partial Q_+}, p_k\right\} + \left\{q_j, \frac{\partial p_k}{\partial Q_+}\right\} = 0\ , \\ &\left\{\frac{\partial p_j}{\partial Q_+}, p_k\right\} + \left\{p_j, \frac{\partial p_k}{\partial Q_+}\right\} = 0\ . \end{aligned}$$

Recalling that q, p do not depend on P_+, and denoting

$$f_j = \frac{\partial q_j}{\partial Q_+}\ , \quad g_j = \frac{\partial p_j}{\partial Q_+}\ ,$$

we write the preceding identities as

$$\frac{\partial f_j}{\partial p_k} - \frac{\partial f_k}{\partial p_j} = 0\ , \quad \frac{\partial f_j}{\partial q_k} - \frac{\partial g_k}{\partial p_j} = 0\ , \quad \frac{\partial g_j}{\partial q_k} - \frac{\partial g_k}{\partial q_j} = 0\ .$$

This implies the (local) existence of a function $F(q, p, Q_+)$ such that

$$\frac{\partial q_j}{\partial Q_+} = \frac{\partial F}{\partial p_j}\ , \quad \frac{\partial p_j}{\partial Q_+} = -\frac{\partial F}{\partial q_j}\ . \tag{2.36}$$

With this function we complete the transformation as in (2.35). The assumed preservation of the fundamental Poisson brackets for all t assures that $\{q_+, q_j\} = \{q_+, p_j\} = 0, \{q_+, p_+\} = 1$; for in view of (2.36) we have

$$\{p_+, q_j\} = \frac{\partial q_j}{\partial Q_+} - \{F, q_j\} = 0 \ , \quad \{p_+, p_j\} = \frac{\partial p_j}{\partial Q_+} - \{F, p_j\} = 0 \ .$$

Therefore the extended transformation is canonical. Replacing the transformation in the Hamiltonian (2.33) and removing the extension of the phase space, the claim follows. *Q.E.D.*

2.4.3 Time-Depending Generating Functions

The method of generating functions is fully effective also for time-dependent transformations. For brevity we discuss only the most common form of the generating function. However, any form can be used, as the reader can verify, with enough patience.

Proposition 2.32: *Let $S(P, q, t)$ be a function satisfying*

$$\det\left(\frac{\partial^2 S}{\partial P_j \partial q_k}\right) \neq 0$$

identically in t. Then the transformation implicitly defined by

$$Q_j = \frac{\partial S}{\partial P_j} \ , \quad p_j = \frac{\partial S}{\partial q_j} \ , \quad 1 \le j \le n$$

is canonical, and the transformed Hamiltonian takes the form

$$K(Q, P, t) = H(q, p, t)\Big|_{q=q(Q,P,t),p=p(Q,P,t)} + \frac{\partial S}{\partial t}(P, q, t)\Big|_{q=q(Q,P,t)} \ . \tag{2.37}$$

Proof. In the extended phase space consider the generating function

$$\tilde{S}(P, q, P_+, q_+) = P_+ q_+ + S(P, q, q_+) \ .$$

The corresponding transformation is

$$\begin{aligned} Q_j &= \frac{\partial S}{\partial P_j} \ , \quad p_j = \frac{\partial S}{\partial q_j} \ , \quad 1 \le j \le n \ , \\ Q_+ &= q_+ \ , \quad\ \ p_+ = P_+ + \frac{\partial S}{\partial q_+} \ . \end{aligned} \tag{2.38}$$

Replacing the transformation in the Hamiltonian (2.33) and removing the extension of the phase space, the claim follows. *Q.E.D.*

2.4.4 Changing the Time Variable

Canonical transformation in the extended $(2n+2)$-dimensional phase space can be used also to change the independent variable. The key remark is that the canonically conjugated variables q_+, p_+ we have introduced are the time and the energy (the value of the function H). Indeed, for the Hamiltonian

$H(q,p,t)$ we have $\dot{H} = \frac{\partial H}{\partial t}$. Actually, the new pair of variables q_+, p_+ (which we could well rename as t, E, if we like) may be considered on the same foot as the other pairs: it is not forbidden to remove the restriction made in the two previous sections, namely that the time variable must remain unchanged. The point not to be forgotten is that if we introduce a new time τ, then the new Hamiltonian K must be the variable conjugated to τ.

Let us limit the discussion to the simplest case. Suppose that we want to replace the time variable t with a new variable τ in the Hamiltonian $H(q,p,t)$, leaving the other variables unchanged. The Hamiltonian may well be an autonomous one. To this end, take a positive smooth function $\psi(t)$ on a suitably chosen time interval (which can well be infinite), and define

$$\tau(t) = \int_{t_0}^{t} \psi(s) ds \,. \tag{2.39}$$

The positiveness of $\psi(s)$ assures that the function $\tau(t)$ is invertible, so that $\tau(t)$ is a diffeomorphism and there is a smooth inverse function $t(\tau)$.

Extending the phase space as explained in Section 1.1.1, let us consider the Hamiltonian

$$F(q,p,q_+,p_+) = H(q,p,q_+) + p_+ \;, \quad q_+ = t \tag{2.40}$$

(we replace t with q_+ only to keep uniform notations in all discussions concerning time-dependent systems; there is no special meaning here). The manifold $F(q,p,q_+,p_+) = h$ is clearly invariant in the extended phase space. On the other hand the translation $\tilde{p}_+ = p_+ - h$, with all the other coordinates left unchanged, is clearly canonical, so that we restrict our attention to the manifold $F(q,p,q_+,p_+) = 0$ which is invariant for the flow.[11] On that manifold we have $p_+ = -H(q,p,q_+)$.

The transformation (2.39) may be seen as a point transformation, which can be extended to a canonical one through the generating function

$$S(P,q,p_+,q_+) = \sum_{j=1}^{n} P_j q_j + p_+ \tau(q_+) \,.$$

The coordinates q,p are clearly unchanged, so let us keep the notation q,p for them. We should only apply the canonical transformation which introduces the new conjugated coordinates τ, p_+, namely

$$\tau = \tau(q_+) \;, \quad p_+ = p_+ \psi(q_+).$$

Inverting the first equation, we find $q_+(\tau)$ as a smooth function, so that the Hamiltonian is changed to

---

[11] Adding a constant to the energy does not change the dynamics.

$$F(q,p,\tau,p_+) = \left[H(q,p,q_+) + p_+\psi(q_+)\right]\Big|_{q_+=q_+(\tau)} .$$

The quantity conjugated to the new time τ is $K(q,p,\tau) = -p_+$, where p_+ is found by solving the equation $F(q,p,\tau,p_+) = 0$. We get

$$K(q,p,\tau) = \frac{1}{\psi(q_+)} H(q,p,q_+)\Big|_{q_+=q_+(\tau)} . \tag{2.41}$$

The canonical equations are thus changed to

$$\frac{dq_j}{d\tau} = \frac{\partial K}{\partial p_j} , \quad \frac{dp_j}{d\tau} = -\frac{\partial K}{\partial q_j} . \tag{2.42}$$

The argument just presented may appear somewhat mysterious, so let us add a little extra explanation. Assuming that q, p depend on t through the new time τ we have

$$\begin{aligned} \frac{\partial H}{\partial p_j} &= \dot{q}_j = \frac{dq_j}{d\tau}\frac{d\tau}{dt} = \psi\frac{dq_j}{d\tau} , \\ -\frac{\partial H}{\partial q_j} &= \dot{p}_j = \frac{dp_j}{d\tau}\frac{d\tau}{dt} = \psi\frac{dp_j}{d\tau} . \end{aligned}$$

By comparing the first and last members of every line, we get

$$\frac{dq_j}{d\tau} = \frac{\partial K}{\partial p_j} , \quad \frac{dp_j}{d\tau} = -\frac{\partial K}{\partial q_j} ,$$

namely the canonical equations (2.42).

2.5 The Hamilton–Jacobi Equation

The integration of the canonical equations can be performed by looking for a generating function of a time-dependent canonical transformation which gives the Hamiltonian an extremely simple form.[12]

Having given the Hamiltonian $H(q,p,t)$, we look for a function S which is a solution of the *Hamilton–Jacobi equation*:

$$H\left(q, \frac{\partial S}{\partial q}, t\right) + \frac{\partial S}{\partial t} = 0 . \tag{2.43}$$

We are actually looking for the generating function of a transformation such that the transformed Hamiltonian is identically zero. The problem is to find

---

[12] The method has been introduced in slightly different forms by William Rowan Hamilton [106][107][108] and by Carl Gustav Jacobi [120][121][122]. The form used here is the one due to Jacobi. For a thorough discussion of the relations between the two formulations see, e.g., [144].

a solution of Eq. (2.43) depending on $q_1, \ldots, q_n, t$ and on n arbitrary parameters $\alpha_1, \ldots, \alpha_n$; this is said to be a *complete integral.*

Proposition 2.33: *Consider the Hamiltonian $H(q, p, t)$, and assume that we are given a complete integral $S(\alpha, q, t)$ of the Hamilton–Jacobi equation (2.43), depending on n arbitrary parameters $\alpha_1, \ldots, \alpha_n$ and satisfying*

$$\det\left(\frac{\partial^2 S}{\partial \alpha_j \partial q_k}\right) \neq 0 \, .$$

Then the solutions of the canonical equations are written in implicit form as

$$\beta_j = \frac{\partial S}{\partial \alpha_j}(\alpha, q, t) \, , \quad p_j = \frac{\partial S}{\partial q_j}(\alpha, q, t) \, , \quad 1 \leq j \leq n \, , \tag{2.44}$$

where $\alpha_1, \ldots, \alpha_n, \beta_1, \ldots, \beta_n$ are constants depending on the initial data.

Proof. The function $S(\alpha, q, t)$ satisfies the conditions of Proposition 2.32; therefore, it is the generating function of a canonical transformation, actually the transformation (2.44). Since the transformed Hamiltonian is identically zero, the corresponding canonical equations are

$$\dot{\alpha}_j = 0 \, , \quad \dot{\beta}_j = 0 \, , \quad 1 \leq j \leq n$$

(i.e., α, β are constants depending on the initial data). By inversion of (2.44) with respect to q, p, one gets functions

$$q = q(\alpha, \beta, t) \, , \quad p = p(\alpha, \beta, t) \, ;$$

this is the desired solutions of the canonical equations. *Q.E.D.*

Example 2.34: *Free particle.* Let the Hamiltonian be

$$H = \frac{1}{2m}(p_x^2 + p_y^2 + p_z^2) \, .$$

The corresponding Hamilton–Jacobi equation is

$$\frac{1}{2m}\left[\left(\frac{\partial S}{\partial x}\right)^2 + \left(\frac{\partial S}{\partial y}\right)^2 + \left(\frac{\partial S}{\partial z}\right)^2\right] + \frac{\partial S}{\partial t} = 0 \, .$$

We use the method of *separation of variables*. We look for a solution of the form

$$S(x, y, z, t) = X(x) + Y(y) + Z(z) + T(t) \, ,$$

so that the equation is rewritten as

$$\frac{1}{2m}\left[\left(\frac{dX}{dx}\right)^2 + \left(\frac{dY}{dy}\right)^2 + \left(\frac{dZ}{dz}\right)^2\right] + \frac{dT}{dt} = 0 \, . \tag{2.45}$$

Therefore, we obtain the equations[13]

$$\frac{dX}{dx} = \alpha_x \ , \quad \frac{dY}{dy} = \alpha_y \ , \quad \frac{dZ}{dz} = \alpha_z \ , \quad \frac{dT}{dt} = -\frac{\alpha_x^2 + \alpha_y^2 + \alpha_z^2}{2m} \ ,$$

with $\alpha_x, \alpha_y, \alpha_z$ arbitrary constants. By integration we construct the generating function

$$S(\alpha_x, \alpha_y, \alpha_z, x, y, z, t) = \alpha_x x + \alpha_y y + \alpha_z z - \frac{\alpha_x^2 + \alpha_y^2 + \alpha_z^2}{2m} t \ ,$$

so that the transformation is

$$\begin{aligned} p_x &= \alpha_x \ , & p_y &= \alpha_y \ , & p_z &= \alpha_z \ , \\ \beta_x &= x - \frac{\alpha_x}{m} t \ , & \beta_y &= y - \frac{\alpha_y}{m} t \ , & \beta_z &= z - \frac{\alpha_z}{m} t \ . \end{aligned}$$

This is the solution of the canonical equations. *E.D.*

Example 2.35: *Mechanical systems with one degree of freedom.* Consider the Hamiltonian of a point mass on a line, subject to a potential $V(x)$, where x is a Cartesian coordinate:

$$H(q, p) = \frac{p^2}{2m} + V(x) \ .$$

The corresponding Hamilton–Jacobi equation is

$$\frac{1}{2m}\left(\frac{\partial S}{\partial x}\right)^2 + V(x) + \frac{\partial S}{\partial t} = 0 \ .$$

We apply again the method of separation of variables, looking for a complete solution in the form $S(x, t) = X(x) + T(t)$. By substitution we get

$$\frac{1}{2m}\left(\frac{\partial X}{\partial x}\right)^2 + V(x) + \frac{\partial T}{\partial t} = 0 \ .$$

Hence we split the equation as

$$\frac{1}{2m}\left(\frac{\partial X}{\partial x}\right)^2 + V(x) = E \ , \quad \frac{\partial T}{\partial t} = -E \ ,$$

having introduced the energy E as an integration constant. The equations are solved as

$$X(x) = \sqrt{2m}\int \sqrt{E - V(x)}\, dx \ , \quad T(t) = -Et \ ,$$

[13] Differentiating (2.45) with respect to x we get $\frac{d^2X}{dx^2} = 0$, so that $\frac{dX}{dx}$ must be constant. Similarly, differentiating with respect to y, z, and t we get that $\frac{dY}{dy}$, $\frac{dZ}{dz}$, and $\frac{dT}{dt}$ are constants, too. In view of (2.45), only three of these constants are arbitrary.

thus giving the generating function

$$S(x,t) = \sqrt{2m} \int \sqrt{E - V(x)}\, dx - Et \ .$$

Calling β the coordinate conjugated to E, we get the canonical transformation in implicit form

$$\beta = \frac{\partial S}{\partial E} = \sqrt{\frac{m}{2}} \int \frac{dx}{\sqrt{E - V(x)}} - t \,,$$
$$p = \frac{\partial S}{\partial x} = \sqrt{2m(E - V(x))} \ .$$

The solution so found coincides with (1.34) and (1.36) found in Section 1.3.1: just identify the integration constant β with $-t_0$, the initial time, and introduce the initial condition $x(t_0) = x_0$ in the integral. E.D.

Example 2.36: *Motion in a central field.* Let us consider the simpler case of a plane motion, which is relevant for the Kepler problem. From Example 1.28, Eq. (1.42) of Section 1.3.4 we know that the Hamiltonian in polar coordinates is

$$H = \frac{1}{2m}\left(p_r^2 + \frac{p_\vartheta^2}{r^2}\right) + V(r).$$

The corresponding Hamilton–Jacobi equation reads

$$\frac{1}{2m}\left[\left(\frac{\partial S}{\partial r}\right)^2 + \frac{1}{r^2}\left(\frac{\partial S}{\partial \vartheta}\right)^2\right] + V(r) + \frac{\partial S}{\partial t} = 0 \ .$$

Still applying the separation of variables, we look for a function in the form $S = R(r) + \Theta(\vartheta) + T(t)$, so that the equation splits into the three separate equations for $R(r)$, $\Theta(\vartheta)$ and $T(t)$

$$\frac{1}{2m}\left[\left(\frac{\partial R}{\partial r}\right)^2 + \frac{\alpha_\vartheta^2}{r^2}\right] + V(r) = \alpha_r \ ,$$
$$\left(\frac{\partial \Theta}{\partial \vartheta}\right)^2 = \alpha_\vartheta^2 \ ,$$
$$\frac{\partial T}{\partial t} = -\alpha_r .$$

Here α_r and α_ϑ are the integration constants. A complete solution is

$$S(r, \vartheta, \alpha_r, \alpha_\vartheta, t) = + \int \sqrt{2m[\alpha_r - V(r)] - \frac{\alpha_\vartheta^2}{r^2}}\, dr + \vartheta \alpha_\vartheta - \alpha_r t$$

and generates the canonical transformation

$$
\begin{aligned}
p_r &= \sqrt{2m[\alpha_r + V(r)] - \frac{\alpha_\vartheta^2}{r^2}}\ , \\
p_\vartheta &= \alpha_\vartheta\ , \\
\beta_r + t &= m \int \frac{dr}{\sqrt{2m[\alpha_r - V(r)] - \frac{\alpha_\vartheta^2}{r^2}}}\ , \\
\beta_\vartheta - \vartheta &= -\int \frac{dr}{r^2\sqrt{2m[\alpha_r - V(r)] - \frac{\alpha_\vartheta^2}{r^2}}}\ .
\end{aligned}
$$

In the notations of Section 1.3.4 the constants here represent the energy $E = \alpha_r$, the angular momentum $\ell = \alpha_\vartheta$, the initial time $t_0 = -\beta_r$ and the initial angle $\vartheta(t_0) = -\beta_\vartheta$. The initial value $r_0 = r(t_0)$ should be used as the lower limit for the integrals. Writing the solution in more explicit form requires the calculation of the integrals for a given potential $V(r)$ and the inversion of the fourth equation. With some patience the reader will realize that the solution so found is the same as the one produced along the lines of Example 1.28. *E.D.*

Exercise 2.37: Extend the calculation of Example 2.36 to the spatial case in spherical coordinates. The Hamiltonian has been calculated in Example 1.4:

$$
H = \frac{1}{2m}\left(p_r^2 + \frac{p_\vartheta^2}{r^2} + \frac{p_\varphi^2}{r^2 \sin^2\vartheta}\right) + V(r)\ .
$$

(Hint: let $S(r, \vartheta, \varphi, \alpha_r, \alpha_\vartheta, \alpha_\varphi, t)) = R(r) + \Theta(\vartheta) + \Phi(\varphi) + T(t)$, using again the separation of variables.) *A.E.L.*

Before closing the chapter, it is worth mentioning that for the Kepler problem there is a particularly significant set of canonical variables, first introduced by Charles Eugène Delaunay [57] and named after him. He introduced these variables with the aim of calculating the motion of the Moon. However, Delaunay variables proved to have a major significance in Quantum Mechanics, and in particular for the problem of the hydrogen atom: the canonical momenta of Delaunay are appropriate action variables for quantization (see the treatise of Max Born [28]). A deduction of Delaunay canonical variables via the Hamilton–Jacobi method may be found in [190]. A deduction following the scheme of Born's treatise will be reported in Chapter 3, Section 3.6.

The latter example is an excellent illustration of a general fact: finding a complete integral of the *Hamilton–Jacobi equation* involves an arbitrary choice of the integration constants (α, in the notations used here), which in turn become half of the new canonical variables. It goes without saying that some choices may be better than others. According to the experience of astronomers, Delaunay's variables are the best replacement in the Hamiltonian framework for the so-called orbital elements of classical Astronomy (which are not canonical variables).

The question of making the best choice, in some sense, of the integration constants is strictly connected with the problem of finding suitable first integrals for a system: this problem will be discussed in Chapter 3.

3
Integrable Systems

The present chapter is devoted to classical integration methods based on use of first integrals. In this sense it is a continuation of the discussion in Section 1.2.2. The underlying purpose is to discover a general method for integrating every mechanical system; this was indeed, after Newton, the dream of great mathematicians till the end of the nineteenth century. The methods discussed in the present chapter, based on use of first integrals, may be seen as a complement and often a useful support to the Hamilton–Jacobi theory presented in Section 2.5.

Nowadays we are well aware that integrable systems are rather exceptional, and this is indeed a good justification of the commonly observed fact that most textbooks contain the same examples; just a few which are the classical and more interesting ones and are well representative of the main features of the dynamics of mechanical systems. The classical examples are the Kepler problem and the motion of a rigid body – the problem of Poinsot and Lagrange's top. Nevertheless, integrable systems represent an excellent first approximation of interesting mechanical systems and are the starting point of classical perturbation theory. It is not far from truth to say that in many cases the goal of perturbation methods is to reduce a system of differential equations to a form as close as possible to an integrable one. It may also be mentioned that the recent developments of perturbation theory reveal that integrable systems are not so rare but are not so easily identified.

Before entering the discussion, we should be in agreement on the meaning of the term *integrable system*. In view of the theorem of existence and uniqueness of the solutions of a system of differential equations, every Hamiltonian system can be said to be integrable provided that some mild regularity conditions are satisfied by the Hamiltonian function. This is useful, of course, if one is interested in computing the orbit corresponding to a given initial datum (e.g., with numerical methods). However, we should bear in mind that the theorem has a *local* character: in our case it assures only the existence of the solution for some time interval. The process of continuation of a given solution may be used to establish the *existence* of the solution for larger time intervals, but it gives essentially no information about the *global behaviour of the orbits.* In addition, most current books of analysis provide theorems assuring the existence of the solution, without taking care to provide also a method to construct it.[1] In the framework of Hamiltonian systems it is customary to assign a more precise meaning to the word *integrability.* In some sense, one asks to be able to write the solution for all times.

The traditional interpretation involves the concept of *integrability by quadratures.* As said at the beginning of Section 1.3, it means that the solution should be found via a finite number of algebraic operations, including inversion of functions, and of computations of integrals of known functions (quadrature). The integration method discussed in Section 1.3 for systems with one degree of freedom is a good example; the question is how to extend it to the case of more degrees of freedom, in a general form.

In the framework of Hamiltonian theory, Liouville's theorem can be considered as the most advanced general result in the desired direction [152]. The paradigmatic model is represented by a Hamiltonian depending only on the momenta $p_1, \ldots, p_n$, that is, $H = H(p_1, \ldots, p_n)$, which is trivially integrable. In short, Liouville's theorem says that if a Hamiltonian system possesses enough independent first integrals, then the first integrals themselves play the role of momenta, and the Hamiltonian can be given the form just presented with a suitable coordinate transformation which is constructed by quadratures.

In recent times more attention is paid to the *global* description of the behaviour of the solutions, with particular attention to the existence of *periods*, or *frequencies.* Thus, many authors concentrate the attention on systems for which the coordinates $q_1, \ldots, q_n$ conjugated to $p_1, \ldots, p_n$ are angles. In such a case the canonical coordinates p, q are called *action-angle variables.* This seems to be an excessively strong restriction: for instance, it excludes the problem of a mass point freely moving on the space.[2] However, such a strong attitude can be justified a posteriori. Indeed, small perturbations of

---

[1] A remarkable exception to this claim is represented by the method of solution by series, known to Newton [181] and reformulated in rigorous terms by Cauchy. However, it applies only to holomorphic vector fields.

[2] A typical situation arising in Mechanics is the study of a system of many particles, either free of moving all around the space under the mutual interactions, as is the case of our planetary system or of an atomic system, or subjected to some constraints, as, e.g., in the case of a rigid body. The first step usually consists in exploiting the conservation of the total momentum by eliminating the motion of the centre of mass, which is a trivial one, being that of a free particle. Then in many cases action-angle variables may be introduced in some approximation.

an integrable system which possesses action-angle variables typically produce a very complicated dynamical behaviour, which is still not completely understood. It is the most intriguing case.

Thus we raise the following question. *Let us consider a Liouville-integrable system. Can we introduce action-angle variables?* The use of action-angle variables was well known in connection with classical problems like the planetary motions and the motion of a rigid body. During the first decades of the twentieth century it has become relevant also in connection with the early developments of quantum theory. A classical and valuable reference is Max Born's treatise [28]. The Arnold–Jost theorem states that integrability in classical mechanics is strongly connected with the existence of action-angle variables.

3.1 Involution Systems

The investigation of integrability is often based on the existence of independent first integrals. In the framework of Hamiltonian systems a relevant role is played by first integrals with vanishing mutual Poisson brackets.

Definition 3.1: *A collection of r functions $\{\Phi_1(q,p),\ldots,\Phi_r(q,p)\}$ is said to be an involution system if the functions are independent, that is,*

$$\operatorname{rank}\left(\frac{\partial(\Phi_1,\ldots,\Phi_r)}{\partial(q_1,\ldots,q_n,p_1,\ldots,p_n)}\right)=r\ ,$$

and the Poisson bracket between any two functions vanishes, that is,

$$\{\Phi_j,\Phi_k\}=0\ ,\quad j,k=1,\ldots,r\ .$$

In the language of symplectic geometry and using the compact notation, the involution property is expressed as: *the symplectic gradients of Φ_j, Φ_k are symplectic orthogonal, that is, $\mathsf{J}\partial_{\mathbf{z}}\Phi_j \angle\, \mathsf{J}\partial_{\mathbf{z}}\Phi_k$.*

3.1.1 Some Useful Lemmas

The following lemmas will be used in the rest of the chapter.

Lemma 3.2: *An involution system contains at most n independent functions, where n is the number of degrees of freedom.*

Proof. It is convenient to use the compact notation. At any point $\mathbf{z}\in\mathscr{F}$ the symplectic gradients $(\mathsf{J}\partial_z\Phi_1,\ldots,\mathsf{J}\partial_z\Phi_r)$ span a r-dimensional subspace which is isotropic, due to the involution property satisfied by the functions. By Lemma B.7 the dimension of such a subspace cannot exceed n. We conclude that $r\leq n$. *Q.E.D.*

Example 3.3: *Using the canonical coordinates.* The most immediate and useful example is provided by the canonical coordinates themselves. Consider any partition J, K of $\{1, \ldots, n\}$; then the n functions $\{q_j\}_{j\in J} \cup \{p_k\}_{k\in K}$ form an involution system. *E.D.*

Other examples may be easily constructed by making reference to known integrable systems. For instance, let the phase space be $\mathbb{R}^3 \times \mathbb{R}^3$, with canonical coordinates x, y, z and momenta p_x, p_y, p_z. The latter three quantities are the components of the momentum and form an involution system. This reminds us the case of a free particle in the ordinary Euclidean space. On the other hand, it is just a particular case of the first example, since it corresponds to a partition which selects only the momenta.

Example 3.4: *Using the angular momentum.* It seems natural to try to construct involution systems by using the three components of the angular momentum, namely

$$M_x = yp_z - zp_y \,,\ M_y = zp_x - xp_z \,,\ M_z = xp_y - yp_x \ .$$

However, it is immediately seen that the latter three quantities are independent, but not in involution: this has been shown in Example 1.18. Replacing some components of the angular momentum with some components of the momentum does not help, for the same reason.
An involution system may be constructed by considering one of the components of the angular momentum; for instance, M_z and the quantity $\Gamma^2 = M_x^2 + M_y^2 + M_z^2$, namely the square of the norm of the angular momentum. The latter two quantities are indeed in involution. As a third function in involution with M_x and Γ^2 we may select, for example,

$$E = \frac{1}{2m}(p_x^2 + p_y^2 + p_z^2) + V(r) \ ,$$

where $r = \sqrt{x^2 + y^2 + z^2}$, and $V(r)$ is an arbitrary (differentiable) function depending only on r.
The same example can be reformulated using spherical coordinates. The phase space is $(0, +\infty) \times (0, \pi) \times \mathbb{T} \times \mathbb{R}^3$, with canonical coordinates $r, \vartheta, \varphi, p_r, p_\vartheta, p_\varphi$. The three functions

$$J = p_\varphi \ , \quad \Gamma^2 = p_\vartheta^2 + \frac{J^2}{\sin^2\vartheta} \ , \quad E = \frac{1}{2m}\left(p_r^2 + \frac{\Gamma^2}{r^2}\right) + V(r)$$

form an involution system. The reader will recognize here the first integrals of the problem of motion under central forces, Example 1.20. *E.D.*

Example 3.5: *Harmonic oscillators.* Let the phase space be $\mathbb{R}^{2n}$, with canonical coordinates x, y. The n functions

$$\Phi_1 = \frac{x_1^2 + y_1^2}{2} \,, \ldots, \, \Phi_n = \frac{x_n^2 + y_n^2}{2}$$

form an involution system.

Although apparently trivial, this example plays a central role in studying the small oscillations of a system in the neighbourhood of a stable equilibrium, since it represents a remarkable first approximation. *E.D.*

Lemma 3.6: *Let $\Phi_1, \ldots, \Phi_n$ be an involution system on the phase space $\mathscr{F}$. Then at every point $P \in \mathscr{F}$ there is a partition J, K of $\{1, \ldots, n\}$ such that*

$$\det \left(\frac{\partial(\Phi_1, \ldots, \Phi_n)}{\partial(q_J, p_K)} \right) \neq 0 \ .$$

The statement of the lemma is just a more general formulation of Lemma 2.30; we proved it by exploiting the fact that the coordinates $q_1, \ldots, q_n$ form a complete involution system; the same argument applies to any complete involution system of functions. The easy adaptation of the proof is left to the reader.

3.1.2 Variational Equations and First Integrals

Consider a system of differential equations, not necessarily Hamiltonian,

$$\dot{x}_j = X_j(x_1, \ldots, x_n) \ , \quad 1 \leq j \leq n \ , \tag{3.1}$$

and let $x(t)$ be an orbit with initial point x_0. Let also $x_0 + \delta x_0$ be a point close to x_0, with an infinitesimal increment δx_0, and let $x(t) + \delta x(t)$ be the corresponding orbit, so that it is a solution of the differential equations

$$\begin{aligned} \frac{d}{dt}(x_j + \delta x_j) &= X_j(x_1 + \delta x_1, \ldots, x_n + \delta x_n) \\ &= X_j(x_1, \ldots, x_n) + \sum_{l=1}^{n} \frac{\partial X_j}{\partial x_l}(x_1, \ldots, x_n)\, \delta x_l + \ldots \ , \end{aligned}$$

where the dots denote terms of higher order in δx. Since $x(t)$ is assumed to be a solution of the equation $\dot{x} = X(x)$, and forgetting nonlinear terms in δx, we immediately get that $\delta x(t)$ obeys the so-called *variational equation*,

$$\frac{d}{dt}\delta x_j = \sum_{l=1}^{n} \frac{\partial X_j}{\partial x_l} \delta x_l \ , \quad 1 \leq j \leq n \ , \tag{3.2}$$

where the functions $\frac{\partial X_j}{\partial x_l}(x_1, \ldots, x_n)$ must be evaluated along the known solution $x(t)$.

In geometric language, the infinitesimal increment δx is a vector in the tangent space, which evolves according to the variational equation.

A similar procedure applies to the Hamiltonian case. Let us do it in detail, recalling that the canonical equations have the rather particular form

$$\dot{q}_j = \frac{\partial H}{\partial p_j} \ , \quad \dot{p}_j = -\frac{\partial H}{\partial q_j} \ , \quad 1 \leq j \leq n \ . \tag{3.3}$$

Let us denote by $\delta q_j, \delta p_j$ the increments with respect to the variables q_j, p_j. Then the variational equations are

$$
(3.4)\qquad
\begin{aligned}
\frac{d}{dt}\delta q_j &= \sum_{l=1}^{n}\left(\frac{\partial^2 H}{\partial p_j \partial q_l}\delta q_l + \frac{\partial^2 H}{\partial p_j \partial p_l}\delta p_l\right),\\
\frac{d}{dt}\delta p_j &= -\sum_{l=1}^{n}\left(\frac{\partial^2 H}{\partial q_j \partial q_l}\delta q_l + \frac{\partial^2 H}{\partial q_j \partial p_l}\delta p_l\right).
\end{aligned}
$$

An interesting relation between first integrals and variational equations is given by the following

Proposition 3.7: *Let Φ be a first integral of the canonical system with Hamiltonian $H(q,p)$. Then a solution of the variational equations (3.4) is*

$$
(3.5)\qquad \delta q_j = \tau\frac{\partial \Phi}{\partial p_j}\ ,\quad \delta p_j = -\tau\frac{\partial \Phi}{\partial q_j}\ ,\quad 1 \le j \le n\ ,
$$

where $\tau \neq 0$ is an arbitrary constant.

Proof. In view of the linearity of the variational equations, it is enough to prove the statement for $\tau = 1$. By differentiating the relation $\{\Phi, H\} = 0$, we immediately get

$$
\left\{\frac{\partial \Phi}{\partial p_j}, H\right\} + \left\{\Phi, \frac{\partial H}{\partial p_j}\right\} = 0\ ,\quad \left\{\frac{\partial \Phi}{\partial q_j}, H\right\} + \left\{\Phi, \frac{\partial H}{\partial q_j}\right\} = 0\ ,\quad 1 \le j \le n\ ,
$$

that is,

$$
\frac{d}{dt}\frac{\partial \Phi}{\partial p_j} = \left\{\frac{\partial H}{\partial p_j}, \Phi\right\}\ ,\quad \frac{d}{dt}\frac{\partial \Phi}{\partial q_j} = \left\{\frac{\partial H}{\partial q_j}, \Phi\right\}\ .
$$

Writing in explicit form the r.h.s. of these equations, we get

$$
\begin{aligned}
\frac{d}{dt}\frac{\partial \Phi}{\partial p_j} &= \sum_{l=1}^{n}\left(\frac{\partial^2 H}{\partial p_j \partial q_l}\frac{\partial \Phi}{\partial p_l} - \frac{\partial^2 H}{\partial p_j \partial p_l}\frac{\partial \Phi}{\partial q_l}\right),\\
\frac{d}{dt}\frac{\partial \Phi}{\partial q_j} &= \sum_{l=1}^{n}\left(\frac{\partial^2 H}{\partial q_j \partial q_l}\frac{\partial \Phi}{\partial p_l} - \frac{\partial^2 H}{\partial q_j \partial p_l}\frac{\partial \Phi}{\partial q_l}\right),
\end{aligned}
$$

which in view of (3.5) coincides with (3.4). *Q.E.D.*

A remark is in order. The proposition essentially says that the increment $(\delta q, \delta p)$ in (3.5) is the linear approximation in τ of the canonical flow due to $\Phi(q,p)$. This is illustrated in Figure 3.1. Denote by $\mathbf{z} = (q,p)$ the initial point and by X_Φ the canonical vector field generated by Φ; that is, $X_\Phi = L_\Phi(q,p)$, or, in compact notation, $X_\Phi = \mathsf{J}\partial_z\Phi$. The small displacement will be correspondingly written as $\delta\mathbf{z} = \tau X_\Phi$. Then the statement of the proposition may be rewritten, forgetting terms of higher order, as

$$
(3.6)\qquad \phi^t\big(\tau X_\Phi(\mathbf{z})\big) = \tau X_\Phi(\phi^t\mathbf{z})\ .
$$

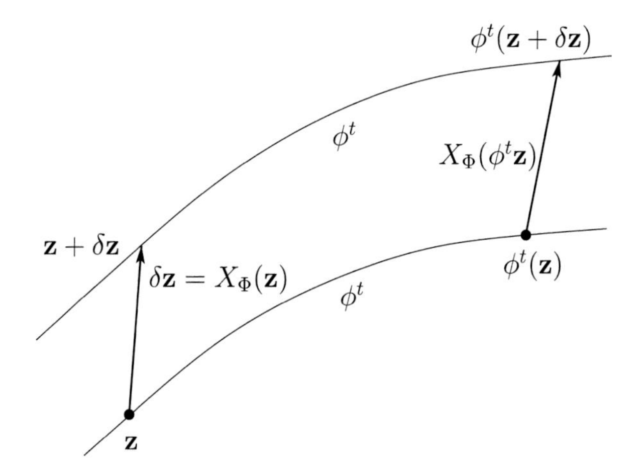

Figure 3.1 Illustrating Proposition 3.7. The displaced orbit with initial point $\mathbf{z}+\delta\mathbf{z}$ is obtained by translating every point of the orbit with initial point $\mathbf{z}$ along the flow generated by $X_\Phi(\phi^t\mathbf{z})$. The figure represents the first-order approximation.

That is, the linear approximation of the increment at the point $\phi^t\mathbf{z}$ is still the approximated flow at time τ of the Hamiltonian field X_Φ generated by Φ; it corresponds to exchanging the flow with the translation along the direction of X_Φ. This raises the question about commutation of flows, which is the subject of the next section.

3.1.3 Commutation of Canonical Flows

Suppose that we are given two functions, $F(q,p)$ and $G(q,p)$. Both of them may be considered as Hamiltonian functions; so we have two different flows that we may denote as ϕ_F and ϕ_G, respectively; Figure 3.2 may help to illustrate the following question. Given an initial point (q,p), let us follow the flow ϕ_F^t of F up to a time t, thus reaching the point $\phi_F^t(q,p)$; then follow the flow ϕ_G^τ of G up to time τ, ending at $\phi_G^\tau \circ \phi_F^t(q,p)$. Then exchange the order of the flows, thus moving from (q,p) to $\phi_G^\tau(q,p)$ along the flow of G and then to $\phi_F^t \circ \phi_G^\tau(q,p)$ along the flow of F, the times t and τ being unchanged. The question concerns the difference between the two final points; in general, one cannot expect them to coincide.

Lemma 3.8: *Let the Hamiltonians F and G be given and consider the canonical flows at infinitesimal times t and τ generated by F and G, respectively, namely*

$$(q_t,p_t) = \phi_F^t(q,p) \ , \quad (q_\tau,p_\tau) = \phi_G^\tau(q,p) \ .$$

Then we have

$$(\phi_F^t\phi_G^\tau - \phi_G^\tau\phi_F^t)(q,p) = \tau t L_{\{G,F\}}(q,p) + \dots \ , \tag{3.7}$$

the dots standing for terms of order higher than 2 in t and τ.

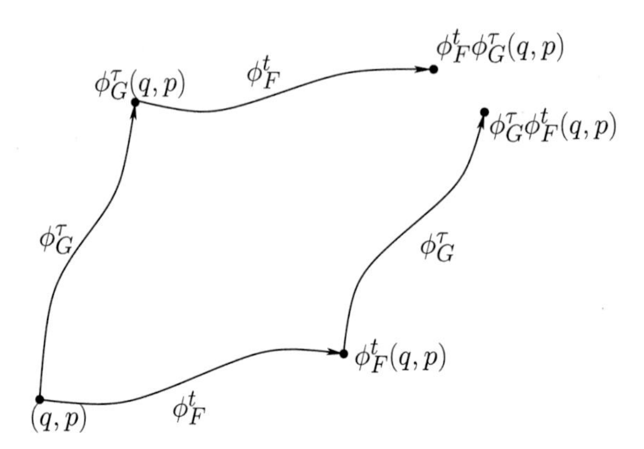

Figure 3.2 Commutation of flows. In general, exchanging the order of the flows changes the final point.

Proof. Taking into account that t and τ are infinitesimal quantities and expressing the time derivatives of coordinates through the Lie derivative L_F, for any point (q,p), we write the Taylor expansion of the flow as

$$\phi_F^t(q,p) = \Bigl(1 + tL_F + \frac{t^2}{2}L_F^2 + \ldots\Bigr)(q,p) \ ,$$

$$\phi_G^\tau(q,p) = \Bigl(1 + \tau L_G + \frac{\tau^2}{2}L_G^2 + \ldots\Bigr)(q,p) \ ,$$

the dots denoting terms of higher order in t and τ. Replacing (q,p) with $\phi_G^\tau(q,p)$ and reordering in powers of t and τ up to the second order we get

$$\begin{aligned}\phi_F^t \circ \phi_G^\tau(q,p) &= \Bigl(1 + tL_F + \frac{t^2}{2}L_F^2 + \ldots\Bigr)\Bigl(1 + \tau L_G + \frac{\tau^2}{2}L_G^2 + \ldots\Bigr)(q,p) \\ &= \Bigl(1 + tL_F + \tau L_G + \frac{t^2}{2}L_F^2 + t\tau L_F L_G + \frac{\tau^2}{2}L_G^2 + \ldots\Bigr)(q,p) \ .\end{aligned}$$

Similarly, exchanging t with τ and F with G, we get

$$\phi_G^\tau \circ \phi_F^t(q,p) = \Bigl(1 + \tau L_G + tL_F + \frac{\tau^2}{2}L_G^2 + \tau t L_G L_F + \frac{t^2}{2}L_F^2 + \ldots\Bigr)(q,p) \ .$$

Hence, forgetting terms of higher order, we have

$$\begin{aligned}\bigl(\phi_F^t \circ \phi_G^\tau - \phi_G^\tau \circ \phi_F^t\bigr)(q,p) &= t\tau\bigl(L_F L_G - L_G L_F\bigr)(q,p) + \ldots \\ &= t\tau L_{\{G,F\}}(q,p) + \ldots \ ,\end{aligned}$$

in virtue of Jacobi's identity (1.25). *Q.E.D.*

Lemma 3.8 provides a condition for the commutation of flows: the right member should be zero. The interesting fact is that *if F and G are in involution, then the final points do coincide.*

Proposition 3.9: *Let the dynamical variables $F(q,p)$ and $G(q,p)$ be in involution, and consider the canonical flows at times t and τ generated by F and G, respectively, that is,*

$$(3.8)\qquad (q_t, p_t) = \phi_F^t(q,p)\ ,\quad (q_\tau, p_\tau) = \phi_G^\tau(q,p)\ .$$

Then the following statements hold true.

(i) The function G is invariant for the canonical flow generated by F; conversely, the function F is invariant for the canonical flow generated by G. That is,

$$F(q_\tau,p_\tau)\Big|_{(q_\tau,p_\tau)=\phi_G^\tau(q,p)} = F(q,p)\ ,\ G(q_t,p_t)\Big|_{(q_t,p_t)=\phi_F^t(q,p)} = G(q,p)\ .$$

(ii) The flows (3.8) do commute, that is, for every (q,p) and for every t and τ, we have

$$\phi_G^\tau \circ \phi_F^t(q,p) = \phi_F^t \circ \phi_G^\tau(q,p)\ .$$

Proof. (i) It is just a different formulation of the claim that G is a first integral for F, and vice versa.
(ii) By Lemma 3.8, if $[L_F, L_G] = 0$ then also the derivatives of $[\phi_F^t, \phi_G^\tau]$ with respect to t an τ are zero at every point (q,p), that is,

$$\frac{\partial}{\partial t}[\phi_F^t, \phi_G^\tau](q,p)\Big|_{t=0} = \frac{\partial}{\partial \tau}[\phi_F^t, \phi_G^\tau](q,p)\Big|_{\tau=0} = 0\ .$$

Since the point (q,p) is arbitrary, we conclude that $[\phi_F^t, \phi_G^\tau](q,p) = 0$ at every point along every path. *Q.E.D.*

A suggestive interpretation of Proposition 3.9 is the following. Consider an orbit of F, and apply the flow ϕ_G^τ to every point of the orbit. The resulting set is again an orbit of F. The roles of F and G may be exchanged, of course. In short: *the flow of G sends orbits of F into orbits of F; and the same holds true by exchanging F and G.*

Remark. Let us recall two facts. The first one is the following obvious property of canonical transformations. Let $(q,p) = \mathscr{C}(Q,P)$ be canonical, and denote by $(\mathscr{C}H)(Q,P) = H(q,p)\big|_{(q,p)=\mathscr{C}(Q,P)}$ the transformed Hamiltonian. Pick a point $(Q,P) = \mathscr{C}(q,p)$ of the phase space. The obvious fact is that the image $\mathscr{C}\big(\Omega(q,p)\big)$ of an orbit of $H(q,p)$ is an orbit of $\big(\mathscr{C}H\big)(Q,P)$. That is: the canonical map $\mathscr{C}$ sends orbits of $H(q,p)$ into orbits of $\big(\mathscr{C}H(Q,P)\big)$. The second fact is that the canonical flow is a canonical transformation, as stated by Proposition 2.15. Therefore, with the notations of Proposition 3.9, the flow ϕ_G sends orbits of ϕ_F into orbits of ϕ_F, and vice versa. This is an alternative proof of claim (ii) of the proposition.

Remark. We have stated the proposition for canonical flows (which are of interest in the present notes); but the statement is true for generic vector fields which commute. Commutation of two vector fields X and Y on a manifold is introduced as in Figure 3.2, with obvious substitutions of symbols. Moreover, the proof of Lemma 3.8 is easily reformulated; one gets that the commutator $Z = [X,Y]$ is defined in coordinates $x_1, \ldots, x_n$ as

$$Z_j = \sum_k \left(X_k \frac{\partial Y_j}{\partial x_k} - Y_k \frac{\partial X_j}{\partial x_k} \right) .$$

The commutator of vector fields replaces the Poisson bracket (see Exercise 3.10). The vector fields commute in case $[X, Y] = 0$.

Exercise 3.10: Check that if X_F and X_G are the canonical vector fields generated by the Hamiltonian functions F and G, then the commutator $[X_F, X_G]$ is the canonical vector field generated by the Hamiltonian function $\{G, F\}$. (Hint: see Proposition 1.12.) A.E.L.

3.1.4 Coordinates on Invariant Manifolds

The result of Proposition 3.9 turns out to be very useful when we are given a complete involution system $\Phi_1(q,p), \ldots, \Phi_n(q,p)$ on a phase space $\mathscr{F}$. Let

$$M_0 = \big\{(q,p) \in \mathscr{F} \ : \ \Phi_1(q,p) = \cdots = \Phi_n(q,p) = 0\big\} \ ; \tag{3.9}$$

the value zero is, of course, arbitrary and may be replaced by any other value: just add a constant to the Φ. Since the functions are independent, M_0 so defined is an n-dimensional invariant manifold (see [189], or just appeal to the implicit functions theorem). At every point of M_0 the functions $\Phi_1(q,p), \ldots, \Phi_n(q,p)$ define n independent canonical vector fields which are tangent to M_0 at every point (q,p), and so n flows $\phi_1, \ldots, \phi_n$; this because every Φ_j is a first integral for all the other functions, so that the flows are confined to M_0. Moreover, the flows do commute, because the functions are in involution.

Forgetting for a moment the canonical character of the flows, let us consider in general an n-dimensional manifold M; let us also assume that at every point $P \in M$ we are assigned n vector fields $X_1, \ldots, X_n$ which are linearly independent, and that these vector fields do commute. We denote again the corresponding flows by $\phi_1, \ldots, \phi_n$. Recall that every flow possesses a local group property; that is, $\phi_1^{t_1} \circ \phi_1^{s_1} = \phi_1^{t_1+s_1}$ whenever the three flows are defined. Since the flows commute, for $t = (t_1, \ldots, t_n)$ and $s = (s_1, \ldots, s_n)$ we have also the local group property $\phi^t \circ \phi^s = \phi^{t+s}$. The interesting fact is that in the neighbourhood of any point $P \in M_0$ the times of the flows $\phi_1, \ldots, \phi_n$ generate a smooth local coordinate system; this is illustrated in Figure 3.3.

Proposition 3.11: *Let M be an n-dimensional manifold on which there are n commuting vector fields $X_1, \ldots, X_n$, tangent to M and linearly independent at every point of M. Then for any $P \in M$ there exists a neighbourhood V of the origin of $\mathbb{R}^n$ and a diffeomorphism ψ mapping V onto a neighbourhood U of P.*

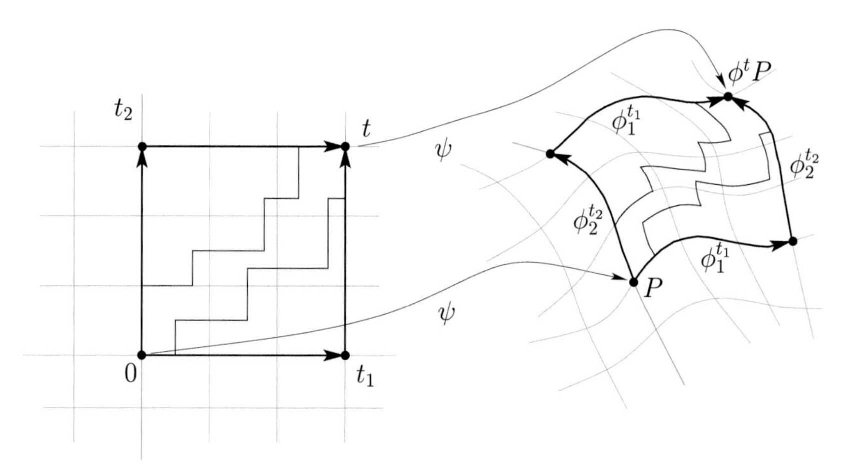

Figure 3.3 Illustrating the construction of local coordinates on a manifold using the flows of commuting vector fields.

Proof. Let $t_1, \ldots, t_n$ be coordinates in V. Apply to the point $P \in M$ the flows $\phi_1^{t_1}, \ldots, \phi_n^{t_n}$. Since the flows do commute we are allowed to apply them in any order, or even to follow any path in V connecting 0 with $t = (t_1, \ldots, t_n)$, as illustrated in Figure 3.3. This means that for any $t \in V$, the point $\psi(t) = \phi^t x$ is uniquely defined. Therefore ψ is a differentiable one-value map from V to M; this follows from known theorems on continuity and differentiability of the flow with respect to initial points. On the other hand, the Jacobian determinant of the function ψ at the origin is not zero, because the rows of the Jacobian matrix are nothing but the vector fields $X_1, \ldots, X_n$, which are independent. Thus, by the implicit function theorem, ψ has a differentiable inverse provided that V is small enough. *Q.E.D.*

3.1.5 Local Coordinates in Phase Space

Let us revert to the Hamiltonian case, so that the manifold M of the previous section is identified with the invariant manifold M_0 determined by $\Phi = 0$. The canonical vector fields generated by $\Phi_1, \ldots, \Phi_n$ on the invariant manifold M_0 satisfy the hypotheses of Proposition 3.11. Hence the times of the flows are local coordinates on M_0 in the neighbourhood of any point; let us highlight the latter fact by renaming these times as coordinates $\alpha = (\alpha_1, \ldots, \alpha_n)$ in a neighbourhood V_α of the origin.

The question now is how to extend the coordinates to a neighbourhood of the manifold M_0, using the Φ themselves as coordinates transversal to M_0. The discussion in this section is just a heuristic analysis of the problem, to put the question in a definite form. The answer, due to Joseph Liouville, will be treated in Section 3.1.6; the impatient reader may skip this section and jump forward to the next one.

Pick a point $(q_0, p_0) \in M_0$, so that $\Phi(q_0, p_0) = 0$, let $U \subset \mathscr{F}$ be a neighbourhood of (q_0, p_0), and let $(q', p') \in U$, so that $\Phi(q', p') \in V_\Phi \subset \mathbb{R}^n$, a neighbourhood of the origin. Recalling (3.9), let

$$M_\Phi = \{(q, p) \in \mathscr{F} : \Phi_1(q, p) = \Phi_1(q', p'), \ldots, \Phi_n(q, p) = \Phi_n(q', p')\} ;$$

then M_Φ is an invariant manifold for the flows $\phi_1^t, \ldots, \phi_n^t$ generated by the Φ, with $t \in V$ (possibly restricting V so that the flows are defined). On every such invariant manifold we may introduce local coordinates $\alpha = (\alpha_1, \ldots, \alpha_n)$, which are the times of the flows; this is just an application of Proposition 3.11.

The question is reformulated as follows: *we look for a local diffeomorphism* $(q, p) = \chi(\alpha, \Phi)$ *mapping* $V_\alpha \times V_\Phi \subset \mathbb{R}^{2n}$ *to* U*; we also want the diffeomorphism to be canonical.* The latter request appears to be a reasonable one, due to the following argument. The functions $\Phi_1(q, p), \ldots, \Phi_n(q, p)$ obey the same involution rules as the fundamental Poisson brackets for the momenta $p_1, \ldots, p_n$ or the coordinates $q_1, \ldots, q_n$. In turn, this fact reminds us that a point transformation is defined by n functions $q_j = q_j(Q_1, \ldots, Q_n)$ which form themselves a complete involution system. This is akin to saying that part of the canonicity conditions expressed via Poisson brackets are satisfied. Therefore, it seems natural to ask whether we can associate to $\Phi_1, \ldots, \Phi_n$ a set of conjugate coordinates, $\alpha_1, \ldots, \alpha_n$ say, so that we are given a canonical transformation. The two cases differ in that the functions Φ may depend on all the $2n$ canonical coordinates, not just on half of them.

The question is partially settled in view of Proposition 3.11, for the coordinates α define a local diffeomorphism $\psi_\Phi(\alpha)$, parameterized by Φ, on every invariant manifold $U \cap M_\Phi$. The critical point is the dependence on Φ, since for every $\Phi \in V_\Phi$ we should choose a point (q_0, p_0) which corresponds to $\alpha = 0$; this is illustrated in Figure 3.4. In more precise terms, we should choose a diffeomorphism $\big(q(\Phi), p(\Phi)\big)$ mapping V_Φ into an n-dimensional manifold $\Sigma_0 \subset U$, transversal to the manifolds $\Phi = \text{constant}$; this determines the desired points $(q_0, p_0) = \chi(0, \Phi)$. Then the complete diffeomorphism χ should be defined so as to satisfy both conditions

$$(q_0, p_0) = \chi(0, \Phi) \in \Sigma_0 , \quad \chi(\alpha, \Phi) = \phi^\alpha \chi(0, \Phi) ; \tag{3.10}$$

remark that in terms of the coordinates (α, Φ) the flow is just a translation on α, leaving Φ invariant. Therefore the coordinates Φ will parameterize a continuous family of local invariant manifolds, while the coordinates α cover each manifold. In view of Proposition 2.15 the flow ϕ^α is canonical, being a composition of canonical flows. However, canonicity can be broken due to an unsuitable choice of the manifold Σ_0 : this point needs careful consideration.

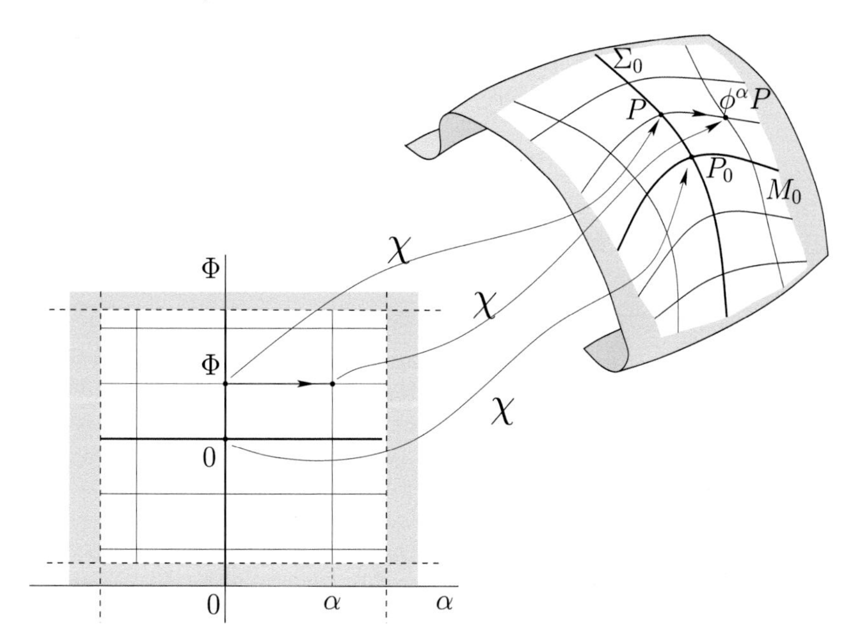

Figure 3.4 Illustrating the local coordinates induced by the flow of the involution system $\Phi_1(q,p),\ldots,\Phi_n(q,p)$. For graphical reasons the notation $P_0 = (q_0,p_0)$ and $P = (q,p)$ is used in the figure.

The comments after the statement of Corollary 3.13 should help to clarify the latter mysterious observation.

3.1.6 Liouville's Canonical Coordinates

The following proposition provides an answer to the questions raised in the previous section. Furthermore it claims that the transformation can be constructed by quadratures.

Proposition 3.12: *Let $\{\Phi_1(q,p),\ldots,\Phi_n(q,p)\}$ be a complete involution system. Then there exists a generating function $S(\Phi,q)$ of a local canonical transformation $(q,p) = \chi(\alpha,\Phi)$, implicitly defined as*

$$\alpha_j = \frac{\partial S}{\partial \Phi_j}\ , \quad p_j = \frac{\partial S}{\partial q_j}\ , \quad j = 1,\ldots,n\ , \tag{3.11}$$

which associates to $\Phi_1,\ldots,\Phi_n$ the new canonically conjugated coordinates $\alpha_1,\ldots,\alpha_n$. With the nonrestrictive hypothesis

$$\det\left(\frac{\partial(\Phi_1,\ldots,\Phi_n)}{\partial(p_1,\ldots,p_1)}\right) \neq 0\ , \tag{3.12}$$

the generating function of the canonical transformation is constructed by quadrature as

$$(3.13) \qquad S(\Phi, q) = \int \sum_j p_j(\Phi, q) dq_j \ ,$$

where $p_1(\Phi, q), \ldots, p_n(\Phi, q)$ *are found by inversion of* $\Phi_1(q,p), \ldots, \Phi_n(q,p)$.

Let us agree to name *Liouville's coordinates* the canonical coordinates so introduced.

Remarks. 1. Referring to the discussion of the previous section, the reader will realize that the functions $p(\Phi, q)$ describe precisely the manifold Σ_0. **2.** Condition (3.12) is not restrictive, for in view of Lemma 3.6 it can always be fulfilled by exchanging some of the coordinates with the conjugated momenta: we should always bear in mind that our construction is a local one. **3.** It should also be noted that in the new coordinates α, Φ the canonical equations for the Hamiltonian Φ_k are $\dot{\alpha}_j = \delta_{j,k}$ and $\dot{\Phi}_j = 0$, with $j = 1, \ldots, n$. Thus, α_k is the time of the flow generated by Φ_k, coherently with the discussion of Section 3.1.5.

Corollary 3.13: *The canonical coordinates of Proposition 3.12 are determined up to a canonical transformation with generating function*

$$(3.14) \qquad W(\overline{\Phi}, \alpha) = \sum_j \overline{\Phi}_j \alpha_j + f(\overline{\Phi}) \ ,$$

where $f(\overline{\Phi})$ *is an arbitrary function. Equivalently, one can add an arbitrary function* $F(\Phi)$ *to the generating function* $S(\Phi, q)$ *defined by (3.13).*

The arbitrary choice of the function $F(\Phi)$ corresponds to changing the choice of the n-dimensional surface Σ_0 corresponding to $\alpha = 0$: the coordinates α may be translated by a quantity depending on Φ. However, the translation is not completely arbitrary: it introduces some arbitrariness in the choice of the n-dimensional manifold corresponding to $\alpha = 0$, which, however, must be compatible with the desired canonicity of the transformation. To this end, it must be written as $\alpha_j = \frac{\partial W}{\partial \Phi_j}$, with arbitrary $W(\Phi)$.

The proof of Corollary 3.13 is trivial, and is left to the reader.

Proof of Proposition 3.12. In view of condition (3.12) we can invert the functions $\Phi_1, \ldots, \Phi_n$ with respect to $p_1, \ldots, p_n$, thus getting the functions $p_1 = p_1(\Phi, q), \ldots, p_n = p_n(\Phi, q)$. In order to construct the generating function $S(\Phi, q)$ in (3.13) we should prove that the differential form $\sum_j p_j(\Phi, q) dq_j$ is closed. To this end, let us differentiate the identity

$$\Phi_j = \Phi_j(q, p)\big|_{p = p(\Phi, q)}$$

with respect to Φ, q, namely taking into account that in the r.h.s. p must be replaced by its expression in terms of Φ, q. This gives

$$d\Phi_j = \sum_{k,l} \frac{\partial \Phi_j}{\partial p_k} \left(\frac{\partial p_k}{\partial \Phi_l} d\Phi_l + \frac{\partial p_k}{\partial q_l} dq_l \right) + \sum_l \frac{\partial \Phi_j}{\partial q_l} dq_l \ .$$

By comparison of the coefficients of dq, $d\Phi$, we get the identities

$$\sum_k \frac{\partial \Phi_j}{\partial p_k} \frac{\partial p_k}{\partial \Phi_l} = \delta_{j,l} \ ,$$

$$\sum_k \frac{\partial \Phi_j}{\partial p_k} \frac{\partial p_k}{\partial q_l} = -\frac{\partial \Phi_j}{\partial q_l} \ , \qquad j,l = 1, \ldots, n \ .$$

Replace now the second of these identities in the relation $\{\Phi_j, \Phi_m\} = 0$, which holds true because the functions are assumed to be in involution. With a few calculations we get

$$\begin{aligned} \{\Phi_j, \Phi_m\} &= \sum_l \left(\frac{\partial \Phi_j}{\partial q_l} \frac{\partial \Phi_m}{\partial p_l} - \frac{\partial \Phi_j}{\partial p_l} \frac{\partial \Phi_m}{\partial q_l} \right) \\ &= -\sum_{l,k} \frac{\partial \Phi_m}{\partial p_l} \frac{\partial \Phi_j}{\partial p_k} \frac{\partial p_k}{\partial q_l} + \sum_{l,k} \frac{\partial \Phi_j}{\partial p_l} \frac{\partial \Phi_m}{\partial p_k} \frac{\partial p_k}{\partial q_l} \\ &= -\sum_l \frac{\partial \Phi_m}{\partial p_l} \sum_k \frac{\partial \Phi_j}{\partial p_k} \left(\frac{\partial p_k}{\partial q_l} - \frac{\partial p_l}{\partial q_k} \right) = 0 \end{aligned}$$

(note that in the second sum on the second line the indexes l and k can be exchanged). By condition (3.12), this implies

$$\frac{\partial p_k}{\partial q_l} - \frac{\partial p_l}{\partial q_k} = 0 \ , \quad l,k = 1, \ldots, n \ ,$$

so that the differential form $\sum_j p_j dq_j$ is closed, as claimed. By integration we construct the generating function (3.13) which, in view of (3.12), satisfies the invertibility condition (2.21) of Proposition 2.23. Therefore, the desired canonical transformation is implicitly defined as in (3.11). *Q.E.D.*

3.1.7 Changing the Involution System

The construction of Liouville's coordinates depends on the choice of the involution system. If a complete involution system $\Psi_1, \ldots, \Psi_n$ different from $\Phi_1, \ldots, \Phi_n$ is considered, then new coordinates $\beta_1, \ldots, \beta_n$ may be constructed by applying again Liouville's procedure. However, there is a simpler way.

Proposition 3.14: *Let $\{\Phi_1(q,p), \ldots, \Phi_n(q,p)\}$ be a complete involution system, and let the functions $\Psi_1(\Phi), \ldots, \Psi_n(\Phi)$ satisfy*

$$\det\left(\frac{\partial\Psi_j}{\partial\Phi_k}\right) \neq 0 \ . \tag{3.15}$$

Then:

(i) the functions $\Psi_1,\ldots,\Psi_n$ form a complete involution system;
(ii) the Liouville coordinates conjugated to Ψ are

$$\beta_j = \sum_{k=1}^{n} \frac{\partial\Phi_k}{\partial\Psi_j}\alpha_k \ , \tag{3.16}$$

where $\alpha_1,\ldots,\alpha_n$ are the Liouville coordinates conjugated to Φ.

Proof. (i) The functions Ψ are independent in view of condition (3.15); moreover, they are obviously in involution, being functions only of Φ.
(ii) Recall the extended point transformation, Example 2.25. The generating function is written as

$$S(\Phi,\beta) = \sum_{j=1}^{n} \Psi_j(\Phi)\beta_j \ ,$$

from which (3.16) immediately follows. *Q.E.D.*

3.2 Liouville's Theorem

For a generic system of differential equations on an n-dimensional manifold a complete integration by quadrature can be performed when $n-1$ independent first integrals are known, n being the dimension of the space. Thus, one expects that in the Hamiltonian case, the dimension of phase space being $2n$, one needs $2n-1$ first integrals. However, the canonical structure allows us to perform the complete integration if only n first integrals are known, provided that they fulfil the further condition of being in involution. The proof exploits Liouville's canonical coordinates introduced by Proposition 3.12.

Theorem 3.15: *Assume that an autonomous canonical system with n degrees of freedom and with Hamiltonian $H(q,p)$ possesses n independent first integrals $\{\Phi_1(q,p),\ldots,\Phi_n(q,p)\}$ forming a complete involution system. Then the system is integrable by quadratures. More precisely, one can construct the generating function $S(\Phi,q)$ of a canonical transformation $(q,p) = \chi(\alpha,\Phi)$ such that the transformed Hamiltonian depends only on the new momenta $\Phi_1,\ldots,\Phi_n$, and the solutions are expressed as*

$$\alpha_j(t) = \alpha_{j,0} + t\frac{\partial H}{\partial\Phi_j}\bigg|_{(\Phi_{12,0},\ldots,\Phi_{n,0})} \ , \quad j = 1,\ldots,n \ , \tag{3.17}$$

with $\alpha_{j,0}$ and $\Phi_{j,0}$ determined by the initial data.

Proof. By Proposition 3.12 there is a canonical transformation $(q,p) = \chi(\alpha,\Phi)$ such that $\Phi_1,\ldots,\Phi_n$ are the new momenta. In view of preservation of

Poisson brackets, we can compute the Poisson bracket $\{H, \Phi_j\}$ with respect to the new variables α, Φ. Since $\Phi_1, \ldots, \Phi_n$ are first integrals, this gives

$$0 = \{H, \Phi_j\} = \frac{\partial H}{\partial \alpha_j} , \quad j = 1, \ldots, n ,$$

so that the transformed Hamiltonian depends only on the momenta, that is, $H = H(\Phi)$. Therefore, the canonical equations are

$$\dot{\alpha}_j = \frac{\partial H}{\partial \Phi_j} , \quad \dot{\Phi}_j = 0 , \quad j = 1, \ldots, n$$

and are trivially integrable, as in (3.17). *Q.E.D.*

3.2.1 *Integration Procedure*

A point not to be underestimated is that Liouville's theorem provides an explicit integration algorithm. The procedure could be reconstructed by isolating the relevant points in the previous sections. However, let us do it, even if it appears to be pedantic.

(i) If necessary, exchange some pairs of canonical variables so that the condition

$$\det\left(\frac{\partial(\Phi_1, \ldots, \Phi_n)}{\partial(p_1, \ldots, p_n)}\right) \neq 0$$

is fulfilled. Then perform an inversion, finding $p = p(\Phi, q)$.

(ii) By a quadrature, construct the generating function

$$S(\Phi, q) = \int \sum_j p_j(\Phi, q) dq_j .$$

(iii) By substitution, determine the transformed Hamiltonian $H(\Phi)$.

(iv) The solutions of the canonical equations are

$$\Phi_j(t) = \Phi_{j,0} , \quad \alpha_j(t) = \left.\frac{\partial H}{\partial \Phi_j}\right|_{\Phi_j = \Phi_{j,0}} t + \alpha_{j,0} , \quad j = 1, \ldots, n ,$$

where Φ_{j_0} and $\alpha_{j,0}$ are the initial values which can be computed from the initial data.

(v) Find the functions $q = q(\Phi, \alpha)$ and $p = p(\Phi, \alpha)$ by inverting the canonical transformation.

(vi) The solutions $q(t), p(t)$ in the original variables are found by substitution of $\Phi(t), \alpha(t)$ given by (3.17).

The following examples demonstrate the procedure of Liouville. The reader may find them annoying, being nothing but a repetition in a slightly different form of examples which have already been discussed in the previous chapters. What should be appreciated is that the knowledge of first integrals for a Hamiltonian system makes the process of solution an automatic one: a kind of engine which produces the result once the correct input is put in.

The relationed with the Hamilton–Jacobi method are also worthy of attention. It may be said that Liouville's method consists in using the first integrals as constants which allow to find a complete solution of the Hamilton–Jacobi equations (see, e.g., [209], § 148). That the final formulæ expressing the solutions are the same as already found should not be surprising, of course. Once more, the impatient reader may skip the examples.

Example 3.16: *Systems with one degree of freedom.* Let us consider the Hamiltonian

$$H(x,p) = \frac{p^2}{2m} + V(x) ,$$

describing the motion of a mass point on a straight line under the action of the potential $V(x)$. The condition at point (i) reduces to $\frac{\partial H}{\partial p} \neq 0$, which is fulfilled for $p \neq 0$. Setting $H(x,p) = E$, we invert the relation with respect to p, getting

$$p = \pm\sqrt{2m[E - V(x)]} . \tag{3.18}$$

The generating function is

$$S(E,x) = \sqrt{2m} \int \sqrt{E - V(x)} dx ,$$

and the canonical transformation in implicit form is written as

$$p = \pm\sqrt{2m[E - V(x)]} , \quad \alpha = \sqrt{\frac{m}{2}} \int \frac{dx}{\sqrt{E - V(x)}} .$$

The first of these relations coincides with (3.18), as expected. The transformed Hamiltonian is, trivially, $H(E) = E$, and the solutions of Hamilton's equations in coordinates α, E are[3]

$$E(t) = E_0 , \quad \alpha(t) = t - t_0 ,$$

[3] Saying that the formula expresses the solution of the canonical equations may seem a bizarre claim to the reader: it seems to say merely that the energy E is constant and the time flows uniformly, which we knew very well in advance. The point is that the canonical transformation allows us to write the solution for the original coordinates, depending on the constants E and t_0.

E_0 and t_0 being the initial values of energy and time, respectively. Therefore, in order to actually compute the solutions we need to compute the integral

$$t - t_0 = \sqrt{\frac{m}{2}} \int_{x_0}^{x} \frac{d\xi}{\sqrt{E_0 - V(\xi)}} , \tag{3.19}$$

where $x_0 = x(0)$ is the initial datum and E_0 is the initial energy. The latter formula actually coincides with (1.36), of Section 1.3.1, as should be expected. *E.D.*

Example 3.17: *Harmonic oscillators.* For the Hamiltonian

$$H = \frac{1}{2} \sum_{l=1}^{n} (y_l^2 + \omega_l^2 x_l^2) , \tag{3.20}$$

the quantities

$$\Phi_l = \frac{1}{2}(y_l^2 + \omega_l^2 x_l^2) , \quad 1 \le l \le n \tag{3.21}$$

form an involution system, and moreover we have

$$H = \sum_l \Phi_l . \tag{3.22}$$

By inversion of (3.21) with respect to y_l, we get

$$y_l = \sqrt{2\Phi_l - \omega_l^2 x_l^2} ,$$

and the generating function is

$$S(\Phi, x) = \sum_{l=1}^{n} \int \sqrt{2\Phi_l - \omega_l^2 x_l^2} dx_l .$$

The canonical transformation is completed by the new coordinates

$$\alpha_l = \frac{\partial S}{\partial \Phi_l} = \int \frac{dx_l}{\sqrt{2\Phi_l - \omega_l^2 x_l^2}} = \frac{1}{\omega_l} \text{arcos} \left(\frac{\omega_l x_l}{\sqrt{2\Phi_l}} \right) .$$

The Hamiltonian is given by (3.22), and the canonical equations

$$\dot{\Phi}_l = 0 , \quad \dot{\alpha}_l = \frac{\partial H}{\partial \Phi_l} = 1$$

have solutions

$$\Phi_l(t) = \Phi_{l,0} , \quad \alpha_l(t) = t - t_0 ,$$

where t_0 is the initial time, and $\Phi_{l,0}$ are constants to be computed by the initial data. Finally, by inversion we obtain the solution

$$x_l = \frac{\sqrt{2\Phi_{l_0}}}{\omega_l} \cos \omega_l (t - t_0) .$$

E.D.

Example 3.18: *Motion under central field on a plane.* As a further example, let us consider the Hamiltonian

$$H = \frac{1}{2m}\left(p_r^2 + \frac{p_\vartheta^2}{r^2}\right) + V(r) ,$$

describing the planar motion of a mass point in a central force field. As first integrals we can use

$$\Gamma = p_\vartheta , \quad \Phi = \frac{1}{2m}\left(p_r^2 + \frac{\Gamma^2}{r^2}\right) + V(r) . \tag{3.23}$$

By inversion, compute

$$p_\vartheta = \Gamma , \quad p_r = \left[2m\big(\Phi - V(r)\big) - \frac{\Gamma^2}{r^2}\right]^{1/2} , \tag{3.24}$$

so that the generating function is

$$S(\Phi,\Gamma,r,\vartheta) = \int \left[2m\big(\Phi - V(r)\big) - \frac{\Gamma^2}{r^2}\right]^{1/2} dr + \int \Gamma d\vartheta .$$

Denoting by φ and γ the canonical variables conjugated to Φ and Γ, respectively, the transformation in implicit form is given by (3.24) and

$$\begin{aligned}\varphi &= \frac{\partial S}{\partial \Phi} = m \int \left[2m\big(\Phi - V(r)\big) - \frac{\Gamma^2}{r^2}\right]^{-1/2} dr ,\\ \gamma &= \frac{\partial S}{\partial \Gamma} = -m\Gamma \int \frac{1}{r^2}\left[2m\big(\Phi - V(r)\big) - \frac{\Gamma^2}{r^2}\right]^{-1/2} dr + \int d\vartheta .\end{aligned}$$

Once $V(r)$ is known, by quadrature we can compute

$$\varphi = f(\Phi,\Gamma,r) , \quad \gamma = g(\Phi,\Gamma,r) + \vartheta - \vartheta_0 , \tag{3.25}$$

where ϑ_0 is given by the initial conditions, and the functions f and g are given by the integrals in the preceding formula. The transformed Hamiltonian is $H = \Phi$, so that the solutions of Hamilton's equations are

$$\begin{aligned}\varphi &= t - t_0 , \quad &\gamma &= \gamma_0 ,\\ \Phi &= \Phi_0 , \quad &\Gamma &= \Gamma_0 .\end{aligned} \tag{3.26}$$

Here, γ_0, Φ_0 and Γ_0 must be computed from the initial data $r_0, \vartheta_0, p_{r,0}, p_{\vartheta,0}$ at time t_0 using (3.23) and (3.25). The solutions $r(t), \vartheta(t), p_r(t), p_\vartheta(t)$ in the original coordinates are computed by inverting (3.25) so as to obtain

$$r = r(\Phi,\Gamma,\varphi) , \quad \vartheta = \vartheta_0 + \gamma - g(\Phi,\Gamma,r)\Big|_{r=r(\Phi,\Gamma,\varphi)} .$$

By substitution of (3.26) in the latter expressions, we get r and ϑ as functions of time and of the initial values, namely

$$r = r(\Phi_0, \Gamma_0, t - t_0) \ , \quad \vartheta = \vartheta_0 + \gamma_0 - g(\Phi_0, \Gamma_0, r)\Big|_{r=r(\Phi_0,\Gamma_0,t-t_0)} \ .$$

Finally, the momenta p_r and p_ϑ as functions of time are computed by replacing (3.26) and the latter expressions in (3.24). This completes the solution of the problem. *E.D.*

Exercise 3.19: Apply Liouville's theorem to the case of a central field of forces in space, reducing it to quadrature. *A.E.L.*

3.2.2 Some Comments on Liouville's Theorem

The general form (3.17) of the solutions for the Hamiltonian in the new variables, as stated in Theorem 3.15, appears to be quite simple. However, the example of the harmonic oscillators shows that all phenomena related to the periodicity of the motion are hidden and show up only when the transformation back to the original variables is performed.

For comparison, if we apply to the Hamiltonian (3.20) the transformation to action-angle variables for the harmonic oscillators, namely

$$x_l = \sqrt{2I_l} \cos\varphi_l \ , \quad y_l = \sqrt{2I_l} \sin\varphi_l \ , \quad l = 1, \ldots, n \ ,$$

we get the transformed Hamiltonian

$$H(I_1, \ldots, I_n) = \sum_l \omega_l I_l \ .$$

The canonical equations then are

$$\dot\varphi_l = \omega_l \ , \quad \dot I_l = 0 \ ,$$

and the evolution of the phases $\varphi_l(t) = \omega_l t + \varphi_{l,0}$ is still uniform with velocity ω. The remarkable difference with respect to (3.17) is that the new coordinates φ are *angles* representing the phases of the oscillators. Therefore, the fact that the evolution is a combination of periodic motions emerges immediately: the velocity ω of the phases is the angular frequency.

The same problem shows up also in the discussion of Example 3.18: the algorithm is well defined, but the fact that, for example, in the Keplerian case the motion is periodic will not be recognized until the complete solution is explicitly calculated.

That this is a general problem is further illustrated by the following.

Example 3.20: *Free rotator.* Let the phase space be $\mathbb{T}\times\mathbb{R}$, with coordinates $q \in \mathbb{T}$ and $p \in \mathbb{R}$, and let the Hamiltonian be[4]

$$H = \frac{p^2}{2} \ . \tag{3.27}$$

The system is actually trivial: the equations are $\dot{q} = p$, $\dot{p} = 0$, with solutions $p(t) = p_0$, $q(t) = p_0 t + q_0$, where p_0 , q_0 are the initial data. Recalling that q is an angle, the motion is immediately seen to be periodic with angular frequency p_0 depending on the initial data. If, however, we forget this fact and apply the procedure suggested by Liouville's theorem, using the Hamiltonian H as a first integral, then we get that the new canonical coordinates are the time t, which flows uniformly, and the energy H, which is constant. We see again that all information concerning the periodicity of the motion is lost and is recovered only after writing the solutions for the original variables. *E.D.*

The common aspect to all these examples is that the choice of the first integrals to be used in order to apply Liouville's procedure is quite arbitrary. However, the examples of the harmonic oscillator and of the free rotator suggest that, at least in these cases, there is a particular choice which is in some sense the best one, and that it is connected with the fact that the coordinates conjugated to the momenta $\Phi_1, \ldots, \Phi_n$ should be *angles*.

3.2.3 Action-Angle Variables for Systems with One Degree of Freedom

The construction of action-angle variables turns out to be particularly simple in the case of a system with one degree of freedom; this is always a Liouville-integrable system, since the Hamiltonian is a first integral. Moreover, the following procedure is instructive, since it suggests how to proceed for higher-dimensional systems, as we shall do later, in Section 3.4. Let $\overline{q}, \overline{p}$ be an extremum for $H(q,p)$, so that it is an equilibrium point, and $q(t) = \overline{q}, p(t) = \overline{p}$ is a solution of Hamilton's equations. Then there is an open interval $\mathscr{E}$ such that for $E \in \mathscr{E}$ the set of points satisfying $H(q,p) = E$ contains a continuous family of closed curves surrounding the point $(\overline{q}, \overline{p})$, which are orbits. Let γ_E be any such curve; it can be described via a coordinate $s \in \mathbb{T}$, in many ways. With such a coordinate it is easy to account for the periodicity of the motion, since a period is completed when s is incremented by 2π. It is quite natural to ask if there exists a canonical momentum I which parameterizes the family γ_E of closed curves.

If such a quantity I exists, it must be constant on every curve γ_E; this implies that it must be a first integral. Thus, let us look for a function $I(q,p)$

---

[4] The Hamiltonian of a rigid body freely rotating around a fixed axis may be put in this form.

which is in involution with the Hamiltonian $H(q, p)$ and satisfies $\frac{\partial I}{\partial p} \neq 0$. If such a function exists, by Proposition 3.12 we can construct a further function

$$S(I, q) = \int p(I, q) dq \ , \tag{3.28}$$

the latter being the generating function of a canonical transformation which defines I as the new momentum. Thus, there is also a coordinate φ, conjugated to I; on the other hand, φ turns out to be periodic, because it is a coordinate on a closed curve, and we can always manage so that the period is 2π. Let us now see how we can construct $I(q, p)$. If φ, I are canonical variables, by Proposition 2.19 we must have

$$\oint_{\gamma_E} p dq = \oint_{\gamma_E} I d\varphi \ .$$

Since $I(q, p)$ must be constant on γ_E and φ is 2π-periodic, the integral on the right-hand side is easily calculated to be $2\pi I$. Therefore, it must be

$$I = \frac{1}{2\pi} \oint_{\gamma_E} p dq \ . \tag{3.29}$$

The latter quantity has been named the *action* of the system. The integral can be computed after expressing p as a function of E and q, and gives a function $I(E)$; replacing $E = H(q, p)$ gives I as a function of q, p, as required. We conclude that $I(q, p)$ can be computed through a quadrature. It will be noticed that this function represents the area enclosed by the curve γ_E passing through the point q, p, divided by 2π.

Since $I(q, p)$ is a first integral, we can apply Liouville's theorem. With a further quadrature we can compute the generating function $S(I, q)$ given by (3.28), and by differentiation we determine the angle φ. The canonical coordinates I, φ are called *action-angle variables.* The transformed Hamiltonian $H(I)$ is independent of φ, and Hamilton's equations read

$$\dot{I} = 0 \ , \quad \dot{\varphi} = \omega(I) \ , \tag{3.30}$$

where $\omega(I) = \frac{\partial H}{\partial I}$. Having fixed the initial conditions $I(0) = I_0, \varphi(0) = \varphi_0$, the corresponding solution is

$$I(t) = I_0 \ , \quad \varphi(t) = \omega_0 t + \varphi^{(0)} \ , \tag{3.31}$$

where $\omega_0 = \omega(I_0)$. The periodicity of the motion is now evident, because φ is an angle, and the period clearly is $T = 2\pi/\omega_0$. The period can be easily computed as

$$T = 2\pi \frac{dI}{dE} \ . \tag{3.32}$$

For, from $H = E$ one has $1 = \frac{dH}{dE} = \frac{dH}{dI}\frac{dI}{dE} = \omega(I)\frac{dI}{dE}$, and the claim follows using $T = 2\pi/\omega(I)$.

Example 3.21: *The harmonic oscillator.* The Hamiltonian

$$H(p, x) = \frac{1}{2}p^2 + \frac{1}{2}\omega^2 x^2$$

has an equilibrium for $p = x = 0$, and for $E > 0$ the curve $H(p, x) = E$ is an ellipse centred on the origin and with semiaxes $\sqrt{2E}/\omega$ and $\sqrt{2E}$ (see Example 1.14). The action is easily computed as $I = E/\omega$, the period is $T = 2\pi/\omega$, and the angle φ represents the phase of the oscillator. The explicit form of the canonical transformation $x = \sqrt{2I}\cos\varphi$, $y = \sqrt{2I}\sin\varphi$ can be obtained by elementary considerations, as in Example 2.14; it is not necessary to proceed to an explicit calculation of the generating function. *E.D.*

Example 3.22: *Oscillations around a stable equilibrium.* Consider the Hamiltonian

$$H(p, x) = \frac{p^2}{2} + V(x)\ , \quad (x, p) \in \mathbb{R}^2\ , \tag{3.33}$$

describing the motion of a point with unitary mass moving on a straight line under the action of a potential $V(x)$. Let $V(x)$ have a point of relative minimum at $x = 0$, and let $V(0) = 0$ (this is not restrictive, of course, since it is always possible to introduce the displacement from equilibrium as a coordinate, and the potential is defined up to a constant). Then there exists an open interval of positive values of E such that the level set of points satisfying $H(p, x) = E$ contains a closed curve around the origin. For such values of E the action is defined as

$$I = \frac{\sqrt{2}}{\pi}\int_{x_{\min}}^{x_{\max}} \sqrt{E - V(x)}dx\ , \tag{3.34}$$

$x_{\min}, x_{\max}$ being the extrema of the interval of oscillation, to be computed as solutions of the equation $E - V(x) = 0$. The period can be computed as

$$T = \sqrt{2}\int_{x_{\min}}^{x_{\max}} \frac{dx}{\sqrt{E - V(x)}}\ . \tag{3.35}$$

The calculation of the action and of the period is thus reduced to a quadrature. If an explicit expression for the angle φ is wanted, then one must compute the generating function

$$S(I, x) = \sqrt{2}\int \sqrt{E - V(x)}dx$$

and replace $E = E(I)$ as computed from (3.34). The canonical transformation is written in implicit form as

$$p = \frac{\partial S}{\partial x} \ , \quad \varphi = \frac{\partial S}{\partial I} \ .$$

Writing the transformation explicitly requires an inversion. E.D.

Example 3.23: *The pendulum.* As a more specific example, consider the Hamiltonian of a pendulum:

$$H(p, \vartheta) = \frac{p^2}{2} - \cos\vartheta \ , \quad (\vartheta, p) \in \mathbb{T} \times \mathbb{R} \ .$$

The Hamiltonian has a minimum for $p = \vartheta = 0$ (recall that ϑ is defined $\bmod 2\pi$), and for $-1 < C < 1$ the equation $H(p, \vartheta) = C$ determines a closed curve around the equilibrium. According to (3.34) the action variable is computed as

$$I = \frac{2\sqrt{2}}{\pi} \int_0^{\vartheta_{\max}} \sqrt{C + \cos\vartheta}\, d\vartheta \ ,$$

where $\vartheta_{\max} = \arccos(-C)$. Remark that the symmetry $V(-\vartheta) = V(\vartheta)$ of the potential has been taken into account in the latter formula. The calculation of the integral can be reduced to that of elliptic integrals as follows. Transform $\vartheta = 2\varphi$, thus getting

$$I = \frac{8}{\pi} \int_0^{\varphi_{\max}} \sqrt{\frac{C+1}{2} - \sin^2\varphi}\, d\varphi \ .$$

Denote now $k^2 = (C+1)/2$, and transform $\sin\varphi = k \sin\psi$; this gives

$$\begin{aligned} I &= \frac{8k^2}{\pi} \int_0^{\pi/2} \frac{\cos^2\psi}{\sqrt{1 - k^2 \sin^2\psi}}\, d\psi \\ &= \frac{16(k^2+1)}{\pi} K(k^2) + \frac{16}{\pi} E(k^2) \ , \end{aligned}$$

where $K(k^2)$ and $E(k^2)$ are the complete elliptic integrals of the first and second kind, namely

$$\begin{aligned} K(k^2) &= \int_0^{\pi/2} \frac{d\psi}{\sqrt{1 - k^2 \sin^2\psi}} \ , \\ E(k^2) &= \int_0^{\pi/2} \sqrt{1 - k^2 \sin^2\psi}\, d\psi \ . \end{aligned}$$

These integrals are computed via a series expansion which is convergent for $|k|^2 < 1$. Therefore, the action I as a function of C is defined only for $C < 1$. The value $C = 1$ corresponds indeed to the separatrix. The transformed

Hamiltonian $H(I)$ can be computed by inversion of the function given by the integral, putting $H = C$. *E.D.*

Exercise 3.24: Show how to calculate the action for the rotating pendulum, that is, when $|C| > 1$. *A.E.L.*

Remark. Before proceeding, let us prepare the next step by pointing out the question that we are going to tackle. The Liouville coordinates are *local* in nature, for they are introduced in the neighbourhood of any point where the vector fields of $\Phi_1, \ldots, \Phi_n$ are non-singular. In the examples in this section it seems that everything is global, but this is untrue: the locality is hidden behind the fact that in all examples we know that the level curves of constant energy are actually *cycles*, and we calculate the actions via a quadrature over the cycle, formula (3.28). However, a careful consideration of the formula shows that we face the same problem seen in the quadrature formulæ of Section 1.3.1 (see in particular note 13 in that section). On the other hand the action-angle variables that we have constructed provide us with a *global* description of the level curve as a torus $\mathbb{T}$; the actions remain local variables. Example 3.23 of the pendulum should make all this clear.

The question now is: *can we extend the procedure of construction of action-angle variables to systems with more than one degree of freedom?* The rest of the chapter answers this question. We proceed in two main steps. The first one is a topological result of general interest concerning a n-dimensional connected manifold where there are n independent vector fields at every point. The second step is concerned precisely with the Hamiltonian case which is the object of the present notes, when we are given a complete involution system; our argument makes wide use of Liouville coordinates, extending them in a global sense through angular variables.

3.3 On Manifolds with Non-Singular Vector Fields

The aim of the present section is to investigate the *global* behaviour of the flow of commuting vector fields on a connected manifold implicitly defined by a complete involution system as

$$M_0 = \{(q,p) \in \mathscr{F} \ : \ \Phi_1(q,p) = 0, \ldots, \Phi_n(q,p) = 0\} \; ; \tag{3.36}$$

the value 0 might be replaced by any suitable value $c \in \mathbb{R}^n$, of course. If more than one component is defined by Eq. (3.36), just select one. Here we focus our attention on the n-dimensional manifold M, letting aside for a while the $2n$-dimensional phase space. The vector fields are also requested to be independent at every point of the manifold; hence they must be non-singular. Our hypothesis is clearly satisfied by the canonical vector fields generated by the first integrals $\Phi_1, \ldots, \Phi_n$: they are independent because the functions are independent; they commute because the functions are in involution. The *local* aspects have been discussed in Section 3.1.4, where it is also pointed out that the canonical character of the vector fields is unessential. So, let us highlight the general interest of the result by using again the notations M for the invariant manifold and $X_1, \ldots, X_n$ for the vector fields.

Proposition 3.25: *Let M be a connected differentiable manifold of dimension n, and let $X_1, \ldots, X_n$ be n vector fields which are independent at every point of M and satisfy $[X_j, X_k] = 0$ on M, for $j, k = 1, \ldots, n$; assume, moreover, that the flows $\phi_1^{t_1}, \ldots, \phi_n^{t_n}$ of $X_1, \ldots, X_n$ can be indefinitely prolonged on M. Then there is a nonnegative $k \leq n$ such that M is diffeomorphic to $\mathbb{T}^k \times \mathbb{R}^{n-k}$.*

Corollary 3.26: *If M is compact, then it is diffeomorphic to an n-dimensional torus.*

3.3.1 The Flow in the Large

It is useful to recall the local result of Proposition 3.11: for every $P \in M$ there are a neighbourhood V of the origin of $\mathbb{R}^n$ and a neighbourhood $U \subset M$ of P which are diffeomorphic; the diffeomorphism is defined through the local group $\phi^t = \phi_1^{t_1} \circ \cdots \circ \phi_n^{t_n}$ generated by commuting flows. The first step of the proof consists in extending the local result by considering the action of the flow on the whole compact manifold. The global property is that any two points can be connected by the action of the group, so that the image of $\mathbb{R}^n$ by the group covers M.

Lemma 3.27: *Under the hypotheses of Proposition 3.25 there exists a n-parameter group of differentiable maps $\phi^t : \mathbb{R}^n \times M \to M$ with the following properties:*

(i) For every $t \in \mathbb{R}^n$ and every $P \in M$, there is a unique $\phi^t P \in M$.

(ii) For every pair P, Q of points of M, there is $t \in \mathbb{R}^n$ such that $\phi^t P = Q$; here t need not be unique.

Proof. (i) By Proposition 3.11 the flow ϕ^t is uniquely defined for $t \in V$, a neighbourhood of the origin in $\mathbb{R}^n$. Since the flows $\phi_1^{t_1}, \ldots, \phi_n^{t_n}$ can be indefinitely prolonged on M, the point $\phi^t P$ is uniquely defined for any $t \in \mathbb{R}^n$. (ii) By Proposition 3.11 there is a neighbourhood U_P of P and a neighbourhood U_Q of Q which are diffeomorphic to neighbourhoods of the origin $V_P \subset \mathbb{R}^n$ and $V_Q \subset \mathbb{R}^n$, respectively. By a suitable restriction we may consider V_P and V_Q to be the same neighbourhood V of the origin (e.g., set $V = V_P \cap V_Q$, which is open). Suppose for a moment that $U_P \cap U_Q \neq \emptyset$, as in Figure 3.5, and let $P' \in U_P \cap U_Q$. Then there are $t \in V$ and $s \in V$ such that both equalities $P' = \phi^t P$ and $P' = \phi^s Q$ hold true; by the group property of the flows the latter also implies that $\phi^{-s} P' = \phi^{-s} \phi^s Q = Q$. Again by composition of flows we finally get $Q = \phi^{-s} P' = \phi^{-s} \phi^t P = \phi^{t-s} P$.

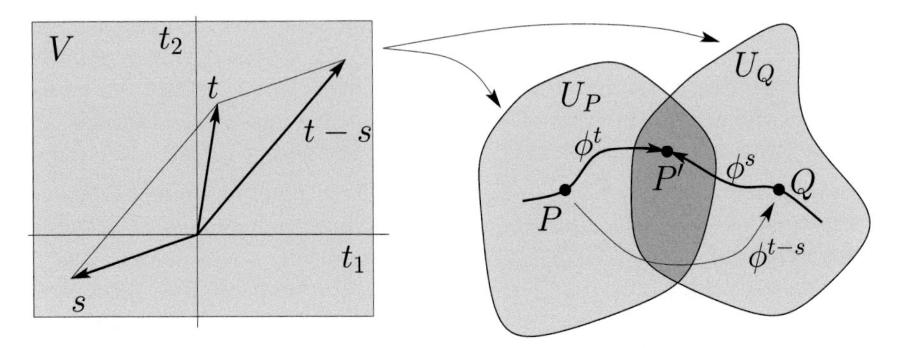

Figure 3.5 A neighbourhood V of the origin maps locally on two different neighbourhoods: U_P of the point P, and U_Q of the point Q, with non-void intersection. The flow ϕ^t moves the point P to $P' \in U_P \cap U_Q$; the flow ϕ^s moves Q to the same point P'. By the group property of the flows, ϕ^{t-s} moves P to Q.

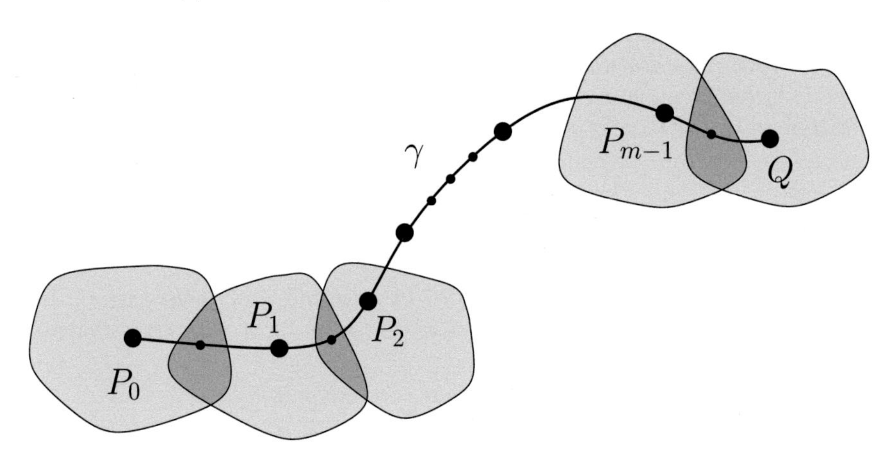

Figure 3.6 Every point Q of the invariant connected manifold M_0 can be reached with the flow ϕ^t, starting from P, through a chain of partial flows connecting two nearby points on a curve γ.

Let now $U_P \cap U_Q = \emptyset$. Since M is connected, there is a continuous curve γ connecting P with Q. To every point $P' \in \gamma$ we can associate a neighbourhood $U_{P'}$ diffeomorphic to a neighbourhood V, thus covering γ with an infinite family $\{U_{P'}\}$ of neighbourhoods. Since γ is compact, we can extract a finite sequence $P_0 = P$, $P_1, \ldots, P_m = Q$ of points of γ with corresponding neighbourhoods $U_{P_0}, \ldots, U_{P_m}$ such that $U_{P_j} \cap U_{P_{j+1}} \neq \emptyset$, as illustrated in Figure 3.6. Therefore there exists a finite sequence $t_1, \ldots, t_m$ such that $P_j = \phi^{t_j} P_{j-1}$ for $j = 1, \ldots, m$. Setting $t = t_1 + \cdots + t_m$, we conclude that $\phi^t P = Q$. Q.E.D.

3.3.2 The Stationary Group

According to Lemma 3.27 we may regard the flows ϕ^t as a differentiable map from $\mathbb{R}^n \times M$ to M, the image of which is the whole manifold M. However, this is not a diffeomorphism, for it may happen that $\phi^\tau P = P$ for some $\tau \in \mathbb{R}^n$ and for some $P \in M$. We call τ a *period.*

Lemma 3.28: *Let $\phi^\tau P = P$ for some nonzero $\tau \in \mathbb{R}^n$ and some $P \in M$. Then we have $\phi^\tau Q = Q$ for all $Q \in M$.*

In short, *a period is a property of the flow, not of a particular point P.*

Proof. By (iii) of Lemma 3.27 we have $Q = \phi^s P$ for some $s \in \mathbb{R}^n$; on the other hand we have also $\phi^\tau Q = \phi^\tau \phi^s P = \phi^{s+\tau} P = \phi^s \phi^\tau P = \phi^s P = Q$, by the group property. *Q.E.D.*

Definition 3.29: *The stationary set of ϕ^t is defined as*

$$G = \{t \in \mathbb{R}^n \mid \phi^t P = P \quad \text{for all } P \in M\} . \tag{3.37}$$

Lemma 3.30: *The stationary set G defined by (3.37) is a non-empty discrete subgroup of $\mathbb{R}^n$ (i.e., a lattice); that is: it possesses the group property, and contains no accumulation points.*

Proof. The set G is not empty, since it contains at least the origin of $\mathbb{R}^n$. If $t \in G$ then $\phi^{-t}P = \phi^{-t}\phi^t P = P$, so that $-t \in G$. If $t_1, t_2 \in G$, then we have $\phi^{t_1+t_2} P = \phi^{t_1}\phi^{t_2} P = \phi^{t_1} P = P$, so that $t_1 + t_2 \in G$. We conclude that G has the group property.
The origin is an isolated point of G, because in a sufficiently small neighbourhood V_0 of the origin ϕ^t is a diffeomorphism, which implies that $G \cap V_0 = \{0\}$. Let now $0 \neq t \in G$, and $V_t = t + V_0 = \{t + s \mid s \in V_0\}$. Let $t' \in G \cap V_t$, that is, $\phi^{t'} P = P$. By the group property, we also have $\phi^{t'-t} P = P$, that is, $t' - t \in G \cap V_0$, which implies $t' - t = 0$. This proves that t is an isolated point. *Q.E.D.*

3.3.3 Angular Coordinates

The stationary group G spans a k-dimensional subspace $\mathrm{span}(G) \subset \mathbb{R}^n$, with $0 < k \leq n$; the dimension k is naturally assigned to G. Now we may proceed to the construction of a diffeomorphism between M and $\mathbb{T}^k \times \mathbb{R}^{n-k}$. To this end we need a statement of algebraic character, namely that a discrete subgroup G of $\mathbb{R}^n$ has a basis. That is: *there exist k independent vectors $e_1, \ldots, e_k$ in G such that*

$$G = \{m_1 e_1 + \cdots + m_k e_k \; : \; (m_1, \ldots, m_k) \in \mathbb{Z}^k\} . \tag{3.38}$$

Such a claim seems to be natural, since a basis of the subspace span$(G) \subset \mathbb{R}^n$ exists. However, using a basis of span(G) in order to construct a corresponding basis for G is not a completely trivial matter: it involves some subtleties due to G being discrete.[5] The proof is a rather long detour and is deferred to Appendix A.

Proof of Proposition 3.25. Take any basis $\{e_1, \ldots, e_k\}$ of G; complete it by adding $n-k$ vectors $u_1, \ldots, u_{n-k}$, such that $\{e_1, \ldots, e_k, u_1, \ldots, u_{n-k},\}$ is a basis of $\mathbb{R}^n$. Writing an arbitrary point $t \in \mathbb{R}^n$ as

$$t = \frac{\vartheta_1}{2\pi} e_1 + \cdots + \frac{\vartheta_k}{2\pi} e_k + \tau_1 u_1 + \cdots + \tau_{n-k} u_k$$

and letting $(\vartheta_1, \ldots, \vartheta_k) \in \mathbb{T}^k$, we establish a global one-to-one correspondence between $\mathbb{T}^k \times \mathbb{R}^{n-k}$ and M by taking an arbitrary point $P \in M$ as corresponding to the origin and setting $P(\vartheta_1, \ldots, \vartheta_k, \tau_1, \ldots, \tau_{n-k}) = \phi^t P$, with t given earlier. In the neighbourhood of any point $Q \in M$ this mapping is a local diffeomorphism; hence M is diffeomorphic to $\mathbb{T}^k \times \mathbb{R}^{n-k}$. *Q.E.D.*

Proof of Corollary 3.26. By Proposition 3.25, M must be diffeomorphic to $\mathbb{T}^k \times \mathbb{R}^{n-k}$ for some k. Since M is compact, then $k = n$. *Q.E.D.*

3.4 Action-Angle Variables

We revert now to considering the $2n$-dimensional phase space; the role of M in the previous section will be taken by an n-dimensional manifold M_0 defined by the first integrals, with the restriction that it must be compact. The restriction is necessary in order to assure that M_0 is diffeomorphic to $\mathbb{T}^n$, since we are going to exploit the angular coordinates on it. Action-angle variables may be introduced in a neighbourhood of M_0 through an elaborate extension of the argument used for $n = 1$ in Section 3.2.3.[6]

Proposition 3.31: *Let $\Phi_1, \ldots, \Phi_n$ be a complete involution system, and assume that the level surface defined by $\Phi_1(q,p) = \cdots = \Phi_n(q,p) = 0$ contains a connected and compact component M_0. Then:*

---

[5] A basis of span(G) is easily found by taking k linearly independent vectors in G. However, this does not imply that it is also a basis of G, for it may happen that applying (3.38) to it reproduces only a subgroup of G. A basis of G must satisfy the further condition span$(G) \cap \mathbb{Z}^n = G$ where $\mathbb{Z}^n$ is to be seen as a subset of $\mathbb{R}^n$. More on this in Appendix A.

[6] The existence of action-angle variables as a general fact for systems that are integrable in Liouville's sense has been first claimed by Arnold [5]. The paper by Res Jost [123] completes the proof proposed by Arnold, also removing an unnecessary assumption hidden in Arnold's proof. (See [177], ch. 3.)

(i) *M_0 is an n-dimensional manifold diffeomorphic to an n-dimensional torus;*

(ii) *in an open neighbourhood $U(M_0)$ one can introduce action-angle variables $I \in \mathscr{G} \subset \mathbb{R}^n$ and $\vartheta \in \mathbb{T}^n$, where $\mathscr{G}$ is a neighbourhood of the origin, via a canonical diffeomorphism*

$$\begin{array}{rcl} \mathscr{A}\colon \mathbb{T}^n \times \mathscr{G} & \to & U(M_0)\,, \\ (\vartheta, I) & \mapsto & (q,p) = \mathscr{A}(\vartheta, I) \end{array}$$

such that $I_1, \ldots, I_n$ depend only on $\Phi_1, \ldots, \Phi_n$.

The action-angle variables so constructed are actually Liouville's canonical coordinates generated by the actions $I = (I_1, \ldots, I_n)$. The remarkable fact, however, is that the angles are globally defined on the invariant tori; the actions remain local coordinates in a neighbourhood of M_0. This marks the difference with respect to a generic set of Liouville coordinates.

The claim (i) is true in view of Proposition 3.25 and Corollary 3.26, since the manifold M_0 is assumed to be compact. The rest of the proof requires two main steps. First, we prove that in a neighbourhood of M_0 there is a family of n-dimensional tori parameterized by $\Phi_1, \ldots, \Phi_n$, actually level sets of the first integrals, and that the periods on the family are smooth functions of $\Phi_1, \ldots, \Phi_n$. Second, we find the canonical transformation $\mathscr{A}$ which introduces a new complete involution system $I_1, \ldots, I_n$ of action variables which parameterize the invariant tori and the conjugated angles $\vartheta_1, \ldots, \vartheta_n$ on an n-dimensional torus, thus replacing the Liouville coordinates $\Phi_1, \ldots, \Phi_n$ and $\alpha_1, \ldots, \alpha_n$. The two steps of the proof are worked out in separate subsections.

3.4.1 Periods in a Neighbourhood of the Torus

We extend our considerations to a neighbourhood of the torus M_0, by letting the values of the first integrals $\Phi_1, \ldots, \Phi_n$ to vary in some neighbourhood of $\Phi = 0$. What we know till now is only that the manifold M_0 is a torus, and that in the neighbourhood of any point $P \in M_0$ there are *local* Liouville coordinates; the nature of the manifolds M_Φ for $\Phi \neq 0$ is unknown.

Our aim is to prove that there is a smooth set of tori $\{M_\Phi\}$ parameterized by $\Phi = (\Phi_1, \ldots, \Phi_n)$ and that the flow ϕ^t on each torus M_Φ possesses periods $\tau_1(\Phi), \ldots, \tau_n(\Phi)$ close to the periods $e_1, \ldots, e_n$ on M_0. To this end we exploit the Liouville coordinates α, Φ of Proposition 3.12 defined in a $2n$-dimensional neighbourhood U_0 of a given point $P_0 \in M_0$ and the corresponding local map $\chi(\alpha, \Phi)$. We also exploit the claim of Lemma 3.27 that the flows of $\Phi_1, \ldots, \Phi_n$ cover the whole manifold M_0. We show that the flow extends the map χ of Liouville coordinates in U_0 so as to cover a full neighbourhood of M_0. What is not immediate is that for $\Phi \neq 0$ the invariant manifolds M_Φ, which exists locally, should still be tori.[7] Using the flow, we prove the existence of periods and angular coordinates on the full neighbourhood of M_0.

Lemma 3.32: *Let the hypotheses of Proposition 3.31 be fulfilled, and M_0 be the connected and compact manifold of point (i) of the proposition. Let $P_0 \in M_0$ be arbitrary but fixed, and let $\varrho > 0$ be finite but large enough so that $M_0 \subset \{\phi^t P_0\,,\ |t| \le \varrho\}$. Then the following statements hold true.*

(i) There is a neighbourhood U_ϱ of P_0 such that for every $Q \in M_0$ the map $\phi^t : U_\varrho \to U_Q = \phi^t U_\varrho$ satisfying $\phi^t P_0 = Q$ exists for $|t| < \varrho$, and is a canonical diffeomorphism.

(ii) The flow ϕ^t is globally defined in a neighbourhood of M_0 and possesses a stationary groups of periods $G(\Phi)$ parameterized by Φ; furthermore, there exist differentiable functions $W_1(\Phi), \ldots, W_n(\Phi)$ such that $G(\Phi)$ possesses a basis

$$\tau_j(\Phi) = e_j + \partial_\Phi W_j\ , \quad j = 1, \ldots, n\ ,$$

where $e_1, \ldots, e_n$ is a basis of the stationary group G of periods on M_0, and $\partial_\Phi W_j = \left(\frac{\partial W_j}{\partial \Phi_1}, \ldots, \frac{\partial W_j}{\partial \Phi_n}\right)$.

The proof occupies the rest of the present section. In view of Lemma 3.27 the flow ϕ^t restricted to M_0 is defined for all $t \in \mathbb{R}^n$; moreover, for any $Q \in M_0$ there is t such that $Q = \phi^t P_0$. The hypothesis on ϱ in the statement of the lemma assures that every point $Q \in M_0$ may be reached in a time $|t| < \varrho$.

Let now $P_0 \in M_0$ be fixed and let $Q \in M_0$ be any point. By Proposition 3.12 both P_0 and Q have a local $2n$-dimensional neighbourhood U_0 and U_Q, respectively, endowed with Liouville coordinates $(\alpha, \Phi) \in V_\alpha \times V_\Phi$ through a local map $P = \chi(\alpha, \Phi) = \phi^\alpha \chi(0, \Phi)$ in U_0 and a similar map χ_Q in U_Q; the relevant fact is that ϕ^α leaves Φ unchanged. Recall also that Liouville coordinates have been introduced by first identifying a manifold $\Sigma_0 = \{\chi(0, \Phi) : \Phi \in V_\Phi\}$, transversal to M_0. Similarly, Liouville coordinates can be introduced in U_Q; however, we must choose another manifold $\Sigma_Q = \{\chi_Q(0, \Phi) : \Phi \in V'_\Phi\} \subset U_Q$, with some open V'_Φ, where $\chi_Q(\alpha, \Phi)$

[7] Consider, for instance, a centre equilibrium of a differential equation in the plane. In the linear approximation, the equilibrium point is surrounded by closed curves; however, adding a nonlinear term may break the curves. For example, the nonlinearity may create a limit cycle which is isolated: the orbits in the neighbourhood of the cycle are not periodic. A well-known example is the equation of Van der Pol. The persistence of closed invariant curves, and so also of periodic orbits, for a canonical system is forced by the invariance of the Hamiltonian. In higher dimensions, as is the present case, the persistence of invariant tori in the neighbourhood of M_0 is forced by the existence of a complete involution system of functions.

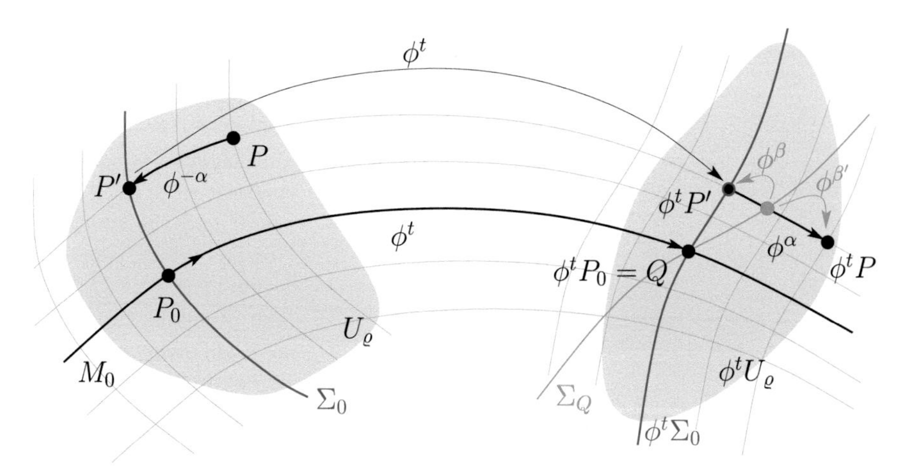

Figure 3.7 The flow ϕ^t generates a local diffeomorphism from a neighbourhood U_0 of $P_0 \in M_0$ to a neighbourhood $\phi^t U_0$ of $Q = \phi^t P_0 \in M_0$. See text. Every point in the neighbourhoods is assigned local Liouville coordinates α, Φ. The local coordinates in $\phi^t U_0$ may be assigned with respect to an arbitrary Σ_Q of with respect to $\phi^t \Sigma_0$. The flow leaves Φ invariant, while the local coordinate α is subject to a translation.

provides a local coordinate map in U_Q. As we have seen, the choice of Σ_Q may be rather arbitrary; Proposition 3.12 suggests a particular one.

Let now $Q \in M_0$, and $\phi^t P_0 = Q$. By continuity and differentiability of the flow with respect to initial data, there is a neighbourhood $U_\varrho \subset U_0$ such that $\phi^t U_\varrho \subset U_Q$, and is a diffeomorphism. Let $P \in U_\varrho$, so that $\phi^t P \in U_Q$. With an arbitrary choice of the manifold Σ_Q we determine the local Liouville coordinates in U_Q, and get $(\beta, \Phi) = \chi_Q^{-1}(\phi^t P')$ and $(\beta', \Phi) = \chi_Q^{-1}(\phi^t P)$, where β and β' are known because they are determined via the local Liouville flow. Thus, with the help of Fig. 3.7, the image $\phi^t P$ through the flow may be constructed through the sequence of operations

$$P \to P' = \phi^{-\alpha} P \to \phi^t P' \to \phi^{-\beta} \phi^t P' \to \phi^{\beta'} \phi^{-\beta} \phi^t P' \in U_Q \ ,$$

where $P' \in \Sigma_0$ and $\phi^t P' \in \phi^t \Sigma_0$. Therefore, we are given a coordinate map from U_ϱ to U_Q, which can be expressed as[8]

$$(\beta', \Phi) = \chi_Q^{-1}\bigl(\phi^t P\bigr) = \chi_Q^{-1}\bigl(\phi^{\beta' - \beta} \circ \phi^t \circ \phi^{-\alpha} \circ \chi(\alpha, P)\bigr) \ . \tag{3.39}$$

[8] Pedantically, reading the expression (3.39) from right to left: find $P = \chi(\alpha, \Phi)$ from the local Liouville coordinates (α, Φ); transport it to $P' \in \phi^{-\alpha}\Sigma_0$ and then to $\phi^t P' \in \phi^t \Sigma_0$ along the flow; using the local Liouville coordinates β and β' on U_Q, move $\phi^t P'$ to $\phi^t P$ via the flow $\phi^{\beta' - \beta}$; finally, the inverse map χ_Q^{-1} gives the local Liouville coordinates (α', Φ) in the local coordinates of U_Q.

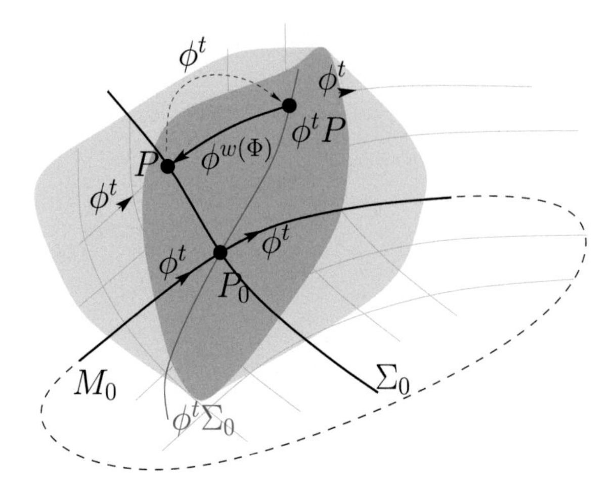

Figure 3.8 Periods in the neighbourhood of M_0. Let $\phi^{e_j} P_0 = P_0$; then the neighbourhood U_ϱ is mapped to $\phi^{e_j} U_\varrho$ (light gray), and the intersection $U_\varrho \cap \phi^{e_j} U_\varrho$ (darker) is non-empty. The manifold Σ_0 maps to a different manifold $\phi^{e_j}\Sigma_0$, and a further flow $\phi^{w(\Phi)}$ restores the period.

Here, by the group property of the flows, we clearly have $\alpha = \beta' - \beta$. The transformation is canonical, being a composition of canonical flows, and is a diffeomorphism.

Now we want to extend to U_Q the Liouville coordinates of U_0. Hence we want to make a particular choice for Σ_Q: we let the flow transport the manifold $\Sigma_0 \subset U_0$ to $\Sigma_Q = \phi^t\Sigma_0 \subset U_Q$, as illustrated in Fig. 3.7. In view of $\phi^t P' \in \Sigma_Q = \phi^t\Sigma_0$ we clearly have $\chi_Q^{-1}(\phi^t P') = (0, \Phi)$; therefore, formula (3.39) is replaced by the obvious one,

$$(\alpha, \Phi) = \chi_Q^{-1}(\phi^t P) = \chi_Q^{-1}\bigl(\phi^\alpha \circ \phi^t \circ \phi^{-\alpha} \circ \chi(\alpha, P)\bigr) \ ,$$

meaning that the local coordinates in U_0 and U_Q are actually the same. Exploiting again the group property of the flow, we extend the coordinate map from U_ϱ to $\phi^t U_\varrho \subset U_Q$ by assigning the Liouville coordinates (t, Φ) to the points of $\phi^t\Sigma_0$, and so also coordinates $(t + \alpha, \Phi)$ to $\phi^t P$ for any $P \in \phi^t U_\varrho$. This removes the need for the local map χ_Q. Moreover, since Q is arbitrary the Liouville coordinates are extended from U_ϱ as global ones in a full neighbourhood of M_0, as claimed at point (i).

Let us now restrict our attention to the periodic points on M_0 by letting t to be one of the periods $e_1, \ldots, e_n$ on M_0. The argument is worked out with the help of Fig. 3.8. Let j, with $1 \leq j \leq n$, be arbitrary but fixed. For $t = e_j$ we have $\phi^{e_j} P_0 = P_0$, so that the flow ϕ^{e_j} is actually a canonical diffeomorphism between neighbourhoods of P_0. By suitably restricting U_ϱ, if needed, we may manage that $\phi^{e_j} U_\varrho \subset U_0$ for $j = 1, \ldots, n$, so that the argument applies to all periods; for example, it is enough to set $U_\varrho = \bigcap_{j=1,\ldots,n} \phi^{-e_j}\bigl(U_0 \cap \phi^{e_j} U_0\bigr)$, which is open, being a finite intersection of open sets. For $P = \chi^{-1}(\alpha, \Phi) \in U_\varrho$ and fixed j we introduce the map

$$
(3.40)\qquad \begin{aligned} \psi\, : \chi^{-1}(U_\varrho) &\to \chi^{-1}(\phi^{e_j} U_\varrho)\ , \quad j = 1, \ldots, n\ , \\ (\alpha, \Phi) &\mapsto \psi(\alpha, \Phi) = \chi^{-1}\bigl(\phi^{e_j + \alpha} \chi(0, \Phi)\bigr)\ . \end{aligned}
$$

The manifolds Σ_0 and $\phi^{e_j}\Sigma_0$ do not need to coincide, except at the point P_0; hence we use again formula (3.39), substituting the coordinate map χ in place of χ_Q. For points $P \in \Sigma_0$ we have $\alpha = 0$; therefore the map ψ satisfies

$$
(3.41)\qquad \psi(0,0) = (0,0)\ , \quad \psi(\alpha, \Phi) = \bigl(\alpha - w(\Phi), \Phi\bigr)\ ,
$$

with a vector function $w(\Phi) = -\chi^{-1}\bigl(\phi^{e_j}\chi(0, \Phi)\bigr)$; the special notation $w(\Phi)$ aims at highlighting that in Liouville coordinates the map ψ reduces to a translation depending on Φ. The map is a composition of canonical diffeomorphisms, hence it is canonical, and $w(\Phi)$ is a differentiable vector. This means that $w(\Phi)$ cannot be fully arbitrary, since it must obey the canonicity conditions. Precisely, writing (3.41) in more explicit form as

$$
\alpha'_k = \alpha_k + w_k(\Phi)\ , \quad \Phi'_k = \Phi_k\ , \quad k = 1, \ldots, n\ ,
$$

the preservation of fundamental Poisson brackets requires in particular

$$
\{\alpha'_k, \alpha'_l\} = \{\alpha_k, w_l\} - \{\alpha_l, w_k\} = \frac{\partial w_l}{\partial \Phi_k} - \frac{\partial w_k}{\partial \Phi_l} = 0\ .
$$

Therefore, there exists a (local) function $W_j(\Phi)$ such that

$$
(3.42)\qquad w(\Phi) = -\partial_\Phi W_j\ , \quad \partial_\Phi W_j\big|_{\Phi=0} = 0\ .
$$

Searching for a period $\tau(\Phi)$ in the neighbourhood of M_0 is now a trivial matter. Let us do it in some detail, though the conclusion may appear to be obvious. A period must satisfy the equation

$$
(3.43)\qquad \phi^{\tau(\Phi)}\chi(0, \Phi) = \chi(0, \Phi)\ ;
$$

the invariance of Φ under the flow implies that the period $\tau_j(\Phi)$, if it exists, depends only on Φ. A look to Fig. 3.8 leads us to set $\tau(\Phi) = e_j + w(\Phi)$; let us (pedantically) check that it is so. Applying the map $\phi^{\tau(\Phi)}$ to $P \in \Sigma_0$, so that $\chi^{-1}(P) = (0, \Phi)$, and using (3.40) and (3.41), we get

$$
\chi^{-1}\bigl(\phi^{e_j + w(\Phi)}\chi(0, \Phi)\bigr) = \psi\bigl(w(\Phi), \Phi\bigr) = (0, \Phi)\ ,
$$

as expected. Therefore, to each period e_j on M_0 we associate a period

$$
\tau_j(\Phi) = e_j + \partial_\Phi W_j\ , \quad j = 1, \ldots, n
$$

on the manifold M_Φ, smoothly dependent on Φ.

We conclude that in a neighbourhood $U(M_0)$ of M_0 the flow ϕ^t is defined for all $t \in \mathbb{R}^n$, and the subgroup of $\mathbb{R}^n$

$$
G(\Phi) = \{m_1\tau_1(\Phi) + \cdots + m_n\tau_n(\Phi)\,,\ (m_1, \ldots, m_n) \in \mathbb{Z}^n\}
$$

is a stationary group of periods of ϕ^t on a compact manifold M_Φ.

A last doubt may arise: we have proven that $G(\Phi)$ is a set of solutions of Eq. (3.43), but we should also prove that it exhausts the set of periods, namely that if $\phi^t\chi(0,\Phi) = \chi(0,\Phi)$ for some $\Phi \in \mathscr{G}$, then $t \in G(\Phi)$. To this end, recall that $0 \in G(\Phi)$ for all $\Phi \in \mathscr{G}$, and that $G(\Phi)$ is a discrete group. Let now t be a period on M_Φ. Then $t + G(\Phi)$ is obviously a set of periods, and in this set there exists $t' = \mu_1\tau_1(\Phi) + \cdots + \mu_n\tau_n(\Phi)$ with $|\mu_j| \leq 1/2$, $(j = 1, \ldots, n)$. By continuity, if Φ is sufficiently close to the origin, then t' must belong to an arbitrarily small neighbourhood of $0 \in G(0)$. Since $G(\Phi)$ is discrete (Lemma 3.30), we conclude $t' = 0$, and so $t \in G(\Phi)$. This concludes the proof of Lemma 3.32.

3.4.2 Global Coordinates and Action-Angle Variables

Now we are ready to complete the proof of Proposition 3.31 by replacing the Liouville coordinates (α, Φ), extended over the full neighbourhood of M_0, with action-angle variables (ϑ, I) which are *globally* one-to-one. Proceeding as we did in Section 3.3.3 for M_0 we define a new mapping

$$\tilde{\chi} : \mathbb{R}^n / G(\Phi) \times \mathscr{G} \to U(M_0) \tag{3.44}$$

as the restriction of χ to the quotient set of $\mathbb{R}^n$ with respect to the group $G(\Phi)$. Recalling the form of the periods in Lemma 3.32, we introduce the angles $\vartheta \in \mathbb{T}^n$ on every invariant manifold M_Φ through the relations

$$\alpha = \frac{1}{2\pi} \sum_k \vartheta_k (e_k + \partial_\Phi W) . \tag{3.45}$$

The transformation is linear, with a matrix depending on Φ. The matrix is not singular in a neighbourhood of $\Phi = 0$, because at $\Phi = 0$ it is the matrix of the periods e_j, which are linearly independent vectors by construction.

In order to find the conjugated actions $I_1, \ldots, I_n$, we introduce the generating function[9]

$$S(\Phi, \vartheta) = \frac{1}{2\pi} \sum_{k=1}^{n} \vartheta_k \big[\langle e_k, \Phi \rangle + W_k(\Phi) \big] . \tag{3.46}$$

The transformation is written in implicit form as

$$\begin{aligned} I_j &= \frac{1}{2\pi} \langle e_j, \Phi \rangle + W_j(\Phi) , \\ \alpha_j &= \frac{1}{2\pi} \sum_k \vartheta_k \left(e_{k,j} + \frac{\partial W_k}{\partial \Phi_j} \right) , \quad j = 1, \ldots, n. \end{aligned} \tag{3.47}$$

[9] Here and immediately after the following notations are used. For vectors or vector-valued functions a, b we write $\langle a, b \rangle = \sum_j a_j b_j$; for the components of the periods we write $e_k = (e_{k,1}, \ldots, e_{k,n})$.

The new functions $I_1, \ldots, I_n$ so defined are a complete involution system: they are independent because for $\Phi = 0$ the Jacobian determinant reduces to the nondegenerate matrix of the vectors e_j, and it remains nonzero in a neighbourhood of $\Phi = 0$; they do not depend on α, and so are in involution. Hence the actions I are well defined in an open set $\mathscr{G} \subset \mathbb{R}^n$, which is the image of the neighbourhood V_Φ of the Φ. On the other hand the angles ϑ are globally defined on the torus $\mathbb{T}^n$.

Recall that the Liouville transformation $(\alpha, \Phi) = \chi(q, p)$ can be globally extended to a whole neighbourhood of M_0. Thus the canonical transformation $(q, p) = \mathscr{A}(\vartheta, I)$ is found by composition of (3.47) with the inverse χ^{-1} of Liouville's transformation. This completes the proof of Proposition 3.31.

3.4.3 Non-uniqueness of the Action-Angle Variables

The action-angle variables of Proposition 3.31 are not unique. Three arbitrary choices have been made, namely: (i) the choice that M_0 corresponds to $\Phi_1 = \cdots = \Phi_n = 0$; (ii) the choice of the point P_0 and of the manifold Σ_0; (iii) the choice of the basis $e_1, \ldots, e_n$ of periods on M_0.

Lemma 3.33: *Let ϑ, I be action-angle variables. New action-angle variables are constructed by composition of the following canonical transformations:*

(i) translation of the action variables

$$I_j = \overline{I}_j + c_j \ , \quad \vartheta_j = \overline{\vartheta}_j \ , \quad 1 \le j \le n \tag{3.48}$$

with $c \equiv (c_1, \ldots, c_n) \in \mathbb{R}^n$;

(ii) translation of the origin of the angles by a quantity depending on the torus, namely

$$\vartheta_j = \overline{\vartheta}_j + \frac{\partial S}{\partial \overline{I}_j}(\overline{I}) \ , \quad I_j = \overline{I}_j \ , \quad 1 \le j \le n \, , \tag{3.49}$$

where $S(\overline{I})$ is an arbitrary differentiable function of the action variables;

(iii) linear transformation of the angles by a unimodular matrix[10] A:

$$\overline{\vartheta} = \mathsf{A}\vartheta \ , \quad I = \mathsf{A}^\top \overline{I} \, . \tag{3.50}$$

Proof. The canonicity of these transformations is easily checked. For instance, for the transformations (i) and (ii) it is enough to verify that the fundamental Poisson brackets are preserved. For the transformation (iii) we just remark that $\overline{\vartheta} = \mathsf{A}\vartheta$ involves only the angles; therefore it is legitimate to apply the method of the extended point transformation, thus writing the generating function as $S = \overline{I}^\top \mathsf{A}\vartheta$.

---

[10] A matrix A is said to be unimodular if it has integer entries, and $\det \mathsf{A} = \pm 1$.

Only the condition that A be a unimodular matrix needs some justification. To this end, we just remark that the angle structure of the torus is preserved if and only if the basis $(e_1, \ldots, e_n)$ of the group of periods is changed by $e'_j = \sum_k A_{jk} e_k$ (where $A_{j,k}$ are the entries of A) into a new basis of the same group. This leads to the condition that A should be unimodular. A detailed proof is given in Appendix A, Lemma A.4. *Q.E.D.*

3.4.4 Explicit Construction of Action-Angle Variables

The construction of action variables in the previous sections appears to be rather elaborated. Traditionally one follows a more direct procedure, based on identifying cycles on a torus, as illustrated in Fig. 3.9, and calculating the actions via path integrals over the cycles. The procedure is justified in view of the existence of the family of periods on the manifolds M_Φ considered in the previous sections.[11]

Denote by $e_1, \ldots, e_n$ the basis of the stationary group on any of the tori parameterized by a given value of Φ. The following properties are immediate:

(i) each period e_j $(j = 1, \ldots, n)$ corresponds on the torus M_Φ to a differentiable closed curve γ_j, which we call a *cycle*, defined as (see Fig. 3.9)

$$\gamma_j = \{\phi^{se_j} P \, , \, 0 \le s < 1\} \; ;$$

(ii) for $j \neq k$ the cycles $\gamma_j \, , \, \gamma_k$ are independent, in the sense that γ_j can not be continuously deformed into γ_k;

(iii) in action-angle variables I, ϑ the continuous set of tori is parameterized by $I_1, \ldots, I_n$, and the cycle γ_j is represented by

$$\vartheta_j \in [0, 2\pi) \; , \quad \vartheta_k = \vartheta_{k,0} \text{ for } k \neq j \; ,$$

$\vartheta_{k,0}$ being constants.

---

[11] An interesting discussion concerning the calculation of the action by choosing suitable cycles is reported in a paper by Albert Einstein on quantization [64]. The method is thoroughly presented in Born's treatise [28]. The introduction of action-angle variables in Born's treatise is connected with the search for *adiabatic invariants*, which were well known to physicists when dealing with one-dimensional oscillators depending on slow-varying parameters. A clever generalization of these concepts to the case of separable systems with many degrees of freedom is made by Born. Essentially, it may be said that Born's book contains a complete treatment of action-angle variables for systems which exhibit a clear separation into one-dimensional systems for which the existence of cycles is immediate. This in turn implies that the orbits lie on invariant tori. What is still missing with respect to the Arnold–Jost theorem is that the existence of invariant tori is a general fact for a wide class of integrable systems.

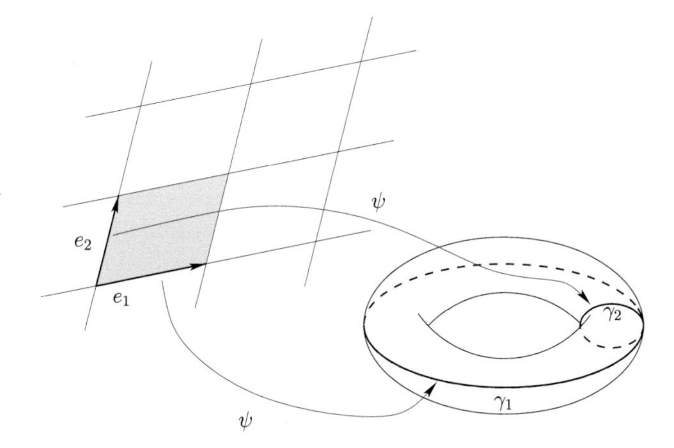

Figure 3.9 The construction of cycles on a torus.

Since the transformation to action-angle variables is canonical, we have

$$\oint_{\gamma_j} \sum_k p_k \, dq_k = \oint_{\gamma_j} \sum_k I_k \, d\vartheta_k \; .$$

On the other hand, by the characterization (iii) of the cycles all actions $I_1, \ldots, I_n$ and all angles ϑ_k with $k \neq j$ are constant on γ_j, so that the integral on the r.h.s. gives

$$\oint_{\gamma_j} \sum_k I_k \, d\vartheta_k = I_j \int_0^{2\pi} d\vartheta_j = 2\pi I_j \; .$$

We are thus led to the traditional expression

$$I_j = \frac{1}{2\pi} \oint_{\gamma_j} \sum_k p_k \, dq_k \; . \tag{3.51}$$

The integral on the right may be calculated using any set of canonical variables; hence the easiest thing is to use the functions Φ themselves. That is, $p_1(\Phi, q), \ldots, p_n(\Phi, q)$ are obtained from $\Phi_1 = \Phi_1(q,p), \ldots, \Phi_n = \Phi_n(q,p)$, by inversion with respect to $\Phi_1, \ldots, \Phi_n$. We emphasize that a continuous deformation of the cycle γ_j does not change the result, so that any determination of the cycles can be used in computing the integral (3.51). The resulting functions $I_1, \ldots, I_n$ depend only on $\Phi_1, \ldots, \Phi_n$, by construction, and are the desired action variables. Their independence is indirectly stated by the argument in Section 3.4.2. We have thus found a complete involution system to which Liouville's procedure can be applied. By construction, the conjugated coordinates ϑ are angles.

Let us resume in explicit terms the algorithm for constructing action-angle variables in a few steps.

(i) Find the cycles γ_j $(j = 1, \dots, n)$.
(ii) Compute the action variables by quadrature, calculating the integrals (3.51). This can be done by possibly introducing some arbitrary angle variables on the cycles and then integrating over them.
(iii) Apply the algorithm of Liouville to the new involution system $I_1, \dots, I_n$ in order to find the angle variables $\vartheta_1, \dots, \vartheta_n$.
(iv) If useful, and if there is any reason to do it, apply any of the transformations of Lemma 3.33 in order to obtain better sets of action-angle variables, depending on the problem at hand.

The hardest part of the algorithm is the first step (i), for it requires an integration of the system via Liouville's algorithm applied to the involution system $\Phi_1, \dots, \Phi_n$. However, in the most commonly considered examples the first integrals have a nice form, so that the cycles are easily determined.

3.5 The Arnold–Jost Theorem

We turn now to the statement of the of Arnold–Jost Theorem.

Theorem 3.34: *Let the Hamiltonian $H(q,p)$ on the phase space $\mathscr{F}$ possess an involution system $\Phi_1, \dots, \Phi_n$ of first integrals (so that it is integrable in Liouville's sense). Let $c = (c_1, \dots, c_n) \in \mathbb{R}^n$ be such that the level surface determined by the equations $\Phi_1(q,p) = c_1, \dots, \Phi_n(q,p) = c_n$ contains a compact and connected component M_c. Then in a neighbourhood U of M_c there are canonical action-angle coordinates I, ϑ mapping $\mathscr{G} \times \mathbb{T}^n$ to U, where $\mathscr{G} \in \mathbb{R}^n$ is an open set, such that the Hamiltonian depends only on $I_1, \dots, I_n$, and the corresponding flow is*

$$\vartheta_j(t) = \vartheta_{j,0} + t\omega_j(I_{1,0}, \dots, I_{n,0}) \ , \quad I_j(t) = I_{j,0} \ , \quad j = 1, \dots, n \ ,$$

where $\vartheta_{j,0}$ and $I_{j,0}$ are the initial data, and $\omega_j = \frac{\partial H}{\partial I_j}$.

The proof is a straightforward application of Proposition 3.31. Just proceed as in the proof of Liouville's theorem, using the actions $I_1, \dots, I_n$ as first integrals.

3.6 Delaunay Variables for the Keplerian Problem

A remarkable application of the Arnold–Jost Theorem is the calculation of action-angle variables for the motion in a central field of force, with particular reference to the case of the Keplerian potential.[12] The latter problem

[12] The action-angle variables for Kepler's problem were discovered by Delaunay. His aim was to replace the orbital elements of a Keplerian orbit, which were used since Lagrange's time in perturbation theory, with an appropriate set of canon-

is known to possess four independent first integrals (see Examples 1.20 and 1.21). A complete involution system of first integrals has been constructed in Example 3.3. Let us recall that in spherical coordinates r, ϑ, φ with the conjugated momenta p_r, p_ϑ, p_φ the functions are

$$J = p_\varphi \ , \quad \Gamma^2 = p_\vartheta^2 + \frac{J^2}{\sin^2\vartheta} \ , \quad E = \frac{1}{2m}\left(p_r^2 + \frac{\Gamma^2}{r^2}\right) + V(r) \ , \tag{3.52}$$

where m is the mass of the point. In Kepler's case the potential $V(r)$ is

$$V(r) = -\frac{k}{r} \ , \tag{3.53}$$

where k is a positive constant. We also recall the expression of the Hamiltonian

$$H = \frac{1}{2m}\left(p_r^2 + \frac{p_\vartheta^2}{r^2} + \frac{p_\varphi^2}{r^2\sin^2\vartheta}\right) + V(r) \ ; \tag{3.54}$$

this actually coincides with the third integral presented earlier when the explicit expressions of Γ^2 and J are substituted.

3.6.1 Determination of Cycles

We consider the canonical flows generated by the three functions (3.52). The discussion here is quite plain, because each function involves only two conjugated variables. This considerably simplifies the construction of cycles.

The function J is a trivially integrable Hamiltonian: the conjugate variable φ is actually an angle which parameterizes the cycle γ_φ.

The function Γ^2 can be considered as the Hamiltonian of a point with unit mass, moving on the segment $(0,\pi)$ under the action of the potential $V(\vartheta) = J^2/\sin^2\vartheta$. For $\Gamma^2 > \Gamma^2_{\rm min} = J^2$ the orbit in the phase plane ϑ, p_ϑ is a closed line, giving the second cycle γ_ϑ (see Fig. 3.10). Note that the construction of the cycles γ_φ and γ_ϑ does not depend on the form of the potential $V(r)$.

The peculiar character of the Keplerian problem shows up when we come to consider the third function. It represents the Hamiltonian of a point moving on the half-line $r > 0$ under the action of the potential

$$V^*(r) = \frac{\Gamma^2}{2mr^2} + V(r) \ .$$

In the Keplerian case, setting $E_{\rm min} = -mk^2/(2\Gamma^2)$, the motion is bounded for $E_{\rm min} < E < 0$, while for $E \geq 0$ it is unbounded. In the former case

ical variables. A deduction of Delaunay variables using the Hamilton–Jacobi method is found in Poincaré's treatises [188] and (more detailed) [190]. The calculation in the present notes reflects the exposition in M. Born's book [28].

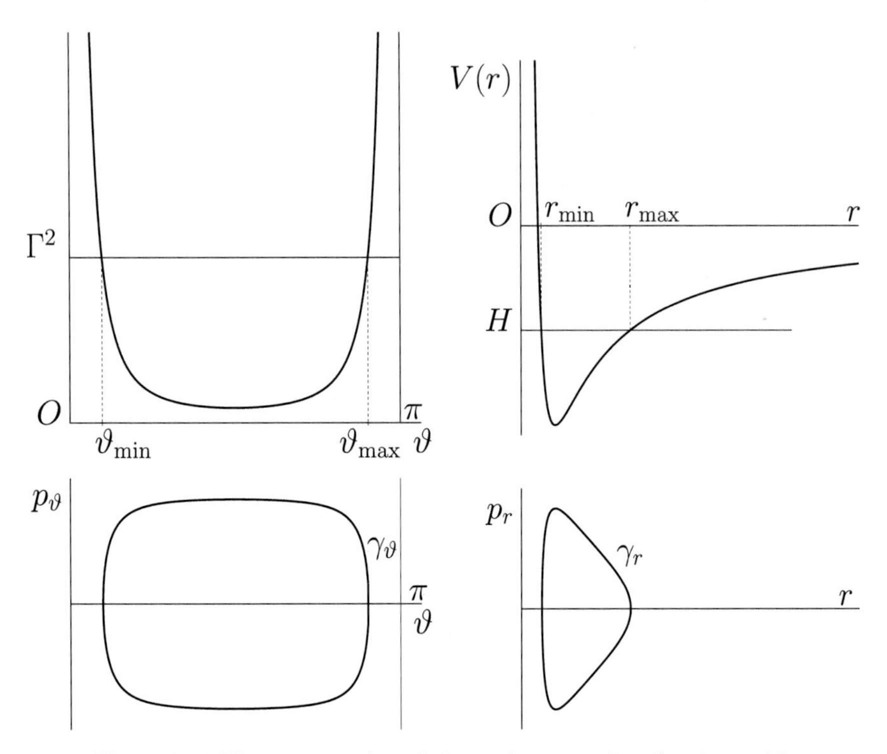

Figure 3.10 The construction of the cycles γ_ϑ and γ_r for the problem of motion in a central field under the Keplerian potential.

the orbit in the phase plane r, p_r is a closed curve, and this gives the third cycle γ_r (see Fig. 3.10). The cycle actually describes the motion of a planet on an elliptic orbit, in full agreement with Kepler's first law. Conversely, no cycle can be found for $E \geq 0$: the radial motion is unbounded, and the invariant surface in phase space for the complete problem is not compact. In the latter case, in agreement with Proposition 3.25, the invariant surface is the product $\mathbb{T}^2 \times \mathbb{R}$, and angular variables can be introduced only for the cycles γ_φ and γ_ϑ. The orbit is either a parabola, for $E = 0$, or a hyperbola, for $E > 0$.

3.6.2 Construction of the Action Variables

Here we restrict our consideration to the Keplerian potential, with the condition $E_{\min} < E < 0$. By inversion of (3.52) we get

$$
(3.55)\qquad
\begin{aligned}
p_r &= \left[2m(E - V(r)) - \frac{\Gamma^2}{r^2}\right]^{\frac{1}{2}} ,\\
p_\vartheta &= \left(\Gamma^2 - \frac{J^2}{\sin^2\vartheta}\right)^{\frac{1}{2}} ,\\
p_\varphi &= J .
\end{aligned}
$$

We should integrate the differential form $p_r\,dr + p_\vartheta\,d\vartheta + p_\varphi\,d\varphi$ over the cycles γ_φ, γ_ϑ and γ_r. This gives the actions I_φ, I_ϑ and I_r as functions of J, Γ and E, namely

$$
(3.56)\qquad
\begin{aligned}
I_\varphi &= \frac{1}{2\pi}\oint_{\gamma_\varphi} p_\varphi \mathrm{d}\varphi = J ,\\
I_\vartheta &= \frac{1}{2\pi}\oint_{\gamma_\vartheta} p_\vartheta \mathrm{d}\vartheta = \Gamma - |J| ,\\
I_r &= \frac{1}{2\pi}\oint_{\gamma_r} p_r dr = -\Gamma + k\sqrt{-\frac{m}{2E}} .
\end{aligned}
$$

The explicit expression as a function of the canonical coordinate is readily found by replacing the expressions of J, Γ, and E in (3.52).

3.6.3 Delaunay variables

By a straightforward inversion of the third part of (3.56) we calculate the Hamiltonian as

$$
(3.57)\qquad H = -\frac{mk^2}{2\left(I_r + I_\vartheta + |I_\varphi|\right)^2} .
$$

It is immediately seen that the Hamiltonian actually depends on the sum of the action variables. This implies that the three frequencies of the system coincide, which justifies the fact that in the Keplerian description of the planetary motion only one frequency does actually appear. A better set of action variables is constructed by introducing the variables of Delaunay L, G, Θ defined by the linear transformation

$$
(3.58)\qquad
\begin{aligned}
L &= I_r + I_\vartheta + |I_\varphi| ,\\
G &= I_\vartheta + |I_\varphi| ,\\
\Theta &= |I_\varphi| .
\end{aligned}
$$

It is immediately clear that G and Θ coincide with Γ and J, respectively. Since the transformation is performed via a unimodular matrix, the corresponding transformation on the angles preserves the periods, as stated by Lemma 3.33. The Hamiltonian in Delaunay's variables takes the form

$$H = -\frac{mk^2}{2L^2} \ . \tag{3.59}$$

Denoting by ℓ, g, h the angles conjugated to the actions G, G, Θ, we can write Hamilton's equations as

$$\dot{\ell} = \frac{mk^2}{L^3} \ , \quad \dot{g} = \dot{h} = \dot{L} = \dot{G} = \dot{\Theta} = 0 \ .$$

Thus, the motion is periodic with a single frequency

$$\omega(L) = \frac{mk^2}{L^3} \ .$$

3.6.4 Construction of the Angle Variables

The canonical transformation should now be completed by constructing the angle variables associated to the actions L, G and Θ. To this end we must first write the generating function

$$\begin{aligned} S &= \int (p_r dr + p_\vartheta d\vartheta + p_\varphi d\varphi) \\ &= \int \sqrt{-\frac{m^2k^2}{L^2} + \frac{2mk}{r} - \frac{G^2}{r^2}}\, dr + \int \sqrt{G^2 - \frac{\Theta^2}{\sin^2\vartheta}}\, d\vartheta + \int \Theta d\varphi \ . \end{aligned}$$

The angle variables are then given by

$$\begin{aligned} \ell &= \frac{\partial S}{\partial L} = \frac{m^2k^2}{L^3} \int \frac{dr}{\sqrt{-\frac{m^2k^2}{L^2} + \frac{2mk}{r} - \frac{G}{r^2}}} \ , \\ g &= \frac{\partial S}{\partial G} = G \int \frac{d\vartheta}{\sqrt{G^2 - \frac{\Theta^2}{\sin^2\vartheta}}} - G \int \frac{dr}{r^2 \sqrt{-\frac{m^2k^2}{L^2} + \frac{2mk}{r} - \frac{G^2}{r^2}}} \ , \\ h &= \frac{\partial S}{\partial \Theta} = -\Theta \int \frac{d\vartheta}{\sin^2\vartheta \sqrt{G^2 - \frac{\Theta^2}{\sin^2\vartheta}}} + \int d\varphi \ . \end{aligned}$$

Thus, the calculation of the angle variables is reduced to a quadrature; the actual calculation presents minor differences with respect to the case of the integrals (3.55).

It may also be useful to recall the relation between the Delaunay actions and the so called *orbital elements*. Without entering into the details, here are the relations:

$$L = \sqrt{mka} \ , \quad G = L\sqrt{1-e^2} \ , \quad \Theta = G\cos\iota \ , \tag{3.60}$$

where a is the semimajor axis, e is the eccentricity, and ι is the inclination of the orbital plane. Concerning the conjugated angles: ℓ is the so-called *mean anomaly*, namely an angle which evolves uniformly, thus averaging in some sense the *true anomaly*, which is the angle giving the actual position of the planet with respect to the Sun; the angles g and h are the *longitude of the perihelion* and the *longitude of the node*, respectively.

3.7 The Linear Chain

In the first half of the eighteenth century, a strong discussion arose about the dynamics of a string with fixed ends, for example, a string of a musical instrument such as a harpsichord. Among the mathematicians involved in the discussion we find Daniel Bernoulli and Jean le Rond D'Alembert. It was in this connection that D'Alembert discovered the string equation known under his name and found the solution which describes the wave propagation along an infinite string. The problem relates to the one of propagation of sound, already raised by Newton ([182], Liber II, sect. VIII). In the case of a string with fixed ends, the problem under discussion was whether the solution could be written as a superposition of sinusoidal stationary waves with shape $\sin\frac{k\pi x}{L}$, where L is the length of the string (see [134], § 25).

Lagrange had written two long memoirs on the subject of propagation of sound [132][133]. Concerning the problem of the string with fixed ends, he exploited Newton's idea of investigating the dynamics of particles. In a further memoir [134] he started his investigations introducing a discrete approximation of the continuous string, representing it as a system of $N+2$ particles on a line, with the two particles at the ends kept fixed and subject to a first-neighbours interaction, as represented in Fig. 3.11. The simplifying hypothesis is assumed that the particles move orthogonally to the rest line. Moreover, the interaction between two neighbouring particles is uniform along the whole chain. Therefore, by symmetry, the equilibrium configuration is the one with the jth particle placed at distance $\frac{jL}{N+1}$ from, say, the left end, where L is the length of the string, and the configuration of the chain is determined by the vertical displacement x_j. Thus, denoting by y_j the canonical momentum conjugated to x_j, the equations may be written as a canonical system with Hamiltonian

$$(3.61)\qquad H(x,y)=\frac{1}{2}\sum_{j=1}^{N}y_j^2+\sum_{j=0}^{N}V(x_{j+1}-x_j)\ ,\quad x_0(t)=x_{N+1}(t)=0\ .$$

The simplest choice, made by Lagrange, consists in assuming an elastic interaction, namely substituting $V(r)=kr^2/2$ in Eq. (3.61), with a constant k which we may set to one, for simplicity. We are thus led to study the quadratic Hamiltonian

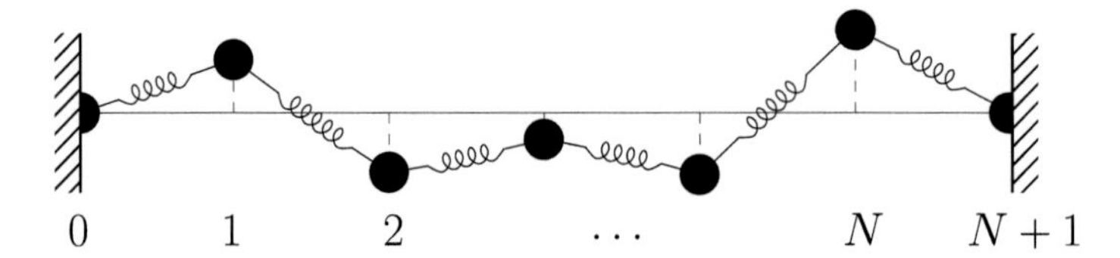

Figure 3.11 The discrete model of a string with fixed ends, according to Lagrange.

$$H(x,y) = \frac{1}{2}\sum_{j=1}^{N} y_j^2 + \frac{1}{2}\sum_{j=0}^{N}(x_{j+1}-x_j)^2 \ , \quad x_0(t) = x_{N+1}(t) = 0 \ . \tag{3.62}$$

3.7.1 A Complete Involution System

The Hamiltonian is the sum of two separate quadratic forms that we may write as

$$H = \frac{1}{2}\langle \mathsf{I}y, y\rangle + \frac{1}{2}\langle \mathsf{L}x, x\rangle \ ,$$

where I is the $N \times N$ identity matrix and L is the $N \times N$ symmetric matrix

$$\mathsf{L} = \begin{pmatrix} 2 & -1 & 0 & 0 & \dots & 0 & 0 \\ -1 & 2 & -1 & 0 & \dots & 0 & 0 \\ 0 & -1 & 2 & -1 & \dots & 0 & 0 \\ \vdots & \vdots & \vdots & \vdots & \dots & \vdots & \vdots \\ 0 & 0 & 0 & 0 & \dots & -1 & 2 \end{pmatrix} . \tag{3.63}$$

The Hamiltonian may be cast into a simpler form by making both quadratic forms diagonal via a linear transformation of variables. This is a known problem in algebra, namely the *simultaneous diagonalization of two quadratic forms*. In the present case, the way out is a rather straightforward one; we emphasize that the following argument applies to any matrix L which is symmetric and positive definite, as in the present case (the potential has a minimum at $x = 0$). We know that the eigenvalues of a symmetric matrix are real and the eigenvectors are orthogonal. Therefore L may be diagonalized through a linear transformation $x = \mathsf{R}\xi$ with an orthogonal matrix R.

Exercise 3.35: Show that the transformation of coordinates $x = \mathsf{R}\xi$ with an orthogonal matrix R is promoted to a canonical one by transforming the momenta as $y = \mathsf{R}\eta$, with the same matrix R. (Hint: recall the extended point transformation, Example 2.25.) *A.E.L.*

The kinetic part $T = \frac{1}{2}\langle \mathsf{I}y, y\rangle$ of the Hamiltonian remains unchanged under the transformation $y = \mathsf{R}\eta$. Furthermore, the hypothesis that L is positive definite entails that the eigenvalues are positive. Hence the Hamiltonian in diagonal form is written as a system of independent harmonic oscillators, namely

$$(3.64)\qquad H(\xi,\eta)=\frac{1}{2}\langle \mathsf{I}\eta,\eta\rangle+\frac{1}{2}\langle\Omega\xi,\xi\rangle\ ,\quad \Omega=\operatorname{diag}(\omega_1^2,\ldots,\omega_n^2)\ .$$

We immediately conclude: *the Hamiltonian (3.64) possesses a complete involution system of first integrals*

$$E_j=\frac{1}{2}(\eta_j^2+\omega_j^2\xi_j^2)\ ,\quad j=1,\ldots,N\ .$$

By transforming back to the original variables

$$\xi=\mathsf{R}^\top x\ ,\quad \eta=\mathsf{R}^\top y,$$

we find the first integrals for the original Hamiltonian (3.62). Therefore the system is integrable in the sense of Arnold–Jost, and we conclude that the motion is a Kronecker flow on a torus $\mathbb{T}^N$ with frequencies $\omega_1,\ldots,\omega_N$ which are the square roots of the eigenvalues of L. The construction of action-angle variables is a straightforward matter: see Example 3.21. The actions are determined as $I_j=E_j/\omega_j$, and the Hamiltonian is

$$H=\sum_{j=1}^N\omega_j I_j\ .$$

3.7.2 The Frequencies of the Linear Chain

In order to complete the discussion of the specific example (3.64) of a linear chain, it remains to calculate the frequencies ω, and the matrix R which diagonalizes the Hamiltonian. To this end, we should solve the characteristic equation

$$\mathsf{L}X=\omega^2X\ ,\quad X\in\mathbb{R}^N\ ,$$

with the matrix L given by (3.63). In more explicit form, the equation including the fixed boundary conditions is written as

$$(3.65)\quad -X_{j+1}+2X_j-X_{j-1}=\omega^2X_j\ ,\quad j=1,\ldots,N\ ,\quad X_0=X_{N+1}=0\ .$$

We accept the suggestion that the motion of the chain should be a superposition of sinusoidal waves; this is quite natural for us but was only a not-yet-proved conjecture in the first half of the eighteenth century, when the problem was under discussion. So, let us suppose, with Lagrange, that $X_j=\sin j\gamma$. By substitution in (3.65) we get the equation for γ:

$$-\sin(j+1)\gamma+2\sin j\gamma-\sin(j-1)\gamma=\omega^2\sin j\gamma\ .$$

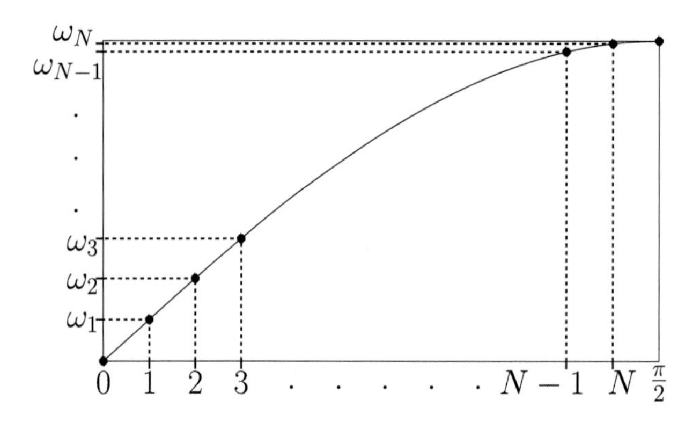

Figure 3.12 Frequency spectrum of the linear chain.

Exploiting the formulæ for sum and bisection of angles in trigonometric functions, we transform the equation as

$$\left(4\sin^2\frac{\gamma}{2} - \omega^2\right)\sin j\gamma = 0\ .$$

This is merely a relation between ω and γ, which we write in more explicit form as

$$\omega = 2\sin\frac{\gamma}{2}\ . \tag{3.66}$$

In order to select appropriate values for γ we should exploit the boundary conditions $X_0 = X_{N+1} = 0$. The first one on X_0 is obviously satisfied; the second one gives $X_{N+1} = \sin(N+1)\gamma = 0$, which selects infinitely many values $\gamma_k = \frac{k\pi}{N+1}$ with integer k. However, putting these values in (3.66),i we find exactly N distinct values corresponding to $k = 1, \ldots, N$; we have thus found the desired eigenvalues

$$\omega_k = 2\sin\frac{k\pi}{2(N+1)}\ , \qquad k = 1, \ldots, N\ . \tag{3.67}$$

Apparently the latter formula seems to provide infinitely many frequencies, letting k be an arbitrary integer. However, it is immediate to check that only N different values are found, as we should expect because N is the number of particles. The complete frequency spectrum is represented in Fig. 3.12.

Finally we calculate the eigenvectors $\mathbf{u}_1, \ldots, \mathbf{u}_N$ by setting $u_{k,j} = a\sin j\gamma_k$ with $k = 1, \ldots, N$ and with a normalization factor a that is easily calculated. The components of every eigenvector are the amplitudes X_j of the particles. For a given integer k we find the eigenvector corresponding to the frequency ω_k, with components

$$u_{k,j} = \sqrt{\frac{2}{N+1}}\sin\frac{jk\pi}{N+1}\ , \tag{3.68}$$

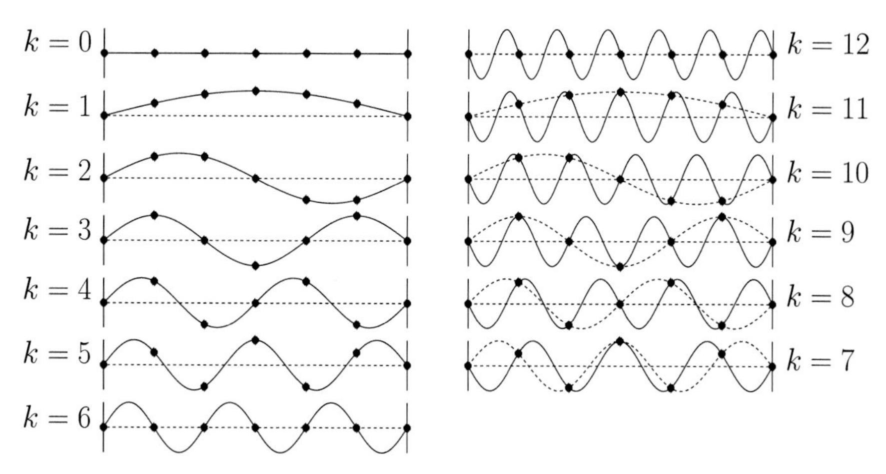

Figure 3.13 Configuration of the normal modes of a linear chain. Only modes $1, \ldots, N$ are significant, as should be expected because the system has N degrees of freedom. Other modes, though included in formula (3.68) of the eigenvectors, are either trivial, with all masses at rest, or reproduce one of the modes $1, \ldots, 6$.

including a normalization factor. Here too, there are apparently infinitely many eigenvectors. However, let us look at Fig. 3.13, which represents the case $N = 5$. The configuration of the particles is found by placing them on a sine curve. It is immediately seen that only N different configurations may be found: for example, the values k and $2(N+1) - k$ correspond to different sine functions but reproduce the same configuration of the particles; the same happens by adding an arbitrary multiple of $2N + 2$ to the pair of values so found. In addition, the values $k = 0,\, N+1,\, 2(N+1), \ldots$ all produce the trivial configuration with all particles at rest. By the way, the phenomenon of multiple configurations corresponding to the same k is a well-known one, named *aliasing*, which shows up when dealing with discrete Fourier transforms

The orthogonal matrix R which diagonalizes the Hamiltonian has the eigenvectors $\mathbf{u}_1, \ldots, \mathbf{u}_N$ thus found on the columns; in this case the matrix turns out to be also symmetric. The new coordinates and momenta introduced via the canonical transformation $\xi = \mathsf{R}^\top x\ ,\quad \eta = \mathsf{R}^\top y$ are usually called *normal modes*.

A last remark is in order. Lagrange found the general solution for a continuous string by letting $N \to \infty$, but keeping constant the density of mass and the density of force all along the chain. His well-known result is that the spectrum consists of a fundamental frequency ω_0, together with infinitely many harmonics $\omega_k = k\omega_0$. The motion associated to the frequency ω_k is a sine wave with $k-1$ nodes, as in the representation of Fig. 3.13 if the particles are removed and the sine function represents the actual configuration of the string. This is the phenomenon which lies at the very basis of the theory of musical harmony, which in fact exploits the integrability of the linear equation of the string.

3.8 The Toda Lattice

A widely studied model of great interest in Physics is obtained by introducing a nonlinearity in the equations for the linear chain. In its simplest formulation the Hamiltonian is written again in the form (3.61), namely

$$H(x,y)=\frac{1}{2}\sum_{j=1}^{N} y_j^2+\sum_{j=0}^{N} V(x_{j+1}-x_j)\ ,$$

with either fixed boundary conditions $x(0)=x(N+1)=0$ or periodic conditions $x(0)=x(N)$. The first-neighbours interaction is described by the potential $V(r)$, which is chosen to be quadratic in the case of the linear chain, but in general is a nonlinear function satisfying $V'(0)=0$. A celebrated model has been numerically studied by Fermi, Pasta and Ulam, who considered an expansion of $V(r)$ around the equilibrium $r=0$ up to terms of degree 3 or 4. Such a system is actually non-integrable.

A particularly interesting case has been proposed by Morikazu Toda [207], using the potential

$$V(r)=\frac{a}{b}e^{-br}+ar+\text{const}\ ,\qquad a,b>0\ ,$$

which has an equilibrium at $r=0$; this is known as the *Toda model.*

Numerical studies suggested that the Toda system could be integrable, due to the lack of chaotic behaviour [69]. First integrals for the Toda model have been eventually constructed by Michel Hénon [111] and by Hermann Flaschka [68] using two different methods. The latter paper exploits a method proposed by Peter David Lax in a seminal and remarkable paper [149], and originally developed in order to study the phenomenon of solitary waves for the celebrated Korteweg–de Vries equation. Here we include a short and simplified exposition concerning the finite-dimensional case, which is enough for the application to the Toda lattice.

3.8.1 *Lax Pairs*

In a finite-dimensional framework, the Lax method consists in finding a pair of time-dependent matrices L, A such that the system of canonical equations for a given Hamiltonian $H(q, p)$ turns out to be equivalent to the so-called *Lax equation*,

$$\frac{d\mathsf{L}}{dt} = [\mathsf{A}, \mathsf{L}] \ , \tag{3.69}$$

where $[\mathsf{A}, \mathsf{L}] = \mathsf{AL} - \mathsf{LA}$ is the commutator. The matrices L, A are said to be a *Lax pair*.

Proposition 3.36: *Let $\mathsf{L}(t)$ and $\mathsf{A}(t)$ be real matrices, with $\mathsf{L}(t)$ symmetric and $\mathsf{A}(t)$ skew-symmetric, satisfying the Lax equation (3.69). Then the eigenvalues of L are independent of time t.*

Proof. Let $\mathsf{U}(t)$ be a matrix obeying the auxiliary equation

$$\frac{d\mathsf{U}}{dt} = \mathsf{AU} \ . \tag{3.70}$$

In view of A being skew-symmetric, we have

$$\frac{d}{dt}(\mathsf{U}^\top \mathsf{U}) = \mathsf{U}^\top (\mathsf{A}^\top + \mathsf{A})\mathsf{U} = 0 \ ;$$

hence if $\mathsf{U}(0)$ is an orthogonal matrix, so is $\mathsf{U}(t)$ for all t. This also implies $\dot{\mathsf{U}}^{-1} = -\mathsf{U}^{-1}\dot{\mathsf{U}}\mathsf{U}^{-1}$. Using both (3.69) and (3.70), calculate

$$\begin{aligned}\frac{d}{dt}(\mathsf{U}^{-1}\mathsf{LU}) &= -\mathsf{U}^{-1}\dot{\mathsf{U}}\mathsf{U}^{-1}\mathsf{L} + \mathsf{U}^{-1}\dot{\mathsf{L}}\mathsf{U} + \mathsf{U}^{-1}\mathsf{L}\dot{\mathsf{U}} \\ &= \mathsf{U}^{-1}(-\mathsf{AL} + \dot{\mathsf{L}} + \mathsf{LA})\mathsf{U} = 0 \ .\end{aligned}$$

This shows that the matrix $\mathsf{U}^{-1}\mathsf{LU}$ is independent of t. By taking the initial condition $\mathsf{U}(0) = \mathsf{I}$, the identity matrix, we get

$$\mathsf{U}(t)^{-1}\mathsf{L}(t)\mathsf{U}(t) = \mathsf{L}(0) \ ;$$

that is, the matrices $\mathsf{L}(t)$ and $\mathsf{L}(0)$ have the same eigenvalues $\lambda_1, \ldots, \lambda_n$ which are also time independent. Since L is assumed to be symmetric, they are real quantities independent of t. *Q.E.D.*

The interesting fact is that if the dynamical equations of a system may be represented through a Lax pair, then finding its first integrals is just a matter of calculating the eigenvalues of L. The difficult part is to find the Lax pair associated to a given system.

3.8.2 First Integrals for the Toda Lattice

As an application of the method of Lax pairs, we sketch the construction of first integrals for the case of a periodic Toda lattice, with parameters $a = b = 1$ in the potential $V(r)$; a complete discussion including also the case of fixed boundary conditions may be found in Section 3.4 of [177]. Thus, we write the Hamiltonian as

$$H(x,y) = \frac{1}{2}\sum_{j=1}^{N} y_j^2 + \sum_{j=0}^{N} e^{-(x_{j+1}-x_j)} \,, \quad x_j = x_{j+N} \;;$$

the dynamical equations are

$$\dot{x}_j = y_j \,, \quad \dot{y}_j = e^{-(x_{j+1}-x_j)} - e^{-(x_j-x_{j-1})} \,, \quad j = 1,\ldots,N \,. \tag{3.71}$$

Following [68], we introduce the new variables

$$a_j = e^{-(x_{j+1}-x_j)/2} \,, \quad b_j = -\frac{1}{2}y_j \,. \tag{3.72}$$

The transformation is not canonical; hence we transform directly the dynamical equations, thus getting

$$\dot{a}_j = a_j(b_{j+1} - b_j) \,, \quad \dot{b}_j = 2(a_j^2 - a_{j-1}^2) \,. \tag{3.73}$$

The latter equations are equivalent to the Lax equation (3.69) with the tridiagonal matrices

$$\mathsf{L} = \begin{pmatrix} b_1 & a_1 & 0 & \ddots & a_N \\ a_1 & b_2 & a_2 & \ddots & 0 \\ 0 & a_2 & b_3 & \ddots & 0 \\ \vdots & \vdots & \ddots & \ddots & a_{N-1} \\ a_N & 0 & \ldots & a_{N-1} & b_N \end{pmatrix}$$

and

$$\mathsf{A} = \begin{pmatrix} 0 & a_1 & 0 & \ddots & -a_N \\ -a_1 & 0 & a_2 & \ddots & 0 \\ 0 & -a_2 & 0 & \ddots & 0 \\ \vdots & \vdots & \ddots & \ddots & a_{N-1} \\ a_N & 0 & \ldots & -a_{N-1} & 0 \end{pmatrix} .$$

In view of Proposition 3.36, the eigenvalues of the matrix L are first integrals of both systems (3.69) and (3.73). Since L is symmetric, its eigenvalues are real. However, we may observe that the eigenvalues are found by solving the secular (or characteristic) equation

$$\det(\lambda\mathsf{I} - \mathsf{L}) = \lambda^N + \Phi_1\lambda^{N-1} + \cdots + \Phi_N \,,$$

where $\Phi_1,\ldots,\Phi_N$ are coefficients depending on a_j, b_j. This is indeed a set of first integrals discovered by Hénon [111], who also gave a proof that they are independent. Other sets of first integrals may be constructed as functions of the eigenvalues $\lambda_1, \ldots, \lambda_N$.

For a complete discussion we should prove that the first integrals so found are independent and in involution; for this part we refer to [177]. The existence of action-angle variables follows from the Arnold–Jost theorem. However, the explicit construction of such variables is not so trivial; it may be found in [114].

4
First Integrals

According to Poincaré, the general problem of dynamics is formulated as follows ([188], Vol. I, §13).

> *Nous sommes donc conduit à nous proposer le problème suivant:*
> *Étudier les équations canoniques*
>
> $$\frac{dx_i}{dt} = \frac{\partial F}{\partial y_i}\ , \quad \frac{dy_i}{dt} = -\frac{\partial F}{\partial x_i}$$
>
> *en supposant que la function F peut se développer suivant les puissances d'un paramètre très petit μ de la manière suivante:*
>
> $$F = F_0 + \mu F_1 + \mu^2 F_2 + \dots\ ,$$
>
> *en supposant de plus que F_0 ne dépend que des x et est indépendent des y; et que F_1, $F_2, \dots$ sont des fonctions périodiques de période 2π par rapport aux y.*

The problem may be restated in terms closer to our current language, with some changes of notation coherent with the rest of the present notes.

> *To investigate the dynamics of a canonical system of differential equations with Hamiltonian*
>
> $$H(p, q, \varepsilon) = H_0(p) + \varepsilon H_1(p, q) + \varepsilon^2 H_2(p, q) + \dots\ , \tag{4.1}$$
>
> *where $p \equiv (p_1, \dots, p_n)\ \in \mathscr{G} \subset \mathbb{R}^n$ are action variables in the open set $\mathscr{G} \subset \mathbb{R}^n$, $q \equiv (q_1, \dots, q_n)\ \in \mathbb{T}^n$ are angle variables, and ε is a small parameter. The Hamiltonian is assumed to be a real analytic function of all its arguments; in particular, we shall assume that it is expanded in power series of ε in a neighbourhood of the origin.*

The smallness of the parameter ε (and the implicit assumption that the functions $H_1, H_2, \dots$ are bounded) means that the system is a small perturbation of an integrable one. This is a typical situation for mechanical systems of interest in Physics and Astronomy. The interest of Poincaré was indeed focused on Celestial Mechanics. However, systems which can be described as perturbations of integrable systems are common in Physics.

For $\varepsilon = 0$ the unperturbed system is integrable in the Arnold–Jost sense and all orbits lie on invariant tori parameterized by the actions $p_1, \ldots, p_n$. This leads us to study the character of periodic and quasiperiodic motions on invariant tori. For $\varepsilon \neq 0$ the question is raised whether the dynamics remains similar to that of the unperturbed system. The natural approach consists in trying to prove that the perturbed system is integrable; for instance, by constructing a complete set of first integrals in involution. However, it has been discovered by Poincaré that the dynamics of the perturbed system can be very complicated, and in general it cannot be reduced to that of the unperturbed system.

Sections 4.1 to 4.4 contain a detailed discussion of the dynamics of the *unperturbed system*, namely the Hamiltonian (4.1) with $\varepsilon = 0$. This includes also the description of resonances and some ergodic properties of the quasiperiodic motions. Section 4.5 deals with the main subject, namely the theorem of Poincaré on non-existence of analytic first integrals for Hamiltonians of the form (4.1).

4.1 Periodic and Quasi-Periodic Motion on a Torus

Let us begin by investigating the dynamics of an integrable system in the Arnold–Jost sense. Referring to the general problem of dynamics as stated by Poincaré, we consider the (subset of the) phase space $\mathscr{F} = \mathscr{G} \times \mathbb{T}^n$, where $\mathscr{G} \subset \mathbb{R}^n$ is an open set, $q = (q_1, \ldots, q_n) \in \mathbb{T}^n$ are angle variables and $p = (p_1, \ldots, p_n) \in \mathscr{G}$ are action variables; the Hamiltonian function takes the form

$$H = H(p_1, \ldots, p_n), \tag{4.2}$$

depending only on the actions p. We know that the canonical equations have the particular form

$$\dot{q}_j = \omega_j(p_1, \ldots, p_n)\ , \quad \dot{p}_j = 0\ , \qquad 1 \leq j \leq n\ , \tag{4.3}$$

where

$$\omega_j(p_1, \ldots, p_n) = \frac{\partial H}{\partial p_j} \tag{4.4}$$

are the frequencies. We also know that the orbits in phase space are given by

$$p_j(t) = p_{j,0}\ , \quad q_j(t) = q_{j,0} + t\omega(p_{1,0}, \ldots, p_{n,0})\ , \tag{4.5}$$

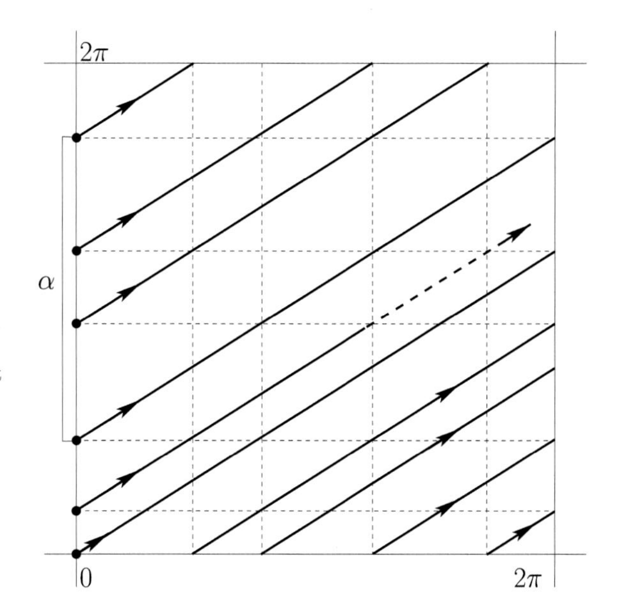

Figure 4.1 The Kronecker flow on the torus $\mathbb{T}^2$. The torus is represented by a square of side 2π. The orbit is a straight line with angular coefficient ω_2/ω_1. All segments inside the square are images of this straight line translated by a multiple of 2π, in both the horizontal and the vertical direction.

where $q_{1,0}, \ldots, q_{n,0}, p_{1,0}, \ldots, p_{n,0}$ are the initial data. Therefore, the orbits lie on invariant n-dimensional tori parameterized by the action variables. The linear flow (4.5) on a torus is named *a Kronecker flow*.

4.1.1 Motion on a Two-Dimensional Torus

In the simple case of a two-dimensional torus, we consider the orbit

$$q_1(t) = q_{1,0} + \omega_1 t \ , \quad q_2(t) = q_{2,0} + \omega_2 t \ , \tag{4.6}$$

where ω_1 and ω_2 are constant frequencies and $q_{1,0}$ and $q_{2,0}$ are the initial phases. As usual, we represent the torus $\mathbb{T}^2$ as a square of side 2π on the plane; that is, the points (q_1, q_2) and (q_1', q_2') in $\mathbb{R}^2$ are said to represent the same point in case $q_1 - q_1'$ and $q_2 - q_2'$ are integer multiples of 2π. By elimination of t in (4.6) we find that the orbit is the straight line

$$\omega_2(q_1 - q_{1,0}) - \omega_1(q_2 - q_{2,0}) = 0 \ ;$$

we simplify the discussion by putting $q_{1,0} = q_{2,0} = 0$, which is tantamount to translating the origin of the angles, so that the orbit is the straight line $\omega_2 q_1 - \omega_1 q_2 = 0$. The image of the orbit on the torus is represented in Fig. 4.1. All segments representing a slice of the orbit are parallel, so that they never intersect transversally. However, the orbit may return to its initial point; in this case the motion is periodic.

In order to investigate the dynamics it is convenient to consider the successive intersections of the orbit with the vertical side of the square corre-

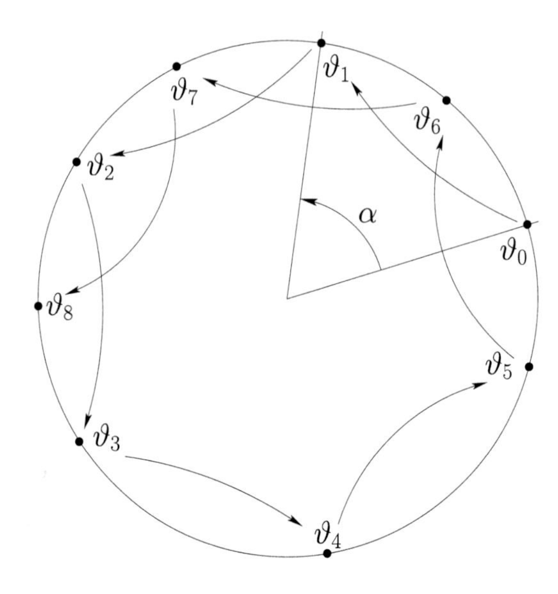

Figure 4.2 The map of a circle into itself defined as a rotation by an angle α.

sponding to $q_1 = 0$. It is easily seen that these intersections are represented by the sequence of points

$$(4.7)\quad 0\,,\ \alpha\,(\mathrm{mod}\,2\pi)\,,\ 2\alpha\,(\mathrm{mod}\,2\pi)\,,\ \ldots,\ s\alpha(\mathrm{mod}\,2\pi),\ \ldots\ ;\qquad \alpha = 2\pi\frac{\omega_2}{\omega_1}\ .$$

Therefore, we are led to investigate the behaviour of a sequence of points on a circle defined by a rotation by a fixed angle α (see Fig. 4.2).

Lemma 4.1: *The sequence $\{s\alpha(\mathrm{mod}\,2\pi)\}_{s\geq 0}$ on the circle is periodic if and only if $\alpha/(2\pi)$ is a rational number. If $\alpha/(2\pi)$ is irrational, then the sequence is everywhere dense on the circle.*

Proof. The sequence is periodic in case there is an integer $s \neq 0$ such that $s\alpha(\mathrm{mod}\,2\pi) = 0$, that is, $s\alpha = 2r\pi$ for some integer $r \neq 0$. Hence, $\alpha/(2\pi) = r/s$, namely a rational number. The condition is clearly necessary and sufficient, and this proves the first claim. Let now $\alpha/(2\pi)$ be irrational; then all points of the sequence are distinct. Since the circle is compact, the sequence has at least one accumulation point.[1] Therefore, for every positive ε there exist integers s, r such that $|(s\alpha - r\alpha)(\mathrm{mod}\,2\pi)| < \varepsilon$. Setting $j = |s-r|$

[1] Dirichlet used the so-called *pigeonhole principle: if $N+1$ (or more) pigeons are placed in N holes, then at least one hole contains more than one pigeon.* (The principle is sometimes called the *Dirichlet box principle* and enunciated in terms of objects in boxes.) In our case we use it as follows: pick N arbitrarily large and divide the circle in N equal intervals of length $\varepsilon = 2\pi/N$; take the first $N+1$ points of the sequence. Then at least one interval contains more than one point.

we also have $|k\alpha - (k+j)\alpha \pmod{2\pi}| < \varepsilon$ for any integer k. The subsequence $\{nj\alpha \pmod{2\pi}\}_{n\geq 0}$ rotates monotonically with steps of length smaller than ε on the circle. Hence, every interval of length ε contains at least one point of the sequence. Since ε is arbitrary, the claim follows. *Q.E.D.*

By a straightforward application of this lemma, we conclude that the orbit (4.6) on the torus $\mathbb{T}^2$ is periodic in case ω_1/ω_2 is rational and is everywhere dense on the torus in case ω_1/ω_2 is irrational.

4.1.2 The Many-Dimensional Case

The case $n > 2$ requires a generalization of the concept of rational ratio of the frequencies. This leads in a natural manner to introducing the concept of resonance module.

The *resonance module* associated to $\omega \in \mathbb{R}^n$ is the set[2]

$$\mathscr{M}_\omega = \{k \in \mathbb{Z}^n \colon \langle k, \omega\rangle = 0\} \ , \tag{4.8}$$

where $\langle k, \omega\rangle = \sum_j k_j\omega_j$. It is an easy matter to check that $\mathscr{M}_\omega$ is a subgroup of $\mathbb{Z}^n$ (i.e., a lattice). A relation $\langle k, \omega\rangle = 0$ is called a *resonance*, and the number $\dim \mathscr{M}_\omega$ is called the *multiplicity of the resonance*. The latter number actually represents the number of independent resonances to which ω is subjected. The extreme cases are $\dim \mathscr{M}_\omega = 0$, which is called the *non-resonant* case, and $\dim \mathscr{M}_\omega = n-1$, the *completely resonant* case.

Example 4.2: *Resonance relations in dimension* $n = 2$. For a frequency vector $\omega \in \mathbb{R}^2$ only two cases may occur.

(i) Complete resonance, $\dim \mathscr{M}_\omega = 1$: the ratio ω_1/ω_2 must be rational; for example, $\omega = (1, 1)$ or $\omega = (\sqrt{3}, 2\sqrt{3})$.

(ii) Non-resonance, $\dim \mathscr{M}_\omega = 0$: the ratio ω_1/ω_2 must be irrational; for example, $\omega = (1, \sqrt{2})$ or $\omega = (\sqrt{3}, \sqrt{5})$. *E.D.*

Example 4.3: *Resonance relations in dimension* $n > 2$. Here are a few examples.

(i) Complete resonance, $\dim \mathscr{M}_\omega = n-1$: all ratios ω_j/ω_1 for $j = 2, \ldots, n$ are rational numbers. This entails that the ratio between any two frequencies is rational, and all frequencies are integer multiples of a common quantity (prove it).

(ii) Partial resonance, $0 < \dim \mathscr{M}_\omega < n - 1$. For example, suppose that $n - 1$ frequencies are rational, and the remaining one is not; then we have $\dim \mathscr{M} = n - 2$. A similar situation occurs if the $n - 2$ ratios

---

[2] Geometrically: consider $\mathbb{Z}^n$ as immersed in $\mathbb{R}^n$, and the $(n-1)$-dimensional plane orthogonal to ω through the origin in $\mathbb{R}^n$. The resonance module $\mathscr{M}_\omega$ is the intersection of this plane with $\mathbb{Z}^n$.

$\omega_2/\omega_1, \ldots, \omega_{n-1}/\omega_1$ are rational, but the last ratio ω_n/ω_1 is not; for example, an elementary case is $\omega = (\sqrt{2}, \ldots, \sqrt{2}, \sqrt{3})$.

(iii) A less trivial example of partial resonance may occur also when all the ratios ω_j/ω_l (with $j \neq l$) are irrational numbers; this does not exclude the existence of a resonance relation. For example, let $\omega = (1, \sqrt{2}, 1 + \sqrt{2})$; then $\omega_2/\omega_1 = \sqrt{2}$, $\omega_3/\omega_1 = 1 + \sqrt{2}$ and $\omega_3/\omega_2 = (2 + \sqrt{2})/2$, which are all irrational numbers, but $\omega_1 + \omega_2 - \omega_3 = 0$; thus we have $\dim \mathscr{M}_\omega = 1$.

(iv) An elementary example of non-resonant frequencies is $\omega = (1, \sqrt{2}, \sqrt{3})$. The reader may try to invent more examples. However, the most interesting fact is that in an open subset $\Omega \subset \mathbb{R}^n$ the relative measure of non-resonant frequencies is one. This is a remarkable result from Diophantine theory: see Appendix A.2.1. *E.D.*

Example 4.4: *The frequencies of a linear chain.* A remarkable example is represented by the frequencies of the normal modes of a linear chain of identical particles with fixed ends discussed in Section 3.7. We have seen that if N is the number of moving particles, then the frequencies are

$$\omega_j = \sin\left(\frac{j\pi}{2(N+1)}\right), \quad j = 1, \ldots, N . \tag{4.9}$$

A known result is the following: the frequencies are non-resonant if and only if $N + 1$ is either a prime number or a power of 2 (see [109]). *E.D.*

4.1.3 Changing the Frequencies on a Torus

Recall that the linear flow on a torus is generated by the system of differential equations

$$\dot{q} = \omega, \quad \omega \in \mathbb{R}^n . \tag{4.10}$$

A linear transformation

$$\overline{q} = \mathsf{M}q \tag{4.11}$$

with a unimodular matrix M changes the system (4.10) into

$$\dot{\overline{q}} = \overline{\omega}, \quad \overline{\omega} = \mathsf{M}\omega , \tag{4.12}$$

so that the change of coordinates induces a change of the frequencies.

Lemma 4.5: *Let $\omega \in \mathbb{R}^n$ be given, and let $\dim \mathscr{M}_\omega > 0$. Then there is a unimodular matrix M such that $\overline{\omega} = \mathsf{M}\omega$ has exactly $\dim \mathscr{M}_\omega$ vanishing components, while the remaining $n - \dim \mathscr{M}_\omega$ components form a non-resonant vector.*

Corollary 4.6: *The dimension $\dim \mathscr{M}_\omega$ of the resonance module is invariant under linear unimodular changes of the angular coordinates.*

Lemma 4.5 means that we can always introduce coordinates such that the torus $\mathbb{T}^n$ is actually the Cartesian product $\mathbb{T}^r \times \mathbb{T}^{n-r}$, where $r = \dim \mathscr{M}_\omega$, with the property that on $\mathbb{T}^r$ the flow is trivial, all points being fixed, while the frequencies on $\mathbb{T}^{n-r}$ are non-resonant. Thus, the actual frequencies of the motion on a torus are not uniquely defined, depending on the chosen coordinates on the torus, while the dimension of the resonance module is uniquely defined, as stated by Corollary 4.6.

The proof of Lemma 4.5 exploits the unimodular transformation (3.50) of Lemma 3.33. Let $r = \dim \mathscr{M}_\omega$ and let $e_1, \ldots, e_r$ be a basis of the resonance module $\mathscr{M}_\omega$. The integer vectors $e_1, \ldots, e_r$ can be arranged in a rectangular $r \times n$ matrix with integer entries. The problem is to complete this matrix by adding $n - r$ rows so that the resulting $n \times n$ matrix is a unimodular one. This is an algebraic problem which is deferred to Appendix A, Lemma A.8. The linear canonical transformation generated by the unimodular matrix M thus found transforms the frequencies according to formula (4.12), so that r of them are zero. In view of this, the completion of the proof of Lemma 4.5 and of Corollary 4.6 is an easy matter, which is left to the reader.

4.1.4 Dynamics of the Kronecker Flow

The main result concerning the dynamics on $\mathbb{T}^n$ is given by the following.

Proposition 4.7: *Consider the Kronecker flow on $\mathbb{T}^n$,*

$$q(t) = q_0 + \omega t \ (\mathrm{mod}\, 2\pi) \ , \quad \omega \in \mathbb{R}^n \ , \tag{4.13}$$

with initial point q_0, and let $\mathscr{M}_\omega$ be the resonance module associated to the frequency vector ω. Then the orbit (4.13) is dense on a torus $\mathbb{T}^{n-\dim \mathscr{M}_\omega} \subset \mathbb{T}^n$.

In particular, in the non-resonant case the orbit is everywhere dense on $\mathbb{T}^n$, and in the completely resonant case, $\dim \mathscr{M}_\omega = n - 1$, the orbit is periodic.

The statement of Proposition 4.7 appears as a somehow natural generalization of the discussion of Section 4.1.1 to the case of a higher-dimensional torus $\mathbb{T}^n$. In some sense it is so, indeed, but a complete proof requires a number of details, which takes some space. The discussion may be separated into two main points.

The first point exploits the result of Lemma 4.5. If $\dim \mathscr{M}_\omega = m$ with $0 < r < n$, then the torus admits a foliation on a continuous family of $(n - r)$-dimensional tori with non-resonant frequencies. Therefore, there are r angles which remain fixed, the ones with zero frequency; the remaining $n-r$ evolve according to the Kronecker flow on a torus $\mathbb{T}^{n-r}$ in the non-resonant case.

The second point consists precisely in studying the case of non-resonant frequencies: if $\dim \mathscr{M}_\omega = 0$, then the orbit is dense on the torus $\mathbb{T}^n$. This is the longest part of the proof, which will be worked out by generalizing the result of Lemma 4.1 to the case of dimension $n > 1$, namely studying the dynamics of a Kronecker map, in Section 4.2.

4.2 The Kronecker Map

The connection between the Kronecker flow and the Kronecker map comes from an appropriate use of Poincaré section.

In the rest of the present section we shall identify $\mathbb{T}^n$ with $\mathbb{R}^n / \mathbb{Z}^n$; the periods on the torus are all equal to one, thus avoiding an unnecessary proliferation of factors 2π. The coordinates on the torus will be denoted by $x = (x_1, \ldots, x_n)$. The Kronecker map is defined as

$$\phi x = x + \alpha \,(\mathrm{mod}\, 1), \tag{4.14}$$

where $\alpha \in \mathbb{R}^n$ are the so-called *rotation angles*. The case $\alpha = 0$ is not forbidden, but the map is trivially the identity; similarly, if one or some components of α are zero, then the map acts on a torus of reduced dimension; therefore, we may assume that α has no zero components. The change of notation from q to x and from ω to α is meant to help avoid confusion with the flow on a 2π-periodic torus which we have considered until now.

The orbit with initial point x is defined as

$$\Omega_x = \bigcup_{s \in \mathbb{Z}} \{\phi^s x\} \,. \tag{4.15}$$

When dealing at the same time with two different maps with different rotation angles, say α and β, we shall use the notation $\Omega_x^{(\alpha)}$ and $\Omega_x^{(\beta)}$ for the corresponding orbits. A special role will be played by the orbit Ω_0 having the origin of the angles as initial point, namely

$$\Omega_0 = \bigcup_{s \in \mathbb{Z}} \{s\alpha \,(\mathrm{mod}\, 1)\} \,. \tag{4.16}$$

4.2.1 General Properties

The following properties exploit the peculiarity of the map of being a translation on the torus; they will be useful in the proofs in this section.

(i) The set Ω_0 is a subgroup of $\mathbb{R}^n$.

(ii) For every $x \in \mathbb{T}^n$ we have

$$\Omega_x = \bigcup_{\xi \in \Omega_0} \{x + \xi \,(\mathrm{mod}\, 1)\} \,. \tag{4.17}$$

(iii) Let $\delta = \xi - \xi'$ with ξ, $\xi' \in \Omega_x$. Then we have $\delta \in \Omega_0$.

(iv) Let $\delta = \xi - \xi'$ as at point (iii), and let $\Omega_x^{(\alpha)}$ and $\Omega_x^{(\delta)}$ be the corresponding orbits for any given x. Then we have $\Omega_x^{(\delta)} \subset \Omega_x^{(\alpha)}$.

Exercise 4.8: Prove the claims (i)–(iv). *A.E.L.*

4.2.2 Resonances

A vector $\alpha \in \mathbb{R}^n$ is said to be k-resonant with a $k \in \mathbb{Z}^n$ in case[3]

$$\langle k, \alpha \rangle \in \mathbb{Q} \ . \tag{4.18}$$

The *resonance module* $\mathscr{M}_\alpha$ associated to α is defined as

$$\mathscr{M}_\alpha = \{ k \in \mathbb{Z}^n \ : \ \langle k, \alpha \rangle \in \mathbb{Q} \} \ . \tag{4.19}$$

Recalling the case of the frequencies of the flow, it is immediately seen that $\mathscr{M}_\alpha$ is a subgroup of $\mathbb{Z}^n$, so it has a dimension. The following cases may occur:

(i) if $\dim \mathscr{M}_\alpha = 0$, i.e., $\mathscr{M}_\alpha = \{0\}$, we say that α is non-resonant;
(ii) if $0 < \dim \mathscr{M}_\alpha < n$ we have a partial resonance;
(iii) if $\dim \mathscr{M}_\alpha = n$ then α is said to be in full resonance.

Exercise 4.9: Prove that the following properties hold true.

(i) If $\alpha = (\alpha_1, \ldots, \alpha_n)$ is non-resonant, then all components $\alpha_1, \ldots, \alpha_n$ are irrational numbers.
(ii) The converse is not true; that is, a vector α with all irrational components may well be resonant; in this case we have $\dim \mathscr{M}_\alpha < n$.
(iii) If all components of α are rational numbers, then α is completely resonant, and $(\alpha_1, \ldots, \alpha_n)$ are multiples of a single quantity. *A.E.L.*

Example 4.10: *Resonances in dimension $n = 2$.* Three cases may occur.

(i) Complete resonance: $\dim \mathscr{M}_\alpha$=2 and $\alpha \in \mathbb{Q}^2$. Write $\alpha = (m_1, m_2)/s$ with a common denominator; then $s\alpha \pmod 1 = 0$; hence $\phi^s x = x$ for any $x \in \mathbb{T}^n$, and all orbits are periodic, as in Fig. 4.3 (a). Conversely, for a periodic orbit we have $\phi^s x - x = s\alpha \pmod 1 = 0$ for some s; hence $\alpha \in \mathbb{Q}^2$. Remark that if one orbit is periodic, then so are all orbits, in view of property (ii) of Section 4.2.1.

---

[3] The reader may be tempted to define a resonance as $\langle k, \alpha \rangle = 0$, as in the case of the flow; but this is immediately seen to be a bad choice. A more spontaneous and natural definition, which comes from considering a map as an appropriate Poincaré section of the flow, leads to say $\langle k, \alpha \rangle \in \mathbb{Z}$. On the other hand, it should be remarked that if $\langle k, \alpha \rangle = r/s$ is a rational number then we have $s\langle k, \alpha \rangle = r$, so that one will conclude that sk corresponds to a resonance, while k does not; this is clearly unreasonable, for in both cases the resonance phenomena are the same, and the distinction between k and sk is unjustified.

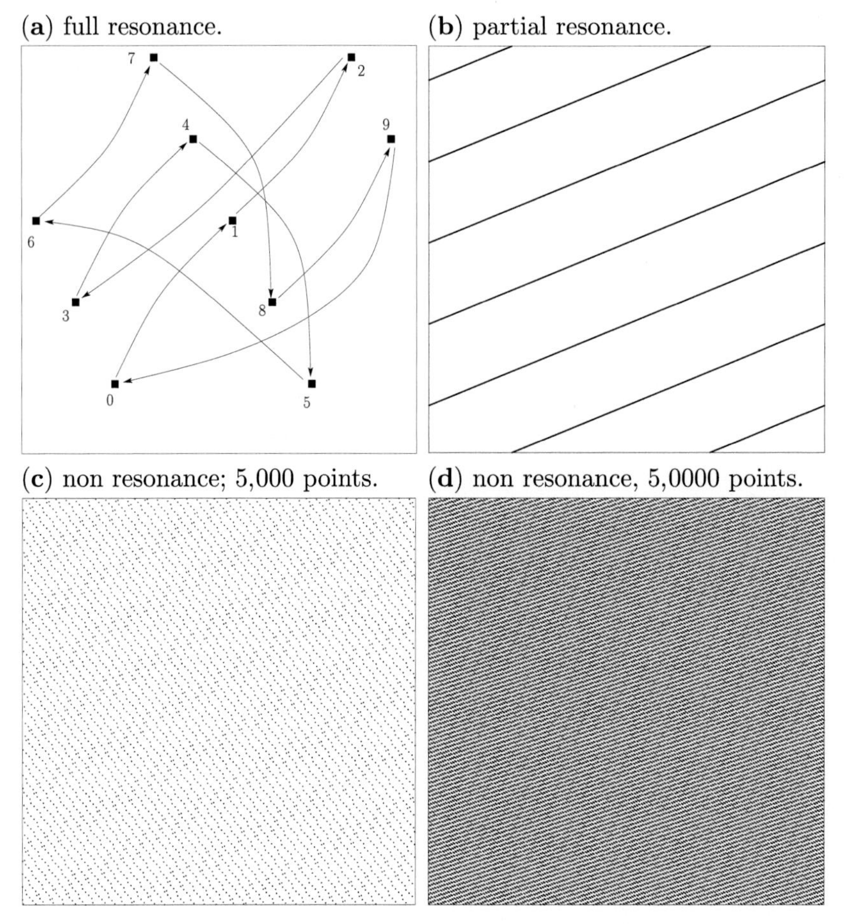

Figure 4.3 Resonances of the Kronecker map on $\mathbb{T}^2$. (**a**) Completely resonant case, $\dim \mathscr{M}_\alpha = 2$, with $\alpha = (\frac{3}{10}, \frac{2}{5})$. (**b**) Partially resonant case, $\dim \mathscr{M}_\alpha = 1$, with $\alpha = (\frac{\sqrt{2}}{2}, \frac{\sqrt{2}}{5})$. (**c**), (**d**) Non-resonant case, $\dim \mathscr{M}_\alpha = 0$, with $\alpha = (\frac{\sqrt{2}}{5}, \frac{\sqrt{5}}{5})$; the two figures show the progressive filling of the torus by increasing the number of points of the orbit.

(ii) Partial resonance, $\dim \mathscr{M}_\alpha = 1$. The orbits are not periodic, hence they contain distinct points. For any pair x, x' of points we have $x - x' = m\alpha$ for some $m \in \mathbb{Z}$. Since $\langle k, \alpha \rangle = r/s$ with non-zero integers r, s we have $\langle k, m\alpha \rangle = mr/s$. On the other hand, $mr/s \pmod 1$ can assume only s distinct values; hence all points belong to s straight equidistant segments in $\mathbb{T}^2$, as in Fig. 4.3 (b).

(iii) Non-resonance: $\dim \mathscr{M}_\alpha = 0$. We want to show that every orbit is dense in $\mathbb{T}^2$, as in Fig. 4.3 (c,d); we do it in Section 4.2.3. The figure illustrates the progressive filling of the torus when the number of points is increased. *E.D.*

4.2.3 Dynamics of the Kronecker Map

We want to prove the following.

Proposition 4.11: *Let $\alpha \in \mathbb{R}^n$ be non-resonant, that is, $\dim \mathscr{M}_\alpha = 0$. Then for every $x \in \mathbb{T}^n$ the orbit Ω_x is dense on $\mathbb{T}^n$.*

Before coming to the proof, let us see how to generate a Kronecker map on $\mathbb{T}^{n-1}$ from a Kronecker flow on $\mathbb{T}^n$. For a given $\alpha \in \mathbb{R}^n$ consider the Kronecker flow $\phi^t x = x + \alpha \,(\mathrm{mod}\, 1)$, with $t \in \mathbb{R}$, with frequencies α which coincide with the rotation angles of the map. We shall denote by Ω^t_x the orbit of the flow on the torus $\mathbb{T}^n$. Remark that the non-resonance condition $\langle k, \alpha\rangle \notin \mathbb{Q}$ implies $\langle k, \alpha\rangle \neq 0$, so that the frequencies of the flow are non-resonant.

Let us write $\alpha = (\tilde{\alpha}, \alpha_n)$ with $\tilde{\alpha} = (\alpha_1, \ldots, \alpha_{n-1}) \in \mathbb{R}^{n-1}$; similarly, let us write $x = (\tilde{x}, x_n) \in \mathbb{T}^n$ for a point on the torus. A useful Poincaré section is constructed by considering the intersection of the orbit of the flow with the plane $x_n = 0$. Set $\tau = \alpha_n^{-1}$ (assuming that $\alpha_n \neq 0$). Taking an initial point $x = (\tilde{x}, 0)$ we clearly have $\phi^\tau x = (x + \tau\tilde{\alpha}\,(\mathrm{mod}\, 1), 0)$, so that the Poincaré section with the plane $x_n = 0$ is the Kronecker map on $\mathbb{T}^{n-1}$ with rotation angles $\tau\tilde{\alpha}$. This map will be used in the proof.

Proof of Proposition 4.11. For $n = 1$ the claim reduces to the statement of Lemma 4.1; therefore, we proceed by induction, assuming that the claim is true up to dimension $n - 1$ and proving it for n. We proceed in two steps:

(i) if the orbits of the Poincaré map ϕ^τ are dense on $\mathbb{T}^{n-1}$, then the orbits of the flow ϕ^t with frequencies α are dense on $\mathbb{T}^n$;

(ii) if α is non-resonant (as assumed in the statement of the proposition), then the orbits of the map with rotation angles α are dense on $\mathbb{T}^n$.

We prove that if α is non-resonant, then the rotation angles

$$\tau\tilde{\alpha} = \left(\frac{\alpha_1}{\alpha_n}, \ldots, \frac{\alpha_{n-1}}{\alpha_n}\right)$$

of the map are non-resonant. By contradiction, let $\tau\langle \tilde{k}, \tilde{\alpha}\rangle = r/s \in \mathbb{Q}$ with $0 \neq \tilde{k} \in \mathbb{Z}^{n-1}$. Then we have $\langle s\tilde{k}, \tilde{\alpha}\rangle - r\alpha_n = 0$, contradicting the hypothesis that α is non-resonant; hence $\tau\tilde{\alpha}$ is non-resonant.
Let $\tilde{x} \in \mathbb{T}^{n-1}$ be a point on the Poincaré section $x_n = 0$. By the induction hypothesis the orbit $\tilde{\Omega}_{\tilde{x}}$ of the Poincaré map ϕ^τ is dense on $\mathbb{T}^{n-1}$; therefore, the orbit $\Omega^t_{(\tilde{x},0)}$ of the flow ϕ^t on $\mathbb{T}^n$ with frequencies α is dense on the torus, being just a translation of $\tilde{\Omega}_{\tilde{x}}$. This obviously implies that for every $x \in \mathbb{T}^n$, the orbit Ω^t_x of the flow $\phi^t x = x + t\alpha\,(\mathrm{mod}\, 1)$ is dense on $\mathbb{T}^n$.

Let us now consider the orbit Ω_x^α of the map $\phi x = x + \alpha \pmod 1$ on $\mathbb{T}^n$. We obviously have $\Omega_x^\alpha \subset \Omega_x^t$; we want to prove that Ω_x^α is dense on $\mathbb{T}^n$.
In view of the non-resonance condition, the orbit Ω_x^α is an infinite set of distinct points on $\mathbb{T}^n$; hence they have an accumulation point; that is, for arbitrary $\varepsilon > 0$ there are two points ξ, $\xi' \in \Omega_x^\alpha$ such that $\mathrm{dist}(\xi, \xi') < \varepsilon/2$, where $\mathrm{dist}(\xi, \xi')$ is a distance on the torus. Set $\delta = \xi - \xi' \pmod 1$, so that $|\delta| < \varepsilon/2$, and consider the orbit Ω_x^δ of the map $\phi x = x + \delta \pmod 1$. By property (iii) of Section 4.2.1 we have $\delta \in \Omega_0$, so that $\delta = s\alpha \pmod 1$ for some $s \neq 0$; moreover, by property (iv) of the same section we have $\Omega_x^\delta \subset \Omega_x^\alpha$. We prove that *for any* $y \in \mathbb{T}^n$ *there is* $\xi \in \Omega_x^\delta$ *such that* $\mathrm{dist}(y, \xi) < \varepsilon$.
Since $\delta = s\alpha \pmod 1$ we have $\langle k, \delta \rangle \notin \mathbb{Q}$; hence δ is non-resonant. Consider the flow $\phi^t x = x + t\delta \pmod 1$. By the argument developed for α in the first part of the proof we know that the orbit of the Poincaré section with rotation angles $(\delta_1/\delta_n, \ldots, \delta_{n-1}/\delta_n)$ is dense on $\mathbb{T}^{n-1}$, and that any orbit Ω_x^t of the flow is dense on $\mathbb{T}^n$; that is, for any $y \in \mathbb{T}^n$ there is $y' \in \Omega_x^t$ such that $\mathrm{dist}(y, y') < \varepsilon/2$. On the other hand, there is a point $\xi \in \Omega_x^{(\delta)}$ such that $\mathrm{dist}(y', \xi) < \varepsilon/2$, because the distance between two consecutive points of the map is $|\delta| < \varepsilon/2$. By the triangle inequality we conclude

$$\mathrm{dist}(y, \xi) \leq \mathrm{dist}(y, y') + \mathrm{dist}(y', \xi) < \varepsilon \ ,$$

as claimed. Since $\varepsilon > 0$ is arbitrary, we conclude that the orbit Ω_x^δ is dense on $\mathbb{T}^n$, and that the same holds true for the orbit Ω_x^α. This completes the induction step. *Q.E.D.*

4.3 Ergodic Properties of the Kronecker Flow

The Kronecker flow on the torus with non-resonant frequencies is a remarkable example of ergodic dynamics. Although ergodicity is not a central point to be discussed in the present notes, it is worth the effort at least to devote a little space to it. The reader already familiar with Ergodic Theory is strongly invited to skip the present section: the exposition is rather informal, strictly referred to the Kronecker flow on a torus.[4]

In our case, and in very rough words, ergodicity means that an orbit fills uniformly the torus $\mathbb{T}^n$. In equivalent but still rough terms, the desired property is that the time average of a dynamical variable must coincide with the phase average of the same variable over the torus. This crucial property has been introduced by Boltzmann as a basic one for the development of classical Statistical Mechanics.

---

[4] The exposition in the present section is widely dependent on Arnold's, [9], § 51. See also [7].

4.3.1 Time Average and Phase Average

Let us consider a function $f: \mathbb{T}^n \to \mathbb{R}$. Using the coordinates $q_1, \ldots, q_n$, the function $f(q_1, \ldots, q_n)$ is 2π periodic in each of the variables. By composition with the Kronecker flow $q(t) = q_0 + \omega t$, where q_0 is the initial point, we get a function $\tilde{f}(t)$ of the time t, namely $\tilde{f}(t) = f(q_0 + \omega t)$.

Let us consider the *time average* of $\tilde{f}(t)$, defined as

$$\overline{f}(q_0) = \lim_{T\to\infty} \frac{1}{T} \int_0^T f(q_0 + \omega t)dt \ , \tag{4.20}$$

assuming that the limit exists. It is an easy matter to check that the result depends only on the orbit, that is, it does not depend on the initial phases.[5]

Let us also define the *phase average* of the function $f(q_1, \ldots, q_n)$ as

$$\langle f \rangle = \frac{1}{(2\pi)^n} \int_0^{2\pi} \cdots \int_0^{2\pi} f(q_1, \ldots, q_n) dq_1 \ldots dq_n \ . \tag{4.21}$$

A system is said to be *ergodic* in case the phase average and the time average of a measurable function do coincide.[6] That the Kronecker flow on the torus is ergodic is the content of the following proposition.

Proposition 4.12: *Let $f: \mathbb{T}^n \to \mathbb{R}$ be a Riemann-integrable function on $\mathbb{T}^n$ and let $\tilde{f}(t) = f(q_0 + \omega t)$. If $\dim \mathscr{M}_\omega = 0$ then the time average $\overline{f}(q_0)$ defined by (4.20) exists everywhere and is constant on $\mathbb{T}^n$, and coincides with the phase average $\langle f \rangle$ defined by (4.21).*

The proof is based on approximation by trigonometric functions.

Lemma 4.13: *Let $f: \mathbb{T}^n \to \mathbb{C}$ be a trigonometric polynomial, that is,*

$$f(q) = \sum_{k\in\mathscr{K}} f_k e^{i\langle k, q\rangle} \ ,$$

where $\mathscr{K} \subset \mathbb{Z}^n$ is a finite set of integer vectors and $f_k \in \mathbb{C}$ is a complex coefficient. Then the time average $\overline{f}$ is constant on $\mathbb{T}^n$ and coincides with the phase average $\langle f \rangle$.

Proof. Let $\mathscr{K}$ contain just one vector k. If $k = 0$, then the function f is constant, namely $f = f_0 \in \mathbb{C}$, and we trivially have $\overline{f} = \langle f \rangle = f_0$. If $k \neq 0$, then we have $f(q) = e^{i\langle k, q\rangle}$ (the actual value of the coefficient is irrelevant) and we may calculate

---

[5] The reader will immediately realize that taking a different initial point q_0' on the orbit with initial point q_0 will not change the result, that is, $\overline{f}(q_0') = \overline{f}(q_0)$. For the difference is just a quantity $\mathcal{O}(1/T)$ which vanishes in the limit $T \to \infty$.

[6] In general we should add 'almost everywhere', that is, except for a set of orbits of zero measure. However, this is not necessary for the Kronecker flow.

$$\int_0^T e^{i\langle k,(q_0+\omega t)\rangle} dt = e^{i\langle k,q_0\rangle} \frac{e^{i(\langle k,\omega\rangle)T} - 1}{i\langle k,\omega\rangle} ;$$

thus the time average is

$$\overline{f}(q_0) = \frac{e^{i\langle k,q_0\rangle}}{i\langle k,\omega\rangle} \lim_{T\to\infty} \frac{e^{i(\langle k,\omega\rangle)T} - 1}{T} = 0$$

for all $q_0 \in \mathbb{T}^n$. On the other hand, by direct integration one easily sees that the phase average vanishes, too. Let now $\mathscr{K}$ contain a finite set of vectors. Since the average is a linear operation we have

$$\overline{f}(q_0) = \frac{1}{T} \sum_{k\in\mathscr{K}} f_k \int_0^T e^{i\langle k,(q_0+\omega t)\rangle} dt = f_0 ,$$

that is, the coefficient of the term with $k = 0$, and the same holds true for the phase average. *Q.E.D.*

Lemma 4.14: *Let f be Riemann-integrable. Then for any $\varepsilon > 0$ there are two trigonometric polynomials $P_1(q_1,\ldots,q_n)$, $P_2(q_1,\ldots,q_n)$ such that for every $q \in \mathbb{R}^n$ one has $P_1(q) < f(q) < P_2(q)$, and moreover*

$$\frac{1}{(2\pi)^n} \int_{\mathbb{T}^n} (P_2 - P_1) dq_1 \ldots dq_n < \varepsilon .$$

Proof. If f is continuous, then by Weierstrass's theorem there is a trigonometric polynomial P which approximates the function within $\varepsilon/2$, that is, $|P(q) - f(q)| < \varepsilon/2$ for any $q \in \mathbb{T}^n$; just set $P_1 = P - \varepsilon/2$ and $P_2 = P + \varepsilon/2$. If f is not continuous but it is Riemann-integrable then there exist two continuous functions f_1, f_2 such that $f_1(q) < f(q) < f_2(q)$ on $\mathbb{T}^n$, and moreover

$$\frac{1}{(2\pi)^n} \int_{\mathbb{T}^n} (f_2 - f_1) dq_1 \ldots dq_n < \frac{\varepsilon}{3} .$$

By approximating f_1 and f_2 with two polynomials $P_1 < f_1, f_2 < P_2$ such that

$$\frac{1}{(2\pi)^n} \int_{\mathbb{T}^n} (P_2 - f_2) dq_1 \ldots dq_n < \frac{\varepsilon}{3} , \quad \frac{1}{(2\pi)^n} \int_{\mathbb{T}^n} (f_1 - P_1) dq_1 \ldots dq_n < \frac{\varepsilon}{3} ,$$

the claim is proven in view of ε being arbitrary. *Q.E.D.*

Proof of Proposition 4.12. For any $T > 0$ the trigonometric polynomials P_1, P_2 of Lemma 4.14 satisfy the relations

$$\frac{1}{T} \int_0^T P_1(q_0 + \omega t) dt < \frac{1}{T} \int_0^T f(q_0 + \omega t) dt < \frac{1}{T} \int_0^T P_2(q_0 + \omega t) dt .$$

By Lemma 4.13, for any $\varepsilon > 0$ there is $T_0(\varepsilon)$ such that for any $T > T_0(\varepsilon)$ the inequality

$$\left| \langle P_i \rangle - \frac{1}{T} \int_0^T P_i(q_0 + \omega t) dt \right| < \varepsilon \ , \quad i = 1, 2$$

holds true, where $\langle P_i \rangle$ is the phase average of the trigonometric polynomial P_i. Furthermore one has $\langle f \rangle - \langle P_1 \rangle < \varepsilon$, $\langle P_2 \rangle - \langle f \rangle < \varepsilon$. Thus, for $T > T_0(\varepsilon)$ one has also

$$\left| \langle f \rangle - \frac{1}{T} \int_0^T f(q_0 + \omega t) dt \right| < 2\varepsilon \ ,$$

which proves the claim. *Q.E.D.*

4.3.2 Quasiperiodic Motions

The result of the previous sections may be used to produce another proof that if the frequencies are non-resonant, then the orbit is dense on the torus. Actually we prove more, but we need a further concept.

Let $D \subset \mathbb{T}^n$ be a measurable subset, and consider the set $\bigcup_{0 \le t \le T} q(t)$ that is, the part of the orbit corresponding to the time interval $[0, T]$. Denote by $\tau_D(T)$ the time spent by the point in the subset D during the latter interval. Letting $T \to \infty$, we may define the average time spent by the orbit in D as $\lim_{T \to \infty} \frac{\tau_D(T)}{T}$, provided the limit exists.

The time τ_D is easily computed by introducing the *characteristic function* χ_D of D defined as $\chi_D(q) = 1$ for $q \in D$ and $\chi_D(q) = 0$ elsewhere. For it is immediately seen that

$$\tau_D(T) = \int_0^T \chi_D(q_0 + \omega t) dt \ . \tag{4.22}$$

Proposition 4.15: *Let $\dim \mathscr{M}_\omega = 0$. Then*

(i) every orbit is dense on the torus $\mathbb{T}^n$;

(ii) the average time spent by the orbit in a measurable subset $D \subset \mathbb{T}^n$ is proportional to the measure of D.

Proof. (i) By contradiction, suppose that there is an open subset $U \subset \mathbb{T}^n$ which is never visited by the orbit, and let $f(q)$ be a continuous function such that $f(q) = 0$ for $q \notin U$ and its phase average is $\langle f \rangle = 1$. Then f is zero valued along the whole orbit, and so the time average $\overline{f}$ is zero, contradicting the statement of Proposition 4.12 that $\overline{f} = \langle f \rangle$.

(ii) By (4.22) the average time spent by the orbit in D is the time average of the characteristic function χ_D of D, that is, $\overline{\chi_D}$. By Proposition 4.12 we also have $\overline{\chi_D} = \langle \chi_D \rangle$, and the phase average of the characteristic function of D is $\langle \chi_D \rangle = (2\pi)^{-n} \mathrm{meas}\,(D)$. *Q.E.D.*

4.4 Isochronous and Anisochronous Systems

Let us now go back to considering a Hamiltonian of the form (4.2), that is, $H(p_1, \ldots, p_n)$ depending only on the action variables, the conjugate variables being angles. The action variables $p_1, \ldots, p_n$ are a complete involution system of first integrals. The question now is whether other first integrals independent of the action variables may exist, although they cannot be in involution with all of them, according to Lemma 3.2.

As we have already remarked, for every $p \in \mathscr{G}$ we have an invariant torus carrying a linear flow with frequencies

$$\omega_j(p) = \frac{\partial H}{\partial p_j}(p) \ , \quad j = 1, \ldots, n \ .$$

It is generally expected that changing the initial value of the actions p causes an actual change of the frequencies. If this happens, then the dynamics on close tori can be very different. For the dynamics basically depends on $\dim \mathscr{M}_{\omega(p)}$, and this quantity can change in a crazy way, depending on the initial conditions. On the other hand, we know that a system of harmonic oscillators is completely isochronous, since the frequencies of the oscillators do not depend on the initial data, and so they are independent of the values of the actions. This leads us to distinguish two extreme cases, namely anisochronous versus isochronous systems.

4.4.1 Anisochronous Non-degenerate Systems

A common hypothesis on the Hamiltonian is that

$$\text{(4.23)} \qquad \det\left(\frac{\partial^2 H}{\partial p_j \partial p_k}\right) \neq 0 \ ;$$

this is called a *non-degeneracy condition* and can be interpreted as the condition that the frequencies ω can be used (at least locally) in place of the actions p as coordinates on $\mathscr{G}$. Under this condition we have the following.

Proposition 4.16: *Let the Hamiltonian $H(p)$ be analytic and non-degenerate, and let $\Phi(q,p)$ be an analytic first integral for H. Then Φ must be independent of the angles.*

Proof. Let us write $\Phi(q,p)$ as a Fourier expansion,

$$\Phi(p,q) = \sum_{k \in \mathbb{Z}^n} \varphi_k(p) e^{i\langle k,q\rangle} \ ,$$

where the coefficients $\varphi_k(p)$ are analytic functions of the actions p only. Since Φ is a first integral it satisfies

$$\{H, \Phi\} = i \sum_k \langle k, \omega(p)\rangle \varphi_k(p) e^{i\langle k,q\rangle} = 0 \; .$$

This implies that for every $k \in \mathbb{Z}^n$ we have either $\langle k, \omega(p)\rangle = 0$ or $\varphi_k(p) = 0$. We prove that $\langle k, \omega(p)\rangle = 0$ cannot happen if $k \neq 0$. For by differentiation we get the linear equation for k

$$\sum_{l=1}^{n} k_l \frac{\partial \omega_l}{\partial p_j} = 0 \; , \quad 1 \le j \le n \; ,$$

which in view of the non-degeneracy condition has the unique solution $k = 0$. We conclude that $\varphi_k(p) = 0$ for $k \neq 0$, and this implies $\Phi = \varphi_0(p)$, namely that Φ is independent of the angles. *Q.E.D.*

Example 4.17: *Uncoupled rotators.* A paradigm example of an anisochronous system which will be used in the following is the system of uncoupled rotators described by the Hamiltonian

$$H = \sum_{j=1}^{n} \frac{p_j^2}{2} \; , \quad p \in \mathbb{R}^n \; , \quad q \in \mathbb{T}^n \; .$$

The system is clearly non-degenerate, and the frequencies are $\omega_j = p_j$. *E.D.*

Example 4.18: *The Kepler problem.* As we have seen in Section 3.6.3, for the Kepler problem we can introduce Delaunay variables L, G, Θ with conjugated angular variables ℓ, g, ϑ and write the Hamiltonian as

$$H_0 = \frac{mk^2}{2L^2} \; ,$$

with constants m and k. Therefore, the unperturbed Hamiltonian of the planetary system is degenerate, which is indeed a major trouble in the study of the problem of three bodies and, more generally, of the dynamics of the solar system. In view of the particular form of the Hamiltonian, we have two further first integrals, namely the angles g and ϑ, which remain constants. *E.D.*

4.4.2 Isochronous Systems

The simplest system which violates the nondegeneracy condition is a system of harmonic oscillators,

$$H = \frac{1}{2} \sum_{j=1}^{n} \omega_j (x_j^2 + y_j^2) \; , \quad (x, y) \in \mathbb{R}^{2n} \; .$$

Changing to action-angle variables via the canonical transformation

$$x_j = \sqrt{2p_j} \cos q_j \; , \quad y_j = \sqrt{2p_j} \sin q_j \; , \quad 1 \le j \le n \; ,$$

the Hamiltonian takes the form

$$H = \sum_{j=1}^{n} \omega_j p_j \ , \quad p \in [0, \infty)^n \ , \quad q \in \mathbb{T}^n \ , \tag{4.24}$$

so that the frequencies ω are actually independent of the initial condition.

Proposition 4.19: *Let H be as in (4.24), and $r = \dim \mathscr{M}_\omega$. Then the Hamiltonian possesses $n + r$ independent first integrals, namely the n actions $p_1, \ldots, p_n$, and r additional first integrals of the form $\Phi_k(q) = \cos\langle k, q\rangle$, where k can take r independent values in $\mathscr{M}_\omega$.*

The proof is just a verification that Φ_k of the form just presented is a first integral for whatever $k \in \mathscr{M}$. By the way, this is connected with the fact that every torus is foliated into a r-parameter family of $(n - r)$-dimensional tori with non-resonant frequencies.

Example 4.20: *Essentially isochronous systems.* Consider an n-dimensional system with Hamiltonian in action-angle variables, $H(p) = F(\langle \omega, p\rangle)$, with $\omega \in \mathbb{R}^n$; that is, the Hamiltonian depends on p only through the expression $\langle \omega, p\rangle$, which is a very special case. The canonical equations are

$$\dot{q}_j = \omega_j \frac{dF}{dp_j} \ , \quad \dot{p}_j = 0 \ , \quad j = 1, \ldots, n \ .$$

We see that the actual frequencies do depend on the initial torus, but the mutual ratio between two frequencies remains constant. Therefore, the geometry of the orbits does not depend on the torus: we may say that changing the initial torus is tantamount to changing the time unit. We call such a system *essentially isochronous.* *E.D.*

Other examples of degenerate systems which possess more than n first integrals (not all in involution) may be imagined; for example, Examples 1.20 (central motion) and 1.21 (Kepler's problem). The case of resonant harmonic oscillators discussed here is the simplest one and will be particularly relevant in the study of dynamics around a stable equilibrium, in Chapter 5.

4.5 The Theorem of Poincaré

A first attempt in investigating the system (4.1) consists in looking for first integrals which are in some sense continuations of the first integrals of the unperturbed system. We first consider the case of a non-degenerate unperturbed Hamiltonian H_0. The underlying idea is that if we find a complete involution system of first integrals, then the system exhibits essentially the characteristics of the unperturbed system: the orbits lie on invariant tori which are close to the unperturbed ones. The relevant non-trivial fact is that the non-existence of first integrals in this case has been proved by Poincaré; his theorem generalizes a previous one, due to Bruns, stating that the problem of three bodies does not admit algebraic first integrals independent of the known ones of the momentum and the angular momentum.

4.5.1 Equations for a First Integral

We look for a first integral as a power expansion in the parameter ε, namely

$$\Phi(p,q,\varepsilon) = \Phi_0(p,q) + \varepsilon\Phi_1(p,q) + \varepsilon^2\Phi_2(p,q) + \ldots \ , \tag{4.25}$$

where $\Phi_0, \Phi_1, \ldots$ are analytic functions of their arguments. To this end, we try to solve the equation

$$\{H, \Phi\} = 0 \ .$$

Replacing the expansions in powers of ε for both H and Φ, and collecting all coefficients of the same power of ε we get the infinite system of equations

$$\begin{aligned} &\{H_0, \Phi_0\} = 0 \, , \\ &\{H_0, \Phi_1\} = -\{H_1, \Phi_0\} \\ &\qquad \ldots\ldots \\ &\{H_0, \Phi_s\} = -\{H_1, \Phi_{s-1}\} - \cdots - \{H_s, \Phi_0\} \\ &\qquad \ldots\ldots \ . \end{aligned} \tag{4.26}$$

Remark that the r.h.s. of the equation for Φ_s depends only on H and on $\Phi_0, \ldots, \Phi_{s-1}$, which are either known or determined by the equations of the preceding steps. Hence we may attempt a recurrent solution of the system.

The first equation simply means that Φ_0 must be a first integral of the unperturbed system. This matter has been discussed in Section 4.4.1 for anisochronous systems and in Section 4.4.2 for isochronous ones. Here the unperturbed Hamiltonian H_0 is assumed to be non degenerate, that is,

$$\det\left(\frac{\partial^2 H_0}{\partial p_j \partial p_k}\right) \neq 0 \ ,$$

so that the system is completely anisochronous. Thus, in view of Proposition 4.16, Φ_0 must be a function of $p_1, \ldots, p_n$ only.

Looking now at the equations for Φ_1, $\Phi_2, \ldots$ we see that at every step we are confronted with the problem of solving an equation (often called the *homological equation*) of the form

$$L_{H_0}\Phi_s = \Psi_s \ , \quad L_{H_0}\cdot = \{\cdot, H_0\}, \tag{4.27}$$

where $\Psi_s(p,q)$ is a known function, computed by calculating all Poisson brackets on the right. Since all functions must be periodic in the angles q, we can consider the Fourier expansion of Ψ_s in complex form, namely

$$\Psi_s(p,q) = \sum_{k\in\mathbb{Z}^n} \psi_k(p)\exp\bigl(i\langle k,q\rangle\bigr) \ ,$$

where the coefficients $\psi_k(p)$ are known and are analytic functions of the actions p. The fact that in our case $\Psi(q,p)$ is a real analytic function implies that the coefficients satisfy the condition

$$\psi_{-k}(p) = \psi_k^*(p) \ , \tag{4.28}$$

as is easily checked. Similarly, we consider the Fourier expansion of $\Phi_s(p,q)$ with unknown coefficients $\varphi_k(p)$. In view of

$$L_{H_0}\Phi_s = i\sum_{k\in\mathbb{Z}^n} \langle k,\omega(p)\rangle\varphi_k(p)\exp\bigl(i\langle k,q\rangle\bigr) \ ,$$

where $\omega(p) = \frac{\partial H_0}{\partial p}$, we get the infinite set of equations

$$i\langle k,\omega(p)\rangle\,\varphi_k(p) = \psi_k(p) \ .$$

The formal solution of such an equation is easily written as

$$\varphi_k(p) = -i\frac{\psi_k(p)}{\langle k,\omega(p)\rangle} \ ,$$

but it is valid only if the denominator $\langle k,\omega(p)\rangle$ does not vanish. Therefore, we are confronted with two problems.

(i) *Consistency:* the coefficient ψ_0 corresponding to $k = 0$, namely the average of the known function $\Psi(p,q)$ over the angles, must vanish. Although this happens in the equation for Φ_1, as is easy to see because only derivatives $\frac{\partial H_1}{\partial q_j}$ do appear in the right member, it is not evident a priori that the same will occur for the right member of all equations (4.26).

(ii) *Null divisors:* for $k \neq 0$ the divisor $\langle k,\omega(p)\rangle$ vanishes in case of resonance among the frequencies. Hence, unless the corresponding coefficient $\psi_k(p)$ is identically zero on the resonant manifold $\langle k,\omega(p)\rangle = 0$, Eq. (4.27) can be solved only in a subset of the action domain $\mathscr{G}$ which excludes the resonant manifolds.

We should remark that even if a divisor is not zero it can assume arbitrarily small values, thus raising doubts on convergence of our series expansion. This is the long-standing problem of *small divisors* which will be discussed in detail in the following chapters.

4.5.2 Nonexistence of First Integrals

The problem of small denominators is actually a major one: Poincaré proved that, under some genericness conditions, no analytic first integrals exist independent of the Hamiltonian[7] [187].

Theorem 4.21: *Let the Hamiltonian (4.1) satisfy the following hypotheses:*

(i) non-degeneracy, that is,

$$\det\left(\frac{\partial^2 H_0}{\partial p_j \partial p_k}\right) \neq 0 \; ;$$

(ii) genericness: no coefficient $h_k(p)$ of the Fourier expansion of $H_1(p,q)$ is identically zero on the manifold $\langle k, \omega(p)\rangle = 0$.

Then there is no analytic first integral independent of H.

Following Poincaré, the proof proceeds in three steps.

i. Due to the non-degeneracy condition on the unperturbed Hamiltonian, Φ_0 must be independent of the angles q.
ii. The condition that Φ be independent of H may be replaced by the condition that Φ_0 be independent of H_0.
iii. Due to the genericness condition, Φ_0 cannot be independent of H_0.

The first claim is nothing but Proposition 4.16; so, we need to prove only the second and the third claims. The proof is split into two separate lemmas.

Lemma 4.22: *If Φ_0 is independent of H_0 then Φ is independent of H. Otherwise, if Φ_0 depends on H_0 but Φ is independent of H, then one can construct a first integral $\Phi' = \Phi'_0 + \varepsilon\Phi'_1 + \ldots$ with Φ'_0 independent of H_0.*

In words, we can always ask Φ_0 independent of H, as claimed at point ii.

Proof. Let Φ_0 be independent of H_0. This means that the Jacobian matrix

$$\frac{\partial(H_0, \Phi_0)}{\partial(p_1, \ldots, p_n)}$$

has rank 2; that is, it contains a 2×2 submatrix whose determinant is not zero. Possibly, with a rearrangement of the order of the variables, we may always assume that

$$\det\begin{pmatrix} \frac{\partial H_0}{\partial p_1} & \frac{\partial H_0}{\partial p_2} \\ \frac{\partial \Phi_0}{\partial p_1} & \frac{\partial \Phi_0}{\partial p_2} \end{pmatrix} \neq 0 \; .$$

---

[7] This section is based on the classical treatise of Poincaré [188], Vol. I, chapt. V. The same proof of Poincaré's theorem can be found also in Whittaker's treatise [209], chapt. XIV, § 165.

Replacing H_0 with $H = H_0 + \varepsilon H_1 + \ldots$ and Φ_0 with $\Phi = \Phi_0 + \varepsilon\Phi_1 + \ldots$, we get the determinant

$$\det\begin{pmatrix} \frac{\partial(H_0+\varepsilon H_1+\ldots)}{\partial p_1} & \frac{\partial(H_0+\varepsilon H_1+\ldots)}{\partial p_2} \\ \frac{\partial(\Phi_0+\varepsilon\Phi_1+\ldots)}{\partial p_1} & \frac{\partial(\Phi_0+\varepsilon\Phi_1+\ldots)}{\partial p_2} \end{pmatrix},$$

which is a continuous function of ε and reduces to the previous one for $\varepsilon = 0$. Since it is not zero for $\varepsilon = 0$, by continuity it must be non-zero for $\varepsilon \neq 0$ small enough, which proves that Φ is independent of H.

Coming to the second part of the statement, let us see how to construct a first integral $\Phi' = \Phi_0' + \varepsilon\Phi_1' + \ldots$ such that Φ_0' is independent of H_0. This requires some considerations.

First we prove that $\Phi_0 = \Phi_0(H_0)$ must be an analytic function of its argument. For in view of the nondegeneracy condition, one at least of the derivatives $\frac{\partial H_0}{\partial p_j}$, with $(j = 1, \ldots, n)$, must be different from zero, and we may assume that it is $\frac{\partial H_0}{\partial p_1}$. Thus, we can invert the relation $H_0 = H_0(p)$, getting $p_1 = p_1(H_0, p_2, \ldots, p_n)$ as an analytic function of its arguments. Replacing this in Φ_0, we construct the function

$$\Phi_0(H_0, p_2, \ldots, p_n) = \Phi_0(p_1, \ldots, p_n)\Big|_{p_1 = p_1(H_0, p_2, \ldots, p_n)},$$

which is analytic, being a composition of analytic functions. On the other hand, Φ_0 was assumed to be function of H_0 only, and so it is an analytic function of H_0, as claimed.

Now we show how to replace Φ with another first integral of the same form. Let us consider the function $\Phi_0(H)$ obtained by replacing H_0 by H in the expression of $\Phi_0(H_0)$. This is clearly a first integral for H, since $\{\Phi_0(H), H\} = 0$. On the other hand it is an analytic function of H, because Φ_0 is an analytic function of its argument, and so we can expand it in power series of ε as $\Phi_0(H) = \Phi_0(H_0) + \varepsilon\tilde{\Phi}_1 + \ldots$, where $\tilde{\Phi}_1, \ldots$ are known functions. By a similar argument, the function $\Psi = \Phi - \Phi_0(H)$ also is a first integral for H, being a difference of first integrals, and is an analytic function of H, so that it can be expanded in power series of ε as $\Psi = \varepsilon\Psi_1 + \varepsilon^2\Psi_2 + \ldots$, because the first term in the expansion of $\Phi_0(H)$ is precisely $\Phi_0(H_0)$. Dividing by ε and setting $\Phi_0' = \Psi_1$, $\Phi_1' = \Psi_2, \ldots$ we get a first integral of the same form of Φ. In particular, $\Phi_0'(p_1, \ldots, p_n)$ cannot depend on the angles.

If Φ_0' so constructed is independent of H_0, we have got the desired first integral. If not, we iterate the procedure, thus finding a new first integral Φ'' of the same form as Φ', and possibly iterate the procedure again if Φ_0'' turns out to depend on H_0, and so on. If Φ is independent of H, then after a finite number of iterations we must end up with a first integral of the same form as Φ, which for $\varepsilon = 0$ is independent of H_0; this is the desired first integral.[8] *Q.E.D.*

Lemma 4.23: *With the hypotheses of Theorem 4.21 the function Φ_0 cannot be independent of H_0.*

Proof. Consider the second equation (4.26), namely

$$L_{H_0}\Phi_1 = \{H_1, \Phi_0\} .$$

The function on the right has zero average, so that the problem of consistency mentioned in Section 4.5.1 does not show up. Denoting by $h_k(p)$ the Fourier coefficients of H_1, we have

$$\{H_1, \Phi_0\} = i \sum_{k\in\mathbb{Z}^n} \left\langle k, \frac{\partial \Phi_0}{\partial p} \right\rangle h_k(p) \exp\bigl(i\langle k, q\rangle\bigr) ,$$

so that the equation for the Fourier coefficients $\varphi_k(p)$ of Φ_1 reads

$$\langle k, \omega(p)\rangle\, \varphi_k(p) = \left\langle k, \frac{\partial \Phi_0}{\partial p}(p) \right\rangle h_k(p) , \quad k \in \mathbb{Z}^n .$$

Pick a $k \neq 0$; on the resonant manifold $\langle k, \omega(p)\rangle = 0$ we can solve the preceding equation only if either condition

$$h_k(p) = 0 \quad \text{or} \quad \left\langle k, \frac{\partial \Phi_0}{\partial p}(p) \right\rangle = 0 \tag{4.29}$$

is fulfilled. Since $h_k(p) = 0$ has been excluded by the genericness hypothesis, the second condition must apply. Recall now that to every $p \in \mathscr{G}$ we can associate the resonance module $\mathscr{M}_{\omega(p)}$ and let p be a completely resonant point, that is, be such that $\dim \mathscr{M}_{\omega(p)} = n - 1$. For this value of p we have

---

[8] In the teaching experience of the author, the latter point may look mysterious; so let us expand it in some more detail. After the first step, let us write $\Phi = \Phi_0(H) + \varepsilon\Phi^{(1)}$ where $\Phi^{(1)} = \Phi_0^{(1)} + \mathcal{O}(\varepsilon)$ is a first integral with $\Phi_0^{(1)}$ as its first term. Thus, $\Phi_0^{(1)}$ must be independent of the angles, and we can check whether it is independent of H_0. If $\Phi_0^{(1)}$ is a function of H_0 then, repeating the same argument, we write $\Phi^{(2)} = \Phi^{(1)} - \Phi_0^{(1)}(H)$, where $\Phi^{(2)} = \varepsilon\Phi_0^{(2)} + \mathcal{O}(\varepsilon)$ is again an analytic first integral. Thus, we have $\Phi = \Phi_0(H) + \varepsilon\Phi_0^{(1)}(H) + \varepsilon^2\Phi^{(2)}$. We repeat this process as many times as we need, say s times, so that we have $\Phi = \Phi_0(H) + \cdots + \varepsilon^{s-1}\Phi_0^{(s-1)}(H) + \varepsilon^s\tilde{\Phi}^{(s)}$, where $\Phi^{(s)} = \Phi_0^{(s)} + \mathcal{O}(\varepsilon)$ is a first integral with its first term $\Phi_0^{(s)}$ independent of H_0, which must happen at some point. For if this does not happen, then we should let $s \to \infty$, so that we end up with the ε expansion of $\Phi = \Phi_0(H) + \varepsilon\Phi_0^{(1)}(H) + \varepsilon^2\Phi_0^{(2)}(H) + \ldots$ as a function of H only, thus contradicting our assumption that Φ is independent of H. The function $\Phi^{(s)}$ is the desired first integral.

$$\left\langle k, \frac{\partial H_0}{\partial p}(p) \right\rangle = 0 \quad \text{and} \quad \left\langle k, \frac{\partial \Phi_0}{\partial p}(p) \right\rangle = 0$$

for $n - 1$ independent vectors k, so that we conclude that every 2×2 submatrix of the Jacobian matrix

$$\frac{\partial(H_0, \Phi_0)}{\partial(p_1, \dots, p_n)} \tag{4.30}$$

has zero determinant. Now, completely resonant points are dense in $\mathbb{R}^n$, and so, by the nondegeneracy of H_0, are dense in $\mathscr{G}$. Therefore, every 2×2 submatrix of the Jacobian matrix in Eq. (4.30) has zero determinant at a set of points which is dense in $\mathscr{G}$, and so, by continuity, it is zero everywhere. This proves that Φ_0 and H_0 cannot be independent, as claimed. *Q.E.D.*

Proof of Theorem 4.21. By Lemma 4.22, if $\Phi(q,p) = \Phi_0(p) + \varepsilon\Phi_1(q,p) + \dots$ is a first integral independent of H, then there is also a first integral of the same form with Φ_0 independent of H_0. Such a first integral must satisfy the system (4.26) of equations, and so it may be constructed by recursively solving that system. But Lemma 4.23 claims that the system may be solved only if Φ_0 is a function of H_0. Thus, a first integral independent of H can not exist. *Q.E.D.*

4.6 Some Remarks on the Theorem of Poincaré

The theorem of Poincaré has sometimes been criticized by mathematicians because the hypotheses have been considered either too strong or at least too difficult to verify for a given system. This section is devoted to a short discussion concerning the condition of genericness. The condition of non-degeneration will be removed in Chapter 5, where a large portion of the discussion will be devoted to the case of isochronous unperturbed Hamiltonians.

4.6.1 On the Genericness Condition

The genericness condition plays a major role in the proof of Poincaré's theorem, but appears as a very strong limitation on the class of Hamiltonians to which the theorem applies. For checking this condition requires a thorough knowledge about the coefficients of the Fourier expansion of the Hamiltonian. For example, does the Hamiltonian of the classical problem of three bodies fulfil the genericness condition? A long discussion on different ways to replace the genericness condition with a weaker one can be found in Poincaré's work.[9]

Following Poincaré, the condition may be weakened, for example, as follows. Let us consider a partition of $\mathbb{Z}^n \setminus \{0\}$ in classes by saying that two vectors k, k' belong to the same class if they are parallel, that is, if $k = \lambda k'$ for some rational number λ. Then replace the genericness condition (ii) in Theorem 4.21 by the following:

For every class there is at least one k such that the coefficient $h_k(p)$ is not identically zero on the manifold $\langle k, \omega(p)\rangle = 0$.

This is clearly enough to conclude that the second condition in (4.29) must hold, so that the rest of the argument goes the same way.

Still, the preceding condition may be further weakened. Let $p \in \mathscr{G}$ and consider the resonance module $\mathscr{M}_{\omega(p)} \subset \mathbb{Z}^n$ associated to $\omega(p)$, as defined in (4.8). Let us say that p is a completely resonant point in case $\mathscr{M}_{\omega(p)} \subset \mathbb{Z}^n$ has multiplicity $n - 1$. Then replace the genericness condition (ii) in Theorem 4.21 by the following:

There is a dense set of completely resonant points $p \in \mathscr{G}$ satisfying the following condition: there are $n - 1$ independent vectors k in $\mathscr{M}_{\omega(p)}$ such that the coefficient $h_k(p)$ is not identically zero on the manifold $\langle k, \omega(p)\rangle = 0$.

Again, this leads us to conclude that (4.30) must hold true for $n - 1$ independent vectors in a dense subset of $\mathscr{G}$; the rest of the argument still applies.

The reader will immediately realize that the weaker conditions are not of great help in deciding whether a given Hamiltonian fulfils the hypotheses of the theorem of Poincaré. We could keep looking for weaker and weaker conditions, of course. We could also end up with the feeling that if we consider a perturbation containing only a finite number of Fourier harmonics – in fact a trigonometric polynomial – then we may escape the strong conclusions of the theorem. But remark also that all the conditions just presented are concerned only with $H_1(p, q)$, the coefficient of the first order in ε. Nothing has been said until now concerning the coefficients of the higher-order terms $H_2(p, q)$, $H_3(p, q)$, ... of the expansion in powers of ε: they could well be zero. Indeed, the proof of the theorem of Poincaré exploits only the first and the second of Eqs. (4.26). Thus, let us see what may happen at higher orders in ε.

---

[9] See [188], chapt. V, §83–85. The discussion in this section shows that the problem is a very complicated one, because it seems to suggest that there are many ways to escape the difficulties of the generic case treated by Poincaré.

4.6.2 A Puzzling Example

Consider the Hamiltonian

$$
\begin{aligned}
H(p,q,\varepsilon) &= H_0(p)+\varepsilon H_1(p,q)\ , \\
H_0(p) &= \frac{1}{2}(p_1^2+p_2^2)\ , \\
H_1(q) &= \cos q_1 + \cos(q_1-q_2) + \cos(q_1+q_2) + \cos q_2\ ,
\end{aligned}
\tag{4.31}
$$

with $p \in \mathbb{R}^2$ and $q \in \mathbb{T}^2$. It should be emphasized that the perturbation H_1 is a trigonometric polynomial of degree 2, so that the genericness condition does not apply. The aim is to show how this condition is eventually recovered, in some sense.

Let us apply the procedure of Section 4.5.1 for a first integral, starting, for example, with $\Phi_0 = p_1$. Using the exponential form for the trigonometric functions, we have

$$
\{H_1, p_1\} = \frac{i}{2}\left[\left(e^{iq_1} + e^{i(q_1-q_2)} + e^{i(q_1+q_2)}\right) - \text{c.c.}\right]\ ,
$$

where c.c. stands for the complex conjugate of the expression in parentheses, so that the resulting function is a real one. In view of $\omega_j(p) = p_j$, we determine Φ_1 as

$$
\Phi_1 = \frac{1}{2}\left[\left(\frac{e^{iq_1}}{p_1} + \frac{e^{i(q_1-q_2)}}{p_1-p_2} + \frac{e^{i(q_1+q_2)}}{p_1+p_2}\right) + \text{c.c.}\right]\ .
$$

The solution is determined up to an arbitrary function of p_1, p_2, which we choose to be zero. This is consistent, provided that we exclude from the domain of definition of Φ_1 (and so also of the first integral Φ) the straight lines $p_1 = 0$, $p_1 - p_2 = 0$ and $p_1 + p_2 = 0$.

Let us now go to the next step. We must consider the equation at order ε^2, namely $\{H_0, \Phi_2\} = \{\Phi_1, H_1\}$. Let us focus our attention on the Fourier modes which are generated. The process is illustrated in Fig. 4.4, where the Fourier modes which appear in the functions Φ_s are represented for orders $s = 1, 2, 3$, by squares of increasing size. The Poisson bracket makes the exponentials to be multiplied; hence $\{H_1, \Phi_1\}$ is a trigonometric polynomial of degree 4 which contains new Fourier modes, including in particular

$$
e^{i(2q_1-q_2)}\ ,\ e^{i(2q_1+q_2)}\ ,\ e^{i(q_1-2q_2)}\ ,\ e^{i(q_1+2q_2)}\ ,
$$

which are not multiples of the previous ones. Therefore, Φ_2 contains the new divisors $2p_1 - p_2$, $2p_1 + p_2$, $p_1 - 2p_2$, $p_1 + 2p_2$ and we must remove the additional resonant manifolds

$$
2p_1 - p_2 = 0\ ,\ 2p_1 + p_2 = 0\ ,\ p_1 - 2p_2 = 0\ ,\ p_1 + 2p_2 = 0\ .
$$

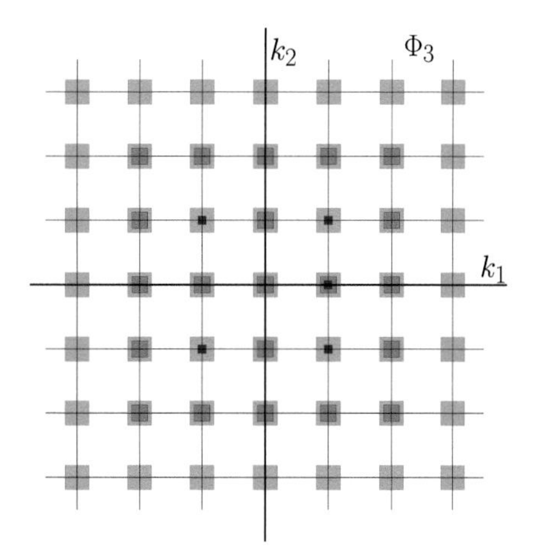

Figure 4.4 The propagation of Fourier modes through the process of construction of a first integral for the Hamiltonian (4.31). Modes from order 1 to 3 are represented with squares of increasing size and with different colours.

At order ε^3 we get the new modes represented in the figure as the largest squares. With a moment's thought, one realizes that at order ε^s the right member $\{H_1, \Phi_{s-1}\}$ contains the Fourier modes $e^{i(k_1 q_1 + k_2 q_2)}$ with $|k_1| + |k_2| \leq 2s$; a finite number, but increasing with s. It is unlikely that these terms will cancel out each other; so, it is reasonable to expect that most of them will appear. Therefore, Φ_s will be defined on a domain excluding all straight lines $k_1 p_1 + k_2 p_2 = 0$, with $|k_1| + |k_2| \leq 2s$. For $s \to +\infty$ the straight lines become dense on $\mathbb{R}^2$, so that the function Φ cannot be defined on an open domain. The unavoidable conclusion is that a first integral independent of H cannot be constructed on an open domain, even formally.

The preceding example is easily generalized. Suppose we are given a non-degenerate Hamiltonian of the form (4.1), with H_s a trigonometric polynomial of degree sK for some positive integer K. That is, H_s is a finite sum of the form

$$H_s = \sum_{|k| \leq sK} h_k^{(s)}(p) e^{i\langle k,q \rangle} , \qquad s \geq 1 .$$

We can easily verify that the r.h.s. of the equation for Φ_s is a known trigonometric polynomial of degree sK. Therefore, we are allowed to determine Φ_s in an open domain, $\mathscr{G}^{(s)}$ say, excluding (generically) all resonances of order up to sK. However, unless there are hidden mechanisms of cancellation, we are not allowed to let $s \to \infty$, because we shall be confronted again with the fact that resonances are dense. So, we see that the process of construction of first integrals is able to generate the resonant terms which do not initially appear in the Hamiltonian.

With a little extra care, the reader will notice that the process of generation of new Fourier components is fully active if the truncated Hamiltonian contains at least a set of Fourier modes k which is a basis of an n-dimensional subgroup of $\mathbb{Z}^n$.

Example 4.24: *An elementary criterion for the existence of first integrals.* Consider a Hamiltonian of the form (4.1), but with the restriction that $H_1, \ldots, H_s, \ldots$ contain only Fourier components on some subgroup $\mathscr{K} \subset \mathbb{Z}^n$, with $\dim \mathscr{K} < n$. Then for any $\alpha \in \mathbb{R}^n$ satisfying $0 \neq \alpha \perp \mathscr{K}$, the function $\Phi = \langle \alpha, p \rangle$ is a first integral. For, denoting generically by $h_k(p)$ the coefficients of the Fourier expansion of H_s, we have

$$\{H_s, \Phi\} = i \sum_{k \in \mathscr{K}} \langle \alpha, k \rangle h_k(p) \exp\bigl(i\langle k, q \rangle\bigr) \ ,$$

which is zero in view of $\langle \alpha, k \rangle = 0$. In such a case we can perform a linear canonical transformation on the angles q so that the Hamiltonian depends only on $\dim \mathscr{K}$ angles. If $\dim \mathscr{K} = 1$, then the system is actually integrable. If $\dim \mathscr{K} > 1$, then we get a reduced system with $n - \dim \mathscr{K}$ degrees of freedom, to which Poincaré's theorem generically applies. *E.D.*

4.6.3 Truncated First Integrals

The preceding examples show that, although first integrals of the form of power series do not exist in general, we can nevertheless construct *approximate* first integrals on domains excluding resonances. For instance, referring to the case of Hamiltonian (4.31), we are able to construct formal expansions for first integrals up to any finite order ε^r, provided that we restrict our attention to a domain, $\mathscr{G}^{(r)}$ say, excluding all straight lines $\langle k, p \rangle = 0$ with $|k| \leq 2r$. That is, we construct n independent functions

$$\Phi^{(j,r)}(p, q, \varepsilon) = \Phi_0^{(j)}(p) + \varepsilon \Phi_1^{(j)}(p, q) + \cdots + \varepsilon^r \Phi_r^{(j)}(p, q) \ , \tag{4.32}$$

with, for example, $\Phi_0^{(j)}(p) = p_j$ for $j = 1, \ldots, n$, which are polynomials of degree r in ε, and satisfy

$$\{H, \Phi^{(j,r)}\} = \mathcal{O}(\varepsilon^{r+1}) \ . \tag{4.33}$$

Let us call such functions *truncated first integrals.*

We can argue that truncated integrals can be very useful in view of the following argument. The time derivative of $\Phi^{(j,r)}$ is $\dot{\Phi}^{(j,r)} = \{\Phi^{(j,r)}, H\}$; for small ε such a quantity is much smaller than the time derivative of $\Phi_0^{(j)} = p_j$, the latter being of order $\mathcal{O}(\varepsilon)$. Therefore, during the evolution of the system we have the bound

$$\bigl|\Phi^{(j,r)}(t) - \Phi^{(j,r)}(0)\bigr| = \mathcal{O}(\varepsilon) \quad \text{for} \quad |t| < T_r = \mathcal{O}(\varepsilon^{-r}) \ . \tag{4.34}$$

A similar result applies also to the unperturbed actions $p_1, \ldots, p_n$ of the system. For, in view of the form (4.32) of the truncated integrals in the domain $\mathscr{G}^{(r)} \times \mathbb{T}^n$, we have

$$|p_j - \Phi^{(j,r)}| = \mathcal{O}(\varepsilon) . \tag{4.35}$$

On the other hand, we have also

$$|p_j(t)-p_j(0)| \le |p_j(t)-\Phi^{(j,r)}(t)|+|\Phi^{(j,r)}(t)-\Phi^{(j,r)}(0)|-|\Phi^{(j,r)}(0)-p_j(0)| .$$

Using (4.34) and (4.35), we conclude

$$|p_j(t) - p_j(0)| = \mathcal{O}(\varepsilon) \quad \text{for} \quad |t| < T_r = \mathcal{O}(\varepsilon^{-r}) . \tag{4.36}$$

This remark lies at the basis of the theory of *complete stability* developed by George David Birkhoff [27]; more on this in Chapter 5.

A fully rigorous development of the remarks made in this section has been worked out by Nikolai Nikolaevich Nekhoroshev [179] [180]. This will be the subject of Chapter 9. The present section should be just considered as a suggestion that Poincaré's theorem should not be taken as the ultimate and hopeless conclusion: that nothing can be said about the general problem of dynamics. Full integrability is off limits, in general, but many interesting results can nevertheless be found. What has been said here is only the beginning of a long story.

5
Nonlinear Oscillations

The aim of this chapter is to investigate the dynamics in the neighbourhood of equilibria of Hamiltonian systems, both in the linear and in the nonlinear case. The argument has strong connections with the general problem of equilibria for systems of differential equations, but the peculiar character of Hamiltonian systems deserves a separate discussion.

Let us recall some facts concerning a generic system of differential equations

$$\dot{x}_j = X_j(x_1, \ldots, x_n) \; , \quad x \in \mathscr{D} \subset \mathbb{R}^n \; ,$$

where $\mathscr{D}$ is open. Assume that this system has an equilibrium point at $(\overline{x}_1, \ldots, \overline{x}_n)$. It is well known that in a neighbourhood of the equilibrium the system may often be approximated by a linear one,

$$\dot{\mathbf{x}} = \mathsf{A}\mathbf{x}, \tag{5.1}$$

where A is a $n \times n$ real matrix with entries $A_{j,k} = \frac{\partial X_j}{\partial x_k}(\overline{x}_1, \ldots, \overline{x}_n)$. The general method for solving such a system of differential equations was discovered by Lagrange [134]; his method is still presented essentially in the same form in many books on Analysis. The complete classification of the equilibria as, for example, *nodes, saddles, foci* and *centres* is due to Poincaré [185], who used the concept of *normal form* for the system. The construction of the normal form is based on finding a suitable linear transformation of coordinates which changes the system into a particularly simple one.

The case of Hamiltonian systems may be treated with the same methods, of course. However, it is interesting to investigate to what extent the concept of normal form may be introduced without losing the canonical character of the equations. The question is given a more precise formulation as follows. Let the Hamiltonian $H(q,p)$ have an equilibrium at $(\overline{q}, \overline{p})$. In a neighbourhood of the equilibrium let us introduce local coordinates $\mathbf{x} = q - \overline{q}$, $\mathbf{y} = p - \overline{p}$. The transformation is canonical, and we can approximate the Hamiltonian with its quadratic part:

$$H(\mathbf{x}, \mathbf{y}) = \sum_{j,k} \bigl(A_{j,k} x_j x_k + B_{j,k} x_j y_k + C_{j,k} y_j y_k \bigr) \ . \tag{5.2}$$

For the contribution $H(\overline{q}, \overline{p})$ can be ignored, being a constant, the linear part vanishes at the equilibrium, and we neglect contributions of higher order. The resulting canonical equations are linear, which takes us back to the general problem of solving a linear equation. However, we may ask for a normal form of the quadratic Hamiltonian (5.2), that is, for a linear transformation to normal form which is canonical. This is the subject of the first part of the chapter.

Having settled the linear problem, we may investigate to which extent the linear approximation describes the dynamics of the complete system. In particular, the question of *stability of the equilibrium* is raised. As a general fact, the theory of Lyapounov states that in most cases the answer can be formulated by analyzing the linear system. This is indeed the case when the equilibria are *foci, nodes or saddle points* in the terminology of Poincaré. These points are also collectively called *hyperbolic equilibrium points.* However, a nontrivial exception to Lyapounov theory is represented by *centre* points, which are the most interesting ones in the Hamiltonian case. These points are also called *elliptic equilibrium points.*

Section 5.2 and the following ones until the end of the chapter are devoted to the study of the nonlinear case. The first step is concerned with the construction of formal first integrals in the neighbourhood of an elliptic equilibrium. At first sight this seems to contradict the claim of the Theorem of Poincaré that such first integrals should not exist. As a matter of fact, it means only that things are definitely more complicated than they appear; suffice it to say, for the moment, that removing the condition of non-degeneracy makes things simpler, at least in appearance. Then a section will be devoted to the so-called *numerical experiments*, as developed starting from 1952; the aim will be to recall how the insurgence of chaotic phenomena became a widely known fact in the scientific community. The final part of the chapter is devoted to discussing the asymptotic character of the formal integrals.

5.1 Normal Form for Linear Systems

In order to better appreciate the peculiarity of the Hamiltonian case, let us first recall the main results which apply to the general system (5.1). Then attention will be focused on the Hamiltonian case.

5.1.1 The Classification of Poincaré

Apply to the system (5.1) the linear transformation

$$\mathbf{x} = \mathsf{M}\boldsymbol{\xi} \ , \quad \det \mathsf{M} \neq 0 \ . \tag{5.3}$$

Then the system is changed to

$$\dot{\boldsymbol{\xi}} = \Lambda\boldsymbol{\xi} \ , \quad \Lambda = \mathsf{M}^{-1}\mathsf{A}\mathsf{M} \ .$$

Lemma 5.1: *Consider the linear system $\dot{\mathbf{x}} = \mathsf{A}\mathbf{x}$ with a real matrix A and assume that the eigenvalues $\lambda_1, \ldots, \lambda_n$ of the matrix A are distinct. Then there exists a complex matrix M such that the linear transformation $\mathbf{x} = \mathsf{M}\boldsymbol{\xi}$ with $\boldsymbol{\xi} \in \mathbb{C}^n$ gives the system the normal form*

$$\dot{\boldsymbol{\xi}} = \Lambda\boldsymbol{\xi} \ , \quad \Lambda = \mathrm{diag}(\lambda_1, \ldots, \lambda_n), \tag{5.4}$$

with $\Lambda = \mathsf{M}^{-1}\mathsf{A}\mathsf{M}$. The matrix M is written as

$$\mathsf{M} = (\mathbf{w}_1, \ldots, \mathbf{w}_n),$$

where the column vectors $\mathbf{w}_1, \ldots, \mathbf{w}_n$ are the complex eigenvectors of A. If the eigenvalues are real, then the matrix M is real, and Eq. (5.4) is restricted to $\mathbb{R}^n$.

The lemma is a standard argument, hence the proof is omitted.

Remark. The matrix M is not uniquely determined, due to the fact that the eigenvectors are determined up to a multiplicative factor. This implies that we can multiply M by a diagonal matrix with non-zero diagonal elements. A general statement is: *if two different matrices M and $\tilde{\mathsf{M}}$ satisfy the relations $\mathsf{M}^{-1}\mathsf{A}\mathsf{M} = \Lambda$ and $\tilde{\mathsf{M}}^{-1}\mathsf{A}\tilde{\mathsf{M}} = \Lambda$, then*

$$\tilde{\mathsf{M}} = \mathsf{M}\mathsf{D} \ , \quad \mathsf{D} = \mathrm{diag}(d_1, \ldots, d_n), \tag{5.5}$$

with non-vanishing $d_1, \ldots, d_n$.

If the matrix A has a pair of complex eigenvalues, then the following lemma will be useful.

Lemma 5.2: *Let $\lambda = \mu + i\omega$ with $\omega \neq 0$ be a complex eigenvalue of the real matrix A, and let $\mathbf{w} = \mathbf{u} + i\mathbf{v}$, with $\mathbf{u}$, $\mathbf{v}$ real vectors, be the corresponding eigenvector, so that $\mathsf{A}\mathbf{w} = \lambda\mathbf{w}$. Then the following statements hold true.*

(i) $\mathbf{w}^* = \mathbf{u} - i\mathbf{v}$ is the eigenvector corresponding to the eigenvalue λ^*, that is: $\mathsf{A}\mathbf{w}^* = \lambda^*\mathbf{w}^*$.

(ii) The real vectors $\mathbf{u}$ and $\mathbf{v}$ are linearly independent.

(iii) We have

$$\mathsf{A}\mathbf{u} = \mu\mathbf{u} - \omega\mathbf{v} \ , \quad \mathsf{A}\mathbf{v} = \omega\mathbf{u} + \mu\mathbf{v} \ .$$

Proof. (i) If $\mathsf{A}\mathbf{w} = \lambda\mathbf{w}$, then by calculating the complex conjugate of both members we have $(\mathsf{A}\mathbf{w})^* = (\lambda\mathbf{w})^*$, which in view of $\mathsf{A}^* = \mathsf{A}$ coincides with $\mathsf{A}\mathbf{w}^* = \lambda^*\mathbf{w}^*$. This proves that $\mathbf{w}^*$ is the eigenvector corresponding to λ^*.

(ii) We know that $\mathbf{w}$ and $\mathbf{w}^*$ are linearly independent over the complex numbers, being eigenvectors corresponding to different eigenvalues λ, λ^* (this is a known general property). By contradiction, let $\mathbf{u} = \alpha\mathbf{v}$ for some real $\alpha \neq 0$. Then we have $\mathbf{w} = \mathbf{u} + i\mathbf{v} = (\alpha + i)\mathbf{v}$ and $\mathbf{w}^* = (\alpha - i)\mathbf{v}$, and so we have also $\mathbf{w} = \frac{\alpha+i}{\alpha-i}\mathbf{w}^*$, which contradicts the linear independence of the eigenvectors over the complex numbers. We conclude that $\mathbf{u}$ and $\mathbf{v}$ must be linearly independent, as claimed.
(iii) Just calculate

$$\mathsf{A}\mathbf{u} = \frac{\lambda\mathbf{w} + \lambda^*\mathbf{w}^*}{2} = \mu\frac{\mathbf{w} + \mathbf{w}^*}{2} + i\omega\frac{\mathbf{w} - \mathbf{w}^*}{2} = \mu\mathbf{u} - \omega\mathbf{v} \ ,$$
$$\mathsf{A}\mathbf{v} = \frac{\lambda\mathbf{w} - \lambda^*\mathbf{w}^*}{2i} = \mu\frac{\mathbf{w} - \mathbf{w}^*}{2i} + i\omega\frac{\mathbf{w} + \mathbf{w}^*}{2i} = \omega\mathbf{u} + \mu\mathbf{v} \ ,$$

and the proof is complete. *Q.E.D.*

Let us now come to the flow. The system (5.4) is easily solved, being diagonal. For, writing it as $\dot{\xi}_1 = \lambda_1\xi_1 \, , \, \ldots \, , \, \dot{\xi}_n = \lambda_n\xi_n$, we have

$$\xi_1(t) = \xi_{1,0}\, e^{\lambda_1 t} \ , \ \ldots \ , \ \xi_n(t) = \xi_{n,0}\, e^{\lambda_n t} \ ,$$

$\xi_{1,0}, \ldots, \xi_{n,0}$ being the initial data. Thus solving the system (5.1) requires calculating the initial data in the new $\boldsymbol{\xi}$ variables, writing the solutions and then performing a transformation back to the original $\mathbf{x}$ variables. In more explicit terms, we should calculate $\boldsymbol{\xi}_0 = \mathsf{M}^{-1}\mathbf{x}_0$, then write the solutions as $\xi(t)$, and finally set $\mathbf{x}(t) = \mathsf{M}\boldsymbol{\xi}(t)$. Some problem may apparently arise when the eigenvalues of the matrix A are complex, but can be overcome by replacing the exponentials with trigonometric functions, as is usually explained in textbooks of Analysis.

A more compact and elegant form of the flow is obtained by introducing a linear *evolution operator* $\mathsf{U}(t)$ as follows. First note that the solution can be written in exponential form as[1] $\mathbf{x}(t) = e^{t\mathsf{A}}\mathbf{x}_0$, where $e^{t\mathsf{A}} = \sum_{k\geq 0} t^k\mathsf{A}^k/k!$. The problem is how to explicitly calculate the exponential of the matrix $t\mathsf{A}$. The answer is easily found if we accept to use complex coordinates ξ because we have

$$e^{t\Lambda} = \mathsf{I} + t\Lambda + \frac{t^2\Lambda^2}{2!} + \frac{t^3\Lambda^3}{3!} + \cdots = \mathrm{diag}\big(e^{\lambda_1 t}, \cdots, e^{\lambda_n t}\big) \ ,$$

where I is the identity matrix. This follows from $\Lambda^k = \mathrm{diag}(\lambda_1^k, \ldots, \lambda_n^k)$. For the real matrix A, using $\mathsf{A} = \mathsf{M}\Lambda\mathsf{M}^{-1}$, we have $\mathsf{A}^0 = \mathsf{I}$ and, by induction,

---

[1] A reader who is not familiar with the meaning of the exponential of a matrix may just check that the expression given here is actually the solution of the equation by formally calculating the time derivative. The question of convergence will be discussed later, in Chapter 6, in the more general context of Lie series. For a reference see, e.g., [8], § 14.

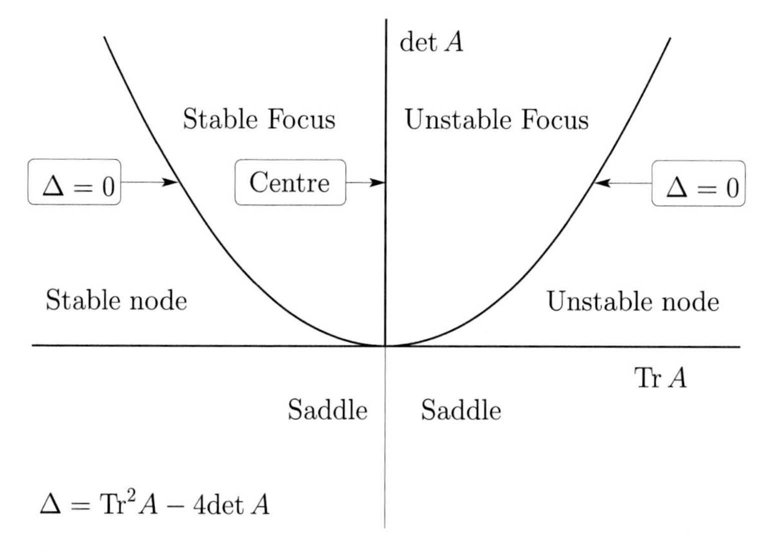

Figure 5.1 The bifurcation diagram for linear differential equations in the plane.

$$\mathsf{A}^k = \mathsf{A}\mathsf{A}^{k-1} = \left(\mathsf{M}\Lambda\mathsf{M}^{-1}\right)\left(\mathsf{M}\Lambda^{k-1}\mathsf{M}^{-1}\right) = \mathsf{M}\Lambda^k\mathsf{M}^{-1} .$$

Remark that A^k is a power of real matrix, hence it is real even if the eigenvalues are complex numbers. Thus we conclude that

$$e^{t\mathsf{A}} = \sum_{k\geq 0} \frac{t^k \mathsf{A}^k}{k!} = \sum_{k\geq 0} \frac{t^k \mathsf{M}\Lambda^k\mathsf{M}^{-1}}{k!} = \mathsf{M}\left(\sum_{k\geq 0} \frac{t^k\Lambda^k}{k!}\right)\mathsf{M}^{-1} = \mathsf{M}e^{t\Lambda}\mathsf{M}^{-1} .$$

The resulting matrix is real. This leads to the following.

Proposition 5.3: *Let the eigenvalues of the real matrix* A *be distinct. Then the flow of the linear system of differential equations* $\dot{\mathbf{x}} = \mathsf{A}\mathbf{x}$ *satisfying the initial condition* $\mathbf{x}(0) = \mathbf{x}_0$ *is*

$$\phi^t\mathbf{x}_0 = \mathsf{U}(t)\mathbf{x}_0 , \quad \mathsf{U}(t) = \mathsf{M}e^{t\Lambda}\,\mathsf{M}^{-1}, \tag{5.6}$$

where $e^{t\Lambda} = \mathrm{diag}\left(e^{\lambda_1 t}, \ldots, e^{\lambda_n t}\right)$ *is constructed using the eigenvalues* $\lambda_1, \ldots, \lambda_n$ *of* A *and* M *is as in Lemma 5.1. The evolution operator* $\mathsf{U}(t)$ *is a real matrix.*

Corollary 5.4: *If the matrix* A *is symmetric, then the eigenvalues are real and the matrix* M *can be chosen to be orthogonal, that is,* $\mathsf{M}^\top\mathsf{M} = \mathsf{I}$.

The corollary is a straightforward consequence of known properties of symmetric matrices. What is relevant here is the analogy with the Hamiltonian case, where the transformation matrix M should be symplectic.

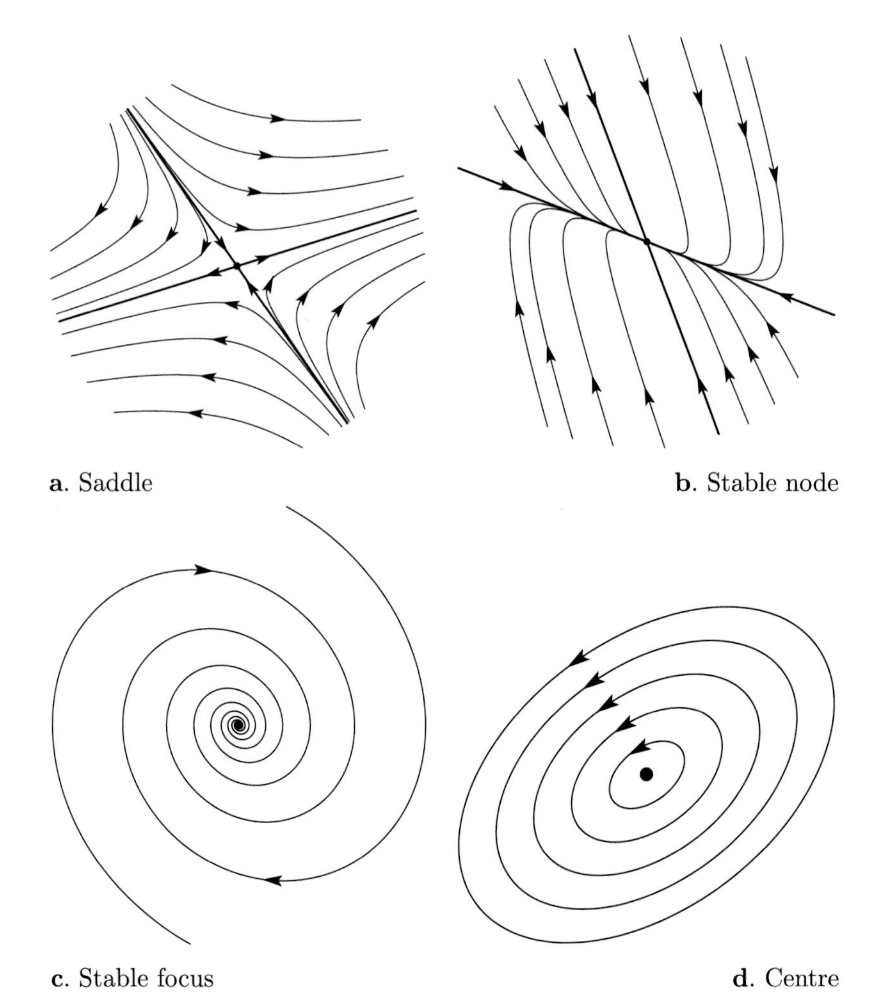

Figure 5.2 Examples of typical phase diagrams corresponding to points on the positive half of the ordinate axis (centre) and on the left half–plane (all the other cases) of the bifurcation diagram of Fig. 5.1.

The qualitative behaviour of the orbits was first investigated by Poincaré [184], paying attention to the case of the plane. The eigenvalues λ_1, λ_2 of the 2×2 matrix A are found by solving the *secular equation*

$$\lambda^2 - \lambda \operatorname{Tr} \mathsf{A} + \det \mathsf{A} = 0 \ , \tag{5.7}$$

with solutions

$$\lambda = \frac{\operatorname{Tr} \mathsf{A} \pm \sqrt{\Delta}}{2} \ , \quad \Delta = \operatorname{Tr}^2 \mathsf{A} - 4 \det \mathsf{A} \ .$$

The classification of Poincaré in the plane case is summarized in the *bifurcation diagram* of Fig. 5.1. Taking $\mathrm{Tr}\,\mathsf{A}$ as the abscissa and $\det \mathsf{A}$ as the ordinate, the plane is divided into different regions separated by the abscissa axis $\det \mathsf{A} = 0$, the positive ordinate half axis $\mathrm{Tr}\,\mathsf{A} = 0$ and the parabola $\Delta = 0$. The phase portrait for the main cases is represented in Fig. 5.2; see, for example, [192] for a detailed discussion. The following cases may occur.

(i) A *saddle point* (Fig. 5.2 a) if the eigenvalues are real and with different signs, which occurs if $\Delta > 0$ and $\det \mathsf{A} < 0$, namely the lower half plane.

(ii) A *node* (Fig. 5.2 b) if the eigenvalues are real with the same sign, which occurs if $\Delta > 0$ and $\det \mathsf{A} > 0$, namely the part of the upper half plane lying below the parabola. The node is *stable* if the eigenvalues are negative, that is, if $\mathrm{Tr}\,\mathsf{A} < 0$, and *unstable* if they are positive, that is, if $\mathrm{Tr}\,\mathsf{A} > 0$.

(iii) A *degenerate node* if the eigenvalues are real and coincident, which occurs if $\Delta = 0$, namely on the parabola. A particular degenerate case occurs when $\mathsf{A} = \mathrm{diag}(\lambda, \lambda)$ is a diagonal matrix (*star node*). The node may be stable or unstable as in the case (ii).

(iv) A *focus* (Fig. 5.2 c) if the eigenvalues are complex conjugate with non-vanishing real part, which occurs if $\mathrm{Tr}\,\mathsf{A} \neq 0$ and $\Delta < 0$, namely the part of the upper half plane lying above the parabola. The focus is *stable* (resp. *unstable*) if the real part of the eigenvalues is negative (resp. positive), which depends on the sign of $\mathrm{Tr}\,\mathsf{A}$ being negative (resp. positive).

(v) A *centre* (Fig. 5.2 d) if the eigenvalues are pure imaginary, which occurs if $\mathrm{Tr}\,\mathsf{A} = 0$ and $\det \mathsf{A} > 0$.

(vi) Degenerate cases on the abscissa axis, except the origin, when one of the eigenvalues is zero. At the origin both eigenvalues are zero.

5.1.2 Normal Form of a Quadratic Hamiltonian

Let us come to the Hamiltonian (5.2). As already pointed out, the problem is to look for a normal form of the linear system of Hamilton's equations keeping the canonical form (i.e., for a quadratic normal form of the Hamiltonian).[2]

---

[2] This is a common topic in textbooks. One is usually led to study a Lagrangian $L(\dot{q}, q) = T(\dot{q}) - V(q)$ where both the kinetic energy T and the potential energy V are quadratic forms in the generalized velocities $\dot{q}$ and in the generalized coordinates q, respectively. In this case the problem reduces to the known one of simultaneous diagonalization of two quadratic forms. The Hamiltonian case is more general because the quadratic Hamiltonian is allowed to include mixed terms in coordinates and momenta. Moreover, we want all transformations to

It is convenient to simplify the notation by using the compact formalism. Thus, denote $\mathbf{z} = (x_1, \ldots, x_n, y_1, \ldots, y_n)^\top$, a column vector, and write the Hamiltonian as a quadratic form,

$$H(z) = \frac{1}{2}\mathbf{z}^\top \mathsf{C}\,\mathbf{z}\,, \quad \mathsf{C}^\top = \mathsf{C}, \tag{5.8}$$

where C is a $2n \times 2n$ real, symmetric and non-degenerate matrix. The corresponding Hamilton's equations are

$$\dot{\mathbf{z}} = \mathsf{J}\mathsf{C}\,\mathbf{z}\,, \quad \mathsf{J} = \begin{pmatrix} 0 & \mathsf{I} \\ -\mathsf{I} & 0 \end{pmatrix}, \tag{5.9}$$

where J is the $2n \times 2n$ antisymmetric matrix defining the symplectic form. Recall that we have $\mathsf{J}^\top = \mathsf{J}^{-1} = -\mathsf{J}$.

As in Section 5.1, let us look for a linear transformation $\mathbf{z} = \mathsf{M}\boldsymbol{\zeta}$ which gives the linear system a diagonal form. We know that if the eigenvalues of the matrix JC are distinct, then the matrix M can be determined, and so we have

$$\mathsf{M}^{-1}\mathsf{JCM} = \Lambda\,. \tag{5.10}$$

The problem is whether the transformation matrix M can be restricted to be symplectic, that is, to satisfy the condition

$$\mathsf{M}^\top \mathsf{JM} = \mathsf{J}\,. \tag{5.11}$$

5.1.3 *The Linear Canonical Transformation*

Let us investigate the properties of the eigenvalues of the matrix JC, recalling that C is assumed to be symmetric.

Lemma 5.5: *The eigenvalues of the matrix* JC *satisfy the following properties:*

(i) if λ is an eigenvalue, so is its complex conjugate λ^*;

(ii) if λ is an eigenvalue, so is $-\lambda$.

That is, the eigenvalues of JC can be organized either in pairs, when they are real or pure imaginary, or in groups of four, as illustrated in Fig. 5.3.

Proof. (i) The characteristic polynomial of a real matrix has real coefficients. Thus, its roots are either real or pairs of complex conjugate numbers.

---

be canonical. Following [205], § 15, the discussion here is limited to the most common case of distinct eigenvalues. For the general case of eigenvalues with multiplicity greater than one see [9] or [210], where results from [211] are used.

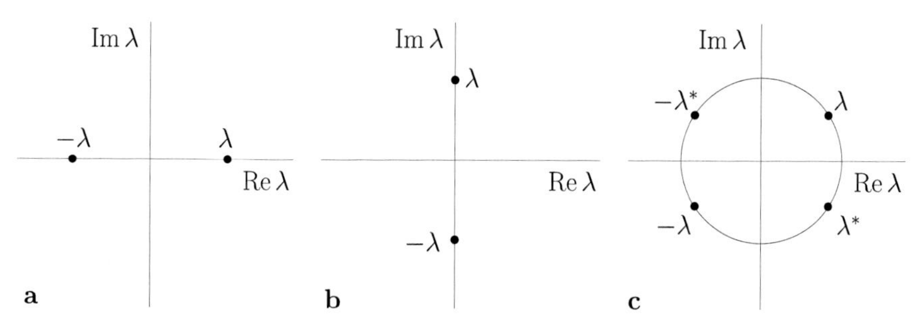

Figure 5.3 The eigenvalues of the matrix JC with C symmetric. **a.** A pair of real eigenvalues, so that the complex conjugates are $\lambda^* = \lambda$ and $-\lambda^* = -\lambda$. **b.** A pair of pure imaginary eigenvalues, namely λ, $-\lambda$; we have $\lambda^* = -\lambda$ and $-\lambda^* = \lambda$. **c.** Four distinct complex eigenvalues λ, $-\lambda$, λ^*, $-\lambda^*$ with real and imaginary parts both non-zero.

(ii) Consider the characteristic equation $\det(\mathsf{JC} - \lambda\mathsf{I}) = 0$ and note that

$$\begin{aligned}(\mathsf{JC} - \lambda\mathsf{I})^\top &= (\mathsf{CJ}^\top - \lambda\mathsf{I}) = -(\mathsf{CJ} + \lambda\mathsf{I}) \\ &= -\mathsf{JJ}^{-1}(\mathsf{CJ} + \lambda\mathsf{I}) = -\mathsf{J}(\mathsf{J}^{-1}\mathsf{CJ} + \mathsf{J}^{-1}\lambda\mathsf{I}) \\ &= \mathsf{J}(\mathsf{JCJ} + \lambda\mathsf{IJ}) = \mathsf{J}(\mathsf{JC} + \lambda\mathsf{I})\mathsf{J}\ .\end{aligned}$$

This shows that

$$\det(\mathsf{JC} - \lambda\mathsf{I}) = \det(\mathsf{JC} + \lambda\mathsf{I})\ ,$$

that is, the characteristic polynomial is symmetric in λ; hence its roots are in pairs $\lambda, -\lambda$, as claimed. *Q.E.D.*

The next lemma may be seen as stating the analogy between symmetric matrices, which have orthogonal eigenvectors, and matrices of the form JC, which have eigenvectors possessing a natural symplectic structure.

Lemma 5.6: *Let the eigenvalues of the matrix* JC *be distinct, and let* λ, λ' *be two eigenvalues corresponding to the eigenvectors* $\mathbf{w}$, $\mathbf{w}'$, *respectively. Then we have*

$$\mathbf{w}^\top \mathsf{J}\,\mathbf{w}' \neq 0 \quad \text{if and only if} \quad \lambda' = -\lambda\ .$$

Proof. Let us show that we have

$$(\lambda + \lambda')\mathbf{w}^\top \mathsf{J}\mathbf{w}' = 0\ .$$

For, in view of $\mathsf{J}^2 = -\mathsf{I}$ and $\mathsf{J}^\top\mathsf{J} = \mathsf{I}$, with a short calculation we get

$$\begin{aligned}(\lambda + \lambda')\mathbf{w}^\top \mathsf{J}\mathbf{w}' &= (\mathsf{JC}\mathbf{w})^\top \mathsf{J}\mathbf{w}' + \mathbf{w}^\top \mathsf{J}(\mathsf{JC}\mathbf{w}') \\ &= \mathbf{w}^\top \mathsf{CJ}^\top \mathsf{J}\mathbf{w}' + \mathbf{w}^\top \mathsf{CJ}^2\mathbf{w}' = \mathbf{w}^\top \mathsf{C}\mathbf{w}' - \mathbf{w}^\top \mathsf{C}\mathbf{w}' = 0\ .\end{aligned}$$

If $\lambda + \lambda' \neq 0$, then we have $\mathbf{w}^\top \mathsf{J}\mathbf{w}' = 0$. If $\lambda + \lambda' = 0$, we must prove that $\mathbf{w}^\top \mathsf{J}\mathbf{w}' \neq 0$, so that $\mathbf{w}$ is symplectic orthogonal to all eigenvectors but $\mathbf{w}'$. By contradiction, assume that also $\mathbf{w}^\top \mathsf{J}\mathbf{w}' = 0$. This would imply that $\mathbf{w}^\top \mathsf{J}\mathbf{w}' = 0$ for all eigenvectors of the matrix JC. Since the eigenvectors are a basis, in view of the non-degeneration of the symplectic form this would imply that $\mathbf{w} = 0$, contradicting the fact that $\mathbf{w}$ is itself an element of the basis. *Q.E.D.*

We now should normalize the eigenvectors so as to be able to construct a symplectic matrix. In view of Lemma 5.5, we can arrange the eigenvalues and the eigenvectors of the matrix JC in the suitable ordering

$$
\begin{array}{cccccc}
\lambda_1\,, & \ldots, & \lambda_n\,, & \lambda_{n+1} = -\lambda_1\,, & \ldots, & \lambda_{2n} = -\lambda_n, \\
\mathbf{w}_1\,, & \ldots, & \mathbf{w}_n\,, & \mathbf{w}_{n+1}\,, & \ldots, & \mathbf{w}_{2n}.
\end{array}
\tag{5.12}
$$

The n quantities

$$
d_j = \mathbf{w}_j^\top \mathsf{J}\mathbf{w}_{j+n}\,, \quad j = 1, \ldots, n
$$

are non-zero in view of Lemma 5.6. Let us construct the matrix

$$
\mathsf{M} = (\mathbf{w}_1/d_1, \ldots, \mathbf{w}_n/d_n,\, \mathbf{w}_{n+1}, \ldots, \mathbf{w}_{2n})
\tag{5.13}
$$

by arranging the eigenvectors in columns in the chosen order.

Lemma 5.7: *The matrix* M *constructed as in (5.13) is symplectic.*

Proof. We must prove that $\mathsf{M}^\top \mathsf{J}\mathsf{M} = \mathsf{J}$. To this end, let us denote its elements by $a_{j,k}$. Then in view of Lemma 5.6 and of the definition of d_j, we have

$$
a_{j,j+n} = \frac{1}{d_j}\mathbf{w}_j^\top \mathsf{J}\mathbf{w}_{j+n} = 1\,, \quad a_{j+n,j} = \frac{1}{d_j}\mathbf{w}_{j+n}^\top \mathsf{J}\mathbf{w}_j = -1\,,
$$

and $a_{j,k} = 0$ in all other cases. *Q.E.D.*

We have thus proven the following.

Proposition 5.8: *The linear transformation* $\mathbf{z} = \mathsf{M}\boldsymbol{\zeta}$ *with the matrix* M *constructed as in (5.13) is canonical and changes the system of linear differential equations (5.9) into its diagonal form*

$$
\dot{\boldsymbol{\zeta}} = \Lambda\boldsymbol{\zeta}\,, \quad \Lambda = \mathrm{diag}(\lambda_1, \ldots, \lambda_n, -\lambda_1, \ldots, -\lambda_n)\,.
$$

The diagonalizing transformation is not unique, for we can multiply it by a symplectic diagonal matrix $\mathsf{R} = \mathrm{diag}(r_1, \ldots, r_{2n})$. This is seen as follows. In view of the remark made after Lemma 5.1, the matrix $\tilde{\mathsf{M}} = \mathsf{MR}$ still gives the system a diagonal form. Thus we should only check that $\tilde{\mathsf{M}}$ is still symplectic, which is true because symplectic matrices form a group.

Let us look for the general form of a diagonal symplectic matrix R. To this end, let us write it in the convenient form

$$\mathsf{R} = \begin{pmatrix} \mathsf{R}_0 & 0 \\ 0 & \mathsf{R}_1 \end{pmatrix} , \quad \mathsf{R}_0 = \mathrm{diag}(r_1, \ldots, r_n) \, , \ \mathsf{R}_1 = \mathrm{diag}(r'_1, \ldots, r'_n) \, .$$

Writing the symplecticity condition, we have

$$\begin{aligned} \begin{pmatrix} \mathsf{R}_0 & 0 \\ 0 & \mathsf{R}_1 \end{pmatrix} \begin{pmatrix} 0 & \mathsf{I} \\ -\mathsf{I} & 0 \end{pmatrix} \begin{pmatrix} \mathsf{R}_0 & 0 \\ 0 & \mathsf{R}_1 \end{pmatrix} &= \begin{pmatrix} \mathsf{R}_0 & 0 \\ 0 & \mathsf{R}_1 \end{pmatrix} \begin{pmatrix} 0 & \mathsf{R}_1 \\ -\mathsf{R}_0 & 0 \end{pmatrix} \\ &= \begin{pmatrix} 0 & \mathsf{R}_0\mathsf{R}_1 \\ -\mathsf{R}_0\mathsf{R}_1 & 0 \end{pmatrix} = \begin{pmatrix} 0 & \mathsf{I} \\ -\mathsf{I} & 0 \end{pmatrix} . \end{aligned}$$

The last line gives us the condition $\mathsf{R}_0\mathsf{R}_1 = \mathsf{I}$. We conclude that *the symplectic matrix* M *which takes the linear canonical system in diagonal form may be replaced with the symplectic matrix* MR*, where* R *is a non-degenerate diagonal matrix of the form*

$$\mathsf{R} = \begin{pmatrix} \mathsf{R}_0 & 0 \\ 0 & \mathsf{R}_0^{-1} \end{pmatrix} , \quad \mathsf{R}_0 = \mathrm{diag}(r_1, \ldots, r_n) \, . \tag{5.14}$$

5.1.4 Complex Normal Form of the Hamiltonian

Let us now come to the transformation of the Hamiltonian. Here it will be convenient to use the compact notation $\boldsymbol{\zeta} = (\boldsymbol{\xi}, \boldsymbol{\eta})$ where $\boldsymbol{\xi} = (\xi_1, \ldots, \xi_n) \in \mathbb{C}^n$ are the coordinates and $\boldsymbol{\eta} = (\eta_1, \ldots, \eta_n) \in \mathbb{C}^n$ are the corresponding conjugate momenta. Furthermore, recalling the ordering (5.12) of the eigenvalues, it will be convenient to write the diagonal matrix Λ as

$$\Lambda = \begin{pmatrix} \Lambda_0 & 0 \\ 0 & -\Lambda_0 \end{pmatrix} , \quad \Lambda_0 = \mathrm{diag}(\lambda_1, \ldots, \lambda_n) \, . \tag{5.15}$$

Proposition 5.9: *With the same hypotheses as in Proposition 5.8, the linear transformation generated by the matrix* M *defined by (5.13) gives the quadratic Hamiltonian (5.8) the form*

$$H(\boldsymbol{\xi}, \boldsymbol{\eta}) = \sum_{j=1}^{n} \lambda_j \xi_j \eta_j \, . \tag{5.16}$$

Proof. By transforming the quadratic form $\frac{1}{2}\mathbf{z}^\top \mathsf{C}\mathbf{z}$, we get $\frac{1}{2}\boldsymbol{\zeta}^\top \mathsf{M}^\top \mathsf{C}\mathsf{M}\boldsymbol{\zeta}$. Recalling that $\mathsf{M}^{-1}\mathsf{J}\mathsf{C}\mathsf{M} = \Lambda$, we calculate

$$\mathsf{M}^\top\mathsf{C}\mathsf{M} = -(\mathsf{M}^\top\mathsf{J})\mathsf{J}\mathsf{C}\mathsf{M} = -(\mathsf{M}^\top\mathsf{J}\mathsf{M})(\mathsf{M}^{-1}\mathsf{J}\mathsf{C}\mathsf{M}) \, , = -\mathsf{J}\Lambda = \begin{pmatrix} 0 & \Lambda_0 \\ \Lambda_0 & 0 \end{pmatrix} .$$

The transformed Hamiltonian is $H(\boldsymbol{\zeta}) = -\frac{1}{2}\boldsymbol{\zeta}^\top \mathsf{J}\Lambda\boldsymbol{\zeta}$, namely (5.16) in complex coordinates $(\boldsymbol{\xi}, \boldsymbol{\eta})$. *Q.E.D.*

5.1.5 First Integrals

The normalized Hamiltonian (5.16) possesses the complete involution system of first integrals

$$\Psi_j(\xi_j, \eta_j) = \xi_j \eta_j \ , \quad j = 1, \ldots, n \ . \tag{5.17}$$

This is easily checked by a straightforward calculation of Poisson brackets $\{\Psi_j, H\}$. Thus the system turns out to be integrable in Liouville's sense.[3] Further first integrals may be found by looking for a solution of the equation $\{\Phi, H\} = 0$, namely

$$\sum_{l=1}^{n} \lambda_l \left(\xi_l \frac{\partial \Phi}{\partial \xi_l} - \eta_l \frac{\partial \Phi}{\partial \eta_l} \right) = 0 \ .$$

Choosing $\Phi = \xi^j \eta^k = \xi_1^{j_1} \cdots \xi_n^{j_n} \eta_1^{k_1} \cdots \eta_n^{k_n}$ (in multi-index notation), we have

$$\langle k - j, \lambda \rangle \xi^j \eta^k = 0 \ ,$$

which can be satisfied only if $\langle k - j, \lambda \rangle = 0$. Thus, further first integrals independent of those in (5.17) do exist only if the eigenvalues λ satisfy a resonance condition. This is indeed the same as Proposition 4.19, specialized to the case of quadratic Hamiltonians in complex variables.

The canonical equations for the Hamiltonian (5.17) are

$$\dot{\xi}_j = \lambda_j \xi_j \ , \quad \dot{\eta}_j = -\lambda_j \eta_j \ , \tag{5.18}$$

and the corresponding solutions with initial datum $\xi_{j,0}, \eta_{j,0}$ are

$$\xi_j(t) = \xi_{j,0} e^{\lambda_j t} \ , \quad \eta_j(t) = \eta_{j,0} e^{-\lambda_j t} \ .$$

This is straightforward, but the structure of the orbits in the real phase space remains hidden until we perform all substitutions back to the original variables, or we write in explicit form the evolution operator $\mathsf{U}(t) = \mathsf{M} e^{t\Lambda} \mathsf{M}^{-1}$ as in Proposition 5.3.

More generally, the drawback of the present section is that the functions so determined have, in general, complex values, while it seems often better to write both the Hamiltonian and the first integrals in real variables. Thus, let us proceed by looking for a normal form in real variables.

5.1.6 The Case of Real Eigenvalues

If the eigenvalues of the matrix JC are real and distinct, then the whole normalization procedure is performed without involving complex quantities.

---

[3] Integrability in the Arnold–Jost sense is not assured because the invariant surfaces determined by the first integrals may be noncompact.

Thus, considering for simplicity the case of a system with one degree of freedom, the normal form of the Hamiltonian is

$$H(\xi, \eta) = \lambda \xi \eta \;, \quad (\xi, \eta) \in \mathbb{R}^2 \;, \tag{5.19}$$

where λ, $-\lambda$ are the eigenvalues of the matrix JC. The equilibrium is a *saddle* and is said to be *hyperbolic.*

It may be interesting, however, to remark that one can also use as a paradigm Hamiltonian the form

$$H(x, y) = \frac{\lambda}{2}(y^2 - x^2) \;. \tag{5.20}$$

This is a typical form of a Hamiltonian describing the motion of a particle on a one-dimensional manifold.

Exercise 5.10: Find a canonical transformation which changes the Hamiltonian (5.20) into (5.19). *A.E.L.*

5.1.7 The Case of Pure Imaginary Eigenvalues

Let us consider again the case of a system with one degree of freedom; we show that the Hamiltonian may be given the normal form

$$H(x, y) = \frac{\omega}{2}(y^2 + x^2) \;, \tag{5.21}$$

which is immediately recognized as the Hamiltonian of a harmonic oscillator.

Let the eigenvalues of the matrix JC be pure imaginary, $\lambda = \pm i\omega$; the equilibrium is a *centre* and is said to be *elliptic.* Then the corresponding eigenvectors are complex conjugated, that is, they may be written as $\mathbf{w} = \mathbf{u} + i\mathbf{v}$ and $\mathbf{w}^* = \mathbf{u} - i\mathbf{v}$. Recalling that the symplectic form is anticommutative, we have

$$\begin{aligned}\mathbf{w}^\top \mathsf{J}\mathbf{w}^* &= (\mathbf{u} + i\mathbf{v})^\top \mathsf{J}(\mathbf{u} - i\mathbf{v}) \\ &= i(-\mathbf{u}^\top \mathsf{J}\mathbf{v} + \mathbf{v}^\top \mathsf{J}\mathbf{u}) = -2i\mathbf{u}^\top \mathsf{J}\mathbf{v} \;,\end{aligned}$$

which is a pure imaginary quantity. We may always order the eigenvalues and the eigenvectors as in (5.12) with the further requirement:

$$d = \mathbf{u}^\top \mathsf{J}\mathbf{v} > 0 \;. \tag{5.22}$$

Such a condition may just require exchanging a pair $(\lambda, \mathbf{w})$ with its conjugate $(-\lambda, \mathbf{w}^*)$. More precisely, we may create the association

$$i\omega \leftrightarrow \mathbf{u} \;, \qquad -i\omega \leftrightarrow \mathbf{v}$$

by choosing the sign of ω so that the condition $d > 0$ is satisfied.

Remark. The sign of ω in the normal form (5.21) depends precisely on condition (5.22), which is dictated by the eigenvectors of the matrix JC: there is no arbitrariness here.

Let us now construct the transformation matrix:

$$\mathsf{T} = (\mathbf{u}/\sqrt{d}\,,\ \mathbf{v}/\sqrt{d})\ . \tag{5.23}$$

This is a symplectic matrix, as is checked by calculating

$$\mathsf{T}^\top \mathsf{J}\mathsf{T} = \frac{1}{d}\begin{pmatrix} \mathbf{u}^\top \mathsf{J}\mathbf{u} & \mathbf{u}^\top \mathsf{J}\mathbf{v} \\ \mathbf{v}^\top \mathsf{J}\mathbf{u} & \mathbf{v}^\top \mathsf{J}\mathbf{v} \end{pmatrix} = \begin{pmatrix} 0 & 1 \\ -1 & 0 \end{pmatrix}\ .$$

By property (iii) of Lemma 5.2, we have

$$\mathsf{J}\mathsf{C}\mathbf{u} = -\omega\mathbf{v}\ ,\quad \mathsf{J}\mathsf{C}\mathbf{v} = \omega\mathbf{u}\ ,$$

and we can write this relation for the columns of the matrix T, thus getting

$$\mathsf{J}\mathsf{C}\mathsf{T} = \mathsf{T}\begin{pmatrix} 0 & \omega \\ -\omega & 0 \end{pmatrix}\ ,$$

that is,

$$\mathsf{T}^{-1}\mathsf{J}\mathsf{C}\mathsf{T} = \mathsf{J}\begin{pmatrix} \omega & 0 \\ 0 & \omega \end{pmatrix}\ .$$

Thus, the equations take the form

$$\dot{x} = \omega y\ ,\quad \dot{y} = -\omega x\ ,$$

and the transformed Hamiltonian is (5.21).

Exercise 5.11: The Hamiltonian of a harmonic oscillator in complex variables is $H(\xi,\eta) = i\omega\xi\eta$. Find a canonical transformation which changes the Hamiltonian (5.21) into this one. *A.E.L.*

5.1.8 The Case of Complex Conjugate Eigenvalues

The simplest case is a system with two degrees of freedom. A good paradigmatic Hamiltonian is

$$H = \mu x_1 y_1 + \mu x_2 y_2 + \omega(x_1 y_2 - x_2 y_1), \tag{5.24}$$

where μ and ω are real parameters. The desired form of the Hamiltonian is found by constructing a real symplectic basis. The equilibrium is a double *focus* and is said to be *hyperbolic*.

Write the four eigenvalues and the corresponding eigenvectors of the matrix $\mathsf{J}\mathsf{C}$ as

$$\begin{aligned} \lambda &= \mu + i\omega\ , & \mathbf{w}_+ &= \mathbf{u}_+ + i\mathbf{v}_+\ , \\ \lambda^* &= \mu - i\omega\ , & \mathbf{w}_+^* &= \mathbf{u}_+ - i\mathbf{v}_+\ , \\ -\lambda &= -\mu + i\omega\ , & \mathbf{w}_- &= \mathbf{u}_- + i\mathbf{v}_-\ , \\ -\lambda^* &= -\mu - i\omega\ , & \mathbf{w}_-^* &= \mathbf{u}_- - i\mathbf{v}_-\ , \end{aligned}$$

the alignment in rows reflecting the correspondence. By Lemma 5.6 we know that $\mathbf{w}_+^\top \mathsf{J}\mathbf{w}_- \neq 0$ and ${\mathbf{w}_+^*}^\top \mathsf{J}\mathbf{w}_-^* \neq 0$, and we can always arrange it so that the vectors $\mathbf{w}_+$ and $\mathbf{w}_-$ satisfy[4]

$$\mathbf{w}_+^\top \mathsf{J}\mathbf{w}_- + {\mathbf{w}_+^*}^\top \mathsf{J}\mathbf{w}_-^* = 0\ . \tag{5.25}$$

Still, by Lemma 5.6 we know also that, omitting four obvious relations, we have

$$\begin{aligned} \mathbf{w}_+^\top \mathsf{J}\mathbf{w}_+^* &= 0\ , & \mathbf{w}_+^\top \mathsf{J}\mathbf{w}_-^* &= 0\ , & \mathbf{w}_-^\top \mathsf{J}\mathbf{w}_-^* &= 0\ ,\\ {\mathbf{w}_+^*}^\top \mathsf{J}\mathbf{w}_+ &= 0\ , & {\mathbf{w}_+^*}^\top \mathsf{J}\mathbf{w}_- &= 0\ , & {\mathbf{w}_-^*}^\top \mathsf{J}\mathbf{w}_- &= 0\ . \end{aligned} \tag{5.26}$$

Lemma 5.12: *With the condition (5.25),* $\mathbf{u}_+$, $\mathbf{u}_-$, $\mathbf{v}_+$, $\mathbf{v}_-$ *satisfy*

$$\mathbf{u}_+^\top \mathsf{J}\mathbf{v}_- = \mathbf{u}_-^\top \mathsf{J}\mathbf{v}_+ \neq 0 \tag{5.27}$$

and

$$\mathbf{u}_+^\top \mathsf{J}\mathbf{v}_+ = \mathbf{u}_-^\top \mathsf{J}\mathbf{v}_- = \mathbf{u}_+^\top \mathsf{J}\mathbf{u}_- = \mathbf{v}_+^\top \mathsf{J}\mathbf{v}_- = 0\ . \tag{5.28}$$

This means that $\mathbf{u}_+$, $\mathbf{u}_-$, $\mathbf{v}_-$, $\mathbf{v}_+$, in this order, form a symplectic basis, still to be normalized.

Proof. From $\mathbf{w}_+^\top \mathsf{J}\mathbf{w}_+^* = 0$, we get

$$(\mathbf{u}_+ + i\mathbf{v}_+)^\top \mathsf{J}(\mathbf{u}_+ - i\mathbf{v}_+) = -2i\mathbf{u}_+^\top \mathsf{J}\mathbf{v}_+ = 0\ .$$

With a similar calculation, from $\mathbf{w}_-^\top \mathsf{J}\mathbf{w}_-^* = 0$ we also get $\mathbf{u}_-^\top \mathsf{J}\mathbf{v}_- = 0$. Thus, two of the relations (5.28) are proven.
Using $\mathbf{w}_+^\top \mathsf{J}\mathbf{w}_-^* = 0$, we get

$$(\mathbf{u}_+ + i\mathbf{v}_+)^\top \mathsf{J}(\mathbf{u}_- - i\mathbf{v}_-) = \bigl(\mathbf{u}_+^\top \mathsf{J}\mathbf{u}_- + \mathbf{v}_+^\top \mathsf{J}\mathbf{v}_-\bigr) - i\bigl(\mathbf{u}_+^\top \mathsf{J}\mathbf{v}_- - \mathbf{v}_+^\top \mathsf{J}\mathbf{u}_-\bigr) = 0\ ,$$

and setting separately to zero the real and the imaginary parts, we get

$$\mathbf{u}_+^\top \mathsf{J}\mathbf{u}_- + \mathbf{v}_+^\top \mathsf{J}\mathbf{v}_-\ , \quad \mathbf{u}_+^\top \mathsf{J}\mathbf{v}_- - \mathbf{v}_+^\top \mathsf{J}\mathbf{u}_- = 0\ . \tag{5.29}$$

Now calculate

$$\begin{aligned} \mathbf{w}_+^\top \mathsf{J}\mathbf{w}_- &= (\mathbf{u}_+ + i\mathbf{v}_+)^\top \mathsf{J}(\mathbf{u}_- + i\mathbf{v}_-)\\ &= \bigl(\mathbf{u}_+^\top \mathsf{J}\mathbf{u}_- - \mathbf{v}_+^\top \mathsf{J}\mathbf{v}_-\bigr) + i\bigl(\mathbf{u}_+^\top \mathsf{J}\mathbf{v}_- + \mathbf{u}_-^\top \mathsf{J}\mathbf{v}_+\bigr)\ . \end{aligned} \tag{5.30}$$

In view of (5.29) we get

$$\mathbf{w}_+^\top \mathsf{J}\mathbf{w}_- = 2\mathbf{u}_+^\top \mathsf{J}\mathbf{u}_- + 2i\mathbf{u}_+^\top \mathsf{J}\mathbf{v}_-.$$

---

[4] For instance, if $\mathbf{w}_+^\top \mathsf{J}\mathbf{w}_- = \varrho e^{i\vartheta}$, it is enough to replace $\mathbf{w}_+$ with $ie^{-i\vartheta}\mathbf{w}_+$, with the corresponding change for $\mathbf{w}_+^*$.

In view of the property (5.25), the left member is a pure imaginary quantity, and since we know that $\mathbf{w}_+^\top \mathsf{J}\mathbf{w}_- \neq 0$, we get

$$\mathbf{u}_+^\top \mathsf{J}\mathbf{u}_- = 0 \ , \quad \mathbf{u}_+^\top \mathsf{J}\mathbf{v}_- \neq 0 \ .$$

By (5.29) we also get

$$\mathbf{v}_+^\top \mathsf{J}\mathbf{v}_- = 0 \ , \quad \mathbf{v}_+^\top \mathsf{J}\mathbf{u}_- \neq 0 \ .$$

Furthermore, by the second part of (5.29) we have $\mathbf{u}_+^\top \mathsf{J}\mathbf{v}_- = \mathbf{v}_+^\top \mathsf{J}\mathbf{u}_-$, which completes the proof. *Q.E.D.*

We are now ready to construct the matrix T which gives the system the normal form (5.24) by suitably adapting the scheme of Section 5.1.7. Since $\mathbf{u}_+^\top \mathsf{J}\mathbf{v}_-$ is a real quantity, we can always arrange it so that

$$\mathbf{u}_+^\top \mathsf{J}\mathbf{v}_- = \mathbf{v}_+^\top \mathsf{J}\mathbf{u}_- = d > 0 \ . \tag{5.31}$$

For, if d is negative, it is enough to exchange $\mathbf{u}_+$ with $\mathbf{u}_-$ and $\mathbf{v}_+$ with $\mathbf{v}_-$. Then we construct the 4×4 matrix

$$\mathsf{T} = \left(\frac{\mathbf{u}_+}{\sqrt{d}}, \frac{\mathbf{v}_+}{\sqrt{d}}, \frac{\mathbf{v}_-}{\sqrt{d}}, \frac{\mathbf{u}_-}{\sqrt{d}} \right) . \tag{5.32}$$

This is a symplectic matrix. For, in view of (5.31) and of lemma 5.12, we have

$$\mathsf{T}^\top \mathsf{J}\mathsf{T} = \frac{1}{d} \begin{pmatrix} \mathbf{u}_+^\top \mathsf{J}\mathbf{u}_+ & \mathbf{u}_+^\top \mathsf{J}\mathbf{v}_+ & \mathbf{u}_+^\top \mathsf{J}\mathbf{v}_- & \mathbf{u}_+^\top \mathsf{J}\mathbf{u}_- \\ \mathbf{v}_+^\top \mathsf{J}\mathbf{u}_+ & \mathbf{v}_+^\top \mathsf{J}\mathbf{v}_+ & \mathbf{v}_+^\top \mathsf{J}\mathbf{v}_- & \mathbf{v}_+^\top \mathsf{J}\mathbf{u}_- \\ \mathbf{v}_-^\top \mathsf{J}\mathbf{u}_+ & \mathbf{v}_-^\top \mathsf{J}\mathbf{v}_+ & \mathbf{v}_-^\top \mathsf{J}\mathbf{v}_- & \mathbf{v}_-^\top \mathsf{J}\mathbf{u}_- \\ \mathbf{u}_-^\top \mathsf{J}\mathbf{u}_+ & \mathbf{u}_-^\top \mathsf{J}\mathbf{v}_+ & \mathbf{u}_-^\top \mathsf{J}\mathbf{v}_- & \mathbf{u}_-^\top \mathsf{J}\mathbf{u}_- \end{pmatrix} = \mathsf{J}.$$

By property (iii) of Lemma 5.2, we have

$$\begin{aligned} \mathsf{JC}\mathbf{u}_+ &= \mu\mathbf{u}_+ - \omega\mathbf{v}_+ \ , & \mathsf{JC}\mathbf{v}_+ &= \omega\mathbf{u}_+ + \mu\mathbf{v}_+ \ , \\ \mathsf{JC}\mathbf{u}_- &= -\mu\mathbf{u}_- + \omega\mathbf{v}_- \ , & \mathsf{JC}\mathbf{v}_- &= -\omega\mathbf{u}_- - \mu\mathbf{v}_- \ . \end{aligned}$$

Using this, we also have

$$\begin{aligned} \mathsf{JCT} &= \left(\frac{\mu\mathbf{u}_+ - \omega\mathbf{v}_+}{\sqrt{d}}, \frac{\omega\mathbf{u}_+ + \mu\mathbf{v}_+}{\sqrt{d}}, \frac{-\mu\mathbf{v}_- - \omega\mathbf{u}_-}{\sqrt{d}}, \frac{\omega\mathbf{v}_- - \mu\mathbf{u}_-}{\sqrt{d}} \right) \\ &= \mathsf{T} \begin{pmatrix} \mu & -\omega & 0 & 0 \\ \omega & \mu & 0 & 0 \\ 0 & 0 & -\mu & -\omega \\ 0 & 0 & \omega & -\mu \end{pmatrix} . \end{aligned}$$

Thus, denoting by (x_1, x_2, y_1, y_2) the new variables, the system (5.9) is transformed to

$$\begin{aligned} \dot{x}_1 &= \mu x_1 - \omega x_2 \ , & \dot{x}_2 &= \omega x_1 + \mu x_2, \\ \dot{y}_1 &= -\mu y_1 - \omega y_2 \ , & \dot{y}_2 &= \omega y_1 - \mu y_2, \end{aligned}$$

which are the canonical equations for the Hamiltonian (5.24).

Exercise 5.13: Find the canonical transformation which changes the Hamiltonian (5.24) into the Hamiltonian in complex variables $H(\xi,\eta) = \lambda\xi\eta$. Hint: this may be used as an example of application of the diagonalizing procedure in Sections 5.1.3 and 5.1.4. *A.E.L.*

5.2 Non-linear Elliptic Equilibrium

As we have seen, the Hamiltonian in a neighbourhood of an elliptic equilibrium can be typically approximated by a quadratic one with the form

$$H_0(x,y) = \frac{1}{2}\sum_l \omega_l(x_l^2 + y_l^2) , \tag{5.33}$$

where $(x,y) \in \mathbb{R}^{2n}$ are the canonical variables, and $\omega = (\omega_1,\ldots,\omega_n) \in \mathbb{R}^n$ is the vector of frequencies, which are assumed not to vanish (see Section 5.1.7). By assuming the Hamiltonian to be analytic and expanding it in power series, we are led to consider a canonical system with Hamiltonian

$$H(x,y) = H_0(x,y) + H_1(x,y) + H_2(x,y) + \ldots, \tag{5.34}$$

where $H_s(x,y)$, for $s \geq 1$, is a homogeneous polynomial of degree $s+2$ in the canonical variables. The power series is assumed to be convergent in a neighbourhood of the origin of $\mathbb{R}^{2n}$. This is a perturbed system of harmonic oscillators, which describes many interesting physical models.

5.2.1 *Use of Action-Angle Variables*

The canonical transformation to action-angle variables,

$$x_l = \sqrt{2p_l}\cos q_l , \quad y_l = \sqrt{2p_l}\sin q_l, \quad 1 \leq l \leq n, \tag{5.35}$$

gives H_0 the form (4.24) of an isochronous system, which fails to satisfy the condition of non-degeneration of Poincaré's theorem.

The lack of a perturbation parameter in the Hamiltonian (5.34) may appear to be a disturbing fact, but it is a trivial matter; the parameter ε is easily replaced by the distance from the origin. Indeed, if we consider the dynamics inside a sphere of radius ϱ centred at the origin, then the homogeneous polynomial $H_s(x,y)$ is of order $\mathcal{O}(\varrho^{s+2})$, so that ϱ plays the role of perturbation parameter. More formally, we may introduce a scaling transformation

$$x_j = \varepsilon x_j' , \quad y_j = \varepsilon y_j' ,$$

which is not canonical but preserves the canonical form of the equations if the new Hamiltonian is defined as

$$H'(x',y') = \frac{1}{\varepsilon^2} H(x,y)\Big|_{x=\varepsilon x' ,\, y=\varepsilon y'}$$

(see Example 2.3). Thus, the Hamiltonian is changed to

$$H'(x',y') = H_0(x',y') + \varepsilon H_1(x',y') + \varepsilon^2 H_2(x',y') + \dots\ ,$$

which introduces the power expansion in ε. This means that the natural reordering of the power series as homogeneous polynomials corresponds precisely to the use of a parameter ε – thus making the parameter useless.

The transformation to action-angle variables is a bit delicate: it introduces a singularity at the origin, which causes a loss of analyticity for $p = 0$. Let us pay a little more attention to this point.

By transforming a homogeneous polynomial of degree s in x, y, we obtain a trigonometric polynomial in the angles q with coefficients depending on the actions p in a particular form. The following rules apply:[5]

(i) The actions p appear only as powers of $\sqrt{p_1}, \dots, \sqrt{p_n}$, namely

$$c_{j_1,\dots,j_n}\, p_1^{j_1/2} \cdots p_n^{j_n/2}\ ,$$

where $c_{j_1,\dots,j_n} \in \mathbb{R}$.

(ii) In every term of the Fourier expansion

$$c_{j_1,\dots,j_n}\, p_1^{j_1/2} \cdots p_n^{j_n/2} e^{i(k_1 q_1 + \ldots + k_n q_n)}\ ,$$

the exponent k_l may take only the values $-j_l,\ -j_l + 2,\ \dots,\ j_l - 2,\ j_l$.

With some patience the reader will be able to check that these properties are preserved by sums, products and Poisson brackets.[6]

The polynomial dependence on $\sqrt{p_1}, \dots, \sqrt{p_n}$ keeps the property of a homogeneous polynomial of degree s at order ϱ^s, so that nothing essential is changed. However, the lack of analyticity may be somewhat annoying when one tries to analyze the convergence. As a matter of fact, all these problems disappear if we choose to work in Cartesian coordinates, which we do in what follows. Let us first develop a formal theory, in the sense that all calculations will be performed disregarding the problem of convergence of the series which will be constructed.

---

[5] The reader will easily check that these rules apply by writing the trigonometric functions in complex form.

[6] The direct verification is straightforward, although tedious; but it can be avoided in view of the following argument. Recall that the transformation to action-angle variables is canonical and that we are transforming polynomials. Since the canonical transformation preserves the Poisson brackets, the calculation in action-angle variables cannot affect the property that in Cartesian variables x, y the result is a polynomial. The same argument, with no use of the canonical structure, applies to the product. The sum is a trivial matter.

5.2.2 A Formally Integrable Case

In view of the degeneration of the unperturbed system we may expect to be able to construct first integrals in our case by just applying the procedure of Section 4.5.1, namely to solve the system (4.26) of equations.

Let us adapt the scheme to our case. We look for a first integral

$$\Phi(x,y) = \Phi_0(x,y) + \Phi_1(x,y) + \ldots , \tag{5.36}$$

where $\Phi_0(x,y) = p_l = \frac{1}{2}(x_l^2 + y_l^2)$ is the action of the lth oscillator, and $\Phi_s(x,y)$ is a homogeneous polynomial of degree $s+2$. By setting $l = 1, \ldots, n$ we may construct n independent first integrals. Thus, splitting the equation $\{\Phi, H\} = 0$ in homogeneous polynomials, we get the system

$$\begin{aligned} &\{\Phi_1, H_0\} = -\{\Phi_0, H_1\}, \\ &\{\Phi_s, H_0\} = -\{\Phi_{s-1}, H_1\} - \cdots - \{\Phi_0, H_s\} \ , \quad s > 0 \ . \end{aligned} \tag{5.37}$$

Looking at the algebraic aspect of these equations makes the problem quite simple. The Lie derivative $L_{H_0}\cdot = \{\cdot, H_0\}$ acts as a linear operator from the linear space $\mathscr{P}^{(r)}$ of homogeneous polynomials of a fixed degree r into itself. Moreover, using the complex canonical coordinates $(\xi, \eta) \in \mathbb{C}^{2n}$ defined by the linear canonical transformation,

$$x_l = \frac{1}{\sqrt{2}}(\xi_l + i\eta_l) \ , \quad y_l = \frac{i}{\sqrt{2}}(\xi_l - i\eta_l) \ , \quad 1 \le l \le n, \tag{5.38}$$

we get

$$H_0 = i \sum_l \omega_l \xi_l \eta_l \ . \tag{5.39}$$

The operator L_{H_0} takes a diagonal form; for, by applying it to a monomial $\xi^j \eta^k \equiv \xi_1^{j_1} \ldots \xi_n^{j_n} \eta_1^{k_1} \ldots \eta_n^{k_n}$, we get

$$L_{H_0} \xi^j \eta^k = i \langle j - k, \omega \rangle \, \xi^j \eta^k \ . \tag{5.40}$$

Define now, as usual, $\mathscr{R}$ as the image of $\mathscr{P}^{(r)}$ by L_{H_0}. We see that Eq. (5.37) can be solved if the r.h.s. belongs to $\mathscr{R}$. On the other hand, by defining the null space (the kernel) $\mathscr{N}$ as $\mathscr{N} = L_{H_0}^{-1}(0)$, the inverse image of the null element by L_{H_0} restricted to $\mathscr{P}^{(r)}$, we get that both $\mathscr{N}$ and $\mathscr{R}$ are linear subspaces of the same space $\mathscr{P}^{(r)}$. The subspaces are disjoint, namely satisfy $\mathscr{N} \cap \mathscr{R} = \{0\}$, and generate $\mathscr{P}^{(r)}$ by direct sum, namely $\mathscr{N} \oplus \mathscr{R} = \mathscr{P}^{(r)}$. It should be remarked that we have $\xi^j \eta^k \in \mathscr{N}$ in case $\langle j - k, \omega \rangle = 0$, that is, $j - k \in \mathscr{M}_\omega$, the resonance module associated to ω; otherwise we have $\xi^j \eta^k \in \mathscr{R}$.

The diagonal form of L_{H_0} in complex variables makes the treatment of the system (5.37) straightforward. For we are requested to solve an equation $L_{H_0}\Phi = \Psi$ with

$$\Phi = \sum_{j,k} \varphi_{j,k} \xi^j \eta^k \ , \quad \Psi = \sum_{j,k} \psi_{j,k} \xi^j \eta^k \ ,$$

where Ψ is a known homogeneous polynomial of a fixed degree, so that the coefficients $\psi_{j,k} \in \mathbb{C}$ are known, while the coefficients $\varphi_{j,k}$ of Φ are the unknowns to be determined. In view of (5.40) we should write

$$\varphi_{j,k} = -i \frac{\psi_{j,k}}{\langle j-k, \omega \rangle} \ ,$$

which is allowed provided $j - k \notin \mathscr{M}_\Omega$. This is tantamount to saying that the system (5.37) can be solved only if the r.h.s. has no component in $\mathscr{N}$. This is the same problem of null divisors pointed out in Sect. 4.5.1.[7]

Dealing with the latter problem is actually not easy, in general. However, a direct solution may be found in a particular but relevant case which occurs in many physical situations.[8]

The formal existence of the first integral is stated by the following.

Proposition 5.14: *Let $H(x,y)$ be as in (5.34) where H_0 has the form (5.33), and assume:*

i. non-resonance: for $k \in \mathbb{Z}^n$ one has $\langle k, \omega \rangle = 0$ if and only if $k = 0$;

ii. reversibility: The Hamiltonian is an even function of the momenta, namely satisfies $H(x,-y) = H(x,y)$.

---

[7] The problem of a degenerate Hamiltonian of the type considered here was first investigated by Whittaker [208]. As a historical remark, it is curious that in his paper Whittaker did not mention the problem of the consistency of the construction. A few years later, Cherry wrote two papers [39][40] where a lot of work is devoted to the consistency problem, but with no definite conclusion.

[8] See [60]. The reader will notice that only the non-resonant case is discussed here; the proof of the proposition cannot be trivially extended to the case of a non-reversible Hamiltonian. The direct construction of first integrals for the resonant case encounters major difficulties due to the non-trivial structure of the kernel $\mathscr{N}$ of L_{H_0}. This is indeed the main concern of the second paper of Cherry [40]: he proposes to determine the arbitrary terms in the solution of the linear equation at a given order so as to remove the unwanted terms in $\mathscr{N}$ from the equation for higher orders. However, Cherry fails to prove that the method works – although his exact method was later implemented on a computer by Contopoulos in order to perform an explicit calculation (see [45], [46] and [47]). First integrals for both the non-reversible case and the resonant one may be constructed using the methods of normal form introduced by Poincaré and widely exploited by Birkhoff. However, the direct approach taken here is definitely simpler. It may be interesting to note that an attempt to justify the method of Cherry, made by Luigi Galgani and the author, has lead to a connection with the algorithm for giving the Hamiltonian a normal form via the Lie transform methods [76]. This will be the object of Chapter 7.

Then for any $\Phi_0 \in \mathscr{N}$ there exists a formal integral Φ of the form (5.36). The first integral is an even function of the momenta y.

Proof. The proof is based on two simple remarks. First, the non-resonance condition implies that any function in $\mathscr{N}$ must be even in the momenta, since it can depend only on the action variables

$$p_1 = i\xi_1\eta_1 = \frac{x_1^2 + y_1^2}{2}, \ \ldots\ , p_n = i\xi_n\eta_n = \frac{x_n^2 + y_n^2}{2},$$

which are quadratic in y; second, the Poisson bracket between functions of the same parity is odd, while the Poisson bracket between functions of different parity is even. Proceeding by induction, one sees that if Φ_s has been determined for $0 \le s \le r$ as an even function of the momenta (which is true for $r = 0$), then the right member of the equation for Φ_{r+1} is an odd function, so that it has no component in $\mathscr{N}$; hence Φ_{r+1} can also be determined and is an even function because $L_{H_0}\Phi_{r+1}$ is odd. Such a solution is unique up to an arbitrary term $\tilde{\Phi}_{r+1}^{(l)} \in \mathscr{N}$, which is an even function. This shows that the construction can be consistently performed at any order. *Q.E.D.*

Proposition 5.15: *Let $H(x, y)$ be as in Proposition 5.14. Then the following statements hold true.*

(i) Any two formal first integrals Φ and Ψ of the form (5.33) are in involution.

(ii) There exist n independent formal integrals $\Phi^{(1)}, \ldots, \Phi^{(n)}$ of the form (5.36) with $\Phi_0^{(l)} = p_l$, the harmonic actions, which form a complete involution system.

Proof. (i) Let $\Upsilon = \{\Phi, \Psi\}$; by Poisson's theorem, it is a first integral and has the form (5.33) with

$$\Upsilon_0 = \{\Phi_0, \Psi_0\}\ , \quad \Upsilon_1 = \{\Phi_0, \Psi_1\} + \{\Phi_1, \Psi_0\}\ , \quad \ldots$$

Hence, by Proposition 5.14, it is an even function. On the other hand, Υ is the Poisson bracket between even functions, hence it must be odd. We conclude that $\Upsilon = 0$, that is, Φ and Ψ are in involution.
(ii) The functions $\Phi_0^{(1)} = p_1, \ldots, \Phi_0^{(n)} = p_n$ are obviously independent, and this entails the independence of $\Phi^{(1)}, \ldots, \Phi^{(n)}$, which can be constructed by Proposition 5.14. By point (i) they are also in involution. *Q.E.D.*

The latter two propositions seem to suggest that the strongly negative result of Poincaré on non-existence of first integrals does not apply to the case of elliptic equilibria. However, we should not forget that our result is a formal one, as has been stressed in all statements, in the sense that the whole construction is performed by merely using algebra, regardless of the convergence of the series so generated. In the same spirit we could say that our system is formally integrable, and that we could apply the method of Liouville and Arnold to build the action-angle variables.[9]

The next question raised is concerned with the convergence properties of the series so generated. Indeed the denominators $\langle k, \omega \rangle$, although non-vanishing, are not bounded from below. On the other hand, there are known examples of series involving small denominators which are convergent. An essentially negative answer to the problem of convergence has been given by Siegel in [202] (see also [171]). Let us go on by showing that the formal first integrals have an asymptotic character, as predicted by Poincaré in the passage quoted at the beginning of Chapter 4. This will allow us to draw interesting conclusions concerning the long-term stability of an elliptic equilibrium.

5.3 Old-Fashioned Numerical Exploration

Before coming to analytical estimates, let us investigate the dynamics in the neighbourhood of an elliptic equilibrium with numerical methods. At the dawn of numerical simulations of dynamics, between 1950 and 1960, two classes of problems concerning Hamiltonian dynamics have been studied.[10]

The first class was initiated around 1952 by Enrico Fermi, John Pasta and Stanislaw Ulam [66]. They wanted to investigate the behaviour of a nonlinear chain of identical particles with the aim of observing the relaxation to equilibrium predicted by Classical Statistical Mechanics. The unexpected lack of tendency to equilibrium observed by the authors originated the so-called *FPU problem* and marked the beginning of a long series of studies which are still continuing.[11] A complete exposition largely exceeds the limits of the present notes; a status report 50 years after the publication of the original work may be found in [73].

---

[9] This is discussed by Whittaker in [209], ch. XVI, §199 (many years before the announcement of the theorem of Arnold and Jost).

[10] It is necessary here to mention that numerical studies of equations governing the circulation in atmosphere had been initiated around 1950 by Edward Norton Lorenz. In 1963 he published a pioneering paper where evidence was produced of the chaotic, unpredictable behaviour of the orbits [165]. His discovery of the so-called *strange attractors* pointed out the existence of an unknown and somehow unexpected dynamical phenomenon. The work of Lorenz is not described in the present notes only because it is not concerned with Hamiltonian flow.

[11] After Fermi's death, the results were collected by the other two authors as an internal report of the Los Alamos Laboratory which has remained unknown for several years. It is worth mentioning the contribution of Mary Tsingou to the work. She had a main role in the computational part, but her contribution is only acknowledged in a short note: see [56].

The second class is connected with the discovery of chaos in the dynamics of apparently simple models. Around 1958 George Contopoulos began to investigate numerically the dynamics of stars in galaxies [45]. He used in particular the method of Poincaré section, applying it to a simple but meaningful model, namely a system of two harmonic oscillators with a cubic nonlinearity. In 1964 Michel Hénon and Carl Eugene Heiles [110] pointed out the existence of a sharp transition from ordered to chaotic behaviour of orbits in a model similar to the ones of Contopoulos.

The discussion in this section is essentially a revisitation of the works of Contopoulos and of Hénon and Heiles. The main – perhaps the sole – difference is that the amazingly increased power of modern computers allows us to push their calculations much further. However, it is necessary to recall that the discovery of chaos in models which look to be very simple ones, such as the Hamiltonian (5.41), has been seen initially with great surprise.[12]

5.3.1 The Galactic Models of Contopoulos and of Hénon and Heiles

As a useful model problem, let us consider the Hamiltonian

$$H = \frac{\omega_1}{2}(x_1^2 + y_1^2) + \frac{\omega_2}{2}(x_2^2 + y_2^2) + x_1^2 x_2 + \frac{\alpha}{3} x_2^3 \,. \tag{5.41}$$

The model is strongly reminiscent of those studied by Contopoulos and by Hénon and Heiles, with a few differences among them.

The model basically describes the motion of a star in a galaxy with axial symmetry, under the action of the average potential due to all other stars (see Exercise 1.5). Recall that in cylindrical coordinates, and setting the mass $m = 1$, the Hamiltonian is written as

$$H = \frac{1}{2}\left(p_r^2 + \frac{p_\vartheta^2}{r^2} + p_z^2\right) + V(r, z) \,. \tag{5.42}$$

Due to the assumption on the physical structure of the galaxy, the dependence of the potential on the radial coordinate r, far from the centre, has the qualitative behaviour of the Keplerian potential, but with a dependence such as r^{-a} with some value $0 < a < 2$; for example, sometimes one sets $a = 1/4$. As to the dependence on z, the potential is assumed to have a minimum for $z = 0$ for every r, corresponding to the symmetry plane of the galaxy. The Hamiltonian possesses the first integral of the angular momentum, namely

[12] We should better speak of *rediscovery* of chaos: the existence of orbits with an unpredictable behaviour had been already pointed out by Poincaré in [187] and [188] (see Chapter 10). However, his discovery remained essentially unknown to physicists and astronomers for more than 70 years. Nowadays it is a well-known and established fact, but is was not so 60 years ago.

$p_\vartheta = \ell$. Proceeding as in the case of the central motion, we are allowed to remove the angle ϑ by introducing an effective potential,

$$V_{\rm eff}(r,z) = V(r,z) + \frac{\ell}{r^2}.$$

The particular form of the potential $V(r,z)$ leads us to recognize that the effective potential $V_{\rm eff}(r,z)$ has a minimum at some point $(\overline{r},0)$, so that close to the minimum the Hamiltonian may be written as a system of two harmonic oscillators with a nonlinear coupling (in fact, unknown). The form (5.41) of the Hamiltonian is a particularly simple model which has been widely used; the choice of the parameters is rather arbitrary.

Contopoulos did not include the cubic term x_2^3, so that $\alpha = 0$ in his studies. Moreover, he considered several different values of the frequencies ω_1, ω_2 in order to take into account the dependence of the frequencies on the distance from the galactic centre. Remarkably, he observed that many orbits had an ordered behaviour, reminiscent of that of an integrable system, which seemed to contradict the non-existence of first integrals stated by Poincaré. Therefore, he exploited the idea of calculating the so-called *third integral* (to be added to the energy and the angular momentum) by a series expansion, as suggested by Edmund Taylor Whittaker [208] and Thomas MacFarland Cherry [39][40] several years before.[13] To this end he implemented a computer program which calculated the coefficients of the series expansion; he also performed a careful study of the effectiveness of the series so constructed, thus showing that increasing the size of nonlinear terms caused a breakdown of the third integral. Unfortunately Contopoulos did not publish the complete figures of Poincaré sections: he rather emphasized the role of the third integral.

Hénon and Heiles choosed to pay particular attention to the mathematical aspect. Thus, they introduced further freedom in the model by adding the cubic term in x_2 with $\alpha = -1$. Moreover, they choosed the strongly resonant values $\omega_1 = \omega_2 = 1$. The value $\alpha = -1$ has no particular meaning for the physical model; the curious peculiarity is that for $\omega_1 = \omega_2 = 1$ it introduces an elegant symmetry in the potential under a rotation of $2\pi/3$ on the plane (x_1, x_2). The strong resonance due to the frequencies being equal makes the phenomenon of transition from order to chaos very impressive; this was indeed the main intention of the authors.[14]

---

[13] See, e.g., [45][46][47][50][54]. An interesting account of what we may call a rediscovery of the third integral by Contopoulos, as seen by the protagonist, may be found in the autobiographic book [53].

[14] It is worth noting that, by setting $\alpha = 1$, the system turns out to be integrable, for by a rotation of angle $2\pi/3$ in the coordinate plane the Hamiltonian separates

5.3.2 Poincaré Sections

The method of Poincaré section is a long-standing one, which was introduced in [187] and [188]; Poincaré called it *théorie des conséquents.*

In order to simplify the discussion, let us assume that both frequencies ω_1, ω_2 are positive, so that the Hamiltonian (5.41) has a minimum at the origin. Therefore, for small positive energy E the surface of invariant energy $H(x,y) = E$ has a compact component which is topologically equivalent to a three-dimensional sphere surrounding the origin of the four-dimensional phase space. The compactness is broken for E bigger than a critical value corresponding to a saddle point of the potential energy

$$V(x_1,x_2) = \frac{\omega_1}{2}x_1^2 + \frac{\omega_2}{2}x_2^2 + x_1^2x_2 + \frac{\alpha}{3}x_2^3 \ .$$

Depending on α, this may happen at the points

$$P_{1,2} = \left(\pm\frac{\sqrt{\omega_1\omega_2 - \alpha\omega_1^2}}{2}, -\frac{\omega_1}{2}\right) \ , \quad P_3 = \left(0, -\frac{\omega_2}{\alpha}\right) \ ;$$

the point P_3 is discarded if $\alpha = 0$, as in the model of Contopoulos; the points $P_{1,2}$ are not real if $\alpha > \omega_2/\omega_1$. The corresponding values of energy are

$$E_1 = E_2 = \frac{\omega_1^2\omega_2}{8} - \frac{\alpha\omega_1^3}{24} \ , \quad E_3 = \frac{\omega_2^3}{6\alpha^2} \ .$$

The smallest of the latter values will be referred to as the *escape energy* and denoted by $E_{\rm esc}$.

Set now $x_1 = 0$ as a section plane for the orbit; the intersection of the plane with the energy surface is the two-dimensional manifold

$$H(0,x_2,y_1,y_2) = \frac{\omega_1}{2}y_1^2 + \frac{\omega_2}{2}(y_2^2 + x_2^2) + \frac{\alpha}{3}x_2^3 = E \ ; \tag{5.43}$$

for $0 < E < E_{\rm esc}$ it has a compact component around the origin, topologically equivalent to a two-dimensional spherical surface. The projection of the surface on the plane (x_2, y_2) delimits a region

$$\frac{\omega_2}{2}(y_2^2 + x_2^2) + \frac{\alpha}{3}x_2^3 \le E \ , \tag{5.44}$$

topologically similar to a circle. Any point (x_2, y_2) inside this region corresponds to a pair of crossing points with opposite directions, namely

$$y_1 = \pm\sqrt{E - \frac{\omega_2}{2}(y_2^2 + x_2^2) - \frac{\alpha}{3}x_2^3} \ . \tag{5.45}$$

The section points may be determined by numerical integration as follows.

---

into two independent parts (see, e.g., [70]). On the other hand, the original model with $\alpha = -1$ has been proven to be non-integrable (see, e.g., [115][119]).

(i) Choose an arbitrary initial point x_2, y_2 inside the allowed region.
(ii) Set $x_1 = 0$ and find y_1 from (5.45) with a definite sign (let us take the positive one).
(iii) Let the orbit evolve until it intersects again the plane $x_1 = 0$ with positive velocity y_1: this is the point named *conséquent* by Poincaré.

These operations actually define a map from the plane onto itself. The sequence of points obtained by iterating the map may be represented on the plane x_2, y_2. The question is: *what do we expect to observe?*

Exercise 5.16: Figure out what will happen in the simplest case of two independent harmonic oscillators, with Hamiltonian

$$H = \frac{\omega_1}{2}(x_1^2 + y_1^2) + \frac{\omega_2}{2}(x_2^2 + y_2^2) .$$

(Hint: the cases of rational and irrational ratio ω_2/ω_2 deserve a different study. A very particular behaviour shows up in the very special case $\omega_1 = \omega_2$, as in the case studied by Hénon and Heiles.) *A.E.L.*

The case of the Hamiltonian (5.41) is rather peculiar.

(i) The limiting curve of the region (5.44) is an orbit with initial point $x_1(0) = y_1(0) = 0$, and $x_2(0), y_2(0)$; however, this orbit should not be naively interpreted as an orbit of the map generated by the Poincaré section: the phase of the oscillator x_1, y_1 is undefined, since the Hamilton equations give $x_1(t) = y_1(t) = 0$ for all t.
(ii) If $x_1(t), x_2(t), y_1(t), y_2(t)$ is an orbit, then

$$\begin{aligned} x_1'(t) &= x_1(-t) \quad , \quad x_2'(t) = x_2(-t) \, , \\ y_1'(t) &= -y_1(-t) \quad , \quad y_2'(t) = -y_2(-t) \end{aligned}$$

also is an orbit (check it). That is, reversing the velocities y, one obtains the same orbit in the coordinate plane, traveled in the opposite sense.

5.3.3 *The Onset of Chaos*

It is now time to exploit the contents of the previous section: let us see the resulting figures. As a model example, let us set

$$\omega_1 = 1 \, , \quad \omega_2 = \frac{\sqrt{2}}{2} \, , \quad \alpha = -1.$$

The choice of the frequencies $\omega_1 = 1 \, , \omega_2 = \sqrt{2}/2$ has the purpose of making the system definitely far from resonance; it is coherent with the theory developed in the present chapter, particularly oriented to the case of non-resonant frequencies. The portrait of the Poincaré section for different values of energy E is represented in Figs. 5.4 to 5.9. The energy may be used as a parameter measuring the strength of nonlinear terms in the equations. So

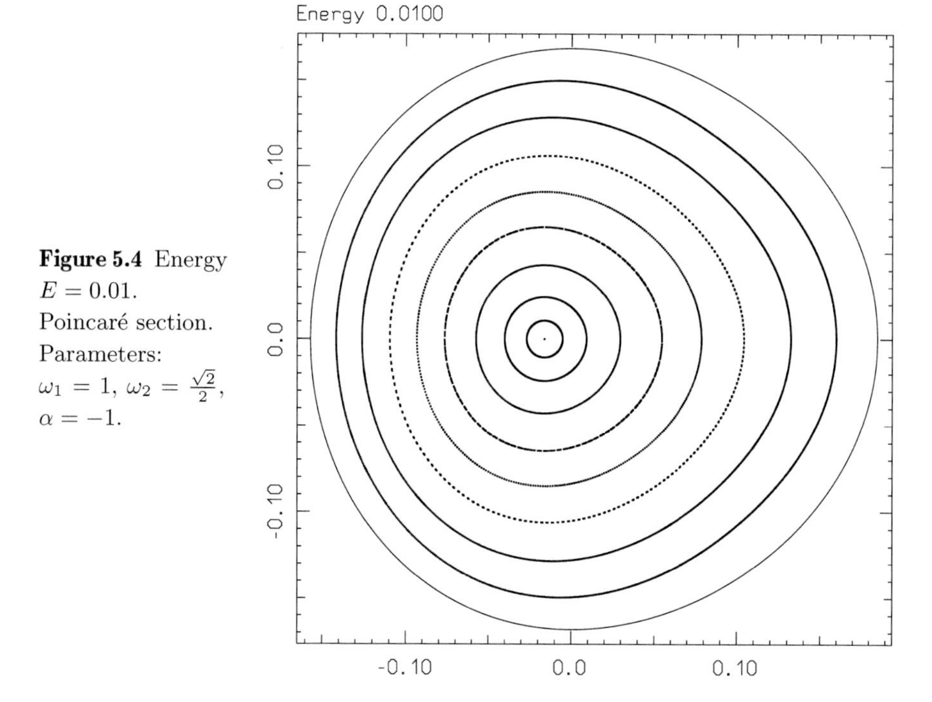

Figure 5.4 Energy $E = 0.01$. Poincaré section. Parameters: $\omega_1 = 1$, $\omega_2 = \frac{\sqrt{2}}{2}$, $\alpha = -1$.

let us see what happens for increasing values of E up to and over the critical value $E_{\text{esc}} = 0.05892556509888\ldots$.

Figure 5.4 represents the portrait of the Poincaré section for $E = 0.01$, which may be considered as a small value (as seen in the figure). The external curve is the border of the allowed region (5.44). The points corresponding to different orbits seem to distribute on curves similar to circles. At the centre there is a single point which represents a periodic orbit which is a deformation of the unperturbed orbit with initial point $x_2 = y_2 = 0$; the existence of such an orbit has been proved by Aleksandr Mikhailovich Lyapounov [166][95]. What we see is what should be expected if the perturbation does not substantially affect the behaviour of the orbit. The reader will notice that the centre of the circle is displaced with respect to the origin of the plane (x_2, y_2); do not forget that what we see is the projection of points which lie on a spherical surface. Moreover, the curves are not perfect circles: they are somewhat deformed, looking as oval curves. Finally, the distribution of points along different curves is not the same. This is because the first effect of the nonlinearity is to induce a slight change of the frequencies: the system is no longer isochronous (recall the nonlinearity of a pendulum). The overall picture suggests that the system could be still integrable: if there are invariant surfaces described by a first integral independent of the energy, then the

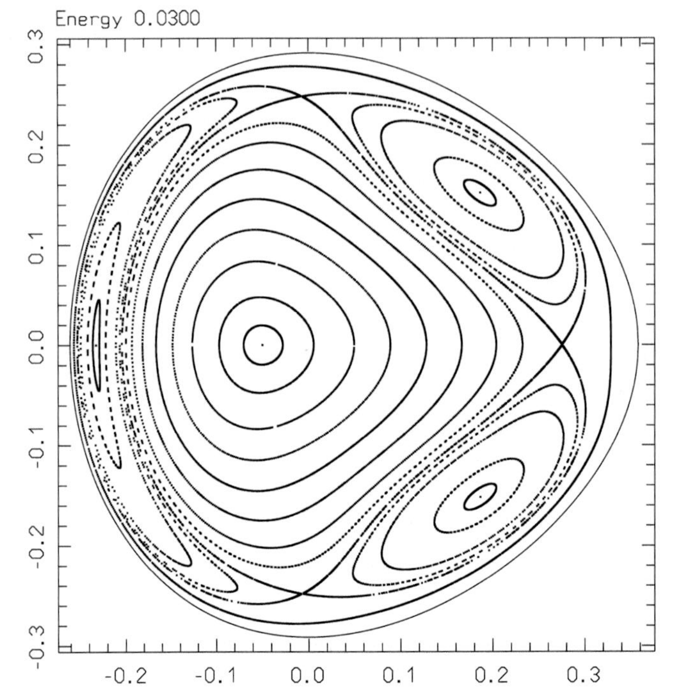

Figure 5.5 Energy $E = 0.03$. Poincaré section. Parameters: $\omega_1 = 1$, $\omega_2 = \frac{\sqrt{2}}{2}$, $\alpha = -1$.

orbits are confined to the constant surface of this integral. In this case the section points must belong to the intersection of the invariant surface with the section plane.

Figure 5.5, for $E = 0.03$, exhibits a first change: in a region close to the limiting curve the ovals of Figure 5.4 are broken. What happens is that the change of frequencies due to the nonlinearity has made them fall into a 3/2 resonance – we have seen in discussing the theorem of Poincaré that resonance is a critical fact. This is the first crucial phenomenon, indeed. In a system integrable in the Arnold–Jost sense the orbits should remain on an invariant torus filled by periodic orbits; in the corresponding sections each orbit should be represented by only three different points. In fact, only a pair of periodic orbits do survive: one of them produces the three points at the centre of the small closed curves; the second one is at the intersection of two curves which resemble the separatrices of a pendulum. The first orbit is stable, as the lower equilibrium of a pendulum; the second one is unstable; the separatrices mark the border between two classes of orbits. Inside the separatrix the orbit librates around the stable periodic orbit, thus forming a kind of tube around it; in the external region the orbits keep the independent behaviour of two oscillators. Nevertheless, the figure does not visually seem to contradict the existence of a first integral independent of the energy.

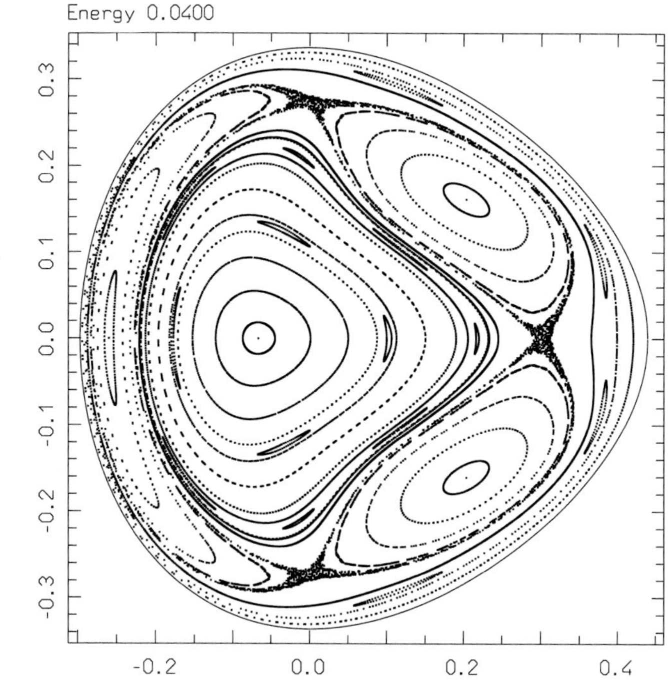

Figure 5.6 Energy $E = 0.04$. Poincaré section. Parameters: $\omega_1 = 1$, $\omega_2 = \frac{\sqrt{2}}{2}$, $\alpha = -1$.

Figure 5.6, for energy $E = 0.04$, yields the surprise. First, the phenomenon of resonance which creates pairs of stable/unstable orbits seems to repeat again. This phenomenon has been predicted by Poincaré: in [191] he claimed precisely that the perturbation acting on a resonant invariant torus is expected to break the periodicity of the orbits, leaving only pairs of stable/unstable ones. In his paper he could not provide a rigorous proof, but he strongly emphasized the overwhelming relevance of that phenomenon. The proof was later completed by Birkhoff [26], and the result is known as the *Poincaré–Birkhoff geometric theorem*; more on this in Section 10.3.1. The second surprise is that the separatrix of Figure 5.5 appears to be unexpectedly fat. It is produced by a single orbit which fills a small region, well visible close to the unstable orbit.

The existence of chaotic orbits close to an unstable periodic orbit is an outcome of the amazing fact discovered by Poincaré in [187] and described at the end of the third volume of *Méthodes Nouvelles*: the existence of homoclinic points which are at the origin of chaotic orbits in the neighbourhood of unstable periodic orbits; a description will be found in Section 10.3.1. It is spontaneous to ask why such a fat chaotic region does not appear in Figure 5.5. The answer is that it is there, indeed, but the width of the chaotic region is too small to be seen on the scale of the figure: one should enlarge

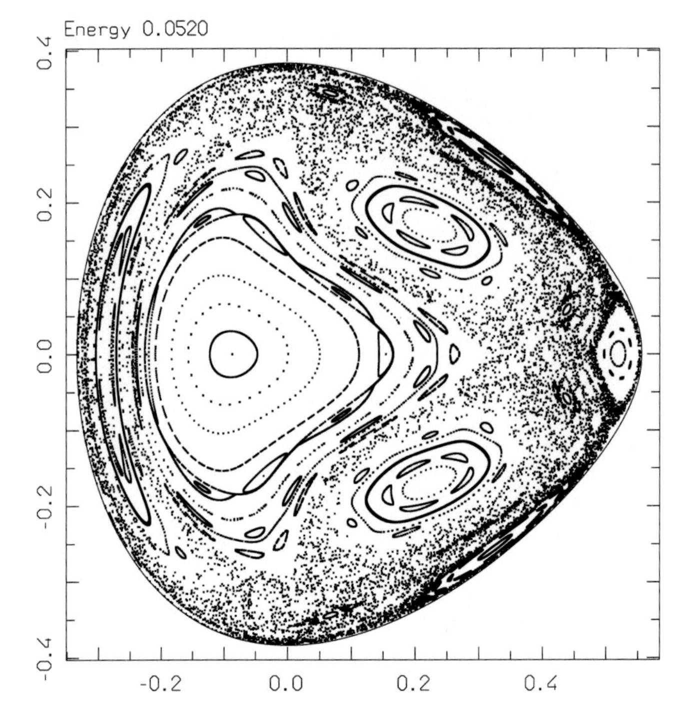

Figure 5.7 Energy $E = 0.052$. Poincaré section. Parameters: $\omega_1 = 1$, $\omega_2 = \frac{\sqrt{2}}{2}$, $\alpha = -1$.

the region around the unstable orbit, as looking at it with a lens. A less evident fact is that the phenomenon described by Poincaré happens generically for *all* unstable orbits; but the size of the chaotic region decreases so fast that it becomes, in fact, invisible.

Figure 5.7, for $E = 0.052$, shows that a further increase of the size of nonlinear terms causes the birth of more and more periodic orbits with their accompanying separatrices. More relevant is the fact that the area covered by chaotic orbits extends considerably and surrounds regions completely controlled by a periodic orbit. With a suggestive image, Hénon and Heiles described the latter phenomenon saying that there are *islands* of ordered motion inside a *chaotic sea*; a language which has soon become familiar to everybody interested in the study of chaotic phenomena.

Figure 5.8 is drawn for $E = 0.058925$, a value lower but very close to the critical value $E_{\rm esc}$. It shows a further increase of the chaotic sea, with smaller and smaller ordered islands which become more and more isolated and seem to be progressively devoured by the sea. Both Figures 5.7 and 5.8 exhibit the remarkable phenomenon of coexistence of ordered and chaotic orbits on the same energy surface.

Finally, Figure 5.9 is drawn for energy $E = 0.07$, exceeding the critical value. The energy surface is no more compact in this case, and it is natural to

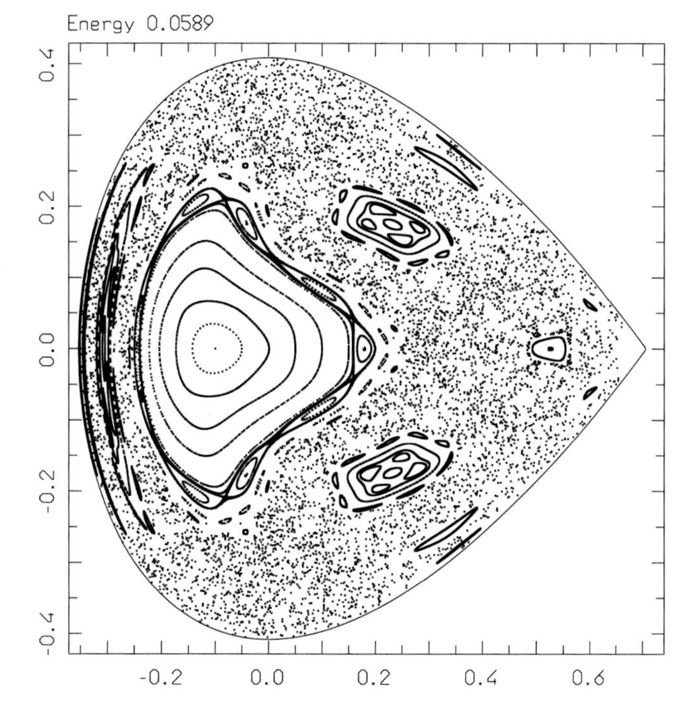

Figure 5.8 Energy $E = 0.058925$. Poincaré section. Parameters: $\omega_1 = 1$, $\omega_2 = \frac{\sqrt{2}}{2}$, $\alpha = -1$.

expect that all orbits should escape to infinity. The figure shows instead that many orbits do escape, indeed, but some islands of order still survive, and the orbits inside them appear to remain confined. In particular, the central periodic orbit seems to remain stable.

It is worthwhile to put, once more, an accent on the fact that the discovery of coexistence of ordered and chaotic phenomena in systems as simple as the Hamiltonian (5.41) caused great astonishment among scientists. It was quite common, at that time, to believe that a statistical behaviour, due to an underlying chaotic dynamics, should be typical of mechanical systems with a large number of bodies, such as atoms or molecules in solids, fluids and gases, or stars in galaxies. Thus, it was with great surprise that Fermi, Pasta and Ulam realized that a model such as the nonlinear chain (Example 4.4), which they had considered as obeying the rules of Statistical Mechanics, unexpectedly failed to tend to statistical equilibrium. Similarly, Contopoulos realized that the orbit of a star in the galaxy, which was expected to be erratic, was instead very ordered. On the other hand, it was also common to believe that systems with few degrees of freedom, for example, the planetary system, should be characterized by ordered motions, actually a combination of periodic motions which we may interpret as typical of an integrable system, in the light of the Arnold–Jost theorem. Thus it was again with great

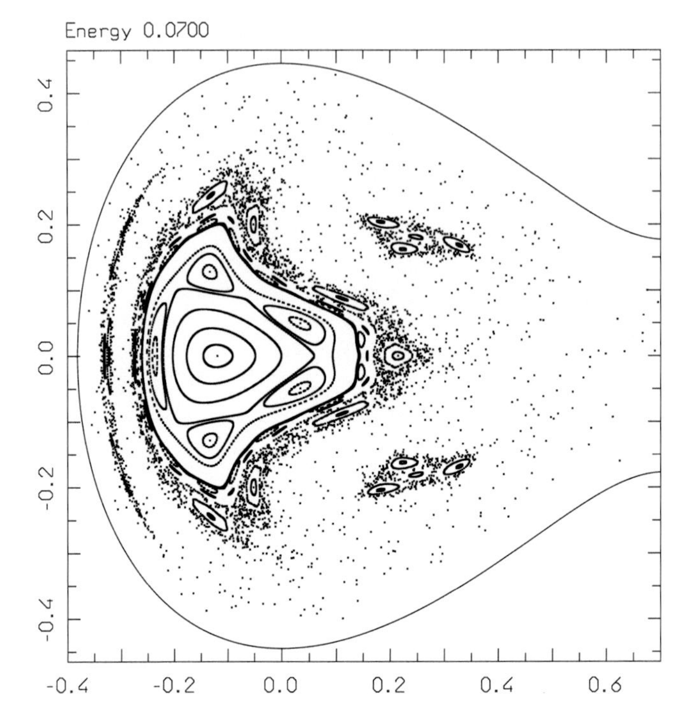

Figure 5.9 Energy $E = 0.07$. Poincaré section. The energy surface is not compact.

surprise that Hénon and Heiles found strong evidence of chaos in a very simple Hamiltonian system. The papers of Fermi–Pasta–Ulam, Contopoulos, Hénon–Heiles and Lorenz marked the beginning of studies concerning chaos in dynamical systems: reading these papers is highly recommended. The interest in chaotic phenomena began to grow amazingly after 1970. A connection with the discoveries of Poincaré and with the theorem of Andrey Nikolaevich Kolmogorov [128] announced in 1954 became progressively known among a wide number of scientists; not only mathematicians, but also astronomers and physicists among others. The amazing increase in the number of papers (and of journals) on the subject of chaos in that period may be seen as a remarkable example of a chaotic system. It is perhaps irreverent, but not far from truth, to say that the pioneering works which have been mentioned up to this point have introduced chaos not just in dynamics – it was already there, waiting to be discovered – but also in the scientific milieu.

5.3.4 Stability of a Non-linear Equilibrium

Stability of an elliptic equilibrium may be intended in (at least) two different senses. The minimal requirement is that orbits which have initial points close to the equilibrium remain confined in a neighbourhood of it; a stronger requirement is that some relevant quantities – typically the actions of the system – keep an almost constant value.

We have already recalled that, in general, an elliptic equilibrium (a product of centre points) is stable in Lyapounov's sense only in the linear approximation; adding a nonlinear perturbation may destroy stability. We could also observe that according to the Lyapounov's theory one is merely interested in confinement of orbits in a neighbourhood of the equilibrium.

In the Hamiltonian case of an elliptic equilibrium, as for the Hamiltonian (5.41), stability in Lyapounov's sense is assured in case all frequencies ω have the same sign, for example, are all positive, for in this case the Hamiltonian has a local minimum at the origin, and the energy surfaces isolate a compact neighbourhood of the origin where the orbits get confined forever. The latter criterion was already known to Dirichlet. An example is the case considered in Section 5.3.3. It also shows that stability in Lyapounov's sense does not imply that the dynamics is of the ordered type: chaos is not incompatible with mere confinement of orbits. However, the Dirichlet's method does not apply to the case of $E > E_{\mathrm{esc}}$: the question of stability cannot be easily settled.

The problem is much more difficult if the frequencies have different signs; for the energy surface $H(x, y) = 0$ has the shape of a saddle around the equilibrium, so that there are no compact surfaces of constant energy $H(x, y) = E$ which surround the equilibrium. Nevertheless, the linear system is stable in Lyapounov's sense, because the two functions (the actions)

$$p_1 = \frac{x_1^2 + y_1^2}{2} , \quad p_2 = \frac{x_2^2 + y_2^2}{2}$$

are independent first integrals (the system is isochronous). The persistence of stability in the nonlinear case is far from being assured, for the actions p_1, p_2 are no longer constant. Therefore, in a neighbourhood of the equilibrium it may well happen that p_1 and p_2, still remaining positive, grow very large without affecting the value of the sole first integral $H(x, y) = E$. The argument is a general one, which applies to more than 2 degrees of freedom; it is indeed a crucial problem in studying the stability of such a system. The following examples show that the problem is not a purely academic one.

Example 5.17: *The triangular equilibria of Lagrange.* It was discovered by Lagrange that the problem of three bodies in the gravitational case has a very particular family of periodic solutions: the three bodies are located at the vertices of an equilateral triangle with fixed side length, so that they rotate uniformly in a plane around the common centre of mass, with constant relative distances. A remarkable fact is that these beautiful solutions are not peculiar to the Newtonian gravitational force: if the attraction between any two bodies depends only on the distance and is proportional to the product of the masses, then the same result is found by assuming very mild smoothness conditions. A widely considered model is the so-called *restricted problem of three bodies*: two masses (*the primaries*) are assumed to move in circular orbits around the common centre of mass; a third much smaller mass (*the planetoid*) moves under the action of the primaries, without affecting their orbit. Considering a reference frame which rotates together with the primaries (so that they are fixed), the planetoid remains at equilibrium at the third vertex of an equilateral triangle; there are two such points, named L_4 and L_5. Applying the normalization method of Section 5.1.2 one finds that the equilibrium is elliptic in case the ratio m_2/m_1 of the masses of the primaries is (moderately) small: the cases Sun–Jupiter and Earth–Moon satisfy this condition, which means that the equilibrium is linearly stable. However, one also finds that two of the frequencies are positive, but the third one is negative: this is indeed the reason why the stability of the equilibrium remains an open problem. The problem is not just academic: there are hundreds of asteroids which librate close to the triangular equilibria of the Sun–Jupiter system. *E.D.*

Example 5.18: *The secular motions of the planets.* A remarkable discovery due to Lagrange (followed by Laplace) is that the perihelia and the nodes of the planetary orbits are not fixed. The elliptic orbits of Kepler are rather subject to slow precessions, named *secular motions* because they have periods of centuries or even more. The calculation shows that the evolution of the pairs of variables eccentricity–perihelion and inclination–node can be described in linear approximation as a system of harmonic oscillators; therefore, Lagrange concluded that the motion will likely be stable. However, the frequencies of the perihelia and those of the nodes have opposite sign. Therefore a perturbation (e.g., the action of Jupiter and Saturn as well as of the other planets), may destroy the stability. For example, the inclination and the eccentricity of a celestial body may both increase until the orbits of two planets intersect those of other planets, thus allowing them to have close encounters or even to collide. Such a process is particularly relevant for the asteroids, as has been discovered in the last, say, 30 years. *E.D.*

The theory of first integrals developed in Section 5.2.2 could help in solving the problem of stability, for we may look for some bound on the evolution of the actions of the system; this would be enough in order to conclude for stability. To this end let us construct first integrals

$$\Phi^{(j)} = \frac{x_j^2 + y_j^2}{2} + \Phi_1^{(j)} + \ldots \ , \qquad j = 1, \ldots, n \ ,$$

which are perturbations of the harmonic actions. In a neighbourhood of the equilibrium the intersection of all the manifolds of constant value of the

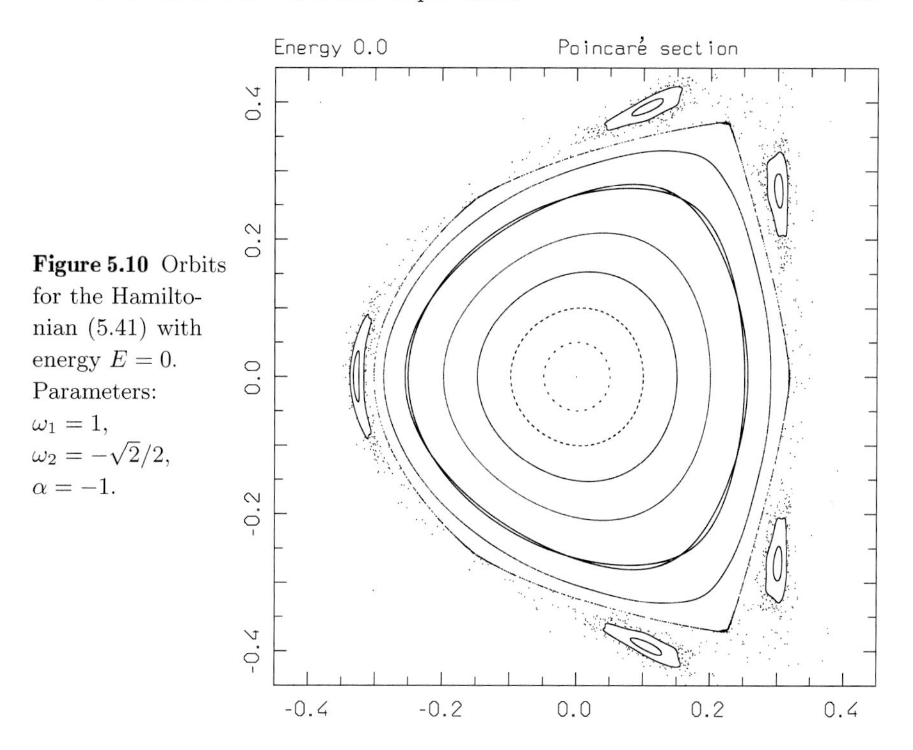

Figure 5.10 Orbits for the Hamiltonian (5.41) with energy $E = 0$. Parameters: $\omega_1 = 1$, $\omega_2 = -\sqrt{2}/2$, $\alpha = -1$.

first integrals is an invariant torus, so that the orbits are confined close to the equilibrium; the argument is the same as Dirichlet's, but the sole Hamiltonian is not enough. This, however, is true in a rigorous sense only if we can prove that the series expansions of the formal first integrals are actually convergent.

In order to investigate the problem numerically, let us change the sign of one of the frequencies in the Hamiltonian (5.41); we set the parameters to $\omega_1 = 1$, $\omega_2 = -\sqrt{2}/2$ and $\alpha = -1$. Here the size of the perturbation increases with the distance from the equilibrium, so that the value of energy has no particular role as a perturbation parameter: we may choose a fixed value for it. The Poincaré section for energy $E = 0$ is represented in Figure 5.10. One sees again orbits which seem to lie on invariant tori (the closed curves), periodic orbits, separatrices emanating from unstable periodic orbits, islands of ordered motions. Chaotic orbits are there, but are not easily seen: they rapidly escape to infinity.

The complexity of the figure emerges with stronger evidence if we enlarge some portions of the figure. A few examples are shown in Figure 5.11 (see details in the caption to the figure). For instance, two successive enlargements of the curve which in Figure 5.10 seems to mark the border of the central ordered region, represented in panels **e** and **f**, reveal a complex structure of

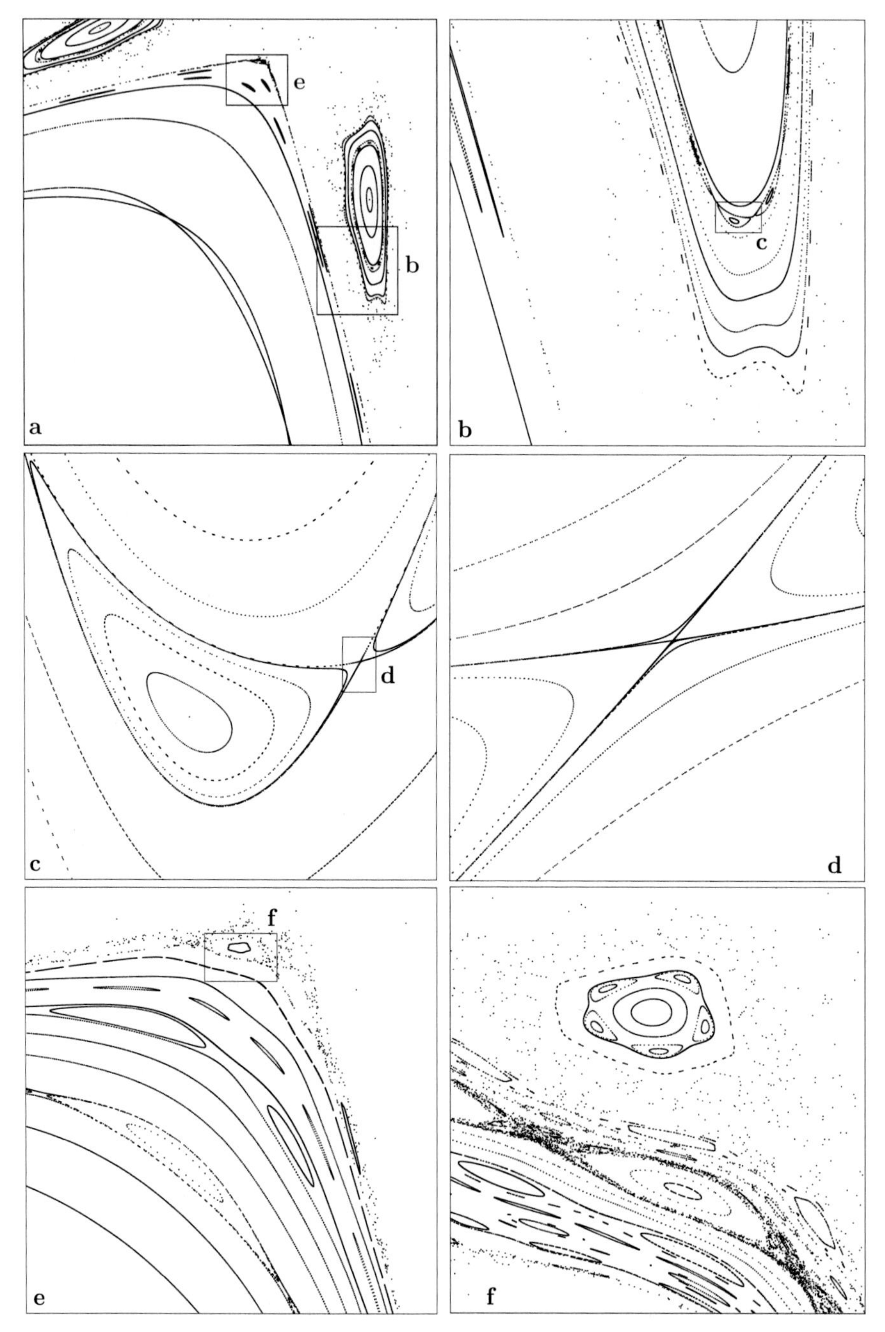

Figure 5.11 Details of the Poincaré section. **a**: blow-up of the square $[0.05, 0.35] \times [0.1, 0.4]$ of Figure 5.10. **b, c, d**: successive blow-ups of the small rectangle **b** in panel **a**. **e, f**: successive blow-ups of the small rectangle **e** in panel **a**.

islands, separatrices and chaos which seems to repeat again and again, ad infinitum.

The spontaneous question is: *can we expect that an analytic first integral is able to provide a global description of such a complex behaviour?*

5.3.5 Exploiting the First Integrals

The scheme of construction of first integrals described in Section 5.2.2 is easily translated into an algorithm which may be implemented on a computer, thus producing the expansion of a formal integral up to some desired order.[15] The results described here have been obtained by truncating the construction of the first integral,

$$\Phi = \Phi_0 + \Phi_1 + \ldots\ , \qquad \Phi_0 = \frac{1}{2}(x_2^2 + y_2^2)\ ,$$

at order 98, namely up to terms of degree 100. The aim is to see what happens when the order increases more and more.

The simplest application consists in comparing the Poincaré section with the level curves of the first integral. We are given a function $\Phi(x_1, x_2, y_1, y_2)$ as a power series truncated at some maximal order. Setting $x_1 = 0$, we get a function $\Phi(0, x_2, y_1, y_2)$ of three variables; on the other hand, for every point (x_2, y_2) we calculate $y_1 = y_1(x_2, y_2, E)$ as the positive root of the equation $H(x, x_2, y_1, y_2) = E$ (in our case $E = 0$). Then we select an orbit with arbitrary initial point $(x_{2,0}, y_{2,0})$ and draw the level lines of the function

---

[15] In the spirit of the whole section, the purpose is to revisit and enhance results which have been obtained already in the sixties and seventies of the past century. The main difference is that the power of today's computers allows us to push the calculation to significantly higher orders. As a comparison, at the beginning of the 1960s Contopoulos usually truncated the expansion at some degree between 6 and 12. In 1966 F. G. Gustavson calculated a first integral up to degree 8 for the model of Hénon and Heiles, on a IBM-7094 computer [105]. Around 1975 Contopoulos and the author could reach degree 15 for a system with 2 degrees of freedom on a CDC-7600 computer, the most powerful one available at that time [49]. Besides the speed of the CPU, the strongest limit was the amount of available memory, at most a few hundred Kbytes: the number of coefficients grows very large with the order. All this gives a pale idea of the growth in power of computers; the expansion calculated here has been performed on a desktop computer with an Intel Core i7-9700K processor and 16 GB of memory (only a small fraction has been actually used). The calculation has been worked out with a self-written program. A description of the methods of computer algebra implemented in the program may be found in [77] or [94].

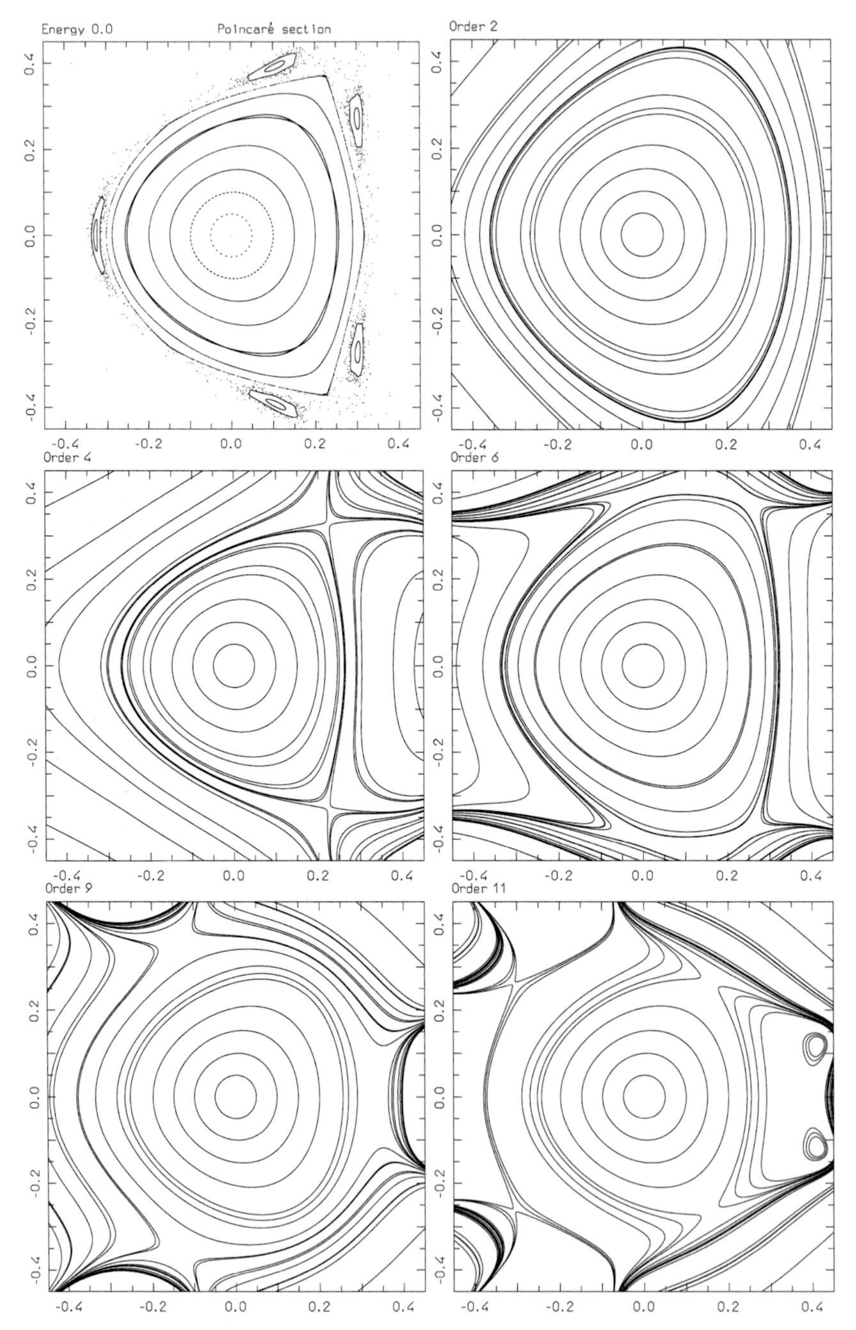

Figure 5.12 Poincaré section of Figure 5.10 and level lines of the formal integral Φ at orders 2, 4, 6, 9 and 11, corresponding to polynomial degrees 4, 6, 11 and 13.

Figure 5.13 Level lines of the formal integral Φ at orders 18, 29, 32, 37, 48 and 98, corresponding to polynomial degrees 20, 31, 34, 39, 50 and 100.

$$\tilde{\Phi}(x_2, y_2) = \Phi(0, x_1, y_1, y_2)\Big|_{y_1=y_1(x_2,y_2,E)} ,$$

depending only on (x_2, y_2), for the value $\tilde{\Phi}(x_{2,0}, y_{2,0})$. If Φ is an exact first integral, then the section points of the orbit should be located on the corresponding level curve.

The results for different truncation orders are reported in Figures 5.12 and 5.13. The portrait of the Poincaré section is drawn again in the upper left corner of Figure 5.12 in order to make the comparison easier; then the level curves of the formal first integral at not-too-high orders are represented. The reader will note that already at low orders there is a visually acceptable correspondence between the level curves and the Poincaré section in the central region. When approaching the border of the central island the correspondence becomes rather poor and is definitely bad in the external region. Such behaviour appears to be reasonable if the series were convergent in some circle around the origin. It also justifies a fact well known to astronomers: since the times of D'Alembert, Clairaut, Euler and Lagrange it has been known that the results obtained via perturbation series were often in good agreement with the observed phenomena; it was also a common conjecture for these authors that the series should converge, so that in the case of unsatisfactory correspondence between calculation and observations it was expected that by increasing the order of approximation, the discrepancy could be removed.[16] The usefulness of the formal expansions in galactic dynamics has also been emphasized by Contopoulos – this was indeed his purpose in calculating the third integral (for a comprehensive report see [50]).

Figure 5.13 shows what happens at higher orders. The overall impression is that the region of apparent convergence of the series shrinks progressively, leaving a smaller and smaller region which resists – like a besieged fortress. It seems reasonable to expect that the radius of apparent convergence will shrink to zero when the order grows to infinity.[17]

---

[16] To quote only an example, the lunar theory was developed independently by D'Alembert and Clairaut around 1749, with a kind of competition aimed at finding better and better approximations for the motion of the apogee of the Moon. A trace of the discussion is reported by D'Alembert in [55], vol. I, ch. XV, Article II, p. 113. For a short summary see, e.g., [100].

[17] The possible asymptotic behaviour of the series has been illustrated in the remarkable work of Jacques Roels and Michel Hénon [194]. The model is an area-preserving map on the plane; they calculated a normal form up to degree 21 on an IBM-7040 computer.

5.4 Quantitative Estimates

It is now time to abandon the numerical investigations and to try to investigate the behaviour of the series with analytical tools. This may be done by introducing an appropriate norm on homogeneous polynomials and translating the formal algorithm of Section 5.2.2 into a scheme of estimates on the norms of the various functions.

5.4.1 *Algebraic and Analytic Setting*

In order to reduce the problem to a simple form let us consider the case of a cubic perturbation, that is, a canonical system with Hamiltonian

$$H(x,y) = H_0(x,y) + H_1(x,y) \ , \tag{5.46}$$

where H_1 is a homogeneous polynomial of degree 3. This is similar to (5.34); the fact that the perturbation is just a polynomial is not relevant.[18] Let us also assume that the Hamiltonian satisfies the hypotheses of non-resonance and reversibility of Proposition 5.14. In view of non-resonance there exists a non-increasing sequence $\{\alpha_s\}_{s\geq 1}$ of positive real numbers such that

$$|\langle k,\omega\rangle| \geq \alpha_s \quad \text{for } k\in\mathbb{Z}^n \ , \quad 0<|k|\leq s+2 \tag{5.47}$$

is satisfied. We have $\alpha_s \to 0$ for $s\to\infty$, of course. Thus, we can perform the formal construction of the first integrals $\Phi^{(l)}(x,y) = p_l + \Phi_1^{(l)} + \ldots$, $1\leq l\leq n$, with $p_l = \frac{1}{2}(x_l^2+y_l^2)$, as described in Sect. 5.2.2. In particular, the solution will be made unique by the condition $\Phi_s^{(l)} \in \mathscr{R}$ for $s\geq 1$.

Equation (5.37) for the Hamiltonian (5.46) takes the simpler form

$$L_{H_0}\Phi_s^{(l)} = \Psi_s^{(l)} := \begin{cases} -\{p_l, H_1\} & \text{for } s=1\ , \\ -\{\Phi_{s-1}^{(l)}, H_1\} & \text{for } s>1\ . \end{cases} \tag{5.48}$$

In order to introduce quantitative estimates we must be able to evaluate the size of a function. To this end, let x, y be real or complex variables, and write a generic homogeneous polynomial $f(x,y)$ of degree s as

$$f(x,y) = \sum_{|j+k|=s} f_{j,k} x^j y^k \ ,$$

with real or complex coefficients $f_{j,k}$. A suitable *polynomial norm* of f is defined as

$$\|f\| = \sum_{j,k} |f_{j,k}| \ . \tag{5.49}$$

---

[18] The general and detailed treatment can be found in [79]. The simpler case discussed here, however, gives full insight into the problem.

Considering a domain

$$\Delta_\varrho = \{(x,y) \in \mathbb{R}^{2n} : x_l^2 + y_l^2 \le \varrho^2,\ 1 \le l \le n\} , \tag{5.50}$$

namely the Cartesian product of n disks of radii ϱ in the coordinate plane x_l, y_l, the size of the homogeneous polynomial $f(x,y)$ is bounded in Δ_ϱ by

$$|f(x,y)| \le \varrho^s \|f\| .$$

5.4.2 Analytical Estimates

The aim now is to translate the recurrent set of equations (5.48) into a set of recurrent estimates on the norms of the polynomials $\Phi_s^{(l)}$ and $\Psi_s^{(l)}$. This is done as follows. First, transform the Hamiltonian to complex variables ξ, η, defined by the canonical transformation (5.38); this will allow us to exploit the diagonal form of the linear operator L_{H_0}. Then build up the expansion in complex variables by recurrently solving (5.48); this involves two basic operations: a Poisson bracket to determine $\Psi_s^{(l)}$, and the inversion of the linear operator $L_{H_0}\cdot = \{H_0, \cdot\}$. Finally, having determined the formal integral up to an arbitrary degree r, transform back to real variables by the inverse of transformation (5.38). We want to know how the norms are propagated through these operations.

(i) The transformation (5.38) to complex variables changes the norm of a homogeneous polynomial of degree s at most by a factor $2^{s/2}$; the same holds true for the inverse.

(ii) The Poisson bracket between two homogeneous polynomials f and g of degree s and r respectively is estimated by

$$\|\{f,g\}\| \le sr\|f\|\,\|g\| . \tag{5.51}$$

(iii) The unique solution $\Phi_s^{(l)} \in \mathscr{R}$ of Eq. (5.48) is estimated in complex variables by

$$\|\Phi_s^{(l)}\| \le \frac{1}{\alpha_s}\|\Psi_s^{(l)}\| , \tag{5.52}$$

with α_s satisfying (5.47).

The proof of (i) is trivial, and (5.52) at point (iii) is a consequence of the diagonal form of L_{H_0} in complex variables.

Concerning (ii), we prove (5.51) by calculating

$$\{f,g\} = \sum_{j,k,j',k'} f_{jk} g_{j'k'} x^{j+j'} y^{k+k'} \sum_{l=1}^{n} \frac{j_l k_l' - j_l' k_l}{x_l y_l} ,$$

and using the definition of the norm to compute[19]

$$\begin{aligned}\|\{f,g\}\| &\leq \sum_{j,k,j',k'} |f_{jk}|\,|g_{j'k'}| \sum_{l=1}^{n} (j_l k'_l + j'_l k_l)\\ &\leq sr\Big(\sum_{j,k} |f_{jk}|\Big)\Big(\sum_{j',k'} |g_{j'k'}|\Big)\\ &= sr\|f\|\|g\| \;.\end{aligned}$$

Using these estimates we readily obtain the following.

Lemma 5.19: *The unique solution $\Phi^{(l)} = p_l + \Phi_1^{(l)} + \ldots$ of Eqs. (5.37) with $\Phi_s^{(l)} \in \mathscr{R}$ satisfies*

$$\big\|\Phi_s^{(l)}\big\| \leq ab^{s-1} \frac{(s+1)!}{\prod_{l=1}^{s} \alpha_l} \quad \textit{for } s \geq 1 \,, \tag{5.53}$$

with positive constants a, b and with a sequence $\alpha_1, \alpha_2, \ldots$ satisfying (5.47).

The proof is based on a few remarks. In order to determine $\Phi_s^{(l)}$ we must solve s times Eq. (5.37). By a recurrent application of the estimates (5.51) and (5.52) we realize that the determination of $\Phi_s^{(l)}$ adds to the estimate of the norm of $\Phi_{s-1}^{(l)}$ a factor $3(s+1)\|H_1\|$ due to the Poisson bracket and a divisor α_s due to the solution of Eq. (5.48). These factors accumulate, which justifies the factorial $(s+1)!$ and the product of divisors α_l. The factor b^{s-1} takes into account the accumulation of constants, including the power of 2 due to the transformation to complex variables at the beginning and the transformation back to real variables at the end of the procedure. Constants which do not accumulate are collected in a.

5.4.3 Truncated Integrals

The estimate (5.53) clearly does not allow us to prove the convergence of the expansions of the first integrals $\Phi^{(l)}$. This, of course, could be merely because we are unable to find better estimates; recall, however, that the non-convergence, in a generic sense, of these first integrals has been proven by Siegel. Nevertheless, we may try to unveil the information hidden in the formal first integrals. Thus, let us suppose that we have performed the procedure up to some arbitrary order r, and consider the truncated polynomial

$$\Phi^{(l,r)} = p_l + \Phi_1^{(l)} + \cdots + \Phi_r^{(l)} \;, \quad l = 1, \ldots, n \,, \tag{5.54}$$

whose time derivatives clearly are

[19] Use $\sum_{l=1}^{n}(j_l k'_l + j'_l k_l) \leq s \sum_{l=1}^{n}(k'_l + j'_l) = sr$, the first inequality being due to $0 \leq j_l \leq s$ and $0 \leq k_l \leq s$.

$$(5.55) \qquad \dot{\Phi}^{(l,r)} = \{\Phi_r^{(l)}, H_1\} \ .$$

Even if we do not know the explicit form of the integrals $\Phi^{(l,r)}$, we can nevertheless obtain significant information from the technical estimates of the Section 5.4.2. Indeed, suppose that, for a real system, we are only able to observe the values of the harmonic actions $p_1, \ldots, p_n$ as functions of time. (This is, essentially, what we do when we compute the osculating elements of the orbit of a planet from the observed positions.) The natural question here is how much these quantities, which are the approximate constants of our problem, can vary in time. Using the fact that the $\Phi^{(l,r)}$ are, hopefully, better conserved than the p_l, we find the bound

$$|p_l(t) - p_l(0)| \le |p_l(t) - \Phi^{(l,r)}(t)| + |\Phi^{(l,r)}(t) - \Phi^{(l,r)}(0)| + |\Phi^{(l,r)}(0) - p_l(0)| \ .$$

In order to simplify the discussion, let us suppose, just for a moment, that the $\Phi^{(l,r)}$ are exact first integrals (this could be true for a particular Hamiltonian, of course), so that the central term in this formula vanishes. The two remaining terms are estimated by noting that for any (x, y) in a domain Δ_ϱ defined by (5.50) one has

$$\left|(\Phi^{(l,r)} - p_l)(x, y)\right| = \left|\Phi_1^{(l)}(x, y) + \cdots + \Phi_r^{(l)}(x, y)\right| < d_r \varrho^3 \ ,$$

with a constant d_r depending, of course, on r, and provided ϱ is small enough. This just expresses the fact that $\Phi^{(l,1)} - p_l$ is a polynomial starting with a term of degree 3. Thus, one has the bound

$$(5.56) \qquad \left|p_l(t) - p_l(0)\right| < 2d_r \varrho^3$$

for any t, provided one can guarantee that the orbit is confined for all times in Δ_ϱ. A simple choice is to assume, for example, $p_l(0) = \varrho^2/3$, and to ask $\left|p_l(t) - p_l(0)\right| < \varrho^2/6$. Thus, comparing the r.h.s. of the latter inequality and of (5.56), we conclude that the orbit is confined forever in the annulus $\varrho^2/6 < p_l(t) < \varrho^2/2$ provided $\varrho < 1/(12 d_r)$. Since we have assumed that the functions $\Phi^{(l,r)}$ are exact first integrals, the variation in time of the harmonic actions $p_1, \ldots, p_n$ clearly appears to be due to a *deformation* of the invariant surfaces. For one has $\Phi^{(l,1)} = \text{const}$ instead of $p_l = \text{const}$. This is illustrated in Fig. 5.14(a): the action $p_l(t)$ exhibits a quasiperiodic oscillation and it is confined in the strip $p_{l,\min} \le p_l(t) \le p_{l,\max}$, whose width is estimated by (5.56).

Let us now take into account the fact that the $\Phi^{(l,r)}$ are not exact first integrals. Thus, by (5.55), the time derivative $\dot{\Phi}^{(l,r)}$ is a homogeneous polynomial of degree $r + 3$, so that for $(x, y) \in \Delta_\varrho$, one has

$$(5.57) \qquad \left|\dot{\Phi}^{(l,r)}(x, y)\right| < C_r \varrho^{r+3} \ ,$$

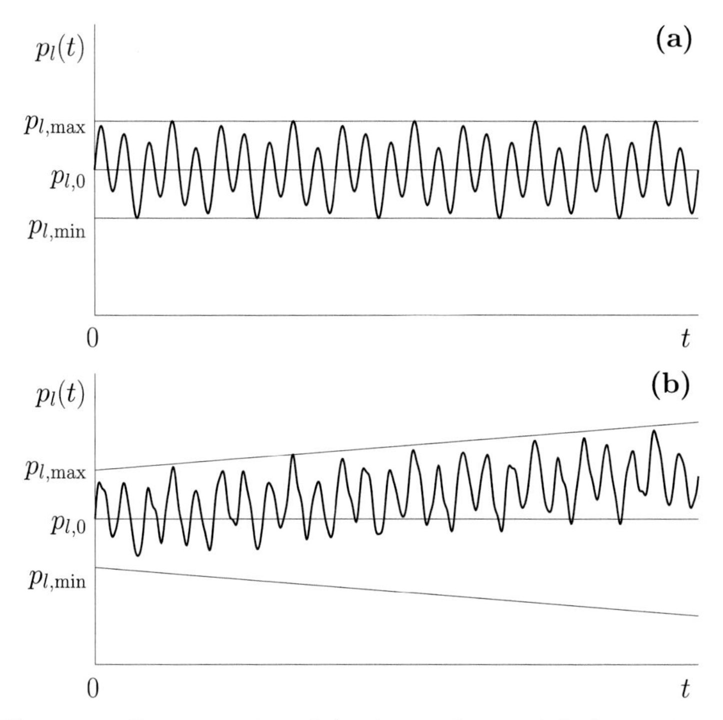

Figure 5.14 Representation of the time evolution of the harmonic action under the effect of *deformation* and of *noise*. (a) The deformation causes a quasiperiodic, bounded oscillation of the action $p_l(t)$. (b) The noise adds further frequencies and may cause a secular variation (drift), at most linear in time. The figure represents the worst case.

with a constant C_r, depending, of course, on r. Thus, superimposed on the deformation, there is a variation in time of $p_l(t)$ due to the dynamical evolution of $\Phi^{(l,r)}$; note, however, that if ϱ is sufficiently small, then such an evolution is hopefully quite slow with respect to the one due to the deformation, due to the estimate (5.57). Following Nekhoroshev, we say that there is a *noise* which causes a slow *drift* of the orbit. The situation is illustrated in Fig. 5.14(b): the disturbing term $\{\Phi_r^{(l)}, H_1\}$ introduces new frequencies and moreover may cause a slow drift of $p_l(t)$. An a priori bound is given by the estimate (5.57), which guarantees that the width of the strip where $p_l(t)$ is confined increases at most linearly with t. The figure represents the worst case, in which the drift is actually linear. Taking into account the effect of the noise, the bound (5.56) must be changed to

$$|p_l(t) - p_l(0)| < 2d_r\varrho^3 + C_r\varrho^{r+3}t \; . \tag{5.58}$$

However, we still have to ensure that $p_l(t) < \frac{1}{2}\varrho^2$ (since all the estimates hold on Δ_ϱ). This, of course, cannot be true for all times, but we can use the fact that the noise has a small effect in order to ensure that it holds for a quite large time. To do this, it is enough, for example, to ask $C_r\varrho^{r+3}t \le d_r\varrho^3$ (i.e.,we do not allow $\left|\Phi^{(l,r)}(t) - \Phi^{(l,r)}(0)\right|$ to be larger than the deformation). Thus, by a little modification of this argument we conclude the following.

Proposition 5.20: *Consider the Hamiltonian (5.46), and assume that the frequencies $\omega_1, \ldots, \omega_n$ satisfy the condition (5.47) for a suitable sequence $\{\alpha_s\}_{s\ge1}$. Then, for any integer $r \ge 1$ there exist constants d_r and C_r such that for ϱ sufficiently small and for any orbit with initial point $(x_0, y_0) \in \Delta_{2\varrho/3}$, we have*

$$\left|p_l(t) - p_l(0)\right| < d_r\varrho^3 \quad \text{for } |t| < T_r := \frac{d_r}{C_r\varrho^r} \ . \tag{5.59}$$

The latter proposition expresses the property of *complete stability* introduced by Birkhoff. As a matter of fact, Birkhoff could not do better because he did not even attempt to estimate the dependence of the constants d_r and C_r on r. Thanks to Lemma 5.19, we do have such an estimate. We have indeed, with possibly a minor change of the constants a and b,

$$C_r = ab^r \frac{(r+2)!}{\alpha_1 \cdots \alpha_r} \ . \tag{5.60}$$

The determination of d_r may require a few more considerations. Observe that

$$\begin{aligned}\left|\Phi_1^{(l)}(x,y) + \cdots + \Phi_r^{(l)}(x,y)\right| &\le \left\|\Phi_1^{(l)}\right\|\varrho^3 + \cdots + \left\|\Phi_r^{(l)}\right\|\varrho^{r+2} \\ &\le \frac{ab}{\alpha_1}\varrho^3 + \cdots + \frac{ab^r}{\alpha_1\cdots\alpha_r}\varrho^{r+2} \ .\end{aligned}$$

Thus, there exists a constant d_r such that for $3 \le s < r$ the coefficient of ϱ^s in the estimate is bounded, for example, by $(d_r/2)^{s-3}$. With such a value for d_r the sum is bounded by $d_r\varrho^3$ provided the condition "ϱ sufficiently small" of Proposition 5.20 is understood as "$\varrho < 1/d_r$."

5.4.4 Exponential Estimates

Let us come back to our first integrals. The estimates of Proposition 5.20 exhibit a strong dependence on r, so that one expects, a priori, that the choice of r has a significant impact on the final result. The unpleasant aspect is that the truncation order r appears to be a quite extraneous element: it is clearly nonsense to introduce the "user-defined parameter r" in the statement of a stability result concerning a physical system. The actual choice of a quite low r is in fact dictated by the practical impossibility of performing an explicit expansion of a first integral up to high orders. However, we have estimates concerning the dependence on r of the constants C_r and d_r in Proposition 5.20. This suggests the possibility of sharpening the theory by looking for a choice of r which is, in some sense, the optimal one. This leads in a quite natural way to exponential estimates of Nekhoroshev's type.

Let's consider the estimate (5.57), with C_r as in (5.60). Recalling that $\{\alpha_s\}_{s\geq 1}$ is a non-increasing sequence, one realizes that the expression $C_r\varrho^r$, considered as a function of r for a fixed ϱ, has a minimum. Indeed, we clearly have

$$C_r\varrho^{r+3} = \frac{b(r+2)\varrho}{\alpha_r}C_{r-1}\varrho^{r-1} \;;$$

hence the estimate is improved by adding the order r if $b(r+2)\varrho/\alpha_r < 1$. Thus, having chosen ϱ so that the initial point of the orbit satisfies, for example, $(x(0), y(0)) \in \Delta_{2\varrho/3}$, we may choose an optimal value $\overline{r}$ of the truncation order r as the largest integer satisfying

$$\overline{r} \leq \frac{\alpha_{\overline{r}}}{b\varrho} - 2 \; . \tag{5.61}$$

Here, "optimal" means that this particular value minimizes the estimate for the time derivative $\dot{\Phi}^{(l,r)}$ in the neighbourhood Δ_ϱ of the equilibrium.

A more explicit analytic estimate can be obtained if one knows more about the sequence $\{\alpha_s\}_{s\geq 1}$. To this end, a natural choice is to make use of the so called *Diophantine condition* of non-resonance,

$$\left|\langle k, \omega\rangle\right| \geq \gamma|k|^{-\tau} \quad \text{for } 0 \neq k \in \mathbb{Z}^n \;,\; \gamma > 0 \;,\; \tau > n-1 \;; \tag{5.62}$$

it is known that the latter condition is satisfied by most frequencies ω in the sense of Lebesgue measure (see Appendix A.2.1 for a proof). Exploiting this fact, let us set

$$\alpha_s = \gamma(s+2)^{-\tau} \;, \quad s > 1,$$

with suitable constants $\gamma > 0$ and $\tau \geq n-1$. Then one is naturally led to choose the optimal truncation order $r_{\rm opt} = r_{\rm opt}(\varrho)$ as

$$r_{\rm opt} = \left(\frac{\varrho^*}{\varrho}\right)^{1/(\tau+1)} - 2 \;, \quad \varrho^* = \frac{\gamma}{b} \; . \tag{5.63}$$

Thus, perturbation theory is useful if $\varrho < \varrho^*$, so that one has $r_{\rm opt} \geq 1$. This allows us to remove r from the estimates of Proposition 5.20 by substituting $r_{\rm opt}(\varrho)$ in place of r, so that the truncation order turns out to be determined by the size of the domain containing the initial data.

Let us now stress only the relevant steps and skip the technical details. Write the r.h.s of the estimate (5.57) in a simple form as $(r!)^a(\varrho/\varrho^*)^r$, and minimize it by setting $r = (\varrho^*/\varrho)^{1/a}$. Then, using $r! \sim r^r e^{-r}$ (by Stirling's formula), we readily get

$$(r!)^a \left(\frac{\varrho}{\varrho^*}\right)^r \sim \exp\left[-\left(\frac{\varrho^*}{\varrho}\right)^{1/a}\right] .$$

Thus, the noise becomes exponentially small with the inverse of the size of the domain. This is the simplest case of an exponential stability estimate of Nekhoroshev's type.[20] The formal statement of the theorem is the following:

Theorem 5.21: *Consider the canonical system with Hamiltonian (5.46), and assume that the harmonic frequencies ω satisfy the non-resonance condition $|\langle k, \omega\rangle| > \gamma\,|k|^{-\tau}$ for $k \in \mathbb{Z}^n$, with real constants $\gamma > 0$ and $\tau > n$. Then there exist positive constants β_1, β_2 and ϱ^* such that the following statement holds true: if*

$$\varrho \le 3^{-(\tau+1)}\varrho^* ,$$

then for any orbit with initial point $(x(0), y(0)) \in \Delta_{2\varrho/3}$, we have

$$\left|p_l(t) - p_l(0)\right| < \beta_1 \varrho^3 \quad \text{for } |t| < T^* := \beta_2 \exp\left[-\left(\frac{\varrho^*}{\varrho}\right)^{1/(\tau+1)}\right] .$$

The proof of the theorem requires a lot of technical estimates which we do not include in the present chapter. We rather defer the complete proof to Chapter 9, where the general case of the Hamiltonian of the general problem of dynamics will be discussed. The proof given there covers also the present case, which is somewhat simpler. For more details the reader may also want to see [79] or [80].

The exponential dependence of T^* on $1/\varrho$ assured by the theorem constitutes a significant improvement with respect to the theory of complete stability of Birkhoff. In the words of Littlewood, "while not eternity, this is a considerable slice of it."

---

[20] The case of an elliptic equilibrium had already been investigated by Moser in [172] and by Littlewood in [153] and [154], with a result similar to Theorem 5.21. The reader may notice the annoying dependence of the estimates on the number n of degrees of freedom, due to $\tau > n - 1$ in the Diophantine condition. This is indeed a major problem for application to large systems.

6
The Method of Lie Series and of Lie Transforms

This chapter should be considered as an *intermezzo*. It has a rather technical character and can be seen as an extension of the chapter on canonical transformations, since it deals with transformations close to identity. The aim is to introduce some useful tools for developing a constructive perturbation scheme based on the algorithms of Lie series and of Lie transforms, which will be used as substitutes for the classical method of constructing near the identity canonical transformations via generating functions in mixed variables, as in Example 2.26.

The usefulness of Lie transform methods relies on two remarks. First, it avoids the cumbersome procedure of inversion of functions and of substitution of variables. Second, completing the formal methods based on Lie transforms with quantitative estimates – as typically required by modern perturbation theory — is not really difficult.

As to the first remark, one may notice that the operations of inversion of functions and substitution of variables are conceptually simple and so introduce no real difficulty from the point of view of Analysis. Moreover, most of the rigorous methods of perturbation theory available in the current literature are based on powerful analytical tools such as the implicit function theorem and fixed point theorems, to quote just the most common ones. The drawback of these method is that they are often not constructive, or at least not easily implemented in explicit calculations, especially if one plans to perform series expansions up to a quite high order in a parameter. For example, the first few steps of an inversion are indeed quite simple – and often faster – than the expansions required by Lie transforms. But soon one realizes that pushing the typical expansions of perturbation theory to higher orders becomes a really cumbersome and time wasting procedure (see [91], sect. 4.3). In contrast, the methods based on Lie transforms allow us to express all series expansions that we need in a very straightforward manner, using only algebraic operations which are easily implemented on computers: sums, products and derivatives.

The second remark is related to an apparent defect of Lie transform methods: they are often used merely as formal tools, skipping all questions related to convergence of the expansions. We could even say that there is a gap between the formal methods of perturbation theory used by astronomers since the times of D'Alembert, Euler and Lagrange and the rigorous methods widely used nowadays in the framework of Analysis. The ambitious attempt of the present notes is to fill, at least partially, this gap.

A few historical notes may be useful. The idea of solving differential equations with series expansions was introduced by Newton [181]. The methods illustrated in the present chapter go back to Marius Sophus Lie. The field of Lie algebras has been developed on the basis of his work; however, this matter will not be discussed in the present notes. The usefulness of Lie series in numerical calculations was pointed out by Wolfgang Gröbner in a series of papers, starting in 1957. The book [104], published in 1967, contains a complete theory including the discussion of the convergence of Lie series for analytic vector fields, when all functions are expressed as power series. The expansions are then applied to numerical integration of differential equations. It should also be mentioned that the approximation of a solution by Lie series has been recently used also as a tool for constructing symplectic integration algorithms which are particularly useful for numerically investigating the dynamics of Hamiltonian systems [23].

The application of Lie series methods to perturbation expansions in Celestial Mechanics was first proposed by Gen-Ichiro Hori [116] and André Deprit [59], who generalized the method by introducing Lie transforms and producing explicit algorithms. Many other papers have been devoted to algorithms based on Lie transforms, mainly for applications in Celestial Mechanics. An attempt to collect several such algorithms may be found in the papers [112] and [113] by Jacques Henrard, where extensive references may be found. However, all these papers were concerned only with the formal aspects, while convergence problems were neglected. An extension of the analytical methods of Gröbner to the Hamiltonian case including use of action-angle variables has been developed in [78].

The exposition here will be concerned only with the application of Lie methods in a Hamiltonian framework, often considering the particular case of action-angle variables. This is indeed what we need in order to establish a working background for the rest of the present notes. However a similar theory may be developed also for general systems of differential equations.

6.1 Formal Expansions

Let us start with a short discussion concerning the meaning to be assigned to the expression *formal calculus*. In Chapter 5 we used it in an intuitive sense – perhaps vague – just saying: "we disregard questions relative to convergence." It is now time to make the last sentence a little more precise, thanks to the assistance of Poincaré.

The easiest and most natural case is concerned with polynomial expansions, or *formal power series*, already used in Chapter 5 in order to construct a first integral for a Hamiltonian in a neighbourhood of an (elliptic) equilibrium. Following Poincaré ([188], Tome II, ch. VIII), we may introduce the concept of formal expansion as follows. Consider the *truncated expansion* $\Phi = \Phi_0 + \cdots + \Phi_r$, that is, a non-homogeneous polynomial of degree $r+2$. We say that *the equation* $\{H, \Phi\} = 0$ *is satisfied in a formal sense if, by replacing the truncated polynomial, we get*

$$\{H, \Phi\} = \mathcal{O}(|(x, y)|^{r+3}) , \tag{6.1}$$

that is, it is of degree at least $r+3$*, which we have identified as order* $r+1$. Let us call the difference *the remainder*. Henceforth we shall denote as $\mathcal{O}(r+1)$ the remainder of order $r+1$, in whatever sense. This is essentially what we did in Chapter 5, and the propositions stated there should be interpreted in this sense.

A similar attitude may be taken when considering expansions in a small parameter ε, as considered in Chapter 4: just interpret order $\mathcal{O}(r+1)$ as meaning *terms which have a power* ε^{r+1} *(or higher) as a factor*. More elaborate cases may occur, as we shall see in the rest of the present notes.

Such an attitude turns out to be useful in view of the fact, known after Poincaré, that most series usually considered in perturbation theory are not convergent. On the other hand there is a long-standing and successful tradition in Celestial Mechanics (or Astronomy) or in Physics that most calculations are based on use of the first-order terms of a series of expansions, disregarding the rest of the series. The problem of convergence is not even mentioned, in most cases.

Such a state of affairs was splendidly described by Poincaré (see [188], tome II, chap. VIII).

> *Il y a entre les géomètres et les astronomes une sorte de malentendu au sujet de la signification du mot* convergence. *Les géomètres, préoccupés de la parfaite rigueur et souvent trop indifférents à la longueur de calculs inextricables dont ils conçoivent la possibilité, sans songer à les entreprendre effectivement, disent qu'une série est convergente quand la somme des termes tend vers une limite déterminée, quand même les premiers termes diminueraient très lentement. Les astronomes, au contraire, ont coutume de dire qu'une série converge quand les vingt premiers termes, par exemple, diminuent très rapidement, quand même les termes suivants devraient croître indéfiniment.*

Ainsi, pour prendre un exemple simple, considérons les deux séries qui ont pour terme général

$$\frac{1000^n}{1 \cdot 2 \cdot 3 \cdots n} \quad \text{et} \quad \frac{1 \cdot 2 \cdot 3 \cdots n}{1000^n} .$$

Les géomètres diront que la première série converge, et même qu'elle converge rapidement, parce que le millionième terme est beaucoup plus petit que le 999 999e; mais ils regarderont la seconde come divergente, parce que le terme général peut croître au delà de toute limite.

Les astronomes, au contraire, regarderont la première série comme divergente, parce que les 1000 premiers termes vont en croissant; et la seconde comme convergente, parce que les 1000 premiers termes vont en décroissant et que cette décroissance est d'abord très rapide.

Les deux règles sont légitimes: la première, dans les récherches théoriques; la séconde, dans les applications numériques. Toutes deux doivent régner, mais dans deux domaines séparés et dont il importe de bien connaître les frontières.

......

Le premier exemple qui a montré clairement la légitimité de certains développements divergentes est l'exemple classique de la série de Stirling. Cauchy a montré que les termes de cette série vont d'abord en décroissant, puis en croissant, de sorte que la série diverge; mais si l'on s'arrête au terme le plus petit, on répresente la function eulérienne avec une approximation d'autant plus grande que l'argument est plus grand.

An example of the usefulness of a divergent series was given in Chapter 5, where an exponential stability estimate was obtained for an elliptic equilibrium. A similar but more general application is the theorem of Nekhoroshev which will be the subject of Chapter 9. In Chapter 8 the method of Lie series will be used to prove the theorem of Kolmogorov on existence of invariant tori for perturbed systems. In all cases it will be essential to implement a scheme of quantitative analytical estimates.

6.2 Lie Series

This section is devoted to the formal aspects of the theory of Lie series, in the sense of the formal calculus. Thus, all questions concerning convergence will be neglected here and deferred to Sections 6.5 and 6.6. Particular care will instead be devoted to the formulation of *constructive* algorithms. In this chapter all functions will be assumed to be analytic.

On a $2n$-dimensional phase space endowed with canonical coordinates p, q, consider an analytic function $\chi(p, q)$, which will be called a *generating function.* The Lie derivative, $L_\chi \cdot = \{\cdot, \chi\}$, will play a basic role throughout the whole chapter.

6.2.1 The Lie Series Operator

The *Lie series* operator is defined as the exponential of εL_χ, namely

$$\exp(\varepsilon L_\chi) = \sum_{s\geq 0} \frac{\varepsilon^s}{s!} L_\chi^s \tag{6.2}$$

(see, e.g., [104]). It represents the time-ε evolution of the canonical flow generated by the autonomous Hamiltonian χ.

The basic use of Lie series is the explicit expression of the canonical flow. Consider $\chi(p', q')$ as a function of the new variables q', p' and write

$$\begin{aligned} p &= \exp(\varepsilon L_\chi)p' = p' - \varepsilon \left.\frac{\partial \chi}{\partial q'}\right|_{p',q'} + \frac{\varepsilon^2}{2} L_\chi \left.\frac{\partial \chi}{\partial q'}\right|_{p',q'} + \ldots \,, \\ q &= \exp(\varepsilon L_\chi)q' = q' + \varepsilon \left.\frac{\partial \chi}{\partial p'}\right|_{p',q'} + \frac{\varepsilon^2}{2} L_\chi \left.\frac{\partial \chi}{\partial p'}\right|_{p',q'} + \ldots \,. \end{aligned} \tag{6.3}$$

This is the explicit expression of the flow at time ε, and is a near the identity canonical transformation in the sense that it depends analytically on ε as a parameter, and for $\varepsilon = 0$ is the identity.[1] The reader will recognize the power expansion in time of the local solution of an holomorphic system of differential equations, as used by Cauchy. It is also an easy matter to write the inverse transformation: exploiting the fact that the flow is autonomous, it is enough to replace χ by $-\chi$ or, equivalently, ε by $-\varepsilon$.

Example 6.1: *Translation and deformation.* Consider the case of action-angle variables $p \in \mathbb{R}^n$ and $q \in \mathbb{T}^n$. The function $\chi = \sum_l \xi_l q'_l$, with $\xi \in \mathbb{R}^n$ generates the one-parameter canonical transformation at time ε

$$p = \exp(\varepsilon L_\chi)p' = p' - \varepsilon\xi \,, \quad q = \exp(\varepsilon L_\chi)q' = q' \,,$$

namely a translation in the action space which leaves the angles unchanged. Similarly, a generating function $\chi = \chi(q')$ (independent of p') generates the transformation

$$p = p' - \varepsilon\frac{\partial \chi}{\partial q'} \,, \quad q = q' \,,$$

namely a q-depending deformation of the action variables. *E.D.*

---

[1] We have already encountered an expression similar to this one in Section 5.1. The flow of a linear system has been expressed as the action of an evolution operator, the exponential of a matrix, i.e., the Lie series operator for a linear vector field.

Lemma 6.2: *The Lie series operator has the following properties.*

(i) Linearity: for any pair of functions f, g and for all $\alpha \in \mathbb{R}$, we have

$$\exp(\varepsilon L_\chi)(f+g) = \exp(\varepsilon L_\chi) f + \exp(\varepsilon L_\chi) g \ ,$$
$$\exp(\varepsilon L_\chi)(\alpha f) = \alpha \exp(\varepsilon L_\chi) f \ .$$

(ii) Distributivity over the product (or preservation of product): for any pair of functions f, g, we have

$$\exp(\varepsilon L_\chi)(f \cdot g) = \exp(\varepsilon L_\chi) f \cdot \exp(\varepsilon L_\chi) g \ .$$

(iii) Distributivity over the Poisson bracket (preservation of the Poisson bracket): for any pair of functions f, g, we have

$$\exp(\varepsilon L_\chi)\{f,g\} = \bigl\{\exp(\varepsilon L_\chi) f, \exp(\varepsilon L_\chi) g\bigr\} \ .$$

The easy proof is left to the reader. Note in particular that the property (iii) means that the Hamiltonian flow is a canonical transformation, a property already known to us from Proposition 2.15.

Having defined a coordinate transformation, we may ask how a function is transformed. Here the considerations of Section 6.1 concerning the formal calculus play a relevant role. Having given a function $f(q,p)$, we should calculate the transformed function $f'(p',q')$ by substitution of the transformation (6.3). In perturbation theory we also need to expand the transformed function in powers of the parameter ε up to a given order, which is clearly a long and boring procedure since we should apply Taylor's formula. A remarkable property is the following.

Lemma 6.3: *Let a generating function $\chi(p,q)$ and a function $f(p,q)$ be given. Then the following equality holds true:*

$$f(p,q)\Big|_{p=\exp(\varepsilon L_\chi)p',\, q=\exp(\varepsilon L_\chi)q'} = \exp(\varepsilon L_\chi) f\Big|_{p=p',\, q=q'} \ . \tag{6.4}$$

This lemma has been named by Gröbner the *exchange theorem.* A few comments are in order.

The claim is that *the series expansion in ε of the transformed function may be calculated by applying the exponential operator of the Lie series directly to the function, with no need of making a substitution of variables.*[2] This

[2] A further comment on the role of the variables may be useful, if perhaps pedantic. The left member of (6.4) produces a function of the new variables q', p'. On the other hand, the right member may be calculated by considering both functions χ and f as depending on the old variables q, p so that the result is a function of q, p. The substitution $p = p'$, $q = q'$ means that the variables q, p are merely renamed. The equality must be interpreted in the sense that we take care only of the form of the function, the name of the variables being irrelevant.

may appear trivial in view of the following considerations. The substitution of variables in the left member of (6.4) produces a function $f'(q', p', \varepsilon)$, where ε is the time of the flow. We may expand the function in Taylor series by calculating the derivatives of f' with respect to ε. The right member says that the derivatives with respect to ε are calculated as Lie derivatives with respect to the flow generated by χ. This seems quite obvious. However, some care is required since we are combining two operations: in the left member we first make the substitution, and then we perform the expansion; in the right member we perform the operations in reverse order. To be rigorous, we should prove that the result is the same. This is evident for a polynomial in view of Properties (i) and (ii) of Lemma 6.2, and for an analytic function follows via approximation by polynomials. The complete proof may be found in [104], § I.2.

6.2.2 The Triangle of Lie Series

An elegant and effective representation of the operation of transforming a function is found as follows. Assume, as is typical in perturbation theory, that the function to be transformed is expanded in power series of the parameter ε, namely $f(p, q, \varepsilon) = f_0(p, q) + \varepsilon f_1(p, q) + \varepsilon^2 f_2(p, q) + \ldots$, and that we want to write the transformed function $g = \exp(\varepsilon L_\chi) f$ as a power series $g_0 + \varepsilon g_1 + \varepsilon^2 g_2 + \ldots$ in ε. Working at a formal level we may use the linearity of the Lie series operator, thus writing

$$g = \exp(\varepsilon L_\chi) f_0 + \varepsilon \exp(\varepsilon L_\chi) f_1 + \varepsilon^2 \exp(\varepsilon L_\chi) f_2 + \ldots .$$

That is: we apply the Lie series to every term of the expansion of f. The action of the operator is nicely represented by a triangular diagram.

$$\begin{array}{cccccc}
g_0 & f_0 & & & & \\
 & \downarrow & & & & \\
g_1 & L_\chi f_0 & f_1 & & & \\
 & \downarrow & \downarrow & & & \\
g_2 & \frac{1}{2!} L_\chi^2 f_0 & L_\chi f_1 & f_2 & & \\
 & \downarrow & \downarrow & \downarrow & & \\
g_3 & \frac{1}{3!} L_\chi^3 f_0 & \frac{1}{2!} L_\chi^2 f_1 & L_\chi f_2 & f_3 & \\
 & \downarrow & \downarrow & \downarrow & \downarrow & \\
\vdots & \vdots & \vdots & \vdots & \vdots & \ddots
\end{array} \tag{6.5}$$

This fact should be kept in mind when we use the algorithm: don't worry about the names of the variables; just transform the functions.

Terms of the same order in ε are aligned on the same line. The calculation may be performed by columns, as indicated by the arrows: if the function f and the generating function χ are known, then every column may be calculated, proceeding top-down until the line corresponding to the desired order in ε is reached. The expression of g is calculated by adding up all terms of the same row, thus finding the expansion of g up to the desired order. Everything not included in the diagram has higher order in ε and need not be calculated. This is precisely what we mean while saying that $g = \exp(L_\chi)f$ in formal sense.

A compact form of the diagram is given by the recurrent formula

$$g_0 = f_0 \ , \quad g_r = \sum_{j=0}^{r} \frac{1}{j!} L_\chi^j f_{r-j} \ . \tag{6.6}$$

6.2.3 Composition of Lie Series

As we have seen, the Lie series defines a near-the-identity transformation, which in our case is a canonical one. A natural question is whether the converse is also true, that is, if every near-the-identity canonical transformation can be expressed as a Lie series.

The question may be formulated in more precise terms. We are actually considering a one-parameter family of transformations. Suppose that we are given such a family in the form

$$\begin{aligned} q &= q' + \varepsilon\varphi_1(p',q') + \varepsilon^2\varphi_2(p',q') + \ldots \ , \\ p &= p' + \varepsilon\psi_1(p',q') + \varepsilon^2\psi_2(p',q') + \ldots \ , \end{aligned} \tag{6.7}$$

where the functions $\varphi_1, \varphi_2, \ldots$ and $\psi_1, \psi_2, \ldots$ are assumed to satisfy the necessary conditions for the transformation to be canonical. The question is whether a generating function $\chi(p',q')$ exists which produces exactly this transformation. The answer in general is in the negative, the limitation being that we are considering the flow of an autonomous system. However, more general transformations can be constructed by *composition of Lie series.*

Suppose we are given a sequence $\chi = \{\chi_1, \chi_2, \chi_3, \ldots\}$ of generating functions, and let the sequence of operators $\{S^{(1)}, S^{(2)}, S^{(3)}, \ldots\}$ be defined by recurrence as

$$S^{(1)} = \exp\bigl(\varepsilon L_{\chi_1}\bigr) \ , \quad S^{(k)} = \exp\bigl(\varepsilon^k L_{\chi_k}\bigr) \circ S^{(k-1)} \quad \text{for } k > 1 \ . \tag{6.8}$$

If we work at formal level, we interrupt the sequence at $k = r$ for some $r > 0$, also truncating all expansions at order r; all the rest of the sequence produces only terms of order $\mathcal{O}(r+1)$. If we can prove that the expansion converges to some limit, then the operator

$$S_\chi = \cdots \circ \exp\bigl(\varepsilon^3 L_{\chi_3}\bigr) \circ \exp\bigl(\varepsilon^2 L_{\chi_2}\bigr) \circ \exp\bigl(\varepsilon L_{\chi_1}\bigr) \tag{6.9}$$

is well defined. This procedure defines the composition of Lie series.

The sequence $\tilde{S}^{(1)}, \tilde{S}^{(2)}, \tilde{S}^{(3)}, \ldots$, defined by recurrence as

$$\tilde{S}^{(1)} = \exp\bigl(-\varepsilon L_{\chi_1}\bigr) \ , \quad \tilde{S}^{(k)} = \tilde{S}^{(k-1)} \circ \exp\bigl(-\varepsilon^k L_{\chi_k}\bigr) \quad \text{for } k > 1 \ , \tag{6.10}$$

produces the inverse operator *in a formal sense*, namely

$$\tilde{S}^{(r)} \circ S^{(r)} = \mathcal{O}(r+1) \ . \tag{6.11}$$

If the sequence $\tilde{S}^{(k)}$ tends to some limit, then we get the inverse of the operator S_χ, namely

$$S_\chi^{-1} = \exp\bigl(-\varepsilon L_{\chi_1}\bigr) \circ \exp\bigl(-\varepsilon^2 L_{\chi_2}\bigr) \circ \exp\bigl(-\varepsilon^3 L_{\chi_3}\bigr) \circ \ldots \ . \tag{6.12}$$

In this case, but only in this case, (6.11) is replaced by $\tilde{S}_\chi \circ S_\chi = \mathrm{Id}$.

Let us pay a little attention to the triangular diagram (6.5), making clear how the calculation should proceed when a composition of Lie series is considered. Keep in mind that all equalities must be considered in a formal sense. Suppose we want to transform the function $f = f_0 + \varepsilon f_1 + \varepsilon^2 f_2 + \ldots$, which means that we want

$$g = \cdots \circ \exp\bigl(\varepsilon^2 L_{\chi_2}\bigr) \circ \exp\bigl(\varepsilon L_{\chi_1}\bigr) f \ .$$

We first calculate $f' = \exp\bigl(\varepsilon L_{\chi_1}\bigr) f = f'_0 + \varepsilon f'_1 + \varepsilon^2 f'_2 + \ldots$ as indicated in the diagram (6.5), just writing f' in place of g and χ_1 in place of χ. Then we calculate $f'' = \exp\bigl(\varepsilon^2 L_{\chi_2}\bigr) f' = f''_0 + \varepsilon f''_1 + \varepsilon^2 f''_2 + \ldots$ through a similar diagram, paying attention to the correct alignment of the powers of ε. With a moment's thought we realize that the diagram should be represented as

$$\begin{array}{ccccccc}
f''_0 & f'_0 & & & & & \\
 & \downarrow & & & & & \\
f''_1 & 0 & f'_1 & & & & \\
 & \downarrow & \downarrow & & & & \\
f''_2 & L_{\chi_2} f'_0 & 0 & f'_2 & & & \\
 & \downarrow & \downarrow & \downarrow & & & \\
f''_3 & 0 & L_{\chi_2} f'_1 & 0 & f'_3 & & \\
 & \downarrow & \downarrow & \downarrow & \downarrow & & \\
f''_4 & \frac{1}{2!} L_{\chi_2}^2 f'_0 & 0 & L_{\chi_2} f'_2 & 0 & f'_4 & \\
 & \downarrow & \downarrow & \downarrow & \downarrow & \downarrow & \\
\vdots & \vdots & \vdots & \vdots & \vdots & \vdots & \ddots \ ,
\end{array}$$

with several null elements, because we proceed by powers of ε^2, not of ε. Similarly, when we go on with our procedure we construct a diagram for $\exp\bigl(\varepsilon^s L_{\chi_s}\bigr)$ which contains in every column one non-zero element out of s. A compact recurrent formula for $f^{(r)} = L_{\chi_r} f^{(r-1)}$ may be given, namely

$$f_s^{(r)} = \sum_{j=0}^{k} \frac{1}{j!} L_{\chi_r}^j f_{s-jr}^{(r-1)} \ , \quad k = \left\lfloor \frac{s}{r} \right\rfloor \ , \tag{6.13}$$

where $\lfloor x \rfloor$ denotes the maximal integer which does not exceed x.

Pedantically, let us see in some more detail how to proceed in a practical calculation, keeping in mind that we shall unavoidably truncate the series to some order ε^r for some $r \geq 1$.

Let us pick a truncated function $f = f_0 + \varepsilon f_1 + \cdots + \varepsilon^r f_r$ and suppose that we want to construct the transformed function up to degree r in ε. With a little attention we realize that *it is enough to construct every diagram until we reach the line corresponding to the power* ε^r*, and in particular we need to know only the generating functions* $\chi_1, \ldots, \chi_r$. For this includes *all* terms up to order ε^r and neglects everything of higher order. Similar considerations apply to the inverse transformation: it is enough to consider only the action of the operator $\tilde{S}^{(r)}$ defined by (6.10).

A fact which may raise some perplexity in an actual calculation is the following. Suppose that we have constructed the generating functions $\chi_1, \ldots, \chi_r$, so that we know how to construct the operators $S_\chi^{(r)}$ and $\tilde{S}_\chi^{(r)}$. Let us calculate the transformation

$$q = S_\chi^{(r)} q' \ , \quad p = S_\chi^{(r)} p'$$

up to degree r in ε. This will give us expressions such as

$$\begin{aligned} q &= q' + \varepsilon\varphi_1(p', q') + \cdots + \varepsilon^r \varphi_r(p', q') \ , \\ p &= p' + \varepsilon\psi_1(p', q') + \cdots + \varepsilon^r \psi_r(p', q') \ , \end{aligned} \tag{6.14}$$

where the functions $\varphi_1(p', q'), \ldots, \varphi_r(p', q')$ and $\psi_1(p', q'), \ldots, \psi_r(p', q')$ may be explicitly calculated. We may then consider the inverse transformation

$$q' = \tilde{S}_\chi^{(r)} q \ , \quad p' = \tilde{S}_\chi^{(r)} p \ .$$

This will give us the expressions

$$\begin{aligned} q' &= q + \varepsilon\tilde{\varphi}_1(p, q) + \cdots + \varepsilon^r \tilde{\varphi}_r(p, q) \ , \\ p' &= p + \varepsilon\tilde{\psi}_1(p, q) + \cdots + \varepsilon^r \tilde{\psi}_r(p, q) \ , \end{aligned} \tag{6.15}$$

which could be explicitly calculated. Suppose now that we substitute the expressions (6.15) into (6.14). By Lemma 6.3 (the exchange theorem) this is equivalent to applying the operator $\tilde{S}_\chi^{(r)}$ to the functions in the right member of (6.14). This is expected to give us the identity, which is definitely true *if we consider the infinite series.* If we work with truncated series at order r, we shall realize that the result is the identity up to a term of order ε^{r+1}.

This is the best we can expect in a practical calculation – and is coherent with the rules of formal calculus.

Lemma 6.4: *Let a near-the-identity canonical transformation be given as in (6.7). Then there exists a sequence* $\chi = \{\chi_1(p',q'), \chi_2(p',q'),\ldots\}$ *of generating functions such that*

$$q = S_\chi q' \ , \quad p = S_\chi p' \ .$$

Proof. Let us see how the sequence χ is constructed, step by step. In the formal approach, we want to determine χ_1 so that

$$\begin{aligned} \exp(\varepsilon L_{\chi_1})q_j - \bigl[q_j + \varepsilon\varphi_{1,j}\bigr] &= \mathcal{O}(2) \ , \\ \exp(\varepsilon L_{\chi_1})p_j - \bigl[p_j + \varepsilon\psi_{1,j}\bigr] &= \mathcal{O}(2) \ . \end{aligned} \tag{6.16}$$

The condition that the transformation (6.7) be canonical (in a formal sense at order 1) is

$$\begin{aligned} \{q_j + \varepsilon\varphi_{1,j}, q_k + \varepsilon\varphi_{1,k}\} &= \{q_j,q_k\} + \varepsilon\{q_j,\varphi_{1,k}\} + \varepsilon\{\varphi_{1,j},q_k\} = \mathcal{O}(2) \ , \\ \{p_j + \varepsilon\psi_{1,j}, p_k + \varepsilon\psi_{1,k}\} &= \{p_j,q_k\} + \varepsilon\{p_j,\psi_{1,k}\} + \varepsilon\{\psi_{1,j},p_k\} = \mathcal{O}(2) \ , \\ \{q_j + \varepsilon\varphi_{1,j}, p_k + \varepsilon\psi_{1,k}\} &= \{q_j,p_k\} + \varepsilon\{q_j,\psi_{1,k}\} + \varepsilon\{\varphi_{1,j},p_k\} = 1 + \mathcal{O}(2) \ . \end{aligned}$$

If the transformation (6.7) is canonical, then the local condition

$$\frac{\partial\varphi_{1,k}}{\partial p_j} - \frac{\partial\varphi_{1,j}}{\partial p_k} = 0 \ , \quad -\frac{\partial\psi_{1,k}}{\partial q_j} + \frac{\partial\psi_{1,j}}{\partial q_k} = 0 \ , \quad \frac{\partial\psi_{1,k}}{\partial p_j} + \frac{\partial\varphi_{1,j}}{\partial q_k} = 0$$

is satisfied; hence there is a (local) function $\chi_1(q,p)$ such that

$$\varphi_{1,j} = \frac{\partial\chi_1}{\partial p_j} \ , \quad \psi_{1,j} = -\frac{\partial\chi_1}{\partial q_j} \ ;$$

the function χ_1 is determined by quadrature, and (6.16) is satisfied, too. Suppose now that we have determined $\{\chi_1,\ldots,\chi_{r-1}\}$ so that

$$\begin{aligned} &\exp(\varepsilon^{(r-1)}L_{\chi_{r-1}}) \circ \cdots \circ \exp(\varepsilon L_{\chi_1})q_j \\ &\qquad - \bigl[q_j + \varepsilon\varphi_{1,j} + \cdots + \varepsilon^{(r-1)}\varphi_{r-1,j}\bigr] - \varepsilon^r\varphi'_{r,j} = \mathcal{O}(r+1) \ , \\ &\exp(\varepsilon^{(r-1)}L_{\chi_{r-1}}) \circ \cdots \circ \exp(\varepsilon L_{\chi_1})p_j \\ &\qquad - \bigl[p_j + \varepsilon\psi_{1,j} + \cdots + \varepsilon^{(r-1)}\psi_{r-1,j}\bigr] - \varepsilon^r\psi'_{r,j} = \mathcal{O}(r+1) \ . \end{aligned}$$

Remark that the expressions $\varphi'_{r,j}$ and $\psi'_{r,j}$ can be explicitly determined. On the other hand we must check that, if (6.7) is canonical, then the local conditions

$$\frac{\partial\varphi'_{r,k}}{\partial p_j} - \frac{\partial\varphi'_{r,j}}{\partial p_k} = 0 \ , \quad -\frac{\partial\psi'_{r,k}}{\partial q_j} + \frac{\partial\psi'_{r,j}}{\partial q_k} = 0 \ , \quad \frac{\partial\psi'_{r,k}}{\partial p_j} + \frac{\partial\varphi'_{r,j}}{\partial q_k} = 0$$

are satisfied. The calculation requires some patience and attention: one must take advantage of the fact that canonicity up to order $r-1$ follows from the construction, for the transformation generated by a Lie series is canonical; we leave it to the reader. Thus, the function χ_r can be determined by quadrature. We conclude that the sequence can be constructed up to any desired order, as claimed. *Q.E.D.*

6.3 Lie Transforms

The algorithm of Lie series may be generalized to the case of a time-dependent generating function $\chi(p,q,t)$, so that the corresponding time evolution is due to the flow of a non-autonomous Hamiltonian. This leads to an algorithm different from (6.2). As already said, several different formulæ have been devised in order to give the Lie transform an algorithmic recurrent form: everybody, of course, has his favourite one. To make a definite choice, the algorithm used here was proposed in [76] and is the favourite of the author. It is related to the "algorithm of the inverse," found by Henrard [112].

Hereafter, particular attention will be paid to the algebraic aspect of the method, leaving somewhat hidden the relation with a canonical flow. In particular, the algorithm will be formulated after setting $\varepsilon = 1$.

Consider a *generating sequence* $\chi = \{\chi_s\}_{s\geq 1}$ of analytic functions on the phase space. The Lie transform operator T_χ is defined as

$$T_\chi = \sum_{s\geq 0} E_s \ , \tag{6.17}$$

where the sequence $\{E_s\}_{s\geq 0}$ of operators is recurrently defined as

$$E_0 = \mathrm{Id} \ , \quad E_s = \sum_{j=1}^{s} \frac{j}{s} L_{\chi_j} E_{s-j} \ . \tag{6.18}$$

A coordinate transformation is written as

$$\begin{aligned} q &= T_\chi q' = q' + L_{\chi_1} q' + \left(\frac{1}{2} L_{\chi_1}^2 q' + L_{\chi_2} q'\right) + \ldots \ , \\ p &= T_\chi p' = p' + L_{\chi_1} p' + \left(\frac{1}{2} L_{\chi_1}^2 p' + L_{\chi_2} p'\right) + \ldots \ . \end{aligned} \tag{6.19}$$

This is a canonical transformation (see Lemma 6.5).

A direct connection with Lie series comes out by considering the case $\chi = \{\chi_1, 0, 0, \ldots\}$, namely a generating sequence containing only the first term. Then the Lie transform generated by χ coincides with the Lie series generated by χ_1, this is, $T_\chi = \exp(L_{\chi_1})$.

Lemma 6.5: *The Lie transform T_χ defined by (6.17) and (6.18) has the following properties.*

(i) Linearity: for any pair f, g of functions and for $\alpha \in \mathbb{R}$, we have

$$T_\chi(f+g) = T_\chi f + T_\chi g \ , \quad T_\chi(\alpha f) = \alpha T_\chi f \ .$$

(ii) Distributivity over the product: for any pair f, g of functions, we have

$$T_\chi(f \cdot g) = T_\chi f \cdot T_\chi g \ .$$

(iii) Distributivity over the Poisson bracket: for any pair f, g of functions, we have

$$T_\chi\{f, g\} = \{T_\chi f, T_\chi g\} \ .$$

For a proof see, for example, [78]. Property (iii) is particularly relevant, since it means that the coordinate transformation (6.19) is canonical.

The reader will notice that the Lie transform and the Lie series have the same formal properties. This is true also for the exchange theorem.

Lemma 6.6: *Let the generating sequence $\chi = \{\chi_s\}_{s\geq 1}$ and a function $f(p,q)$ be given. We have*

$$f(p,q)\Big|_{p=T_\chi p',\, p=T_\chi q'} = T_\chi f\Big|_{p=p',\, q=q'} \ . \tag{6.20}$$

The proof may be obtained using the linearity and the conservation of products. This makes the claim evident for polynomials, and the result may be extended to analytic functions; the remarks made for the Lie series apply.

6.3.1 The Triangular Diagram for the Lie Transform

The calculation of a Lie transform has a nice graphical representation similar to the triangle of Lie series. Suppose that we are given a function $f = f_0 + f_1 + \ldots$ and denote by $g = g_0 + g_1 + \ldots$ its transformed function $g = T_\chi f$. Using the linearity of the Lie transform, we can apply T_χ separately to every term of f. It is useful to rearrange terms according to the triangular diagram

$$\begin{array}{cccccc}
g_0 & f_0 & & & & \\
 & \downarrow & & & & \\
g_1 & E_1 f_0 & f_1 & & & \\
 & \downarrow & \downarrow & & & \\
g_2 & E_2 f_0 & E_1 f_1 & f_2 & & \\
 & \downarrow & \downarrow & \downarrow & & \\
g_3 & E_3 f_0 & E_2 f_1 & E_1 f_2 & f_3 & \\
 & \downarrow & \downarrow & \downarrow & \downarrow & \\
\vdots & \vdots & \vdots & \vdots & \vdots & \ddots ,
\end{array} \tag{6.21}$$

where terms of the same order appear on the same line. Remark that the operator T_χ acts by columns, as indicated by the arrows: the knowledge of f_j and of the generating sequence allows one to construct the whole column below f_j. Thus, we get $g_0 = f_0$ from the first line, $g_1 = E_1 f_0 + f_1$ from the second line, and so on. This shows how to practically perform the transformation.[3] Again, truncating the diagram at line r allows us to calculate all contributions up to order r and forget everything of higher order, which is coherent with the rules of formal calculus.

Finding the inverse of the Lie transform seems to be definitely more complicated than for Lie series: changing the sign of the generating sequence (as for a Lie series) is not enough. However, with a bit of attention we may realize that it is just a matter of using in a skillful manner the triangular diagram (6.21). Assume that g is given and f is unknown. Then, the first line gives $f_0 = g_0$; having determined f_0, all the column below f_0 can be constructed, and the second line gives immediately $f_1 = g_1 - E_1 f_0$; having determined f_1, all the corresponding column can be constructed, so that f_2 can be determined from the third line as $f_2 = g_2 - E_2 f_0 - E_1 f_1$, and so on.

There is also an explicit formula for the inverse, namely

$$T_\chi^{-1} = \sum_{s \geq 0} D_s \ , \tag{6.22}$$

---

[3] The scheme may look mysterious to a reader who is not familiar with the methods of expansion of perturbation theory, so it may be convenient to clarify this matter. Introduce an expansion parameter, ε say, and write $\varepsilon\chi_1, \varepsilon^2\chi_2, \ldots$ in place of $\chi_1, \chi_2, \ldots$. Then the Lie transform applied to a generic function f (independent of the expansion parameter ε) generates in a natural way a transformed function expanded in power series of ε. For, in view of $L_{\varepsilon\chi_1} = \varepsilon L_{\chi_1}$, $L_{\varepsilon^2\chi_2} = \varepsilon^2 L_{\chi_2}$, ... one easily sees that the operator E_s produces a factor ε^s. Let now f itself be a series in ε, namely $f = f_0 + \varepsilon f_1 + \varepsilon^2 f_2 + \ldots$, and look for the ε-expansion $g = g_0 + \varepsilon g_1 + \varepsilon^2 g_2 + \ldots$ of its transformed function $g = T_\chi f$. Then the triangular diagram (6.21) is easily constructed by putting on the same line the functions which have the same power of ε as the coefficient. This should make the whole procedure natural. At this point, just set $\varepsilon = 1$ and leave everything else in its place, noting that the indexes play the role of the exponents of ε. Indeed, this is not just a formal game. We may consider the index s of a function, in our notation, as meaning that the function is *small of order s* in some sense. The quantitative theory will be responsible for giving the latter expression a definite meaning, for example, by assuring that with increasing values of the index s the size of the function decreases in some regular manner.

where

$$D_0 = \mathrm{Id}\ , \quad D_s = -\sum_{j=1}^{s} \frac{j}{s} D_{s-j} L_{\chi_j}\ . \tag{6.23}$$

However, the latter expression is actually useless for a practical computation: the algorithm just described is much more effective. Nevertheless, the explicit recurrent formula is useful for quantitative estimates.

The property that makes the Lie transform quite useful is expressed by the following.

Lemma 6.7: *Let a near-the-identity canonical transformation be given in the form (6.7), namely*

$$\begin{aligned} q &= q' + \varepsilon\varphi_1(p',q') + \varepsilon^2\varphi_2(p',q') + \ldots\ , \\ p &= p' + \varepsilon\psi_1(p',q') + \varepsilon^2\psi_2(p',q') + \ldots\ . \end{aligned} \tag{6.24}$$

Then for any $r > 0$ there exists a finite sequence $\chi^{(r)} = \left\{\varepsilon^s\chi_s(p',q')\right\}_{s=1,\ldots,r}$ such that we have

$$q = T_{\chi^{(r)}} q' + \mathcal{O}(r+1), \quad p = T_{\chi^{(r)}} p' + \mathcal{O}(r+1)\ . \tag{6.25}$$

The generating sequence is determined by solving at every order s the $2n$ equations, for $j = 1, \ldots, n$,

$$L_{\chi_1} q_j = \varphi_{1,j}\ , \quad L_{\chi_1} p_j = \psi_{1,j} \tag{6.26}$$

for $s = 1$, and, for $s > 1$,

$$L_{\chi_s} q_j = \varphi_{s,j} - \sum_{l=1}^{s-1} \frac{l}{s} L_{\chi_l} E_{s-l} q_j\ , \quad L_{\chi_s} p_j = \psi_{s,j} - \sum_{l=1}^{s-1} \frac{l}{s} L_{\chi_l} E_{s-l} p_j\ . \tag{6.27}$$

The lemma is stated in a form which points out the formal character of the construction. The complete transformation is recovered by letting $r \to \infty$, which is true in a formal sense.

Proof. Let us write Eqs. (6.25) in the more explicit form, omitting primes (recall that the name of the variables is irrelevant),

$$\begin{aligned} T_{\chi^{(r)}} q_j &= q_j + \varphi_{1,j} + \cdots + \varphi_{r,j} + \mathcal{O}(r+1)\ , \\ T_{\chi^{(r)}} p_j &= p_j + \psi_{1,j} + \cdots + \psi_{r,j} + \mathcal{O}(r+1)\ . \end{aligned} \tag{6.28}$$

Recalling the definition of T_χ, at every order $r > 0$ we get the equations for the generating sequence

$$E_r q_j = \varphi_{r,j}\ , \quad E_r p_j = \psi_{r,j}\ , \quad j = 1, \ldots, n. \tag{6.29}$$

More explicitly, for $r = 1$ the equations are

$$L_{\chi_1} q_j = \varphi_{1,j}\ , \quad L_{\chi_1} p_j = \psi_{1,j}\ , \tag{6.30}$$

namely (6.26). For $r > 1$ we use the expression (6.18) of E_r and separate the last term of the sum, thus writing the equations as

$$(6.31)\quad L_{\chi_s} q_j + \sum_{l=1}^{r-1} \frac{l}{r} L_{\chi_l} E_{r-l} q_j = \varphi_{r,j} \ , \quad L_{\chi_r} p_j + \sum_{l=1}^{r-1} \frac{l}{r} L_{\chi_l} E_{r-l} p_j = \psi_{r,j} \ ,$$

namely (6.27). We should now prove that all the latter equations may be integrated, so as to find the generating functions χ_r at every desired order. This follows from the canonical character of the transformation, as we are going to see in detail.

Let us first give an explicit form to the canonicity conditions for the transformation which are satisfied by hypothesis. We have

$$\begin{aligned}
&\{q_j + \varepsilon\varphi_{1,j} + \varepsilon^2\varphi_{2,j} + \ldots, q_k + \varepsilon\varphi_{1,k} + \varepsilon^2\varphi_{2,k} + \ldots\} = 0 \ ,\\
&\{p_j + \varepsilon\psi_{1,j} + \varepsilon^2\psi_{2,j} + \ldots, p_k + \varepsilon\psi_{1,k} + \varepsilon^2\psi_{2,k} + \ldots\} = 0 \ ,\\
&\{q_j + \varepsilon\varphi_{1,j} + \varepsilon^2\varphi_{2,j} + \ldots, p_k + \varepsilon\psi_{1,k} + \varepsilon^2\psi_{2,k} + \ldots\} = \delta_{j,k} \ .
\end{aligned}$$

Separating the part of order $r = 1$ we get

$$(6.32)\qquad \begin{aligned}
&\{q_j, \varphi_{1,k}\} + \{\varphi_{1,j}, q_k\} = 0 \ ,\\
&\{p_j, \psi_{1,k}\} + \{\psi_{1,j}, p_k\} = 0 \ ,\\
&\{q_j, \psi_{1,k}\} + \{\psi_{1,j}, p_k\} = 0 \ ,
\end{aligned}$$

while for order $r > 1$ we get

$$(6.33)\qquad \begin{aligned}
&\{q_j, \varphi_{r,k}\} + \{\varphi_{r,j}, q_k\} + \sum_{l=1}^{r-1} \{\varphi_{l,j}, \varphi_{r-l,k}\} = 0 \ ,\\
&\{p_j, \psi_{r,k}\} + \{\psi_{r,j}, p_k\} + \sum_{l=1}^{r-1} \{\psi_{l,j}, \psi_{r-l,k}\} = 0 \ ,\\
&\{q_j, \psi_{r,k}\} + \{\varphi_{r,j}, p_k\} + \sum_{l=1}^{r-1} \{\varphi_{l,j}, \psi_{r-l,k}\} = 0 \ .
\end{aligned}$$

From (6.32), by calculating the Poisson brackets, we find

$$\frac{\partial \chi_1}{\partial p_j} = \varphi_{1,j} \ , \quad -\frac{\partial \chi_1}{\partial q_j} = \psi_{1,j} \ ,$$

so that χ_1 can be determined by integrating equations (6.30). For $r > 1$ we proceed by induction. Remark that the sum in the right members of Eqs. (6.31) is determined by the truncated sequence $\chi^{(r-1)} = \{\chi_1, \ldots, \chi_{r-1}\}$, for we have more explicitly

$$(6.34)\qquad \begin{aligned}
T_{\chi^{(r-1)}} q_j &= q_j + \varphi_{1,j} + \cdots + \varphi_{r-1,j} + F_{r,j} + \mathcal{O}(r+1) \ ,\\
T_{\chi^{(r-1)}} p_j &= p_j + \psi_{1,j} + \cdots + \psi_{r-1,j} + G_{r,j} + \mathcal{O}(r+1) \ ,
\end{aligned}$$

where

$$F_{r,j} = \sum_{l=1}^{r-1} \frac{l}{r} L_{\chi_l} E_{r-l} q_j \ , \quad G_{r,j} = \sum_{l=1}^{r-1} \frac{l}{r} L_{\chi_l} E_{r-l} p_j$$

are the expressions $E_r q_j$ and $E_r p_j$, respectively, generated by $\chi^{(r-1)}$ with $\chi_r = 0$. Hence the equations read

$$L_{\chi_r} q_j = \varphi_{r,j} - F_{r,j} \ , \quad L_{\chi_r} p_j = \psi_{r,j} - G_{r,j} .$$

We should check the local integrability conditions:

$$\begin{aligned} \frac{\partial}{\partial p_j}(\varphi_{r,k} - F_{r,k}) - \frac{\partial}{\partial p_k}(\varphi_{r,j} - F_{r,j}) &= 0 \ , \\ -\frac{\partial}{\partial q_j}(\psi_{r,k} - G_{r,k}) + \frac{\partial}{\partial q_k}(\psi_{r,j} - G_{r,j}) &= 0 \ , \\ \frac{\partial}{\partial p_j}(\psi_{r,k} - G_{r,k}) + \frac{\partial}{\partial q_k}(\varphi_{r,j} - F_{r,j}) &= 0 \ . \end{aligned} \tag{6.35}$$

Recalling that $T_{\chi^{(r-1)}}$ generates a canonical transformation and in view of (6.34), we have

$$\begin{aligned} \{q_j, F_{r,k}\} + \{F_{r,j}, q_k\} + \sum_{l=1}^{r-1} \{\varphi_{l,j}, \varphi_{r-l,k}\} &= 0 \ , \\ \{p_j, G_{r,k}\} + \{G_{r,j}, p_k\} + \sum_{l=1}^{r-1} \{\psi_{l,j}, \psi_{r-l,k}\} &= 0 \ , \\ \{q_j, G_{r,k}\} + \{F_{r,j}, p_k\} + \sum_{l=1}^{r-1} \{\varphi_{l,j}, \psi_{r-l,k}\} &= 0 \ . \end{aligned}$$

By subtracting the latter equations from (6.33) as appropriate, we get

$$\begin{aligned} \{q_j, \varphi_{r,k} - F_{r,k}\} + \{\varphi_{r,j} - F_{r,j}, q_k\} &= 0 \ , \\ \{p_j, \psi_{r,k} - G_{r,k}\} + \{\psi_{r,j} - G_{r,j}, p_k\} &= 0 \ , \\ \{q_j, \psi_{r,k} - G_{r,k}\} + \{\varphi_{r,j} - F_{r,j}, p_k\} &= 0 \ , \end{aligned}$$

namely the desired conditions (6.35). *Q.E.D.*

It is necessary here to emphasize once more that the sequence of generating functions for the Lie transform does not coincide with the sequence which generates the same transformation by composition of Lie series. This is because Lie derivatives do not commute, in general.

6.4 Analytic Framework

It is now time to introduce the technical tools which will allow us to investigate the convergence or the asymptotic properties of perturbation series. The methods exposed here are in fact a *variazione* over the classical method of majorants due to Cauchy.

6.4.1 *Cauchy estimates*

Consider an open disk $\Delta_\varrho(0)$, with $\varrho > 0$, centred at the origin of the complex plane $\mathbb{C}$. Consider a function f analytic in the open disk $\Delta_\varrho(0)$ and bounded on the closure of the disk. The *supremum norm* $|f|_\varrho$ of f in the domain $\Delta_\varrho(0)$ is defined as

$$|f|_\varrho = \sup_{z\in\Delta_\varrho(0)} |f(z)| \; . \tag{6.36}$$

The estimate of Cauchy for the derivative f' of f at the origin states that

$$|f'(0)| \leq \frac{1}{\varrho}|f|_\varrho \; .$$

More generally, for the sth derivative $f^{(s)}$ one has the estimate

$$\bigl|f^{(s)}(0)\bigr| \leq \frac{s!}{\varrho^s}|f|_\varrho \; .$$

For instance, let $\varrho = 1$, and consider the function $f(z) = z^s$. It is an easy matter to check that $f^{(s)} = s!$ and $|f|_1 = 1$, so that Cauchy's estimate gives $|f^{(s)}(0)| \leq s!$, which shows that the estimate cannot be improved in general. The proof of the inequalities just given is an easy consequence of Cauchy's formula:

$$f^{(s)}(z) = \frac{s!}{2\pi i}\oint \frac{f(\zeta)}{(\zeta - z)^{s+1}}\,d\zeta \; .$$

For, writing the contour of the disk as $\zeta = \varrho e^{i\vartheta}$, a straightforward calculation gives

$$\bigl|f^{(s)}(0)\bigr| \leq \frac{s!}{2\pi}\left|\oint \frac{|f(\zeta)|}{\varrho^{s+1}}\,d\zeta\right| \leq \frac{s!}{2\pi\varrho^s}|f|_\varrho \int_0^{2\pi} d\vartheta = \frac{s!}{\varrho^s}|f|_\varrho \; .$$

The case of n variables requires a straightforward extension. Let the domain $\Delta_\varrho(0)$ be the polydisk of radius ϱ centred at the origin of $\mathbb{C}^n$, namely

$$\Delta_\varrho(0) = \{z \in \mathbb{C}^n \;:\; |z| < \varrho\} \; , \tag{6.37}$$

where $|z| = \max_j |z_j|$ is the l_∞ norm on $\mathbb{C}^n$. This is nothing but the Cartesian product of complex disks of radius ϱ in the complex plane. Define the supremum norm of an analytic function f as in (6.36). Then Cauchy's estimate reads

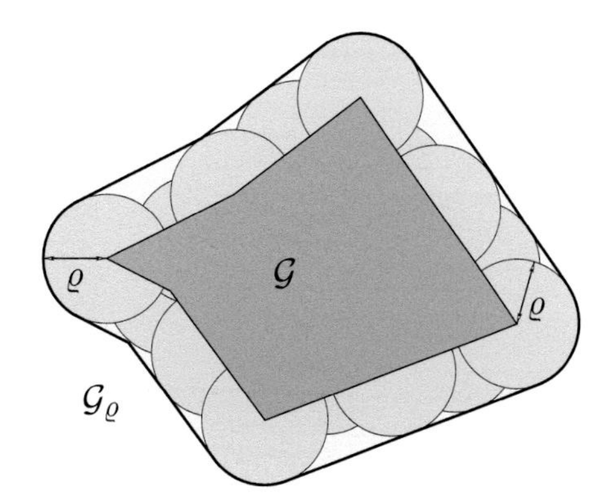

Figure 6.1 The complex extension of a domain $\mathscr{G} \subset \mathbb{C}^2$. The extended domain $\mathscr{G}_\varrho \subset \mathbb{C}^2$ is the union of complex disks of radius ϱ around every point of $\mathscr{G}$. The complex extension may be applied also to a real domain.

$$\left| \frac{\partial f}{\partial z_j}(0) \right| \le \frac{1}{\varrho} |f|_\varrho \ , \quad 1 \le j \le n \ .$$

The corresponding estimate for higher-order derivatives is

$$\left| \frac{\partial^{k_1+\cdots+k_n} f}{\partial p_1^{k_1} \ldots \partial p_n^{k_n}}(0) \right| \le \frac{k_1! \cdot \cdots \cdot k_n!}{\varrho^{k_1+\cdots+k_n}} |f|_\varrho \ .$$

6.4.2 Complexification of Domains

Consider the common case of a phase space $\mathscr{G} \times \mathbb{T}^n$, where $\mathscr{G} \subset \mathbb{R}^n$, endowed with action-angle variables $p \in \mathscr{G}$ and $q \in \mathbb{T}^n$.

In order to use Cauchy's estimates, we need to introduce a *complexification* of domains. For $p \in \mathscr{G}$ consider the open complex polydisk $\Delta_\varrho(p)$ of radius $\varrho > 0$ with centre p defined as in (6.37). The complexification $\mathscr{G}_\varrho$ of $\mathscr{G}$ is defined as the extension in the complex

$$\mathscr{G}_\varrho = \bigcup_{p \in \mathscr{G}} \Delta_\varrho(p) \ , \tag{6.38}$$

as illustrated in Fig. 6.1. It may be remarked that the complex extension may be applied to a domain $\mathscr{G}$ which can be either real or complex. With the same process the complexification $\mathbb{T}^n_\sigma$ for $\sigma > 0$ of the n-torus turns out to be defined as

$$\mathbb{T}^n_\sigma = \{ q \in \mathbb{C}^n \ : \ |\operatorname{Im} q| < \sigma \} \ , \tag{6.39}$$

namely the Cartesian product of strips of width σ around the real axis on the complex plane. From now on, the phase space will be the domain in action-angle variables

$$\mathscr{D}_{\varrho,\sigma} = \mathscr{G}_\varrho \times \mathbb{T}^n_\sigma \ , \quad \mathscr{G} \subset \mathbb{R}^n \ ,$$

as illustrated in Fig. 6.2.

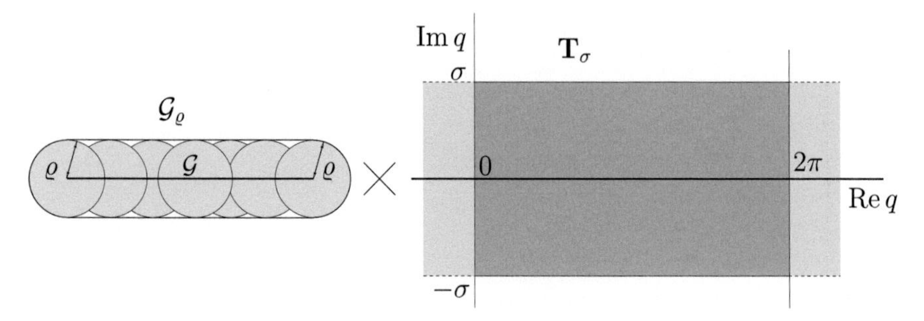

Figure 6.2 The complex extension of a real domain $\mathscr{G} \times \mathbb{T}$, in the case $n = 1$. The extended domain $\mathscr{G}_\varrho$ is the union of disks of radius ϱ around every point of $\mathscr{G}$. The extended torus $\mathbb{T}_\sigma$ is a strip around the real torus.

Introducing a norm on the space of analytic functions on the domain $\mathscr{D}_{\varrho,\sigma}$ is now straightforward. For instance, for a function f analytic on $\mathscr{G}_\varrho$ the supremum norm is

$$|f|_\varrho = \sup_{p\in\mathscr{G}_\varrho} |f(p)| \ . \tag{6.40}$$

Similarly, for a function g analytic on $\mathscr{D}_{\varrho,\sigma}$ the supremum norm $|g|_{\varrho,\sigma}$ is

$$|g|_{\varrho,\sigma} = \sup_{(p,q)\in\mathscr{D}_{\varrho,\sigma}} |g(p,q)| \ . \tag{6.41}$$

In order to apply Cauchy estimates, we must consider a family of complexified domains $\mathscr{G}_{\varrho-\delta}$ parameterized with $0 \le \delta < \varrho$. This point deserves particular attention. In the case of action-angle variables we shall similarly consider a family of domains $\mathscr{D}_{(1-d)(\varrho,\sigma)}$ parameterized by $0 \le d < 1$.

Consider for simplicity a function f analytic and bounded[4] on the complex domain $\mathscr{G}_\varrho$. Consider now a point $p \in \mathscr{G}_{\varrho-\delta}$ for $0 < \delta \le \varrho$ (i.e., the union of polydisks of radius $\varrho-\delta$ centred at every point of $\mathscr{G}$). Note that the polydisk $\Delta_\delta(p)$ is a subset of $\mathscr{G}_\varrho$, so that f is analytic and bounded on $\Delta_\delta(p)$, and we have the estimate $|f(p')| \le |f|_\varrho$ for all $p' \in \Delta_\delta(p)$. By Cauchy's estimate for a derivative we get

$$\left| \frac{\partial f}{\partial p_j}(p) \right| \le \frac{1}{\delta} |f|_\varrho \ , \quad 1 \le j \le n \ .$$

Since this is true for every point $p \in \mathscr{G}_{\varrho-\delta}$, we conclude

$$\left| \frac{\partial f}{\partial p_j} \right|_{\varrho-\delta} \le \frac{1}{\delta} |f|_\varrho \ , \quad 1 \le j \le n \ .$$

[4] Hereafter by "analytic and bounded in a complexified domain," $\mathscr{D}_\varrho$ say, we mean analytic in the interior and bounded on the closure of $\mathscr{D}_\varrho$.

It is relevant here that we have an explicit bound on the derivative of a function, but it is not for free: we must pay this information with a restriction of the complex domain. The family of restricted domains has been introduced precisely for this reason.

Analogous formulæ are obtained for the derivatives of a function $g(p,q)$ using the supremum norm as in (6.41). Using the positive parameter $d < 1$, for $j = 1, \ldots, n$ we have

$$\left|\frac{\partial g}{\partial p_j}\right|_{(1-d)(\varrho,\sigma)} \leq \frac{1}{d\varrho}|g|_{\varrho,\sigma} \, , \qquad \left|\frac{\partial g}{\partial q_j}\right|_{(1-d)(\varrho,\sigma)} \leq \frac{1}{d\sigma}|g|_{\varrho,\sigma} \, . \tag{6.42}$$

This is easily understood by paying due attention to the fact that the derivative with respect to one of the p variables does not affect the analyticity domain for the q variables, and vice versa. The corresponding estimate for higher-order derivatives is

$$\left|\frac{\partial^{k_1+\cdots+k_n+l_1+\cdots+l_n} g}{\partial p_1^{k_1}\ldots\partial p_n^{k_n}\partial q_1^{l_1}\ldots\partial q_n^{l_n}}\right|_{(1-d)(\varrho,\sigma)} \leq \frac{k_1!\cdot\cdots\cdot k_n!\, l_1!\cdot\cdots\cdot l_n!}{(d\varrho)^{k1+\cdots+k_n}(d\sigma)^{l_1+\cdots+l_n}}|g|_{\varrho,\sigma} \, . \tag{6.43}$$

6.4.3 Generalized Cauchy Estimates

The next step is concerned with the generalization of Cauchy estimates, taking into account that we are interested in Lie derivatives (or Poisson brackets).

Remark. In the following estimates the reader should pay a particular attention to the role of the complex domains, in particular when multiple derivatives are considered. When we estimate Poisson brackets, it turns out to be impractical (though not impossible: see, e.g., [104] or [78]) to use Cauchy's estimates for higher-order derivatives, and we are somewhat forced to proceed by recurrence. The key is that every Lie derivative – or Poisson bracket – is estimated by restricting the domain. A recurrent estimate of a multiple derivative requires a multiple restriction, but we must assure that the sum of the restrictions does not cause the final domain to become empty. This is one of the common sources of error in using the quantitative tools based on Cauchy's estimates.

A first result concerns the Poisson bracket between two functions.

Lemma 6.8: *Let g be analytic and bounded in $\mathscr{D}_{\varrho,\sigma}$ and f be analytic and bounded in $\mathscr{D}_{(1-d')(\varrho,\sigma)}$ for some $0 \leq d' < 1$, so that $|g|_{\varrho,\sigma}$ and $|f|_{(1-d')(\varrho,\sigma)}$ are finite. Then for $0 < d < 1 - d'$, we have*

$$\left|\{f,g\}\right|_{(1-d'-d)(\varrho,\sigma)} \leq \frac{1}{d(d+d')\varrho\sigma}|f|_{\varrho,\sigma}|g|_{(1-d')(\varrho,\sigma)} \, . \tag{6.44}$$

Proof. We take advantage of the fact that $L_g f = \{f, g\}$ is the time derivative of the evolution of f under the flow generated by g. Pick any point $(p,q) \in \mathscr{D}_{(1-d'-d)(\varrho,\sigma)}$ and consider the canonical vector field $\left(\frac{\partial g}{\partial p}, -\frac{\partial g}{\partial q}\right)(p,q)$ generated by g at the point (p,q). Define the analytic function

$$F(\tau) = f\left(p - \tau\tfrac{\partial g}{\partial q}, q + \tau\tfrac{\partial g}{\partial p}\right) ,$$

which for fixed (p,q) depends only on τ. Then we have

$$\bigl(L_g f\bigr)(p,q) = \frac{d}{d\tau}F(\tau)\Big|_{\tau=0} .$$

In view of Cauchy estimates at the point (p,q), we have

$$\left|\frac{\partial g}{\partial q}(p,q)\right| \le \frac{|g|_{\varrho,\sigma}}{(d+d')\sigma} , \quad \left|\frac{\partial g}{\partial p}(p,q)\right| \le \frac{|g|_{\varrho,\sigma}}{(d+d')\varrho} ;$$

this is the estimate (6.42) where the restriction of the domain is $d + d'$. Let now τ satisfy both conditions,

$$|\tau|\frac{|g|_{(1-d')(\varrho,\sigma)}}{(d+d')\sigma} \le d\varrho , \quad |\tau|\frac{|g|_{(1-d')(\varrho,\sigma)}}{(d+d')\varrho} \le d\sigma$$

(in fact the same condition). Then we have $\left(p - \tau\frac{\partial g}{\partial q}, q + \tau\frac{\partial g}{\partial p}\right) \in \mathscr{D}_{(1-d')(\varrho,\sigma)}$, so that the function $F(\tau)$ is analytic and bounded on the complex disk

$$|\tau| \le \frac{d(d+d')\varrho\sigma}{|g|_{\varrho,\sigma}} .$$

By Cauchy's estimate we have

$$\left|\frac{dF}{d\tau}(0)\right| \le \frac{|f|_{\varrho,\sigma}|g|_{(1-d')(\varrho,\sigma)}}{d(d+d')\varrho\sigma} .$$

Therefore, the Poisson bracket $\bigl|\{f,g\}\bigr|$ is bounded at the point (p,q) by the right member of the latter inequality. But this is true for every point $(p,q) \in \mathscr{D}_{(1-d'-d)(\varrho,\sigma)}$, hence the same bound applies to the supremum norm $\bigl|\{f,g\}\bigr|_{(1-d'-d)(\varrho,\sigma)}$. *Q.E.D.*

The next step concerns the estimate of multiple Poisson brackets. Consider an expression of the form $L_{g_s} \circ \cdots \circ L_{g_1} f$. We can estimate it by iterating s times the estimate of Lemma 6.8. However, we must appropriately handle the restrictions of domains. To this end, having fixed the final restriction d, we set $\delta = d/s$ and estimate the first j Lie derivatives on the domain restricted by the factor $1 - j\delta$. The final result is stated by

Lemma 6.9: *Let $g_1,\ldots,g_s$ and f be analytic on the domain $\mathscr{D}_{\varrho,\sigma}$, with finite norms and s positive. Then for every positive $d<1$, one has*

$$(6.45)\quad |L_{g_s}\circ\cdots\circ L_{g_1}f|_{(1-d)(\varrho,\sigma)} \le \frac{s!}{e^2}\left(\frac{e^2}{d^2\varrho\sigma}\right)^s |g_1|_{\varrho,\sigma}\cdot\ldots\cdot|g_s|_{\varrho,\sigma}|f|_{\varrho,\sigma}\ .$$

Remark. The operators L_{g_j} do not commute, in general, but the right member of (6.45) does not depend on the order; with a little abuse, we may say that the estimates do commute. Furthermore, the coefficient in front of the estimate depends only on how many operators L_{g_j} are applied to f. However, note that this is true provided the restrictions of the domains are handled the good way: the assumption that all functions g are bounded in the same domain is essential.

Proof of Lemma 6.9. For $s=1$ this is just Lemma 6.8. For $s>1$ let $\delta=d/s$. A straightforward application of Lemma 6.8 gives

$$|L_{g_1}f|_{(1-\delta)(\varrho,\sigma)} \le \frac{1}{\delta^2\varrho\sigma}|g_1|_{\varrho,\sigma}|f|_{\varrho,\sigma}\ .$$

For $j=2,\ldots,s$, recalling that $L_{g_j}\circ\cdots\circ L_{g_1}f$ can be estimated on the domain restricted by the factor $1-j\delta$, use again Lemma 6.8 with δ in place of d and $(j-1)\delta$ in place of d', and get the recurrent estimate

$$\bigl|L_{g_j}\circ\cdots\circ L_{g_1}f\bigr|_{(1-j\delta)(\varrho,\sigma)} \le \frac{1}{j\delta^2\varrho\sigma}|g_j|_{\varrho,\sigma}\bigl|L_{g_{j-1}}\circ\cdots\circ L_{g_1}f\bigr|_{(1-(j-1)\delta)(\varrho,\sigma)}\ .$$

By recurrent application of the latter formula and recalling $\delta=d/s$, we get[5]

$$|L_{g_s}\circ\cdots\circ L_{g_1}f|_{(1-d)(\varrho,\sigma)} \le \left(\frac{1}{d^2\varrho\sigma}\right)^s \frac{s^{2s}}{s!}|g_1|_{\varrho,\sigma}\cdot\ldots\cdot|g_s|_{\varrho,\sigma}|f|_{\varrho,\sigma}\ .$$

In view of the inequality[6] $s^s\le e^{s-1}s!$ for $s\ge 1$, (6.45) follows. *Q.E.D.*

In the following we shall use also a more refined (and boring) estimate by assuming that $|f|_{(1-d')(\varrho,\sigma)}$ is known on a smaller domain.

---

[5] The divisor $j\delta^2\varrho\sigma$ in the latter estimate deserves a bit of attention, for one may naively expect to have $\delta^2\varrho\sigma$, without the factor j. Such a naive estimate would result in a failure to prove the convergence of Lie series. The point is that the norm $|g_j|_{\varrho,\sigma}$ is assumed to be evaluated on the whole domain $\mathscr{D}_{\varrho,\sigma}$, and the derivatives of g_j should be estimated on the domain $\mathscr{D}_{(1-j\delta)(\varrho,\sigma)}$, with a restriction $j\delta$. Only a restriction δ is allowed instead for $|L_{g_{j-1}}\circ\cdots\circ L_{g_1}f|_{(1-(j-1)\delta)(\varrho,\sigma)}$.

[6] By induction: for $s=1$ it is true; for $s>1$ we have $s^s=s\left(\frac{s}{s-1}\right)^{s-1}(s-1)^{s-1}=s\left(1+\frac{1}{s-1}\right)^{s-1}(s-1)^{s-1}<se\cdot(s-1)!e^{s-2}=s!e^{s-1}$.

6.5 Analyticity of Lie Series

The technical tools developed in the previous section allow us to investigate the convergence of Lie series. The aim is to find sufficient conditions which guarantee the analyticity of a Lie series and, more interestingly, of the composition of Lie series.

6.5.1 Convergence of Lie Series

The first step is to prove that the transformation generated by Lie series is analytic in some domain. Recalling that the Lie series applied to coordinates is the expansion in power series of time of the local flow, the reader will immediately notice that we are actually concerned with the (local) existence and uniqueness of the solutions of differential equations in the analytic case, as was done by Cauchy. Here we put the accent on the transformation generated by the flow. The method of proof used here is essentially a revisitation of Cauchy's method of majorants.

Proposition 6.10: *Let the generating vector field χ be analytic and bounded on a domain $\mathscr{D}_{\varrho,\sigma}$. Then for every positive $d < 1/2$ the following statement holds true: if $|\chi|_{\varrho,\sigma}$ is small enough, for example, if*

$$\frac{|\chi|_{\varrho,\sigma}}{d^2\varrho\sigma} \leq \frac{1}{2e^2}\,, \tag{6.46}$$

then for every $|\varepsilon| \leq 1$ and for $(p,q) \in \mathscr{D}_{(1-d)(\varrho,\sigma)}$, the Lie series $\exp(\varepsilon L_\chi)$ defines an analytic canonical diffeomorphism

$$p' = \exp(\varepsilon L_\chi)p\ , \quad q' = \exp(\varepsilon L_\chi)q \tag{6.47}$$

satisfying

$$\begin{aligned}\bigl|\exp(\varepsilon L_\chi)p_j - p_j\bigr|_{(1-d)(\varrho,\sigma)} &\leq \varepsilon d\varrho\ ,\\ \bigl|\exp(\varepsilon L_\chi)q_j - q_j\bigr|_{(1-d)(\varrho,\sigma)} &\leq \varepsilon d\sigma\ .\end{aligned} \tag{6.48}$$

The canonical map defined in the domain $\mathscr{D}_{(1-d)(\varrho,\sigma)}$ satisfies

$$\mathscr{D}_{(1-d-\varepsilon d)(\varrho,\sigma)} \subset \exp(\varepsilon L_\chi)\bigl(\mathscr{D}_{(1-d)(\varrho,\sigma)}\bigr) \subset \mathscr{D}_{(1-d+\varepsilon d)(\varrho,\sigma)}\ . \tag{6.49}$$

Let us spend a few words on the inclusion relations (6.49). For simplicity, let us set $\varepsilon = 1$ and focus our attention on a disk Δ_ϱ in the action variables, as illustrated in Fig. 6.3. The extension of the argument to $\mathscr{D}_{\varrho,\sigma}$ is an easy matter, since it is the union of disks. Every point is moved by the flow, so that the disk is deformed both by the direct transformation (the forward flow) and the inverse transformation (the backward flow). We should choose

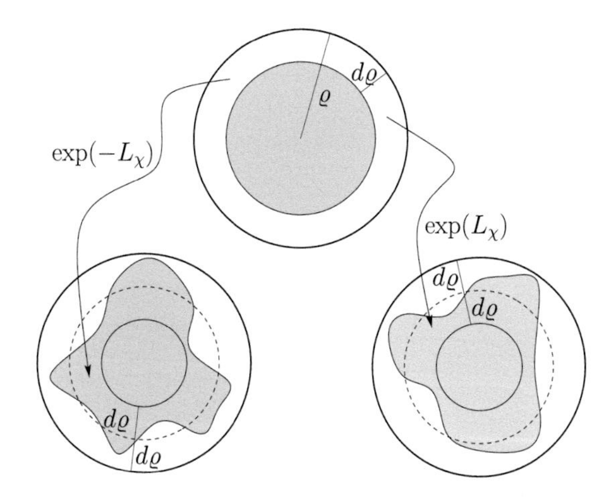

Figure 6.3 The deformation induced on a disk by the near-the-identity transformation of Proposition 6.10. The restriction $d\varrho$ creates a safety zone, so that both the direct and the inverse images of $\Delta_{(1-d)\varrho}$ contain $\Delta_{(1-2d)\varrho}$ and are contained in Δ_ϱ. The centre of the disk belongs to the image of the map in view of $d < 1/2$.

the set of initial data so as to avoid two bad occurrences, namely: (i) a point of the initial set leaves the disk Δ_ϱ where the vector field is defined, and (ii) the centre of the disk does not belong to the image of the disk. In order to avoid such unpleasant facts we restrict the deformation to be a fraction $d\varrho$ of the radius of the disk, with $d < 1/2$ as requested, so that the flow is well defined if we choose the initial set to be a disk $\Delta_{(1-d)\varrho}$. The relations (6.48) are the formal statement of this fact. On the other hand, the value of d cannot be arbitrary: it must be related to the strength G of the vector field, so that a condition such as, for example, (6.46) must be imposed.

Proof of Proposition 6.10. Recall that for $j = 1, \ldots, n$ we have

$$\begin{aligned}
\exp\bigl(\varepsilon L_\chi\bigr) q_j &= q_j + \sum_{s>0} \frac{\varepsilon^s}{s!} L_\chi^{s-1} \frac{\partial \chi}{\partial p_j} \ , \\
\exp\bigl(\varepsilon L_\chi\bigr) p_j &= p_j - \sum_{s>0} \frac{\varepsilon^s}{s!} L_\chi^{s-1} \frac{\partial \chi}{\partial q_j} \ ,
\end{aligned}$$

where we have taken into account that $L_\chi q_j = \frac{\partial \chi}{\partial p_j}$ and $L_\chi p_j = -\frac{\partial \chi}{\partial q_j}$. By (6.42), for $0 < \delta < 1$ we have

$$\begin{aligned}
\left| \frac{\partial \chi}{\partial p_j} \right|_{(1-\delta)(\varrho,\sigma)} &\le \frac{|\chi|_{\varrho,\sigma}}{\delta\varrho} = \frac{|\chi|_{\varrho,\sigma}}{\delta^2 \varrho\sigma} \delta\sigma \ , \\
\left| \frac{\partial \chi}{\partial q_j} \right|_{(1-\delta)(\varrho,\sigma)} &\le \frac{|\chi|_{\varrho,\sigma}}{\delta\sigma} = \frac{|\chi|_{\varrho,\sigma}}{\delta^2 \varrho\sigma} \delta\varrho \ .
\end{aligned}$$

Using the latter estimates in Lemma 6.9, we get

$$\bigl|L_\chi^s p_j\bigr|_{(1-d)(\varrho,\sigma)} \le \frac{s!}{e^2}\left(\frac{e^2|\chi|_{\varrho,\sigma}}{d^2\varrho\sigma}\right)^s d\varrho\ ,$$

$$\bigl|L_\chi^s q_j\bigr|_{(1-d)(\varrho,\sigma)} \le \frac{s!}{e^2}\left(\frac{e^2|\chi|_{\varrho,\sigma}}{d^2\varrho\sigma}\right)^s d\sigma\ .$$

For the whole series we get the estimate

$$\bigl|\exp(\varepsilon L_\chi)p_j - p_j\bigr|_{(1-d)(\varrho,\sigma)} \le \frac{d\varrho}{e^2}\sum_{s>0}\left(\frac{\varepsilon e^2|\chi|_{\varrho,\sigma}}{d^2\varrho\sigma}\right)^s\ ,$$

$$\bigl|\exp(\varepsilon L_\chi)q_j - q_j\bigr|_{(1-d)(\varrho,\sigma)} \le \frac{d\sigma}{e^2}\sum_{s>0}\left(\frac{\varepsilon e^2|\chi|_{\varrho,\sigma}}{d^2\varrho\sigma}\right)^s\ .$$

The series is absolutely convergent for $|\chi|_{\varrho,\sigma}$ small enough, so it defines an analytic map on the domain $\mathscr{D}_{(1-d)(\varrho,\sigma)}$. The map is canonical in view of the formal properties of Lie series. With the condition (6.46), the sum of the series does not exceed 2ε, so that the inequalities (6.48) follow. The inclusion relations (6.49) easily follow from the inequalities (6.48). *Q.E.D.*

The transformation of a function $f(p,q)$ analytic in the domain $\mathscr{D}_{\varrho,\sigma}$ may be estimated as follows. Let

$$g(p',q') = f(p,q)\Big|_{p=\exp(L_\chi)p',\, q=\exp(L_\chi)p'}$$

be the transformed function. In view of Proposition 6.10 it is an analytic function on the restricted domain $\mathscr{D}_{(1-d)(\varrho,\sigma)}$, being a composition of analytic functions, and it is bounded by $|g|_{(1-d)(\varrho,\sigma)} \le |f|_{\varrho,\sigma}$. This is enough in order to have a bound on g.

As an additional remark, we may recall that in view of the exchange theorem we have

$$g = \exp(L_\chi)f = \sum_{s\ge 0}\frac{1}{s!}L_\chi^s f\ .$$

It is often useful to have a direct estimate based on the latter expression, thus avoiding any reference to the coordinate transformation. This can be done along the lines of the proof of Proposition 6.10. Let us state this formally as

Lemma 6.11: *Let the generating function χ and the function f be analytic and bounded on a domain $\mathscr{D}_{\varrho,\sigma}$. Assume that χ satisfies the convergence hypothesis (6.46), namely*

$$\frac{|\chi|_{\varrho,\sigma}}{d^2\varrho\sigma} \le \frac{1}{2e^2}\ .$$

Then for every positive $d < 1$ the function $\exp(L_\chi)f$ is analytic and bounded in $\mathscr{D}_{(1-d)(\varrho,\sigma)}$, and we have

$$\left|\exp(L_\chi)f\right|_{(1-d)(\varrho,\sigma)} \le \frac{1+e^2}{e^2}\,|f|_{\varrho,\sigma}\ .$$

Proof. Applying Lemma 6.9 we have

$$\left|\exp(L_\chi)f\right|_{(1-d)(\varrho,\sigma)} \le |f|_{\varrho,\sigma} + \frac{1}{e^2}\sum_{s>0}\left(\frac{e^2|\chi|_{\varrho,\sigma}}{d^2\varrho\sigma}\right)|f|_{\varrho,\sigma}\ .$$

With the convergence hypothesis the sum of the whole expression on the right does not exceed $|f|_{\varrho,\sigma}(1+1/e^2)$. Hence the series for $\exp(L_\chi)$ is absolutely convergent in $\mathscr{D}_{(1-d)(\varrho,\sigma)}$ and defines there an analytic bounded function satisfying the inequality claimed in the statement. *Q.E.D.*

6.5.2 Analyticity of the Composition of Lie Series

We look now for sufficient conditions for the analyticity of the composition of Lie series. Given a sequence $\chi = \{\chi_1, \chi_2, \chi_3, \ldots\}$ we consider the sequences of operators $S_\chi^{(r)}$ and $\tilde{S}_\chi^{(r)}$ defined by (6.9) and (6.12), namely

$$\begin{aligned} S^{(1)} &= \exp\bigl(L_{\chi_1}\bigr)\ , & S^{(r)} &= \exp\bigl(L_{\chi_r}\bigr)\circ S^{(r-1)}\ , \\ \tilde{S}^{(1)} &= \exp\bigl(-L_{\chi_1}\bigr)\ , & \tilde{S}^{(r)} &= \tilde{S}^{(r-1)}\circ\exp\bigl(-L_{\chi_r}\bigr) \quad \text{for } r>1\ . \end{aligned} \tag{6.50}$$

Proposition 6.12: *Let the vector fields $\chi_1, \chi_2, \chi_3, \ldots$ be analytic and bounded on a domain $\mathscr{D}_{\varrho,\sigma}$. Then for every positive $d < 1/3$, the following statement holds true: if χ and $d > 0$ satisfy the convergence condition*

$$\frac{1}{d^2\varrho\sigma}\sum_{j\ge 1}|\chi_j|_{\varrho,\sigma} \le \frac{1}{2e^2}, \tag{6.51}$$

then the maps $S_\chi : \mathscr{D}_{(1-2d)(\varrho,\sigma)} \to \mathscr{D}_{\varrho,\sigma}$ and $S_\chi^{-1} : \mathscr{D}_{(1-2d)(\varrho,\sigma)} \to \mathscr{D}_{\varrho,\sigma}$ defined as the limit of the sequences of maps (6.50) are analytic and canonical diffeomorphisms, and satisfy

$$\left|S_\chi p_j - p_j\right|_{(1-2d)(\varrho,\sigma)} < d\varrho\ ,\quad \left|S_\chi^{-1} p_j - p_j\right|_{(1-2d)(\varrho,\sigma)} < d\sigma\ . \tag{6.52}$$

The canonical maps defined in $\mathscr{D}_{(1-2d)(\varrho,\sigma)}$ satisfy

$$\begin{aligned} \mathscr{D}_{(1-3d)(\varrho,\sigma)} &\subset S_\chi\bigl(\mathscr{D}_{(1-2d)(\varrho,\sigma)}\bigr) \subset \mathscr{D}_{(1-d)(\varrho,\sigma)}\ , \\ \mathscr{D}_{(1-3d)(\varrho,\sigma)} &\subset S_\chi^{-1}\bigl(\mathscr{D}_{(1-2d)(\varrho,\sigma)}\bigr) \subset \mathscr{D}_{(1-d)(\varrho,\sigma)}\ . \end{aligned} \tag{6.53}$$

Proof. Define the sequence of positive numbers

$$d_r = d\,\frac{|\chi_r|_{\varrho,\sigma}}{\sum_{j\ge 1}|\chi_j|_{\varrho,\sigma}}\ ,\qquad r\ge 1\ ,$$

so that $\sum_{r\ge 1} d_r = d$. In view of (6.51), for every $r \ge 1$ the generating function χ_r satisfies condition (6.46), so that Proposition 6.10 applies to $\exp(L_{\chi_r})$. Setting $\varepsilon = d_r/d$ in (6.48), for every $(p,q) \in \mathscr{D}_{(1-d)(\varrho,\sigma)}$, we have

$$
\begin{aligned}
\bigl|\exp(L_{\chi_r})p_j - p_j\bigr|_{(1-d)(\varrho,\sigma)} &\le d_r \varrho \ ,\\
\bigl|\exp(L_{\chi_r})q_j - q_j\bigr|_{(1-d)(\varrho,\sigma)} &\le d_r \sigma \ .
\end{aligned}
\tag{6.54}
$$

Let now the sequence $\delta_1, \delta_2, \ldots$ be defined by recurrence as

$$\delta_1 = d_1 \ , \quad \delta_r = \delta_{r-1} + d_r \ ,$$

so that $\delta_r \to d$ for $r \to \infty$. With a further restriction d, we prove by induction that for every $(p,q) \in \mathscr{D}_{(1-2d)(\varrho,\sigma)}$ we have $S^{(r)}(p,q) \in \mathscr{D}_{(1-2d+\delta_r)}$ and $\tilde{S}^{(r)}(p,q) \in \mathscr{D}_{(1-2d+\delta_r)}$. This is true for $r = 1$ in view of (6.54), since $\mathscr{D}_{(1-2d)(\varrho,\sigma)} \subset \mathscr{D}_{(1-d)(\varrho,\sigma)}$. If the claim is true for $r-1$, then (6.54) gives

$$
\begin{aligned}
\bigl|\bigl(S^{(r)} - S^{(r-1)}\bigr)p_j\bigr|_{(1-2d+\delta_{r-1})} &\le d_r \varrho \ ,\\
\bigl|\bigl(S^{(r)} - S^{(r-1)}\bigr)q_j\bigr|_{(1-2d+\delta_{r-1})} &\le d_r \sigma \ ,
\end{aligned}
$$

so that the claim for r follows. The proof for $\tilde{S}^{(r)}$ requires a minor change which we leave to the reader. Now, for any $l \ge 1$ we have

$$
\begin{aligned}
\bigl|\bigl(S^{(r+l)} - S^{(r)}\bigr)p_j\bigr|_{(1-2d+\delta_{r-1})} &\le (\delta_{r+l} - \delta_r)\varrho \ ,\\
\bigl|\bigl(S^{(r+l)} - S^{(r)}\bigr)q_j\bigr|_{(1-2d+\delta_{r-1})} &\le (\delta_{r+l} - \delta_r)\sigma \ ,
\end{aligned}
$$

and a similar inequality for $\tilde{S}^{(r)}$. Therefore, in view of $\sum_{r\ge 1} d_r = d$, the sequences $S^{(r)}(p,q)$ and $\tilde{S}^{(r)}(p,q)$ are absolutely and uniformly convergent for every $(p,q) \in \mathscr{D}_{(1-2d)(\varrho,\sigma)}$, and the limits are analytic diffeomorphisms satisfying the claim (6.52). The maps are canonical, being compositions of canonical maps. The inclusion relations (6.53) are a straightforward consequence of (6.52). *Q.E.D.*

6.6 Analyticity of the Lie Transforms

We finally state sufficient conditions for the analyticity of the Lie transform. Here we must provide a direct proof of convergence of the series, because the definition itself of the Lie transform is not based on iteration.

6.6.1 Convergence of the Lie Transforms

The result is stated by the following.

Proposition 6.13: *Let the generating sequence $\chi = \{\chi_s\}_{s\geq 1}$ be analytic on the domain $\mathscr{D}_{\varrho,\sigma}$, and assume*

$$|\chi_s|_{\varrho,\sigma} \leq \frac{b^{s-1}}{s} G \tag{6.55}$$

for some $b \geq 0$ and $G > 0$. Then for every positive $d < 1/2$ the following statement holds true: if the convergence condition

$$\frac{e^2 G}{d^2 \varrho\sigma} + b \leq \frac{1}{2} \tag{6.56}$$

is satisfied, then the operator T_χ and its inverse T_χ^{-1} define an analytic canonical transformation on the domain $\mathscr{D}_{(1-d)(\varrho,\sigma)}$ with the properties

$$\begin{aligned} \mathscr{D}_{(1-2d)(\varrho,\sigma)} &\subset T_\chi \mathscr{D}_{(1-d)(\varrho,\sigma)} \subset \mathscr{D}_{\varrho,\sigma} \ , \\ \mathscr{D}_{(1-2d)(\varrho,\sigma)} &\subset T_\chi^{-1} \mathscr{D}_{(1-d)(\varrho,\sigma)} \subset \mathscr{D}_{\varrho,\sigma} \ . \end{aligned} \tag{6.57}$$

The proof of Proposition 6.13 is based on the following.

Lemma 6.14: *Let f and the generating sequence $\chi = \{\chi_s\}_{s\geq 1}$ be analytic on the domain $\mathscr{D}_{\varrho,\sigma}$, and assume that $|f|_{\varrho,\sigma}$ is finite, and that the generating sequence satisfies (6.55). Then the series $T_\chi f$, $T_\chi^{-1} f$, $T_\chi p$, $T_\chi^{-1} p$, $T_\chi q$ and $T_\chi^{-1} q$ are absolutely convergent in $\mathscr{D}_{(1-d)(\varrho,\sigma)}$, and we get the following estimates:*

(i) for any function f the operators E_s and D_s are estimated by

$$\begin{aligned} |E_s f|_{(1-d)(\varrho,\sigma)} &\leq \left(\frac{e^2 G}{d^2 \varrho\sigma} + b\right)^{s-1} \frac{G}{d^2 \varrho\sigma} |f|_{\varrho,\sigma} \ , \\ |D_s f|_{(1-d)(\varrho,\sigma)} &\leq \left(\frac{e^2 G}{d^2 \varrho\sigma} + b\right)^{s-1} \frac{G}{d^2 \varrho\sigma} |f|_{\varrho,\sigma} \ ; \end{aligned} \tag{6.58}$$

(ii) the transformation of a function is estimated by

$$|T_\chi f|_{(1-d)(\varrho,\sigma)} \leq 2|f|_{\varrho,\sigma} \ , \quad |T_\chi^{-1} f|_{(1-d)(\varrho,\sigma)} \leq 2|f|_{\varrho,\sigma} \ ; \tag{6.59}$$

(iii) the remainder of a transformation truncated at order $r > 0$ is estimated by

$$\begin{aligned} \left| T_\chi f - \sum_{s=0}^{r} E_s f \right|_{(1-d)(\varrho,\sigma)} &\leq \frac{2G}{d^2 \varrho\sigma} \left(\frac{e^2 G}{d^2 \varrho\sigma} + b\right)^{r+1} |f|_{\varrho,\sigma} \ , \\ \left| T_\chi^{-1} f - \sum_{s=0}^{r} D_s f \right|_{(1-d)(\varrho,\sigma)} &\leq \frac{2G}{d^2 \varrho\sigma} \left(\frac{e^2 G}{d^2 \varrho\sigma} + b\right)^{r+1} |f|_{\varrho,\sigma} \ ; \end{aligned} \tag{6.60}$$

(iv) the transformation of coordinates is estimated by

$$\begin{aligned} \left|T_\chi p_j - p_j\right|_{(1-d)(\varrho,\sigma)} &\leq \tfrac{1}{2e} d\varrho \ , \quad \left|T_\chi^{-1} p_j - p_j\right|_{(1-d)(\varrho,\sigma)} \leq \tfrac{1}{2e} d\varrho \ , \\ \left|T_\chi q_j - q_j\right|_{(1-d)(\varrho,\sigma)} &\leq \tfrac{1}{2e} d\sigma \ , \quad \left|T_\chi^{-1} q_j - q_j\right|_{(1-d)(\varrho,\sigma)} \leq \tfrac{1}{2e} d\sigma \ . \end{aligned} \tag{6.61}$$

Proof. We start by proving that

$$(6.62)\quad |E_s f|_{(1-d)(\varrho,\sigma)} \le B_s |f|_{\varrho,\sigma}\ , \quad |D_s f|_{(1-d)(\varrho,\sigma)} \le B_s |f|_{\varrho,\sigma}\ , \quad s \ge 1\ ,$$

where the real sequence $\{B_s\}_{s\ge 1}$ is defined as

$$(6.63)\quad \begin{aligned} B_1 &= \frac{1}{d^2\varrho\sigma} G\ ,\\ B_s &= \frac{e^2}{d^2\varrho\sigma}\sum_{j=1}^{s-1}\frac{s-j+1}{s} b^{j-1} G B_{s-j} + \frac{1}{sed^2\varrho\sigma} b^{s-1} G\ . \end{aligned}$$

For $s = 1$ it is enough to apply the estimate (6.44) of Lemma 6.8 with $d' = 0$. For $s > 1$ we look for a recurrent estimate. We shall do it for the operators E_s, but the same argument applies word by word also to the operators D_s. The whole argument is based on the remark that E_s can be written as $E_s = \sum_{\alpha\in\mathscr{I}_s} c_\alpha F_\alpha$, where $\mathscr{I}_s$ is a set of indexes, $\{c_\alpha\}_{\alpha\in\mathscr{I}_s}$ a set of real coefficients, and $\{F_\alpha\}_{\alpha\in\mathscr{I}_s}$ a set of linear operators, each F_α being a composition of at most s operators L_{χ_j}. This is evident from the recurrent definition (6.18) of E_s. By Lemma 6.9 we have $|F_\alpha f|_{(1-d)(\varrho,\sigma)} \le A_\alpha |f|_{\varrho,\sigma}$, with some set of positive constants $\{A_\alpha\}_{\alpha\in\mathscr{I}_s}$, and so also $|E_s f|_{(1-d)(\varrho,\sigma)} \le \sum_{\alpha\in\mathscr{I}_s} |c_\alpha| A_\alpha |f|_{\varrho,\sigma}$. This information suffices in order to establish a recurrent estimate. To this end, assume we know that $\sum_{\alpha\in\mathscr{I}_r} |c_\alpha| A_\alpha \le B_r$ for $1 \le r < s$; this is true for $s = 2$, because we know B_1. Recalling the recurrent definition (6.18) of E_s, let us look for an estimate of $L_{\chi_j} E_{s-j} = \sum_{\alpha\in\mathscr{I}_{s-j}} c_\alpha L_{\chi_j} F_\alpha$. By Lemma 6.17 we get

$$\bigl|L_{\chi_j} F_\alpha f\bigr|_{(1-d)(\varrho,\sigma)} \le \frac{e^2(s-j+1)}{d^2\varrho\sigma} |\chi_j|_{\varrho,\sigma} A_\alpha |f|_{\varrho,\sigma}$$

(see the remark after the statement of Lemma 6.9, recalling also that F_α contains at most $s-j$ operators L_{χ_j}; also note that the same estimate applies also to D_s because the estimate is not affected by commutation of operators). Thus, we get

$$\begin{aligned} \bigl|L_{\chi_j} E_{s-j} f\bigr|_{(1-d)(\varrho,\sigma)} &\le \frac{e^2(s-j+1)}{d^2\varrho\sigma} |\chi_j|_{\varrho,\sigma} \sum_{\alpha\in\mathscr{I}_{s-j}} |c_\alpha| A_\alpha |f|_{\varrho,\sigma}\\ &\le \frac{e^2(s-j+1)}{d^2\varrho\sigma} B_{s-j} |\chi_j|_{\varrho,\sigma} |f|_{\varrho,\sigma}\ , \end{aligned}$$

where the induction hypothesis has been used. Using the recurrent definition (6.18) of E_s and hypothesis (6.55), we immediately get (6.63).

We come now to the proof of (6.58). For $s = 1, 2$ this is directly checked. For $s > 2$ isolate the term $j = 1$ in the sum in (6.63) and change the summation index j to $j-1$, thus getting

$$\begin{aligned}B_s &= \frac{e^2G}{d^2\varrho\sigma}B_{s-1} + b\sum_{j=1}^{s-2}\frac{(s-1)-j+1}{s}b^{j-1}GB_{s-1-j} + \frac{1}{sed^2\varrho\sigma}b^{s-1}G\\ &\le \frac{e^2G}{d^2\varrho\sigma}B_{s-1} + \frac{s-1}{s}bB_{s-1}\ ;\end{aligned}$$

here, the definition of B_{s-1} in (6.63) has been used. Thus we get, for $s \ge 2$,

$$B_s < \left(\frac{e^2G}{d^2\varrho\sigma} + b\right)B_{s-1}\ ;$$

and (6.58) immediately follows.
(ii) By (6.62) we have

$$\sum_{s>0}|E_s f|_{(1-d)(\varrho,\sigma)} \le \frac{G}{d^2\varrho\sigma}\sum_{s>0}\left(\frac{e^2G}{d^2\varrho\sigma} + b\right)^{s-1}|f|_{\varrho,\sigma}\ ;$$

in view of condition (6.56) the series in the right-hand side of the latter inequality is a geometric series with sum not exceeding 2. This implies that the series $T_\chi f$ is absolutely convergent on the domain $\mathscr{D}_{(1-d)(\varrho,\sigma)}$ and is estimated by (6.59), as claimed.
(iii) The estimates (6.60) follow by simply evaluating the sum of the geometric series just presented, starting from $s = r+1$ and taking into account the condition (6.56).
All arguments used until now, with very minor changes, apply to the operators D_s which define the inverse T_χ^{-1}: just use the fact that the estimates do not depend on the order of application of Lie derivatives.
(iv) Concerning the estimate on the transformation of coordinates (p,q) a similar argument applies, with a minor difference: the terms $L_{\chi_l}p_j$ and $L_{\chi_l}q_j$ must be estimated using (6.42) of Lemma 6.8. Proceeding as earlier, we get (recall that $E_1 = L_{\chi_1}$)

$$\begin{aligned}|E_1p_j|_{(1-d)(\varrho,\sigma)} &\le \frac{G}{d\sigma}\ , \quad |E_sp_j|_{(1-d)(\varrho,\sigma)} \le \left(\frac{e^2G}{d^2\varrho\sigma} + b\right)^{s-1}\frac{G}{d\sigma}\ ,\\ |E_1q_j|_{(1-d)(\varrho,\sigma)} &\le \frac{G}{d\sigma}\ , \quad |E_sq_j|_{(1-d)(\varrho,\sigma)} \le \left(\frac{e^2G}{d^2\varrho\sigma} + b\right)^{s-1}\frac{G}{d\varrho}\ ,\end{aligned}$$

and similar estimates for the operators D_s. Then the same convergence argument is used, giving

$$\begin{aligned}|T_\chi p_j - p_j|_{(1-d)(\varrho,\sigma)} &\le \frac{G}{d\sigma}\sum_{s>0}\left(\frac{e^2G}{d^2\varrho\sigma} + b\right)^{s-1} \le \frac{G}{d\sigma} < \frac{1}{e^2}\left(\frac{e^2G}{d^2\varrho\sigma} + b\right)d\varrho\ ,\\ |T_\chi q_j - q_j|_{(1-d)(\varrho,\sigma)} &\le \frac{G}{d\varrho}\sum_{s>0}\left(\frac{e^2G}{d^2\varrho\sigma} + b\right)^{s-1} \le \frac{G}{d\varrho} < \frac{1}{e^2}\left(\frac{e^2G}{d^2\varrho\sigma} + b\right)d\sigma\ ,\end{aligned}$$

and (6.61) immediately follows by condition (6.56). Checking the remaining estimates requires only minor changes with respect to our argument. The same argument applies to $T_\chi^{-1} f$. *Q.E.D.*

Proof of Proposition 6.13. By Lemma 6.14, the series of analytic functions defining the canonical transformation are absolutely convergent in $\mathscr{D}_{(1-d)(\varrho,\sigma)}$. Thus, the same series are uniformly convergent on every compact subset of $\mathscr{D}_{(1-d)(\varrho,\sigma)}$. By Weierstrass's theorem, this implies that the sums of the series are analytic functions on $\mathscr{D}_{(1-d)(\varrho,\sigma)}$, as claimed. The statement concerning the inclusions of the domains follows from the estimates (6.61). *Q.E.D.*

A straightforward consequence of Proposition 6.13 is the following: if $f(p,q)$ is analytic in $\mathscr{D}_{\varrho,\sigma}$, then the transformed function $f\bigl(T_\chi p, T_\chi q\bigr)$ is analytic in $\mathscr{D}_{(1-d)(\varrho,\sigma)}$, being a composition of analytic functions. However, as remarked at the end of Section 6.2, the transformed function is nothing but $T_\chi f$. The first estimate in (6.60) gives in fact a direct proof of the analyticity of $T_\chi f$. Moreover, taking into account that an explicit computation must be truncated at some order r, it gives an estimate of the error.

6.7 Weighted Fourier Norms

As a matter of fact, the supremum norm is somehow unhandy when dealing with problems involving small denominators. A more convenient norm is the so-called *weighted Fourier norm*, which can be introduced to exploit the properties of analytic Fourier series. The weighted Fourier norm will be used in the following chapters. The aim of this section is to provide the analytical tools which allow us to work with this particular norm. The reader will find that this section is only a *variazione* of Section 6.4. The scope is only technical, indeed.

6.7.1 Analytic Fourier Series

It is known that an analytic function f on the domain $\mathscr{D}_{\varrho,\sigma'}$ admits the Fourier expansion

$$f(p,q) = \sum_{k\in\mathbb{Z}^n} f_k(p) e^{i\langle k,q\rangle} \ .$$

It is easy to show that the coefficients $f_k(p)$ of the Fourier expansion decay exponentially with k. More precisely, we have

$$|f_k|_\varrho \le |f|_{\varrho,\sigma'} e^{-|k|\sigma'} \ ,$$

where $|k| = |k_1| + \cdots + |k_n|$ (the l_1 norm in $\mathbb{Z}^n$). For consider the simple case of a function $f(q)$ of one variable $q \in \mathbb{T}$ and recall that the coefficients are computed as

$$f_k = \frac{1}{2\pi}\int_0^{2\pi} f(q)e^{-ikq}dq \ .$$

If k is positive, move the contour of integration by $i\sigma'$ and get

$$f_k = \frac{1}{2\pi}\int_0^{2\pi} f(q+i\sigma')e^{-ikq}e^{-k\sigma'}dq \ .$$

The inequality then follows in view of $|f(q+i\sigma')| \leq |f|_{\sigma'}$ (by definition of supremum norm) and $|e^{-ikq}| = 1$. For negative k, just move the contour by $-i\sigma'$. In the case $q \in \mathbb{T}^n$ the coefficients f_k, $k \in \mathbb{Z}^n$ are defined as

$$f_k = \frac{1}{(2\pi)^n}\int_0^{2\pi}\cdots\int_0^{2\pi} f(q)e^{-i\langle k,q\rangle}dq_1\,dq_n \ ,$$

and a similar argument applies. In the general case $f(p,q)$ just repeat the same argument for every $p \in \mathscr{G}_\varrho$, and the claim is proved by just replacing the supremum norm $\bigl|f_k(p)\bigr|$ with $\bigl|f_k(p)\bigr|_\varrho$, in the complexified domain.

The exponential decay of coefficients suggests to define the weighted Fourier norm $\|f\|_{\varrho,\sigma}$ for $\sigma < \sigma'$ as[7]

$$\|f\|_{\varrho,\sigma} = \sum_{k\in\mathbb{Z}^n} |f_k|_\varrho e^{|k|\sigma} \ . \tag{6.64}$$

The condition $\sigma < \sigma'$ ensures that the series defining the norm converges and the norm is finite. It is also an easy matter to check that the supremum norm is bounded by the weighted Fourier norm, that is,

$$|f|_{\varrho,\sigma} \leq \|f\|_{\varrho,\sigma} \ .$$

6.7.2 Generalized Cauchy Estimates

The next step is concerned with the generalization of Cauchy estimates, taking into account that we are interested in Lie derivatives (or Poisson brackets) and that we are going to use the weighted Fourier norm instead of the supremum norm.

A first result concerns the Poisson bracket between two functions, and the derivative of a function with respect to one of the canonical coordinates (i.e., the Poisson bracket with the conjugate one).

Lemma 6.15: *Let f be analytic on the domain $\mathscr{D}_{\varrho,\sigma}$, and g be analytic in $\mathscr{D}_{(1-d')(\varrho,\sigma)}$ for some $0 \leq d' < 1$; assume, moreover, that $\|f\|_{\varrho,\sigma}$ and $\|g\|_{(1-d')(\varrho,\sigma)}$ are finite. Then:*

---

[7] Hereafter the supremum norm in a domain $\mathscr{D}_{\varrho,\sigma}$ will be denoted as $|\cdot|_{\varrho,\sigma}$, and the symbol $\|\cdot\|_{\varrho,\sigma}$ will be reserved for the weighted Fourier norm.

i. for $0 < d < 1$ and for $1 \le j \le n$, we have

$$(6.65)\qquad \left\|\frac{\partial f}{\partial p_j}\right\|_{(1-d)(\varrho,\sigma)} \le \frac{1}{d\varrho}\|f\|_{\varrho,\sigma}\ , \qquad \left\|\frac{\partial f}{\partial q_j}\right\|_{(1-d)(\varrho,\sigma)} \le \frac{1}{ed\sigma}\|f\|_{\varrho,\sigma}\ ;$$

ii. for $0 < d < 1 - d'$, we have

$$(6.66)\qquad \bigl\|\{f,g\}\bigr\|_{(1-d'-d)(\varrho,\sigma)} \le \frac{2}{ed(d+d')\varrho\sigma}\|f\|_{\varrho,\sigma}\|g\|_{(1-d')(\varrho,\sigma)}\ .$$

The following corollary improves the estimate for the average over the angles of a Poisson bracket between two functions.

Corollary 6.16: *With the same hypotheses of Lemma 6.15, we have*

$$(6.67)\qquad \bigl\|\overline{\{f,g\}}\bigr\|_{(1-d'-d)(\varrho,\sigma)} \le \frac{1}{ed\varrho\sigma}\|f\|_{\varrho,\sigma}\|g\|_{(1-d')(\varrho,\sigma)}\ .$$

Proof of Lemma 6.15. Using the Fourier expansion, compute

$$\frac{\partial f}{\partial p_j} = \sum_k \frac{\partial f_k}{\partial p_j}(p)e^{i\langle k,q\rangle}\ .$$

According to the definition of the norm, we have

$$\left\|\frac{\partial f}{\partial p_j}\right\|_{(1-d)(\varrho,\sigma)} = \sum_k \left|\frac{\partial f_k}{\partial p_j}\right|_{(1-d)\varrho} e^{(1-d)|k|\sigma}\ .$$

By Cauchy's estimate, and forgetting d in the exponent, the right-hand side of the latter expression is bounded by

$$\frac{1}{d\varrho}\sum_k |f_k|_\varrho e^{|k|\sigma}\ ,$$

and the first of (6.65) follows in view of the definition of the norm. Coming to the second one, compute

$$\frac{\partial f}{\partial q_j} = i\sum_k k_j f_k(p)e^{i\langle k,q\rangle}\ ,$$

and use again the definition of the norm in order to estimate

$$\left\|\frac{\partial f}{\partial q_j}\right\|_{(1-d)(\varrho,\sigma)} = \sum_k |k_j|\,|f_k|_{(1-d)\varrho}e^{(1-d)|k|\sigma} \le \sum_k |f_k|_\varrho e^{|k|\sigma}|k|e^{-d|k|\sigma}\ .$$

By the general inequality[8]

$$x^{\alpha} e^{-\delta x} \leq \left(\frac{\alpha}{e\delta}\right)^{\alpha} \quad \text{for positive } \alpha,\, x,\, \delta \tag{6.68}$$

with $\alpha = 1$ and putting $|k|$ and $d\sigma$ in place of x and δ, respectively, we immediately get $|k|e^{-d|k|\sigma} \leq 1/(ed\sigma)$; the second of (6.65) follows from definition of the norm. The proof of (6.66) requires some more work. Compute

$$\{f, g\} = i \sum_{k,k'} \left[\sum_{l=1}^{n} \left(k_l \frac{\partial g_{k'}}{\partial p_l} f_k - k'_l \frac{\partial f_k}{\partial p_l} g_{k'} \right) \right] e^{i\langle k,q\rangle} e^{i\langle k',q\rangle} \,, \tag{6.69}$$

and use the definition of the norm to estimate

$$\begin{aligned} \|\{f,g\}\|_{(1-d'-d)(\varrho,\sigma)} < \sum_{k,k'} \Biggl[\sum_{l=1}^{n} \Biggl(|k_l| \left| \frac{\partial g_{k'}}{\partial p_l} \right|_{(1-d'-d)\varrho} |f_k|_{\varrho} \\ + |k'_l| \left| \frac{\partial f_k}{\partial p_l} \right|_{(1-d'-d)\varrho} |g_{k'}|_{(1-d')\varrho} \Biggr) \Biggr] e^{(1-d'-d)|k+k'|\sigma} \,. \end{aligned}$$

By Cauchy estimates, the right member of the latter expression does not exceed

$$\begin{aligned} \frac{1}{d\varrho} \sum_{k,k'} |g_{k'}|_{(1-d')\varrho} e^{(1-d')|k'|\sigma} |f_k|_{\varrho} e^{|k|\sigma} \sum_{l=1}^{n} |k_l| e^{-(d'+d)|k|\sigma} \\ + \frac{1}{(d'+d)\varrho} \sum_{k,k'} |g_{k'}|_{(1-d')\varrho} e^{(1-d')|k'|\sigma} |f_k|_{\varrho} e^{|k|\sigma} \sum_{l=1}^{n} |k'_l| e^{-d|k'|\sigma} \,. \end{aligned}$$

Note now that the only terms depending on the index l are $\sum_l |k_l| = |k|$ and $\sum_l |k'_l| = |k'|$; moreover, using the inequality (6.68), we get

$$|k|e^{-(d'+d)|k|\sigma} \leq \frac{1}{(d'+d)e\sigma} \,, \quad |k'|e^{-d|k'|\sigma} \leq \frac{1}{de\sigma} \,.$$

Reordering terms, this expression is found to be smaller than

$$\frac{2}{ed(d'+d)\varrho\sigma} \sum_{k'} |g'_k|_{(1-d')\varrho} e^{(1-d')|k'|\sigma} \cdot \sum_{k} |f_k|_{\varrho} e^{|k|\sigma} \,,$$

and the conclusion follows in view of the definition of the norms $\|f\|_{\varrho,\sigma}$ and $\|g\|_{(1-d')(\varrho,\sigma)}$. *Q.E.D.*

Proof of Corollary 6.16. The average $\overline{\{f,g\}}$ is the component $k = 0$ of the Fourier expansion, which results from the Poisson bracket of the terms

[8] Keep $\alpha > 0$ and $\delta > 0$ fixed and let $x \geq 0$. The function $x^{\alpha}e^{-\delta x}$ has a maximum at $x = \frac{\alpha}{\delta}$, and its value at this point is $\left(\frac{\alpha}{e\delta}\right)^{\alpha}$. This immediately gives the desired inequality.

$f_k(p)e^{i\langle k,q\rangle}$ and $g_{-k}(p)e^{i\langle -k,q\rangle}$ of the Fourier expansion of f and g; we recast (6.69) as

$$\overline{\{f,g\}} = i\sum_k\sum_{l=1}^{n} k_l\left(\frac{\partial g_{-k}}{\partial p_l}f_k + \frac{\partial f_k}{\partial p_l}g_{-k}\right) = i\sum_k\sum_{l=1}^{n} k_l\frac{\partial}{\partial p_l}f_k g_{-k}\ .$$

Since $\overline{\{f,g\}}$ depends only on p, its Fourier norm reduces to the supremum norm, namely $\left\|\overline{\{f,g\}}\right\|_{(1-d'-d'')(\varrho,\sigma)} = \left|\overline{\{f,g\}}\right|_{(1-d-d')\varrho}$. Hence we estimate

$$\begin{aligned}
\left|\overline{\{f,g\}}\right|_{(1-d-d')\varrho} &\le \sum_k\sum_{l=1}^{n}|k_l|\left|\frac{\partial}{\partial p_l}f_k g_{-k}\right|_{(1-d-d')\varrho}\\
&\le \sum_k\sum_{l=1}^{n}|k_l|\frac{|f_k|_\varrho|g_{-k}|_{(1-d')\varrho}}{d\varrho}\\
&\le \frac{1}{d\varrho}\left(\sum_k|k|e^{-|k|\sigma}|f_k|_\varrho e^{|k|\sigma}\right)\left(\sum_{k'}|g_{-k'}|_{(1-d')\varrho}e^{|k'|(1-d')\sigma}\right)\\
&\le \frac{1}{ed\varrho\sigma}\|f\|_{\varrho,\sigma}\,\|g\|_{(1-d')(\varrho,\sigma)}\ ;
\end{aligned}$$

here we used the inequality (6.68) with $\alpha = 1$, $x = |k|$ and $\delta = \sigma$. *Q.E.D.*

The next step concerns the estimate of multiple Poisson brackets. Consider an expression of the form $L_{g_s}\circ\cdots\circ L_{g_1}f$. We can estimate it by iterating s times the estimate of Lemma 6.15. However, we must appropriately handle the restrictions of domains. To this end, having fixed the final restriction d, we choose s positive quantities $\delta_1,\cdots,\delta_s$ the sum of which is d, and estimate $L_\chi^j f$ on the domain restricted by the factor $1-\delta_1-\cdots-\delta_j$. The simplest choice is $\delta_1=\cdots=\delta_s=d/s$. The final result is given by

Lemma 6.17: *Let $g_1,\ldots,g_s$ and f be analytic on the domain $\mathscr{D}_{\varrho,\sigma}$, with finite norms and s positive. Then for every positive $d<1$, one has*

$$\begin{aligned}
&\|L_{g_s}\circ\cdots\circ L_{g_1}f\|_{(1-d)(\varrho,\sigma)}\\
&\qquad\le \frac{s!}{e^2}\left(\frac{2e}{d^2\varrho\sigma}\right)^s\|g_1\|_{\varrho,\sigma}\cdot\cdots\cdot\|g_s\|_{\varrho,\sigma}\|f\|_{\varrho,\sigma}\ .
\end{aligned} \tag{6.70}$$

Remark. As for the supremum norm, the estimate (6.70) actually does not depend on the order of Lie operators: with a little abuse, we could say again that the estimates do commute. Furthermore, the coefficient in front of the estimate depends only on how many operators L_{g_j} are applied to f.

Proof of Lemma 6.17. The proof is a repetition of Lemma 6.9; let us recall it. For $s=1$ this is just Lemma 6.15. For $s>1$ let $\delta=d/s$. A straightforward application of Lemma 6.15 gives

$$\|L_{g_1} f\|_{(1-\delta)(\varrho,\sigma)} \le \frac{2}{e\delta^2\varrho\sigma}\|g_1\|_{\varrho,\sigma}\|f\|_{\varrho,\sigma} .$$

For $j = 2,\ldots,s$, recalling that $L_{g_j}\circ\cdots\circ L_{g_1} f$ can be estimated on the domain restricted by the factor $1-j\delta$, use again Lemma 6.15 with δ in place of d and $(j-1)\delta$ in place of d', and get the recurrent estimate

$$\begin{aligned}&\bigl\|L_{g_j}\circ\cdots\circ L_{g_1} f\bigr\|_{(1-j\delta)(\varrho,\sigma)}\\ &\qquad\le \frac{2}{ej\delta^2\varrho\sigma}\|g_j\|_{\varrho,\sigma}\bigl\|L_{g_{j-1}}\circ\cdots\circ L_{g_1} f\bigr\|_{(1-(j-1)\delta)(\varrho,\sigma)} .\end{aligned}$$

By applying this formula s times and recalling that $\delta = d/s$, we get

$$\|L_{g_s}\circ\cdots\circ L_{g_1} f\|_{(1-d)(\varrho,\sigma)} \le \left(\frac{2}{ed^2\varrho\sigma}\right)^s \frac{s^{2s}}{s!}\|g_1\|_{\varrho,\sigma}\cdot\cdots\cdot\|g_s\|_{\varrho,\sigma}\|f\|_{\varrho,\sigma} .$$

In view of the inequality $s^s \le e^{s-1}s!$ for $s\ge 1$, (6.70) follows. *Q.E.D.*

6.7.3 Quantitative Estimates for Lie Series

The following technical lemma is a reformulation of Lemma 6.11 in terms of weighted Fourier norm. It will be generally useful for producing quantitative estimates in the framework of normal form theory.

Lemma 6.18: *Let the generating function χ and the function f be analytic and bounded on a domain $\mathscr{D}_{\varrho,\sigma}$. Assume that χ satisfies the convergence hypothesis*

$$\frac{\|\chi\|_{\varrho,\sigma}}{d^2\varrho\sigma} \le \frac{1}{4e} . \tag{6.71}$$

Then for every positive $d<1$, the function $\exp(L_\chi)f$ is analytic and bounded in $\mathscr{D}_{(1-d)(\varrho,\sigma)}$, and we have

$$\|\exp(L_\chi)f\|_{(1-d)(\varrho,\sigma)} \le \frac{e^2+1}{e^2}\,\|f\|_{\varrho,\sigma} . \tag{6.72}$$

Furthermore, for the remainder of order $r\ge 1$, we have

$$\Bigl\|\exp(L_\chi)f - \sum_{s=0}^{r-1}\frac{1}{s!}L_\chi^s f\Bigr\|_{(1-d)(\varrho,\sigma)} \le \frac{2\|f\|_{\varrho,\sigma}}{e^2}\left(\frac{2e\|\chi\|_{\varrho,\sigma}}{d^2\varrho\sigma}\right)^r . \tag{6.73}$$

In particular, for $r=1$ we have

$$\bigl\|\exp(L_\chi)f - f\bigr\|_{(1-d)(\varrho,\sigma)} \le \frac{4\|\chi\|_{\varrho,\sigma}}{ed^2\varrho\sigma}\|f\|_{\varrho,\sigma} . \tag{6.74}$$

Proof. Recall that by the definition of Lie series we have

$$\exp(L_\chi)f = f + \sum_{s>0}\frac{1}{s!}L_\chi^s f .$$

In view of Lemma 6.17, we estimate

$$\left\|\exp(L_\chi)f\right\|_{(1-d)(\varrho,\sigma)} \le \|f\|_{\varrho,\sigma} + \frac{\|f\|_{\varrho,\sigma}}{e^2}\sum_{s>0}\left(\frac{2e\|\chi\|_{\varrho,\sigma}}{d^2\varrho\sigma}\right)^s = \frac{e^2+1}{e^2}\|f\|_{\varrho,\sigma}\ ,$$

because in view of the convergence hypothesis the sum does not exceed one; this gives (6.72). Coming to (6.73), we calculate

$$\begin{aligned}\left\|\exp(L_\chi)f - \sum_{s=0}^{r-1}\frac{1}{s!}L_\chi^s f\right\|_{(1-d)(\varrho,\sigma)} &\le \frac{\|f\|_{\varrho,\sigma}}{e^2}\sum_{s\ge r}\left(\frac{2e\|\chi\|_{\varrho,\sigma}}{d^2\varrho\sigma}\right)^s\\ &= \frac{\|f\|_{\varrho,\sigma}}{e^2}\left(\frac{2e\|\chi\|_{\varrho,\sigma}}{d^2\varrho\sigma}\right)^r\sum_{s\ge 0}\left(\frac{2e\|\chi\|_{\varrho,\sigma}}{d^2\varrho\sigma}\right)^s\\ &\le \frac{2\|f\|_{\varrho,\sigma}}{e^2}\left(\frac{2e\|\chi\|_{\varrho,\sigma}}{d^2\varrho\sigma}\right)^r\ ,\end{aligned}$$

as claimed. Finally, (6.74) comes by setting $r=1$ in (6.73). *Q.E.D.*

7
The Normal Form of Poincaré and Birkhoff

The method of normal form was introduced by Poincaré in his investigations of the dynamics in the neighbourhood of an equilibrium of a nonlinear system of differential equations – not necessarily a canonical one [184][185]. In the Hamiltonian case it is known as the *Birkhoff normal form* [27]. The basic idea is to give the system a particularly simple form, possibly an integrable one. For instance, Poincaré's first attempt was to conjugate the nonlinear system to its linear approximation; such an attempt fails in the Hamiltonian case. But the name *normal form* may be used in a definitely wider sense. The obvious question concerns the relations with the search of first integrals which we have extensively discussed in Chapter 5, in the canonical case. Suffice it to say, for the moment, that the method of normal form proves to be more powerful than the direct search for first integrals – despite the personal preference of the author for the latter; for example, the method of normal form allows us to deal with the case of resonant frequencies.

The exposition in the present section is limited to the formal construction of a normal form. A heuristic discussion illustrating the possible mechanism of divergence of the series is included at the end of the chapter in the last Section 7.4. A full quantitative scheme of estimates is deferred to the following chapters.

7.1 The Case of an Elliptic Equilibrium

The case of the elliptic equilibrium has been widely examined in Chapter 5, looking for a direct method of constructing first integrals. However, some questions have been left open. First, the possibility of performing a formal construction has been proved only in the case of non-resonant frequencies with the additional condition of reversibility; second, the case of resonant frequencies could not be considered. The method of normal form, though being indirect, provides a formal answer to both questions.

7.1.1 The Formal Algorithm

Let us recall that in a neighbourhood of an elliptic equilibrium the Hamiltonian may generally be written as a power series

$$
\begin{aligned}
H(x,y) &= H_0(x,y) + H_1(x,y) + H_2(x,y) + \ldots \;, \\
H_0(x,y) &= \frac{1}{2}\sum_{l=1}^{n} \omega_l (x_l^2 + y_l^2) \;,
\end{aligned}
\tag{7.1}
$$

where $\omega = (\omega_1, \ldots, \omega_n) \in \mathbb{R}^n$ are the frequencies in the linear approximation, and $H_s(x,y)$ is a homogeneous polynomial of degree $s+2$ in the canonical variables $(x,y) \in \mathbb{R}^{2n}$.

The linear operator $L_{H_0} = \{\cdot, H_0\}$ will play a crucial role in the discussion, so let us introduce the special notation $\partial_\omega = L_{H_0}$. Let us also recall some relevant properties from Section 5.2.2.

(i) With the canonical transformation to complex variables

$$
x_l = \frac{1}{\sqrt{2}}(\xi_l + i\eta_l) \;, \quad y_l = \frac{i}{\sqrt{2}}(\xi_l - i\eta_l) \;, \quad l = 1, \ldots, n \;,
\tag{7.2}
$$

the unperturbed Hamiltonian H_0 takes the form

$$
H_0 = \sum_{l=1}^{n} \omega_l p_l \;, \quad p_l = i\xi_l\eta_l \;,
\tag{7.3}
$$

while the polynomials $H_s(\xi, \eta)$ are still homogeneous.

(ii) The linear operator ∂_ω is diagonal over the complex basis $\xi^j\eta^k$ of monomials (in multi-index notation), for

$$
\partial_\omega \xi^j \eta^k = i\langle j-k, \omega\rangle\, \xi^j \eta^k \;.
$$

(iii) The kernel $\mathscr{N}_\omega$ and the range $\mathscr{R}_\omega$ of ∂_ω are defined as

$$
\mathscr{N}_\omega = \partial_\omega^{-1}(\{0\}) \;, \quad \mathscr{R}_\omega = \partial_\omega(\mathscr{P}) \;,
$$

where $\mathscr{P}$ may be taken as the vector space of homogeneous polynomial of a given degree.

(iv) Both $\mathscr{N}_\omega$ and $\mathscr{R}_\omega$ are subspaces of the same space $\mathscr{P}$, and we have

$$
\mathscr{N}_\omega \cap \mathscr{R}_\omega = \{0\} \;, \quad \mathscr{N}_\omega \cup \mathscr{R}_\omega = \mathscr{P} \;.
$$

(v) The operator ∂_ω restricted to $\mathscr{R}_\omega$ is uniquely inverted.

The definition of what is meant by *normal form* is a non-trivial matter: it strongly depends on what we are looking for. The most immediate, and also the most frequently used one, is:

the Hamiltonian $Z = Z_0 + Z_1 + Z_2 + \ldots$ is said to be in Poincaré–Birkhoff normal form in case $\partial_\omega Z = 0$.

The symbol Z will be used in place of H in order to stress that the Hamiltonian is in normal form.

The problem is:

find a near-the-identity transformation which gives the Hamiltonian (7.1) the normal form.

The transformation may be performed, for example, by using a generating function in mixed variables, as explained in Example 2.26; however, this method requires substitutions and inversions which can be avoided using either a composition of Lie series or a single Lie transform. Here we proceed with the algorithm of the Lie transform.

We solve the equation $T_\chi Z = H$, the unknowns being the normal form Z itself and the generating sequence $\chi = \{\chi_1, \chi_2, \ldots\}$, with χ_s a generating function of order s, namely a homogeneous polynomial of degree $s+2$. The formal algorithm is found by recalling the definition of the operator T_χ, namely

$$T_\chi = \sum_{s\geq 0} E_s \ , \quad E_0 = 1 \ , \quad E_s = \sum_{j=1}^{s} \frac{j}{s} L_{\chi_j} E_{s-j} \ . \tag{7.4}$$

Let us collect the algorithm in a short scheme.

For $s \geq 1$ find χ_s and Z_s by recurrently solving the homological equation

$$Z_s - \partial_\omega \chi_s = \Psi_s \ , \quad s = 1, \ldots, r \ , \tag{7.5}$$

where

$$\begin{aligned} &\Psi_1 = H_1 \ , \\ &\Psi_s = H_s - \sum_{j=1}^{s-1} \frac{j}{s} \left(L_{\chi_j} H_{s-j} + E_{s-j} Z_j \right) \quad \text{for } 2 \leq s \leq r \ . \end{aligned} \tag{7.6}$$

In order to obtain these formulæ, it is convenient to recall the triangle for the Lie transform.

$$\begin{array}{cccccc} H_0 & Z_0 & & & & \\ & \downarrow & & & & \\ H_1 & E_1 Z_0 & Z_1 & & & \\ & \downarrow & \downarrow & & & \\ H_2 & E_2 Z_0 & E_1 Z_1 & Z_2 & & \\ & \downarrow & \downarrow & \downarrow & & \\ \vdots & \vdots & \vdots & \vdots & \ddots & \\ H_r & E_r Z_0 & E_{r-1} Z_1 & E_{r-2} Z_2 & \cdots & Z_r . \end{array} \tag{7.7}$$

The triangle contains everything which must be calculated in order to give the Hamiltonian a normal form up to an arbitrary order r. The sth row of the triangle gives Z_s as

$$Z_s = H_s - E_s Z_0 - E_{s-1} Z_1 - E_{s-2} Z_2 - \cdots - E_1 Z_{s-1} \;;$$

this comes from the last line of (7.7), naming s the running index for the order. Here all quantities $E_{s-1}Z_1$, $E_{s-2}Z_2$, ..., $E_1 Z_{s-1}$ on the right member depend only on $\chi_1, \ldots, \chi_{s-1}$ and $Z_1, \ldots, Z_{s-1}$, determined in the previous steps; the dependence on χ_s comes only from $E_s Z_0$. Using the explicit expression of the operator E_s, namely

$$E_s Z_0 = L_{\chi_s} Z_0 + \sum_{j=1}^{s-1} \frac{j}{s} L_{\chi_j} E_{s-j} Z_0 \;,$$

we find an expression for Ψ_s similar to (7.5), namely

$$\Psi_s = H_s - \sum_{j=1}^{s-1} \frac{j}{s} L_{\chi_j} E_{s-j} Z_0 - \sum_{j=1}^{s-1} E_{s-j} Z_j \;.$$

Here it is convenient to remove the dependence on Z_0 using Eq. (7.5). To this end, replace in the first sum $E_{s-j}Z_0 = H_{s-j} - \sum_{l=1}^{s-j} E_{s-j-l} Z_l$ (which comes again from the last row of the Lie triangle (7.7) with $s-j$ in place of r); split the second sum as

$$\sum_{j=1}^{s-1} \cdots = \sum_{j=1}^{s-1} \frac{j}{s} \cdots + \sum_{j=1}^{s-1} \frac{s-j}{s} \cdots \;;$$

then use the recurrent definition $E_{s-j} = \sum_{l=1}^{s-j} \frac{l}{s-j} L_{\chi_l} E_{s-j-l}$ and get

$$\begin{aligned} \Psi_s = H_s &- \sum_{j=1}^{s-1} \frac{j}{s} L_{\chi_j} H_{s-j} - \sum_{j=1}^{s-1} \frac{j}{s} E_{s-j} Z_j \\ &+ \sum_{j=1}^{s-1} \sum_{l=1}^{s-j} \frac{j}{s} L_{\chi_j} E_{s-j-l} Z_l - \sum_{j=1}^{s-1} \sum_{l=1}^{s-j} \frac{l}{s} L_{\chi_l} E_{s-j-l} Z_j \;. \end{aligned}$$

The first line gives (7.6). The contribution of the second line is seen to vanish by exchanging the sums in the last term, with the change of limits

$$\left. \begin{array}{l} 1 \le j \le s-1\,, \\ 1 \le l \le s-j \end{array} \right\} \quad \longleftrightarrow \quad \left\{ \begin{array}{l} 1 \le l \le s-1\,, \\ 1 \le j \le s-l\,. \end{array} \right.$$

7.1.2 The Solution of Poincaré and Birkhoff

Having seen how to obtain the algorithm expressed by formulæ (7.5) and (7.6), let us see where it leads us. The reader will note that the *homological equation* (7.5) is similar to (5.37), which we have found in trying to construct first integrals. The relevant difference is the new unknown Z_s which appears here: it provides us with a lot of freedom in solving the equation. Here we exploit the properties (i)–(v) recalled at the beginning of Section 7.1.1.

The straightforward solution, as proposed by Poincaré and Birkhoff, is the following. Project the right-hand side of (7.5) on $\mathscr{N}_\omega$ and $\mathscr{R}_\omega$, that is, (with the obvious meaning of the superscripts)

$$\Psi_s = \Psi_s^{(\mathscr{N})} + \Psi_s^{(\mathscr{R})} , \quad \Psi_s^{(\mathscr{N})} \in \mathscr{N}_\omega , \quad \Psi_s^{(\mathscr{R})} \in \mathscr{R}_\omega .$$

Then set

$$Z_s = \Psi_s^{(\mathscr{N})} , \quad \chi_s = \partial_\omega^{-1} \Psi_s^{(\mathscr{R})} ,$$

so that χ_s is uniquely defined with the condition $\chi_s \in \mathscr{R}_\omega$. An arbitrary term $\tilde{\chi}_s \in \mathscr{N}_\omega$ may be added to χ_s, but usually it is not necessary.

Let us be more explicit. In complex Cartesian variables, write the homogeneous polynomial Ψ_s as

$$\Psi_s = \sum_{|j+k|=s} \psi_{j,k} \xi^j \eta^k .$$

If $\langle j-k, \omega \rangle = 0$, then add $\psi_{j,k}\xi^j\eta^k$ to Z_s ; else, add $-i\frac{\psi_{j,k}}{\langle j-k,\omega\rangle}\xi^j\eta^k$ to χ_s .

Exploiting the algorithm just given, we conclude with the following.

Proposition 7.1: *The Hamiltonian in complex variables* ξ, η

$$H = H_0 + H_1 + \ldots , \quad H_0 = \sum_l i\omega_l \xi_l \eta_l$$

may be cast formally in the normal form of Poincaré and Birkhoff

$$Z = Z_0 + Z_1 + Z_2 + \ldots , \quad \partial_\omega Z = 0 .$$

Thus, we have

$$Z_s(\xi,\eta) = \sum_{j-k\in\mathscr{M}_\omega} c_{j,k}\xi^j\eta^k .$$

7.1.3 The Canonical Transformation

The generating sequence χ of the previous section is constructed without even mentioning that there is an underlying canonical transformation of coordinates. This is due to the property of the Lie transform expressed by the exchange theorem: forget the coordinates, and work on functions.

Writing the transformation of coordinates is an easy matter, indeed: just recall the exchange theorem. We have found the generating sequence by solving the equation $T_\chi Z = H$. In more explicit terms, let us denote by x, y the original coordinates which we use in writing the Hamiltonian H and by x', y' the coordinates of the normal form. The exchange theorem says that

$$Z(x', y')\Big|_{x'=T_\chi x,\, y'=T_\chi y} = (T_\chi Z)(x, y) = H(x, y). \tag{7.8}$$

Hence the coordinate transformation and its inverse are written as

$$x' = T_\chi x = x + \frac{\partial \chi_1}{\partial y} + \ldots\,, \qquad y' = T_\chi y = y - \frac{\partial \chi_1}{\partial x} + \ldots \tag{7.9}$$

and

$$x = T_\chi^{-1} x' = x' - \frac{\partial \chi_1}{\partial y'} + \ldots\,, \qquad y = T_\chi^{-1} y' = y' + \frac{\partial \chi_1}{\partial x'} + \ldots\,. \tag{7.10}$$

Two remarks are necessary here. First, these two formulæ might be misleading, since it seems that the inverse T_χ^{-1} is found by merely changing the sign of the generating sequence: this is quite wrong. The inverse must be constructed as explained in Section 6.3.1. Second, one must keep in mind that the identity $T_\chi^{-1} T_\chi = 1$ is true only if we consider the infinite expansion of the series; if the expansions are truncated at an arbitrary order r, then we have only $T_\chi^{-1} T_\chi = 1 + \mathcal{O}(r+1)$, consistent with the rules of formal calculus.

Similar considerations apply if we use complex coordinates: just write (ξ, η) in place of x, y and (ξ', η') in place of x', y'; for the Lie transform T_χ is defined in terms of Poisson brackets, which are preserved by canonical changes of coordinates.

A further question is whether the generating sequence which produces the normal form is unique or not. The answer is in the negative, for at every step of the procedure we may add an arbitrary kernel term, $\overline{\chi}_s \in \mathscr{N}_\omega$ say, to the generating function χ_s. Such an addition affects all the rest of the construction. The overall effect is more easily seen if we think in terms of action-angle variables; so let us postpone some considerations to Section 7.2.

7.1.4 First Integrals

The Lie transform formalism allows us to put the issue of searching for first integrals in different terms. The process is an indirect one, since it goes through the construction of a normal form. The advantage is that it provides a complete answer to the questions raised in Chapter 5. The basic claim is the following.

Proposition 7.2: *Let Ψ be a first integral for the normal form Z. Then $T_\chi \Psi$ is a first integral for H.*

Proof. By the properties of the Lie transform operator T_χ, Lemma 6.5, and in view of $T_\chi Z = H$, we have

$$\{T_\chi \Psi, H\} = \{T_\chi \Psi, T_\chi Z\} = T_\chi \{\Psi, Z\} = 0 \ . \qquad \text{Q.E.D.}$$

Thus, the problem is to find suitable first integrals for Z. It is relevant that the structure of the kernel $\mathscr{N}_\omega$ strongly depends on the existence of resonances among the frequencies. Here is convenient to discuss in more detail the algebraic aspect of the problem. In the space $\mathscr{P}$ of homogeneous polynomials of a given degree we may distinguish the following subspaces, depending on the frequencies ω:

(i) the *range* $\mathscr{R}_\omega$ and the kernel $\mathscr{N}_\omega$ which we have already defined;
(ii) the *non-resonant kernel* $\mathscr{N}_0$ defined as the kernel in the non-resonant case $\dim \mathscr{M}_\omega = 0$;
(iii) the *commutant* of the kernel, that is, the subspace of $\mathscr{N}_\omega$, defined as

$$\mathscr{N}_\omega^c = \bigl\{\Phi \,:\, \{\Phi, \Psi\} = 0 \text{ for all } \Psi \in \mathscr{N}_\omega\bigr\} \, . \tag{7.11}$$

Lemma 7.3: *The following properties hold true:*

(i) $\mathscr{R}_\omega \cap \mathscr{N}_\omega = \{0\}$ *and* $\mathscr{R}_\omega \cup \mathscr{N}_\omega = \mathscr{P}$*, the union meaning direct sum of subspaces.*
(ii) If $\dim \mathscr{M}_\omega = 0$ *(i.e., non-resonant frequencies) then* $\mathscr{N}_\omega^c = \mathscr{N}_0 = \mathscr{N}_\omega$*, and* $\mathscr{N}_0$ *contains* n *independent functions* $p_l = i\xi_l\eta_l$*,* $(l = 1, \ldots, n)$*, which form a complete involution system.*
(iii) If $\dim \mathscr{M}_\omega > 0$ *(i.e., resonant frequencies) then* $\mathscr{N}_\omega^c \subset \mathscr{N}_0 \subset \mathscr{N}_\omega$*, the inclusion to be intended in strict sense.*
(iv) $\mathscr{N}_\omega$ *contains* $n + \dim \mathscr{M}_\omega$ *independent monomials* $\xi^j\eta^k$ *with* $j - k \in \mathscr{M}_\omega$ *(including the monomials of* $\mathscr{N}_0$ *with* $j = k$*).*
(v) $\mathscr{N}^c$ *contains* $n - \dim \mathscr{M}_\omega$ *independent functions in* $\mathscr{N}_\omega^c$ *written as*

$$\Phi_0 = \langle \mu, p\rangle = i\sum_{l=1}^{n} \mu_l \xi_l \eta_l \ , \quad 0 \neq \mu \perp \mathscr{M}_\omega \ . \tag{7.12}$$

(vi) The Poisson bracket between pairs of functions in the subspaces obeys the following table.

(7.13)

$\{\cdot,\cdot\}$	$\mathscr{N}^c$	$\mathscr{N}_0$	$\mathscr{N}_\omega$	$\mathscr{R}_\omega$
$\mathscr{N}^c$	0	0	0	$\mathscr{R}_\omega$
$\mathscr{N}_0$	0	0	$\mathscr{N}_\omega$	$\mathscr{R}_\omega$
$\mathscr{N}_\omega$	0	$\mathscr{N}_\omega$	$\mathscr{N}_\omega$	$\mathscr{R}_\omega$
$\mathscr{R}_\omega$	$\mathscr{R}_\omega$	$\mathscr{R}_\omega$	$\mathscr{R}_\omega$	$\mathscr{P}$

Let us stress that the property that $\mathcal{N}_0$ contains a complete involution system is independent from the existence of resonances. Moreover, recalling that in real variables we have $p_l = (y_l^2 + x_l^2)/2$, the p are action variables, as constructed by applying Proposition 3.31. What is not true in the resonant case is that these functions are in involution with the Hamiltonian $Z \in \mathcal{N}_\omega$.

Exercise 7.4: Prove Lemma 7.3. Hint: using complex variables, the proof is almost direct. A.E.L.

Proposition 7.5: *The Hamiltonian*

$$H = H_0 + H_1 + \ldots \; , \qquad H_0 = \sum_l i\omega_l \xi_l \eta_l$$

possesses $n - \dim \mathcal{M}_\omega$ formal first integrals of the form

$$\Phi = T_\chi \Phi_0 \; , \quad \Phi_0 = \sum_{l=1}^{n} \mu_l \xi_l \eta_l \; , \quad 0 \neq \mu \perp \mathcal{M}_\omega \; . \tag{7.14}$$

Moreover:

(i) in the non-resonant case the actions $p_1 = i\xi_1\eta_1, \ldots, p_n = i\xi_n\eta_n$ form a complete system of first integrals in involution, and the Hamiltonian depends on them;

(ii) in the resonant case the $n - \dim \mathcal{M}_\omega$ first integrals so found are in involution, and the Hamiltonian is generically a further first integral.

Proof. By Proposition 7.2 and Lemma 7.3 any $\Phi_0 \in \mathcal{N}_\omega^c$ generates a first integral for Z. In general, no more first integrals for Z may be found, unless Z is a very special function. Since $\{\langle \mu, p \rangle, \langle \mu', p \rangle\} = 0$ for any pair μ, μ', we have $\{T_\chi \langle \mu, p \rangle, T_\chi \langle \mu', p \rangle\} = T_\chi \{\langle \mu, p \rangle, \langle \mu', p \rangle\} = 0$, hence the first integrals are in involution.

(i) By the normal form condition, $\partial_\omega Z = 0$, Z may depend only on $p_1, \ldots, p_n$.

(ii) If Z contains a monomial $\xi^j \eta^k$ with $0 \neq j - k \in \mathcal{M}_\omega$, then Z is independent of $p_1, \ldots, p_n$. If Z contains $\dim \mathcal{M}_\omega$ such monomials which are independent, then there are no further first integrals of Z. Q.E.D.

7.1.5 Back to the Direct Construction of First Integrals

As we have seen, the method of normal form provides an elegant solution to the problem of constructing formal first integrals by series expansion, though they are found via an *indirect* procedure. On the other hand, it is immediately clear that the *direct* method of Section 5.2.2 is definitely faster, since it does not require the construction of a generating sequence and of the normal form. The question is whether we can improve the result stated in Proposition 5.14.

Recall that the direct construction of a first integral is based on solving the recurrent system (5.37), which we may rewrite as

$$
\begin{aligned}
&\partial_\omega \Phi_0 = 0 \ , \\
&\partial_\omega \Phi_s = \Psi_s \ , \quad \Psi_s = -\sum_{l=0}^{s-1} \{\Phi_j, H_{s-j}\} \ , \quad s \geq 1 \ .
\end{aligned}
\tag{7.15}
$$

The problem is that this equation may be solved only if $\Psi_s \in \mathscr{R}_\omega$. Here we have to deal separately with the non-resonant and the resonant case

Let us start with the non-resonant case, choosing as Φ_0 one of the actions p; this is enough for a complete discussion. In Proposition 5.14 the hypothesis of reversibility of the Hamiltonian is used to prove that the condition $\Psi_s \in \mathscr{R}_\omega$ is satisfied. Now we show that the hypothesis of reversibility can be removed; hence the direct construction is always legitimate. However, the proof requires a little more analysis.

In order to compare the two methods let us denote by $\Phi = \Phi_0 + \Phi_1 \ldots$ and $\tilde{\Phi} = T_\chi \Phi_0 = \Phi_0 + E_1 \Phi_0 + \ldots$ the formal series obtained via the direct and indirect method, respectively. At first glance one could imagine that the two formal series so found are the same, since both start with the same term Φ_0; but this is untrue. The point is that the solution of the homological equation $\partial_\omega \Phi_s = \Psi_s$ is determined up to an arbitrary term $\overline{\Phi}_s \in \mathscr{N}_\omega$. In the direct method we have made it unique by forcing $\Phi_s \in \mathscr{R}$, namely setting $\overline{\Phi}_s = 0$, but with the indirect method $E_s \Phi_0$ is completely determined by the generating functions $\chi_1, \ldots, \chi_s$, so that kernel terms may be introduced. The point is a puzzling one, so let us examine it in more detail.

At odd orders (i.e., for s odd) it is immediately evident that $E_s \Phi_0 \in \mathscr{R}_\omega$; for in view of non-resonance we have $\xi^j \eta^k \in \mathscr{N}_\omega$ if and only if $j = k$, so that the monomial has even degree. The following example shows that kernel terms are typically generated at even orders.

Example 7.6: *Kernel terms in $E_2 \Phi_0$.* Let us consider the simplest case of one degree of freedom. Let $H_0 = i\omega\xi\eta$, $H_1 = \xi^3 + \eta^3$ and $\Phi_0 = i\xi\eta$. We want to calculate

$$
E_2 \Phi_0 = \frac{1}{2} L_{\chi_1}^2 \Phi_0 + L_{\chi_2} \Phi_0 \ .
$$

It is easy to check that $L_{\chi_2} \Phi_0 \in \mathscr{R}_\omega$, hence we focus attention on $L_{\chi_1}^2 \Phi_0$. We determine χ_1 by solving the homological equation $Z_1 - \partial_\omega \chi_1 = H_1 \in \mathscr{R}_\omega$ (it is odd order), so we have $Z_1 = 0$ and $\chi_1 = -\partial_\omega^{-1} H_1$ (recall that ∂_ω^{-1} is well defined on $\mathscr{R}_\omega$). We have thus

$$
\chi_1 = -\frac{i}{3\omega} (\xi^3 - \eta^3) \ , \quad L_{\chi_1} \Phi_0 = -\frac{1}{\omega} (\xi^3 + \eta^3) \ ,
$$

and finally

$$L^2_{\chi_1}\Phi_0 = -\frac{6i}{\omega}\xi^2\eta^2 \ ;$$

this is a kernel term which is produced in $E_2\Phi_0$. *E.D.*

The example can be generalized to any even degree; there is no need to spend more time on it. It is more interesting to examine how adding a kernel term at an arbitrary order may affect the construction.

Lemma 7.7: *The Hamiltonian*

$$H = H_0 + H_1 + \ldots \ , \qquad H_0 = \sum_l i\omega_l \xi_l \eta_l$$

possesses a formal first integral $\Phi = \Phi_0 + \Phi_1 + \ldots$ constructed as

$$\Phi = T_\chi\bigl(W_0 + W_2 + W_4 + \ldots\bigr) \ , \quad W_0, W_2, W_4, \ldots \in \mathscr{N}_\omega$$

and obeying $\Phi_s \in \mathscr{R}_\omega$ for $s \geq 1$.

Corollary 7.8: *Proposition 5.14 holds true with the sole condition of non-resonance. The reversibility condition may be removed.*

Proof of Lemma 7.7. Since $W \in \mathscr{N}_\omega$ and $\dim \mathscr{M}_\omega = 0$, there cannot be odd-order terms in W. By the properties of T_χ, Lemma 6.5, we have (formally)

$$T_\chi W = T_\chi W_0 + T_\chi W_2 + T_\chi W_4 + \ldots,$$

and in view of $W_s \in \mathscr{N}_\omega$ we have $\{T_\chi W, H\} = T_\chi\{W, Z\} = 0$, hence $T_\chi W$ is a first integral. The term of degree s is $[T_\chi W]_s = W_s + E_2 W_{s-2} + \cdots + E_s W_0$. By recurrence, let us assume that $\Phi_l = [T_\chi W]_l \in \mathscr{R}_\omega$ for $l = 1, \ldots, s-1$; this is true for $s = 2$, and we may well take W as a polynomial of degree $s-1$. Let $[T_\chi W]_s = \Phi_s + \overline{\Phi}_s$ with $\Phi_s \in \mathscr{R}_\omega$ and $\overline{\Phi}_s \in \mathscr{N}_\omega$. If $\overline{\Phi}_s \neq 0$, replace W with $W' = W - \overline{\Phi}_s$. Then $T_\chi W'$ is still a first integral which coincides with $T_\chi W$ up to order $s-1$, and $\Phi_s = [T_\chi W']_s \in \mathscr{R}_\omega$, as required, because we have subtracted the unwanted kernel part. *Q.E.D.*

Proof of Corollary 7.8. The formal first integral Φ of Lemma 7.7 solves (formally) the equation $\{\Phi, H\} = 0$, and so also the system of equations

$$\begin{aligned}
&\{\Phi_1, H_0\} = -\{\Phi_0, H_1\},\\
&\{\Phi_s, H_0\} = -\{\Phi_{s-1}, H_1\} - \cdots - \{\Phi_0, H_s\} \ , \quad s > 0 \ .
\end{aligned}$$

Since $\{\Phi_s, H_0\} \in \mathscr{R}_\omega$, the right member of the equation must belong to $\mathscr{R}_\omega$ for every s. *Q.E.D.*

Thus, the consistency problem is removed in the non-resonant case. With a moment's thought the reader will realize that the construction of a first integral via the direct method remains consistent also if an arbitrary kernel term is added at every (even) order.

Let us now come to the resonant case. The reader might be tempted to imagine that the consistency problem disappears also in this case; but this is untrue. Lemma 7.7 does not apply because the condition $W_s \in \mathscr{N}_\omega$ does not imply that W_s is a first integral for Z, and so $T_\chi W_s$ need not be a first integral for H: the stronger condition $W_s \in \mathscr{N}_\omega^c$ is necessary. Therefore a first integral $\Phi = \Phi_0 + \Phi_1 + \ldots$ satisfying $\Phi_s \in \mathscr{R}_\omega$ cannot be constructed, in general.

Historically, as mentioned in Chapter 5, Cherry proposed to proceed as follows: at step s of the construction, introduce an arbitrary kernel term $\overline{\Phi}_s$ undetermined; later, determine it so that it removes the unwanted kernel terms in the known member of the equations at higher orders. However, he failed to prove that his procedure was effective. Proposition 7.5 states that Cherry's procedure must be effective provided Φ_0 has some special form, and the expression of $T_\chi\Phi_0$ contains appropriate kernel terms which make the construction consistent.

7.2 Action-Angle Variables for the Elliptic Equilibrium

The dynamics of the normalized Hamiltonian is better described in action-angle variables $p = (p_1, \ldots, p_n) \in \mathbb{R}^n_+$ and $q = (q_1, \ldots, q_n) \in \mathbb{T}^n$. Thus, let us rewrite the Hamiltonian for the elliptic equilibrium in action-angle variables by transforming

$$x_l = \sqrt{2p_l}\cos q_l \ , \quad y_l = \sqrt{2p_l}\sin q_l \ , \quad l = 1, \ldots, n \ . \tag{7.16}$$

From complex variables we easily write the Hamiltonian by using the exponential form of trigonometric functions, namely

$$\xi_l = \sqrt{p_l}\, e^{iq} \ , \quad \eta_l = i\sqrt{p_l}\, e^{-iq} \ , \quad l = 1, \ldots, n \ . \tag{7.17}$$

The action variables are thus

$$p_l = i\xi_l\eta_l = \frac{1}{2}\left(y_l^2 + x_l^2\right) ,$$

and the unperturbed Hamiltonian becomes linear in the actions, since

$$H_0 = \langle\omega, p\rangle \ , \quad \partial_\omega = \left\langle\omega, \frac{\partial}{\partial q}\right\rangle \ .$$

From the detailed discussion of Section 5.2.1 we know that a homogeneous polynomial $f(\xi,\eta) = \sum_{|j+k|=s} f_{j,k}\xi^j\eta^k$ of degree s is changed into a trigonometric polynomial of the same degree, which we may write as

$$f(q,p) = \sum_{|k|\le s} c_k(p)\exp(i\langle k, q\rangle)$$

with coefficients $c_k(p)$ which are homogeneous polynomials in $p^{1/2}$. From Section 5.2.1 we also know that the dependence on the angles q is subject to strong limitations.

7.2.1 The Normal Form

The normal form Z is expanded as a series of trigonometric polynomials which contain only Fourier harmonics $\langle k, q\rangle$ with $k \in \mathscr{M}_\omega$, that is,

$$Z_s(q,p) = \sum_{k\in\mathscr{M}_\omega\,,\,|k|\le s} c_k(p)\exp\bigl(i\langle k,q\rangle\bigr),$$

where, again, $c_k(p)$ are homogeneous polynomials of degree s in $p^{1/2}$. The momenta p are perturbations of the actions p' of the unperturbed Hamiltonian H_0; we have indeed $p = T_\chi p'$.

It may be remarked that the formal construction of the normal form could well be performed using action-angle variables. The unpleasant fact is that the calculation of Poisson brackets involves derivatives of $p^{1/2}$, which seem to produce coefficients with powers of $p^{-1/2}$ in the Fourier expansions. However, exploiting the properties in Section 5.2.1 we may check that terms with such coefficients do compensate each other, so they do not show up in the result. Such a check is not really necessary; indeed, the whole procedure of construction of normal form is based on calculation of Poisson brackets, which are preserved by canonical transformation. The final result could be obtained by working with polynomials and then transforming to action-angle variables, so that inverse powers of p cannot appear.

7.2.2 The Non-resonant Case

In the non-resonant case the normal form turns out to be a power series depending only on the actions $p_1,\ldots,p_n$, taking the particular form

$$Z = \langle\omega,p\rangle + Z_2(p) + Z_4(p) + \ldots\ . \tag{7.18}$$

Odd-order terms cannot appear because they must contain odd powers of $p^{1/2}$, which in turn entail a dependence on the angles; this is in view of the detailed discussion of Section 5.2.1.

The usual description of the dynamics applies in this case. The phase space is foliated into invariant tori parameterized by $p_1,\ldots,p_n$, carrying quasiperiodic motions with frequencies

$$\Omega(p) = \omega + \frac{\partial Z_2}{\partial p}(p) + \frac{\partial Z_4}{\partial p}(p) + \ldots.$$

Thus, generically, the dynamics of the normal form is not isochronous.

A last remark concerns the non-uniqueness of the generating sequence. Consider the simpler case of a Lie series with generating function $\xi(p)$ depending only on the actions. With a straightforward calculation we get the finite expressions

$$\exp(L_\xi)q = q + \frac{\partial \xi}{\partial p} \ , \quad \exp(L_\xi)p = p \ .$$

Thus we see that the action p remains unchanged, while the angles q are translated by a p-dependent quantity. The same happens if we perform a Lie transform with a generating sequence $\xi(p)$ depending only on p, as the reader will easily check. Therefore, adding an arbitrary kernel term to the generating function at some step changes the phases of the angles but does not affect the form of Z, for Z depends only on p.

7.2.3 The Resonant Case

The resonant case, $0 < \dim \mathscr{M}_\omega = r < n$, is more intriguing. As we have seen, the normal form depends on the actions p and on combinations of the angles $\langle k, q\rangle$ with $k \in \mathscr{M}_\omega$, that is,

$$Z(p,q) = \langle \omega, p\rangle + Z_1\bigl(p_1, \ldots, p_n, \langle k^{(1)}, q\rangle, \ldots, \langle k^{(r)}, q\rangle\bigr) + \ldots .$$

For $r = 1$ the Hamiltonian Z possesses $n-1$ independent first integrals which are linear combinations of the actions p. Moreover, the Hamiltonian itself is a first integral, and if $Z(p,q)$ does depend on q, then it is independent of the previous first integrals. Therefore, the system is still Liouville-integrable.

A resonant system with $\dim \mathscr{M}_\omega = r > 1$ is not expected to be integrable, except for very particular cases. However, the first integrals may be used to reduce the number of degrees of freedom by r. A general procedure takes advantage of canonical transformations which change the angles (see Lemma 3.33). Let us recall the procedure. Find a basis $k^{(1)}, \ldots, k^{(r)}$ for $\mathscr{M}_\omega$; that is, choose r integer vectors in $\mathscr{M}_\omega$ which are independent, and satisfy the further property that $\mathrm{span}\bigl(k^{(1)}, \ldots, k^{(r)}\bigr) \cap \mathbb{Z}^n = \mathscr{M}_\omega$. Denote by

$$\begin{pmatrix} k_{1,1} & k_{1,2} & \ldots & k_{1,n} \\ \vdots & \vdots & \ldots & \vdots \\ k_{r,1} & k_{n-r,2} & \ldots & k_{r,n} \end{pmatrix}$$

the matrix whose lines are the vectors of the basis. Then, as shown in appendix A, we can complete it by adding $n-r$ rows so as to find a unimodular matrix

$$\mathsf{M} = \begin{pmatrix} k_{1,1} & k_{1,2} & \ldots & k_{1,n} \\ \vdots & \vdots & \ldots & \vdots \\ k_{r,1} & k_{n-r,2} & \ldots & k_{r,n} \\ m_{1,1} & m_{1,2} & \ldots & m_{1,n} \\ \vdots & \vdots & \ldots & \vdots \\ m_{n-r,1} & m_{n-r,2} & \ldots & m_{n-r,n} \end{pmatrix} , \quad \det \mathsf{M} = \pm 1 .$$

With a canonical transformation

$$\varphi = \mathsf{M}q , \quad p = \mathsf{M}^\top I ,$$

with generating function $S(I, q) = \langle I, \mathsf{M}q \rangle$, the Hamiltonian is changed to

$$H_0(I) = \langle \omega', I \rangle , \quad Z_s(I, \varphi) = Z_s(I_1, \ldots, I_n, \varphi_1, \ldots, \varphi_r) ,$$
$$\omega' = \mathsf{M}\omega = (0, \ldots, 0, \omega'_{n-r+1}, \ldots, \omega'_n) .$$

The Hamiltonian turns out to depend only on the *resonant angles* $\varphi_1, \ldots, \varphi_r$. Hence the actions $I_{r+1}, \ldots, I_n$ are first integrals which may be considered as parameters, and we may forget $H_0(I_{r+1}, \ldots, I_n)$, which is constant. We conclude that the dynamics is determined by the family of reduced systems of $r < n$ degrees of freedom with Hamiltonian

$$Z(I_1, \ldots, I_r, c_1, \ldots, c_{n-r}, \varphi_1, \ldots, \varphi_r)$$
$$= Z_1(I_1, \ldots, I_r, c_1, \ldots, c_{n-r}, \varphi_1, \ldots, \varphi_r) + \ldots$$

parameterized by the initial values of the constants $I_{r+1} = c_1, \ldots, I_n = c_{n-r}$. A system of this form needs a separate discussion, which should be adapted to the case under investigation.

The normal form of the Hamiltonian may be simplified, or changed according to some personal preference, by adding arbitrary kernel terms to the generating sequence. The considerations made at the end of Section 7.2.2 do not apply here: the expression of the normal form is actually changed, though not in an arbitrary manner. We do not enter such a long discussion.

A last caveat: we should never forget that *all claims made until now are just formal*: we still lack a discussion of the (non)convergence of the normal form.

7.2.4 An Example of Resonant Hamiltonian

As an example which does not involve long calculations, let us consider again the Hamiltonian (5.41) with parameters $\alpha = 0$, as in the models of

Contopoulos, and with a 2 : 1 resonance between the frequencies,[1] setting $\omega_1 = 1/2$ and $\omega_2 = 1$. Thus, the Hamiltonian is written as

$$H = \frac{1}{4}\left(y_1^2 + x_1^2\right) + \frac{1}{2}\left(y_2^2 + x_2^2\right) + x_1^2 x_2 \ . \tag{7.19}$$

The aim is to study the dynamics of the normal form after one step of normalization, though in an apparently crude approximation. This will turn out to be useful at least for low energies, which means small perturbation. A Poincaré section for energy $E = 0.01$ is represented in Fig. 7.1.

It is convenient to perform the calculation in action-angle variables, with complex representation of trigonometric functions. With the transformation

$$x_l = \sqrt{\frac{p_l}{2}}\left(e^{iq_l} + e^{-iq_l}\right) \ , \quad y_l = -i\sqrt{\frac{p_l}{2}}\left(e^{iq_l} - e^{-iq_l}\right) \ , \tag{7.20}$$

the Hamiltonian is expanded as

$$\begin{aligned} H_0 &= \frac{p_1}{2} + p_2 \ , \\ H_1 &= \frac{p_1\sqrt{p_2}}{2\sqrt{2}}\left(e^{i(2q_1+q_2)} + e^{-i(2q_1+q_2)}\right) \\ &\qquad + \frac{p_1\sqrt{p_2}}{2\sqrt{2}}\left(e^{i(2q_1-q_2)} + e^{-i(2q_1-q_2)}\right) + \frac{p_1\sqrt{p_2}}{\sqrt{2}}\left(e^{iq_2} + e^{-iq_2}\right) \ . \end{aligned}$$

According to (7.5) and (7.6), the first normalization step consists in solving the equation

$$Z_1 - \partial_\omega \chi_1 = H_1 \ , \quad \partial_\omega = \sum_{l=1}^{n} \omega_l \frac{\partial}{\partial q_l} \ .$$

The resonant terms are easily identified, since they depend on the combination $2q_1 - q_2$ of the angles. Thus, following Poincaré and Birkhoff, we set

$$\begin{aligned} Z_0 &= H_0 \ , \\ Z_1 &= \frac{p_1\sqrt{p_2}}{2\sqrt{2}}\left(e^{i(2q_1-q_2)} + e^{-i(2q_1-q_2)}\right) \ , \end{aligned} \tag{7.21}$$

and calculate the generating function χ_1 as

$$\chi_1 = \frac{ip_1\sqrt{p_2}}{4\sqrt{2}}\left(e^{i(2q_1+q_2)} - e^{-i(2q_1+q_2)}\right) + \frac{ip_1\sqrt{p_2}}{\sqrt{2}}\left(e^{iq_2} - e^{-iq_2}\right) \ .$$

---

[1] The choice of a 2 : 1 resonance simplifies the calculation, because a resonant term in the normal form appears already at order 1. Another classical example is the model of Hénon and Heiles [110], with a 1 : 1 resonance. However, in this case the first resonant term in the normal form appears at order 2, thus making the calculation longer. For a study of the normal form for the model of Hénon and Heiles see, for example, [29].

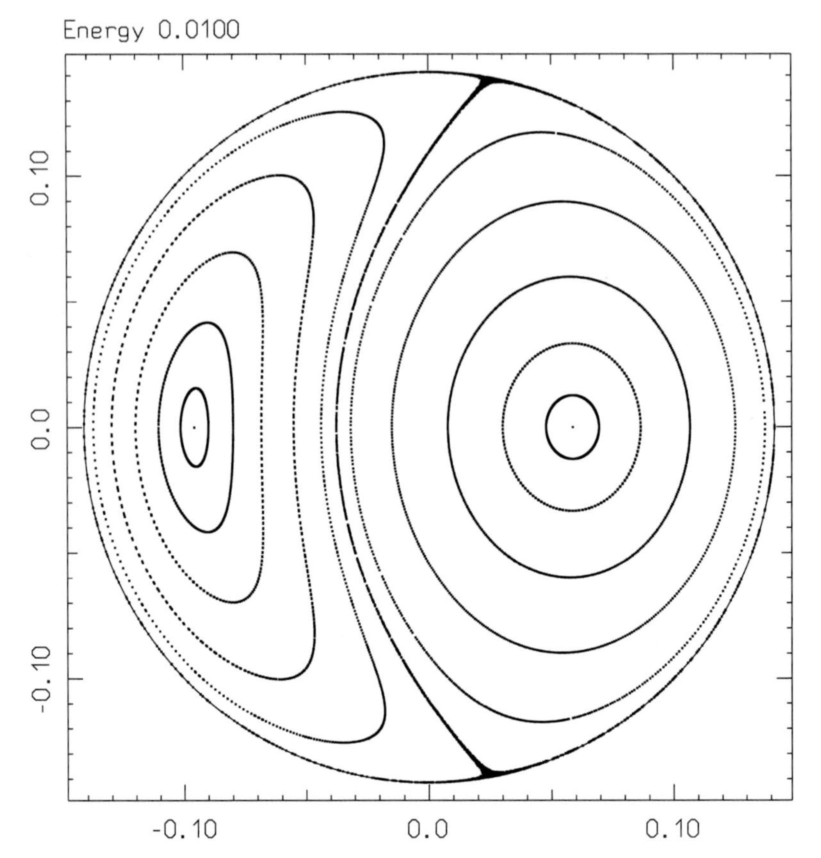

Figure 7.1 Poincaré section for the Hamiltonian (7.19) on the energy surface $E = 0.01$.

Since we truncate our expansions at order 1 there is no need here to calculate the Lie transform of H. Let us write the Hamiltonian $Z = Z_0 + Z_1$, restoring the usual notation for trigonometric functions; thus, we write

$$Z = \frac{p_1}{2} + p_2 + \frac{p_1\sqrt{p_2}}{\sqrt{2}}\cos(2q_1 - q_2) \; .$$

Following the discussion at the end of Section 7.2.3, we change the angles by applying the canonical transformation

$$\varphi_1 = q_1 \; , \quad \varphi_2 = q_2 - 2q_1 \; , \tag{7.22}$$

with inverse

$$q_1 = \varphi_1 \; , \quad q_2 = \varphi_1 + 2\varphi_2 \; . \tag{7.23}$$

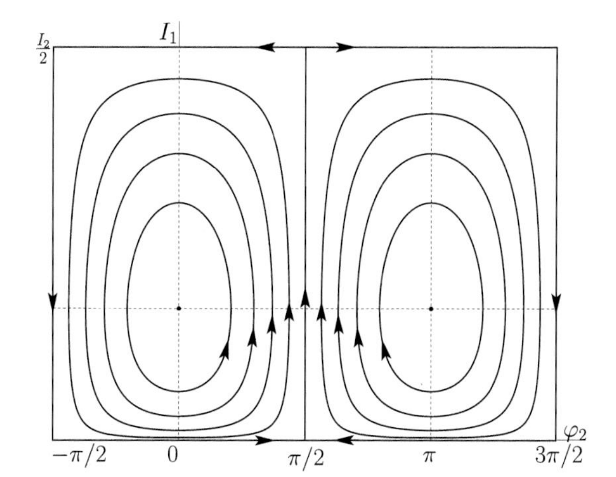

Figure 7.2 Phase portrait for the Hamiltonian (7.25).

The generating function $S = I_1 q_1 + I_2(q_2 - 2q_1)$ allows us to write the transformation of the momenta and its inverse

$$p_1 = I_1 - 2I_2 \ , \quad p_2 = I_2 \ ; \qquad I_1 = p_1 + 2p_2 \ , \quad I_2 = p_2 \ . \tag{7.24}$$

Here we should take into account that, by definition, we have $p_1 \geq 0$ and $p_2 \geq 0$; hence the relations give the constraint $0 \leq I_2 \leq I_1/2$. The Hamiltonian is thus changed to

$$Z = \frac{I_1}{2} + \frac{(I_1 - 2I_2)\sqrt{I_2}}{\sqrt{2}} \cos\varphi_2 \ ; \tag{7.25}$$

the corresponding canonical equations are

$$\begin{aligned} \dot\varphi_1 &= \frac{1}{2} + \sqrt{\frac{I_2}{2}} \cos\varphi_2 \ , \qquad \dot I_1 = 0 \ , \\ \dot\varphi_2 &= \frac{(I_1 - 6I_2)}{2\sqrt{2I_2}} \cos\varphi_2 \ , \quad \dot I_2 = \frac{(I_1 - 2I_2)\sqrt{I_2}}{\sqrt{2}} \sin\varphi_2 \ . \end{aligned} \tag{7.26}$$

The new action I_1 is a first integral; therefore, we may consider the Hamiltonian (7.25) as describing a system with one degree of freedom, depending on the parameter I_1 besides the total energy. The typical phase portrait is represented in Fig. 7.2, which deserves some further comments.

The phase space φ_2, I_2 is represented as a cylinder $[-\pi/2, 3\pi/2] \times [0, I_1/2]$. Looking at the equations (7.26) we find two equilibrium points at the intersection of the dashed vertical lines $\varphi = 0, \pi$ with the horizontal dashed line $I_2 = I_1/3$; these two points are elliptic and correspond to periodic orbits which survive the perturbation. Moreover, there are further equilibrium points at the intersections of the vertical lines $\varphi_2 = -\pi/2$, $\pi/2$ and $3\pi/2$ with the horizontal borders $I_2 = 0$ (the lower border) and $I_2 = I_1/2$

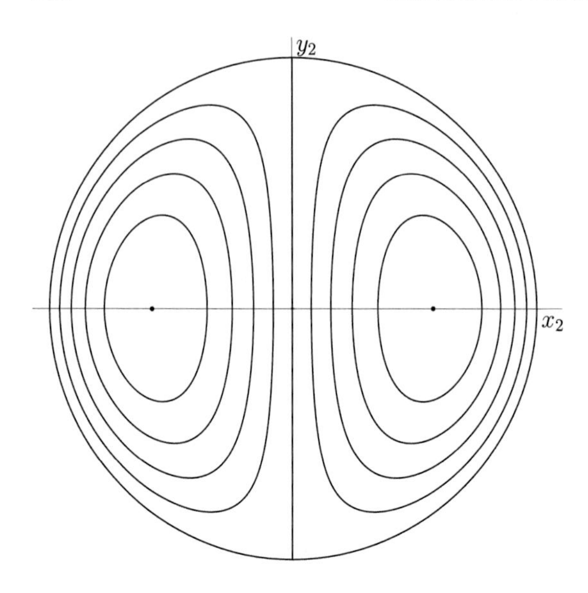

Figure 7.3 The phase portrait of Fig. 7.2 in Cartesian coordinates.

(the upper border); all these points are hyperbolic. The separatrices emanating from these points appear to connect pairs of such equilibria. But here is the puzzling matter. Looking at equations (7.26) we realize that $\dot{\varphi}_2$ is infinite, due to the divisor $\sqrt{I_2}$. A look at the transformations (7.22) shows that we have also $p_2 = 0$, so that the angle q_2 is undefined at that point. The upper border seems to behave better – no time derivatives going to infinity are involved; but the transformations (7.22) tell us that $p_1 = 0$, so that the angle q_1 is undefined there. We should rather consider the upper and lower border as the poles of a spherical surface.

The figure may be better understood by reverting to Cartesian coordinates. By applying the inverse of (7.20) to the expression (7.21) of Z in action-angle variables, we get the Hamiltonian

$$Z = \frac{1}{4}\left(y_1^2 + x_1^2\right) + \frac{1}{2}\left(y_2^2 + x_2^2\right) + \frac{1}{2}x_1 y_1 y_2 - \frac{x_2}{4}\left(y_1^2 - x_1^2\right) .$$

Such an expression does not coincide with the original Hamiltonian (7.19): the cubic term is changed according to the construction of the normal form. However, we should not forget that the transformation to normal form involves a change of variables, which we may write as $x' = T_\chi x$, $y' = T_\chi y$; the expression of Z should be written in variables x', y'. Figure 7.3 represents the level lines of the first integral for the Hamiltonian Z; it may be understood with the help of the change of variables (7.22) and (7.24). A comparison with the Poincaré section of Fig. 7.1 may be made, but the reader should not forget that the coordinates are different, due to the canonical transformation to normal form.

7.3 The General Problem

Let us now turn to investigate how far we can apply the method of normal form to the general problem of dynamics, namely a Hamiltonian

$$H(p,q) = H_0(p) + \varepsilon H_1(p,q) + \varepsilon^2 H_2(p,q) + \dots\ , \tag{7.27}$$

where $(p,q) \in \mathscr{G} \times \mathbb{T}^n$ are action-angle variables.

In view of the theorem of Poincaré on non-existence of first integrals, Section 4.5, we cannot expect to be able to give the system a full normal form. The present section aims at describing how the problem can be reformulated so as to be able to prove at least partial but relevant results. In this sense, the present section may be seen as a preparation for the following chapters.

7.3.1 The Naive But Inconclusive Approach

Working at a purely formal (and candid) level, the algorithm developed in Section 7.1.1 and collected in equations (7.4), (7.5) and (7.6) may be rewritten as is; let us rewrite the formulæ, for convenience. The Lie transform operator is defined as

$$T_\chi = \sum_{s\geq 0} E_s\ ,\quad E_0 = 1\ ,\quad E_s = \sum_{j=1}^{s} \frac{j}{s} L_{\chi_j} E_{s-j}\ . \tag{7.28}$$

The homological equation at every order is

$$Z_s - \partial_\omega \chi_s = \Psi_s\ ,\quad s = 1,\dots,r\ , \tag{7.29}$$

where

$$\begin{aligned} \Psi_1 &= H_1\ ,\\ \Psi_s &= H_s - \sum_{j=1}^{s-1} \frac{j}{s}\left(L_{\chi_j} H_{s-j} + E_{s-j} Z_j\right) \quad \text{for } 2 \leq s \leq r\ . \end{aligned} \tag{7.30}$$

The minor difference concerns the method of solution of the homological equation (7.29), the sole part which depends on the coordinates.

Since we work in action-angle variables, we can write the right member of (7.29) as

$$\Psi(p,q) = \sum_{k\in\mathbb{Z}^n} \psi_k(p) e^{i\langle k,q\rangle}\ , \tag{7.31}$$

where the label s has been suppressed because the argument applies at every order. In view of the generic form of the unperturbed Hamiltonian H_0, the operator $\partial_\omega\cdot = \{\cdot, H_0\}$ is written as

$$\partial_\omega = \sum_{l=1}^{n} \omega_l(p) \frac{\partial}{\partial q_l} \ , \quad \omega_l(p) = \frac{\partial H_0}{\partial p_l} . \tag{7.32}$$

The operator ∂_ω is still diagonal on the Fourier complex basis $e^{i\langle k,q\rangle}$, since

$$\partial_\omega e^{i\langle k,q\rangle} = i\langle k, \omega(p)\rangle e^{i\langle k,q\rangle} \ .$$

Here comes the candid part. As in Section 7.1.1, we may be tempted to say that the Hamiltonian $Z(p,q)$ is in Poincaré–Birkhoff normal form in case $\partial_\omega Z = 0$. Assuming, with Poincaré, that H_0 is non-degenerate, in view of Proposition 4.16 we conclude that $Z(p)$ must be independent of the angles q; hence we try to solve Eq. (7.29) with $\Psi(p,q)$ as in (7.31) by setting

$$Z(p) = \psi_0(p) \ , \quad \chi(p,q) = i \sum_{0 \neq k \in \mathbb{Z}^n} \frac{\psi_k(p)}{\langle k, \omega(p)\rangle} e^{i\langle k,q\rangle} \ . \tag{7.33}$$

Here we fall back on the same troubles that we have encountered in the proof of Poincaré's theorem Section 4.5: unless we introduce strong constraints on the expansion of H_1, $H_2, \ldots$ there is a dense set of points $p \in \mathscr{G}$ where at least one of the coefficients of the formal solution $\chi(p,q)$ goes to infinity. A rearrangement of the proof of Poincaré's theorem shows that our candid procedure fails.

7.3.2 The Case of Constant Frequencies

The discussion concerning the case of elliptic equilibria suggests considering the case of an unperturbed Hamiltonian which is degenerate and has constant frequencies, namely

$$H_0 = \langle \omega, p\rangle \ , \quad \omega \in \mathbb{R}^n \ . \tag{7.34}$$

The discussion in this case follows very closely that of the elliptic case. Denoting here by $\mathscr{P}$ the space of formal Fourier series, the kernel $\mathscr{N}_\omega$ and the range $\mathscr{R}_\omega$ are defined again as $\mathscr{R}_\omega = \partial_\omega(\mathscr{P})$ and $\mathscr{N}_\omega = \partial_\omega^{-1}(\{0\})$; they are disjoint subspaces of the same space $\mathscr{P}$, and $\mathscr{N}_\omega \cup \mathscr{R}_\omega = \mathscr{P}$. Thus, we may split Ψ into its kernel and range components and solve Eq. (7.29) by setting

$$Z(p,q) = \sum_{k \in \mathscr{M}_\omega} \psi_k(p) e^{i\langle k,q\rangle} \ , \quad \chi(p,q) = i \sum_{k \in \mathbb{Z}^n \setminus \mathscr{M}_\omega} \frac{\psi_k(p)}{\langle k, \omega\rangle} e^{i\langle k,q\rangle} \ . \tag{7.35}$$

The problem of zero divisors here is removed. Some doubts may remain concerning the convergence of the Fourier series for χ; this is a minimum request which cannot be ignored by merely invoking the formal character of the series.

The answer to the latter question requires an in-depth look at the concept of resonance, which will be discussed in Chapter 9. Suffice it, for the moment, to recall an argument due to Poincaré.[2] We have seen in Section 6.7.1 that the coefficients of a analytic Fourier series decay exponentially; more precisely, writing for simplicity $\Psi = \sum_{k\in\mathbb{Z}^n} \psi_k e^{i\langle k,q\rangle}$ with coefficients $\psi_k \in \mathbb{C}^n$, we have $|\psi_k| \sim e^{-|k|\sigma}$ with some positive constant σ and with the order of the Fourier component given by $|k| = |k_1| + \cdots + |k_n|$. On the other hand, from the theory of approximation of real quantities with rational ones it was known that for infinitely many real vectors ω one has $\bigl|\langle k,\omega\rangle\bigr| \sim |k|^{-\tau}$ for some $\tau \sim n$; a more precise quantitative estimate is given in Section A.2.1. Thus, we have

$$\left|\frac{\psi_k}{\langle k,\omega\rangle}\right| \sim |k|^{\tau} e^{-|k|\sigma} ,$$

and the Fourier series with such coefficients is still analytic. However, this does not prove the convergence of the process of giving the Hamiltonian a normal form: it is only the first step. The biggest problem lies in the recurrent nature of the system (7.29) and (7.30) of equations: at every step the solution of the homological equations introduces a new divisor $\langle k,\omega\rangle$ on coefficients which already contain the divisors of the previous steps. The critical problem is precisely the *accumulation* of the so-called *small divisors.*

The example of Poincaré shows that the problem of constructing a normal form has a solution, in the formal sense, if the frequencies ω obey an appropriate condition of strong non-resonance. There are frequencies which do not obey the latter condition, still being non-resonant in the strict sense: it may well happen, for example, that $\bigl|\langle k,\omega\rangle\bigr| \sim e^{-|k|}$ or even worse; in this case the frequencies are critically close to resonance. In addition, there are, of course, resonant frequencies with a corresponding resonance module $\mathscr{M}_\omega$. Concerning the resonant case, the condition $\partial_\omega Z = 0$ for the normal form is still valid, as in the case of elliptic equilibria. In the case of frequencies critically close to resonance, a suitable modification of the condition of normal form must be introduced; for example, we may decide to include in Z also range terms with divisors which are too small: there is a lot of freedom in solving the homological equation (7.29).

Example 7.9: *An integrable case.* Let us consider the very special Hamiltonian in action-angle variables,

$$H(p,q) = \langle \omega, p\rangle + \varepsilon H_1(q) ,$$

---

[2] The problem of convergence of perturbation series with small divisors is a long-standing one, and so it was in Poincaré's time. An example of converging series had been proposed by Heinrich Bruns [31]; he proved the convergence of the double series $\sum_{j=1}^{\infty}\sum_{k=1}^{\infty} \frac{q_1^j q_2^k}{j-kA}$, where q_1 and q_2 are proper fractions and $A > 0$ is an irrational algebraic number (see also [209], ch. XVI, § 198). Poincaré's example is more general.

and assume that the frequencies are strongly non-resonant. For definiteness let us assume that the frequencies satisfy the Diophantine condition

$$\bigl|\langle k,\omega\rangle\bigr| \ge \frac{\gamma}{|k|^{\tau}}\ ,\qquad \gamma>0\ ,\quad \tau>n-1\ .$$

Let us construct the normal form; it is actually enough to use the formalism of Lie series with a single generating function χ_1. Forgetting the constant average $\overline{H_1}$, write

$$H_1 = \sum_{0\neq k\in\mathbb{Z}^n} h_k e^{i\langle k,q\rangle}\ ,\qquad h_k\in\mathbb{C}\ ;$$

then the homological equation $-\partial_\omega \chi_1 = H_1$ is solved as

$$\chi_1 = i\sum_{k\neq 0} \frac{h_k}{\langle k,\omega\rangle} e^{i\langle k,q\rangle}\ .$$

Therefore, we get the linear normal form $Z(p)=\langle\omega,p\rangle$. The canonical transformation introduces the new variables q',p' as

$$\begin{aligned} q'_j &= \exp(L_{\varepsilon\chi_1})q_j = q_j\,,\\ p'_j &= \exp(L_{\varepsilon\chi_1})p_j = p_j + \varepsilon\sum_{k\neq 0}\frac{k_j h_k}{\langle k,\omega\rangle} e^{i\langle k,q\rangle}\ . \end{aligned}$$

That is, the transformation leaves the angles invariant and changes the action by introducing a q–dependent deformation. If $H_1(q)$ is analytic then, by the argument of Poincaré already recalled, the series for χ_1 defines a analytic function in view of the Diophantine condition on the frequencies. *E.D.*

The degenerate case of constant frequencies may appear as a very exceptional one, possibly interesting only in case of elliptic equilibria. However, it is not forbidden to expand a general Hamiltonian in a form which isolates the linear part of $H_0(p)$; a remarkable example is provided by the theorem of Kolmogorov on persistence of strongly non-resonant tori. This will be the subject of Chapter 8.

7.3.3 Expansion in Trigonometric Polynomials

A second attempt to overcome the difficulty of small divisors is based on Example 4.24, namely, in a more generic formulation, a Hamiltonian such as

$$H(p,q) = H_0(p) + \varepsilon H_1(p,q)\ ,\quad H_1(p,q) = \sum_{|k|\le K} h_k(p) e^{i\langle k,q\rangle}, \tag{7.36}$$

that is, $H_1(p,q)$ is a *trigonometric polynomial.* The example may appear as rather particular, but it is not so, as we are going to see.[3] Let us put the matter in a rather general form.

In the following a special role will be played by functions which have a finite Fourier representation.

Definition 7.10: *For a non-negative integer K, denote by $\mathscr{F}_K$ the distinguished class of functions $f(p,q)$ which are trigonometric polynomials of degree K in the angles q, with coefficients depending on the action p.*

That is, we may write any function $f(p,q) \in \mathscr{F}_K$ as a finite sum

$$f(p,q) = \sum_{|k| \le K} f_k(p) e^{i\langle k,q\rangle} ,$$

and in (7.36) we have $H_1 \in \mathscr{F}_K$. In particular, $f \in \mathscr{F}_0$ means that $f(p)$ is independent of the angles.

Lemma 7.11: *Let $f \in \mathscr{F}_K$ and $g \in \mathscr{F}_{K'}$; then:*

$$f + g \in \mathscr{F}_{\max(K,K')} , \quad fg \in \mathscr{F}_{K+K'} , \quad \{f,g\} \in \mathscr{F}_{K+K'} .$$

Lemma 7.12: *Let $K > 0$ be arbitrary but fixed. Referring to the triangular diagram for Lie series (6.5), let $\chi \in \mathscr{F}_{rK}$ for some $r > 0$, and let $f_s \in \mathscr{F}_{sK}$ for $s \ge 0$. Then we have $L_\chi^j f_s \in \mathscr{F}_{(jr+s)K}$.*

Lemma 7.13: *Let $K > 0$ be arbitrary but fixed. Referring to the triangular diagram for the Lie transform (6.21), let $\chi_s \in \mathscr{F}_{sK}$ and $f_s \in \mathscr{F}_{sK}$ for $s \ge 0$. Then we have $E_j f_s \in \mathscr{F}_{(j+s)K}$.*

Briefly, the algebraic scheme of Lie series and of Lie transforms is fully compatible with the request that at every order r all functions are of class $\mathscr{F}_{rK}$. This is a very useful property in order to set up an effective algorithm. The

---

[3] Needless to say, the difficulty connected with infinite Fourier expansions is usually a minor one from the viewpoint of Analysis. But in practical applications the problem is by no means irrelevant: the question concerning how many terms should be calculated in the expansion of the equations for, for example, the motion of the moon or of the planets was a major one for D'Alembert, Euler, Clairaut, Lagrange, Laplace and for all the great mathematicians of the nineteenth century who studied the Mechanics of the Solar System: they could rely only on the correspondence of their results with observations. Is some cases, as in the theory of Moon, it was actually found that increasing the number of terms could improve the correspondence. How to provide a satisfactory answer to such a question? This problem had been also pointed out by Poincaré (see [188], Ch. XIII, § 147), who suggested the way out; it is essentially the way exploited in the present notes. In addition the following discussion aims at explaining why the problem has a particular relevance also for quantitative perturbation methods, at least in the form discussed in the present notes.

properties stated in the three lemmas 7.11, 7.12 and 7.13 actually extend to the case of trigonometric polynomials the scheme that we have used for expansions in homogeneous polynomials. The reader may object that the trigonometric polynomials are not homogeneous; but writing the homogeneous polynomials in action-angle variables, it is immediately seen that the resulting trigonometric polynomials are not homogeneous, too.

Exercise 7.14: Prove the three Lemmas 7.11, 7.12 and 7.13. *A.E.L.*

The question now is how to exploit the arrangement in classes in the framework of the general problem of dynamics. Recall that the size of the coefficients of the Fourier expansion of a analytic function decreases exponentially with the degree, as shown in Section 6.7.1. The minor nuisance is that we should get rid of the method of expansion in a parameter used so far. Thus, having given a function

$$f(p,q) = \sum_{k\in\mathbb{Z}^n} f_k(p)e^{i\langle k,q\rangle}\,,$$

we split it as, for example,

$$f(p,q) = \underbrace{\sum_{0\le|k|\le K} f_k(p)e^{i\langle k,q\rangle}}_{f_1(p,q)} + \underbrace{\sum_{K<|k|\le 2K} f_k(p)e^{i\langle k,q\rangle}}_{f_2(p,q)} + \ldots\,, \tag{7.37}$$

so that $f_s \in \mathscr{F}_{sK}$. Such a way of splitting is not unique, of course: we may well invent other criteria provided the condition $f_s \in \mathscr{F}_{sK}$ is satisfied.

Coming back to the general problem of dynamics the straightforward way would be to forget the expansion in a parameter ε, thus writing the Hamiltonian as $H(p,q) = H_0(p) + H'(p,q)$, where $H'(p,q)$ contains the sum of the series in ε representing the perturbation; then we split H' as in (7.37). The drawback is that we should know the full ε expansion of the Hamiltonian, which is actually unknown or at least unpractical in most cases. But in this case we may begin with the expansion $H = H_0 + \varepsilon H_1 + \ldots + \varepsilon^r H_r$ truncated at a suitable order r and split every function H_s as, for example, $H_s = H_{s,1} + H_{s,2} + \ldots$ according to (7.37). Then we rearrange the Hamiltonian as

$$\begin{aligned} H = H_0 &+ \underbrace{\varepsilon H_{1,1}}_{H_1'} + \underbrace{\varepsilon H_{1,2} + \varepsilon^2 H_{2,1}}_{H_2'} + \cdots \\ &+ \underbrace{\varepsilon H_{1,r} + \varepsilon^2 H_{2,r-1} + \cdots + \varepsilon^r H_{r,1}}_{H_r'} + \mathcal{O}(r+1)\,. \end{aligned} \tag{7.38}$$

Such an expansion is consistent with the rules of formal calculus, provided we assign a proper meaning to the symbol $\mathcal{O}(r+1)$, that is, to the expression "of order $r+1$." The first point to be clarified is the role of the parameter K, and how to choose it. In view of the form of the expansion we may be tempted to set $K \sim -\log\varepsilon$, the perturbation parameter; but such a choice has two negative consequences. The first consequence is rather obvious: since K is an integer, it cannot be defined as an analytic function of ε; hence the analyticity in ε of the perturbation expansion should be proved independently, which is an unpleasant task. The second consequence becomes evident in the framework of exponential estimates of Nekhoroshev type, which will be discussed in Chapter 9; so it is hard to provide a satisfactory explanation at this point. It is remarkable indeed that the best choice is to set K to a constant, independent of ε, which does not need to be a large one; for example, setting $K \sim 1/\sigma$ is often enough. The reader may also object that the splitting (7.38) makes the expansion in ε unpractical: one should rather consider using more than one parameter. However, everything works fine if one forgets the parameter and pays attention only to the size of the various terms, in some norm: the formalism of Lie series and of Lie transforms works anyway. Chapters 8 and 9 will serve as complete examples.

7.3.4 Some Analytical Estimates

In order to make the argument more quantitative, and hopefully more clear, let us produce some analytical estimates for the expansion. Here we use the weighted Fourier norm, introduced in Section 6.7.

Lemma 7.15: *Assume that $f(p,q)$ is analytic in $\mathscr{G}_\varrho \times \mathbb{T}^n_{2\sigma}$, and that*

$$|f|_{\varrho,2\sigma} < \infty .$$

Expanding f as in (7.37), we have

$$\|f_s\|_{\varrho,\sigma} \le e^{-(s-1)K\sigma/2}\left(\frac{1+e^{-\sigma/2}}{1-e^{-\sigma/2}}\right)^n |f|_{\varrho,2\sigma}\ ,\qquad s\ge 1\ .$$

Proof. The proof is an immediate consequence of the estimate

$$|f_k|_\varrho \le |f|_{\varrho,2\sigma}\, e^{-2|k|\sigma}$$

for the Fourier coefficients $f_k(p)$ (see Section 6.7.1). For, by definition of the norm, we have

$$\|f_s\|_{\varrho,\sigma} \le \varepsilon\, |f|_{\varrho,2\sigma} \sum_{(s-1)K<|k|} e^{-|k|\sigma}\ .$$

For any positive integer N, we estimate

$$\sum_{|k|\ge N} e^{-|k|\sigma} \le e^{-N\sigma/2}\left(\sum_{j\in\mathbb{Z}} e^{-|j|\sigma/2}\right)^n = e^{-N\sigma/2}\left(\frac{1+e^{-\sigma/2}}{1-e^{-\sigma/2}}\right)^n ,$$

so that the claim follows. *Q.E.D.*

Corollary 7.16: *Assume that in the ε expansion of the Hamiltonian*

$$H(p,q) = H_0(p) + \varepsilon H_1(p,q) + \varepsilon^2 H_2(p,q) + \dots\ ,$$

we have

$$\left\|H_s\right\|_{\varrho,2\sigma} \le \beta^{s-1}E'\ , \quad s>0\ .$$

Choose K such that $2\varepsilon\beta < e^{-K\sigma/2}$. Then the functions in (7.38) satisfy

$$\left\|H'_r\right\|_{\varrho,\sigma} \le h^{r-1}\varepsilon E\ , \quad s>0 \tag{7.39}$$

with

$$h = e^{-K\sigma/2}\ , \quad E = 2E'\left(\frac{1+e^{-\sigma/2}}{1-e^{-\sigma/2}}\right)^n . \tag{7.40}$$

Proof. From (7.38) write

$$H'_r = \varepsilon \sum_{j=0}^{r-1} \varepsilon^j H_{j+1,r-j}\ .$$

By Lemma 7.15 we have the estimate

$$\left\|H'_r\right\|_{\varrho,\sigma} \le \left(\frac{1+e^{-\sigma/2}}{1-e^{-\sigma/2}}\right)^n \varepsilon E' \sum_{j=0}^{r-1} (\varepsilon\beta)^j e^{-(r-j)K\sigma/2}\ .$$

By the choice of K the sum is estimated as

$$\sum_{j=0}^{r-1} (\varepsilon\beta)^j e^{-(r-j)K\sigma/2} = \frac{e^{-rK\sigma/2} - (\varepsilon\beta)^r}{e^{-K\sigma/2} - \varepsilon\beta} \le 2e^{-K\sigma/2}\ ,$$

so that the claim follows. *Q.E.D.*

7.3.5 Truncated Normal Form

The expansion of the Hamiltonian in trigonometric polynomials introduced in Section 7.3.3 is useful if one plans to construct the normal form up to a finite order r. In this case the solution (7.33) of the homological equation is well defined in a particular open domain $\mathcal{V} \subset \mathscr{G}$ with the property that for $p \in \mathcal{V}$ there are no resonances $\langle k, \omega(p)\rangle = 0$ for $0 < |k| \le rK$. Such a domain may be constructed, as one can check by considering the model of perturbed rotators of Section 4.6.2. Similar considerations apply if the open domain $\mathcal{V}$ contains only some resonances.

The construction of a truncated normal form is a basic tool in the proof of the theorem of Nekhoroshev on exponential stability. We do not enter a detailed discussion here: the argument will be treated in full detail in Chapter 9.

7.3.6 Using Composition of Lie Series

The procedure for constructing a normal form may be performed also using the method of composition of Lie series. This is rather obvious, because we have seen in Chapter 6 that the two methods of composition of Lie series and of Lie transforms are equivalent, at least from the viewpoint of the formal calculus.[4] The algorithm should be properly reformulated, of course. We should not forget that for a given transformation the generating sequence of the Lie transform and the generating sequence of the composition of Lie series are not the same: they coincide only at the first order.

Here we write the algorithm for the composition of Lie series, with the aim of making it explicit, so that the reader may get convinced that it can be effectively applied in all cases which can be dealt with using the Lie transform algorithm.

The aim is to find a generating sequence χ and a normal form Z satisfying

$$\ldots \exp\bigl(L_{\chi_r}\bigr) \circ \cdots \circ \exp\bigl(L_{\chi_1}\bigr) H = Z. \tag{7.41}$$

Here we must proceed by iteration, hence the algorithm is less compact than in the case of Lie transforms. Write the Hamiltonian after r normalization steps as

$$H^{(r)} = H_0 + Z_1 + \cdots + Z_r + \sum_{s>r} H_s^{(r)} \ , \tag{7.42}$$

where $Z_1(\xi,\eta),\ldots,Z_r(\xi,\eta)$ are in normal form, and $H_s^{(r)}$ is a remainder not yet normalized. To be definite, in the case of the elliptic equilibrium Z_s and $H_s^{(r)}$ are homogeneous polynomials of degree $s{+}2$; in the case of action-angle variables they are trigonometric polynomials of degree sK. For $r = 0$ the Hamiltonian is already in the desired form, with no functions Z.

Assume that the Hamiltonian has been given a normal form (7.42) up to order $r-1$, so that $H^{(r-1)}$ is known. Here is the algorithm.

The generating function χ_r and the normal form Z_r are determined by solving the homological equation[5]

$$L_{H_0}\chi_r + Z_r = H_r^{(r-1)} \ . \tag{7.43}$$

[4] The difference between the methods of composition of Lie series and of Lie transforms does not show up with evidence even when we translate the algorithm into a scheme of quantitative estimates for the norms. The actual difference becomes relevant when we study in great detail the behaviour of small divisors; in this case the composition of Lie series seems to be in a better position. More on this point in Chapter 8.

[5] The sign of $L_{H_0}\chi_1$ is different than in Eq. (7.5) because here, according to (7.42), we apply the composition of Lie series to H, while in (7.5) we wrote the equation as $T_\chi Z = H$.

The transformed Hamiltonian is expanded as follows:

(7.44)
$$H^{(r)}_{sr+m} = \frac{1}{s!}L^s_{\chi_r} Z_m + \sum_{p=0}^{s-1} \frac{1}{p!} L^p_{\chi_r} H^{(r-1)}_{(s-p)r+m}$$
$$\text{for } r \geq 2\,,\ s \geq 1 \text{ and } 1 \leq m < r\,,$$
$$H^{(r)}_{sr} = \frac{1}{(s-1)!} L^{s-1}_{\chi_r} \left(\frac{1}{s} Z_r + \frac{s-1}{s} H^{(r-1)}_r \right) + \sum_{p=0}^{s-2} \frac{1}{p!} L^p_{\chi_r} H^{(r-1)}_{(s-p)r}$$
$$\text{for } r \geq 1 \text{ and } s \geq 2\,.$$

Justifying this algorithm is just a matter of rearranging terms in the expansion of $\exp(L_{\chi_r})H^{(r-1)}$. Considering first H_0 and $H^{(r-1)}_r$ together, we have

$$\begin{aligned} \exp(L_{\chi_r})\big(H_0 + H^{(r-1)}_r\big) &= H_0 + L_{\chi_r} H_0 + \sum_{s\geq 2} \frac{1}{s!} L^s_{\chi_r} H_0 \\ &\quad + H^{(r-1)}_r + \sum_{s\geq 1} \frac{1}{s!} L^s_{\chi_r} H^{(r-1)}_r\,. \end{aligned}$$

Here, H_0 is the first term in the transformed Hamiltonian $H^{(r)}$ in (7.42). In view of Eq. (7.43) we have $L_{\chi_r} H_0 + H^{(r-1)}_r = Z_r$, which kills the unwanted term $H^{(r-1)}_r$ and replaces it with the normalized term Z_r. The two sums may be collected and simplified by calculating

$$\begin{aligned} &\sum_{s\geq 2} \frac{1}{s!} L^s_{\chi_r} H_0 + \sum_{s\geq 1} \frac{1}{s!} L^s_{\chi_r} H^{(r-1)}_r \\ &\quad = \sum_{s\geq 2} \frac{1}{(s-1)!} L^{s-1}_{\chi_r} \left[\frac{1}{s} \left(L_{\chi_r} H_0 + H^{(r-1)}_r \right) + \frac{s-1}{s} H^{(r-1)}_r \right] \\ &\quad = \sum_{s\geq 2} \frac{1}{(s-1)!} L^{s-1}_{\chi_r} \left(\frac{1}{s} Z_r + \frac{s-1}{s} H^{(r-1)}_r \right)\,. \end{aligned}$$

Here, both $L^{s-1}_{\chi_r} Z_r$ and $L^{s-1}_{\chi_r} H^{(r-1)}_r$ are of order sr and are added to $H^{(r)}_{sr}$ in the second equation of (7.44).

Proceed now by transforming the functions $Z_1, \ldots, Z_{r-1}$ which are already in normal form. Recall that no such term exists for $r = 1$. For $r > 1$ calculate

$$\exp(L_{\chi_r}) Z_m = Z_m + \sum_{s\geq 1} \frac{1}{s!} L^s_{\chi_r} Z_m\,, \quad \text{for } 1 \leq m < r\,.$$

The term Z_m is copied into $H^{(r)}$ in (7.42). The term $L^s_{\chi_r} Z_m$ is of order $sr + m$ and is added to $H^{(r)}_{sr+m}$ in the first equation of (7.44).

Finally, consider all terms $H_s^{(r-1)}$ with $s > r$, which are written as $H_{lr+m}^{(r-1)}$ with $l \geq 1$ and $0 \leq m < r$, the case $l = 1$, $m = 0$ being excluded. We get

$$\exp(L_{\chi_r})H_{lr+m}^{(r-1)} = \sum_{p\geq 0} \frac{1}{p!} L_{\chi_r}^p H_{lr+m}^{(r-1)} ,$$

where $L_{\chi_r}^p H_{lr+m}^{(r-1)}$ is of order $(p+l)r+m$. Collecting all terms of the same order with $m = 0$, $l \geq 2$ and $p + l = s \geq 2$ one gets $\sum_{p=0}^{s-2} \frac{1}{p!} L_{\chi_r}^p H_{(s-p)r}^{(r-1)}$, which is added to $H_{sr}^{(r)}$ in the second equation of (7.44). Similarly, collecting all terms of the same order with $0 < m < r$, $l \geq 1$ and $p + l = s \geq 1$, one gets $\sum_{p=0}^{s-1} \frac{1}{p!} L_{\chi_r}^p H_{(s-p)r+m}^{(r-1)}$, which is added to $H_{sr+m}^{(r)}$ in the first equation of (7.44). The latter case does not occur for $r = 1$. This completes the justification of the formal algorithm.

Thus, the problem is how to solve the homological equation (7.43) for the generating function and the normal form. There is nothing new here with respect to the method of Lie transforms: we know how to do it.

7.4 The Dark Side of Small Divisors

As said at the beginning of the present chapter, we defer the quantitative study of the analytical properties of the normal form to the following chapters, in connection with the theorems of Kolmogorov and of Nekhoroshev. However, it may be interesting to devote some space to a heuristic analysis aimed at pointing out the critical points which may cause divergence of the perturbation series. In particular, we pay attention to the action of small divisors.

7.4.1 The Accumulation of Small Divisors

We follow the algorithm of the Lie transform, which is more immediate. For definiteness we refer to the case of a Hamiltonian around an elliptic equilibrium, with non-resonant frequencies. We consider the simplified Hamiltonian, in complex variables, written as

$$H = H_0 + H_1 \ , \quad H_0 = i \sum_{l=1}^{n} \omega_l \xi_l \eta_l \ , \tag{7.45}$$

with H_1 a homogeneous polynomial of degree 3 in ξ, η. The model is simplified so as to include all the crucial elements. Moreover, the argument applies with very minor modifications to the case of a Hamiltonian in action-angle variables with $H_0 = \langle \omega, p \rangle$ linear in the angles and a perturbation H_1 which is a trigonometric polynomial.

The point to be discussed is concerned with the introduction of small divisors in the process of construction of the normal form. To this end let us introduce the numerical sequence $\{\alpha_s\}_{s\geq 1}$ as

$$\alpha_s = \min_{0\neq|k|\leq s+2} \bigl|\langle k,\omega\rangle\bigr| \ . \tag{7.46}$$

That is, α_s is the smallest divisor which may appear in the solution of the homological equation at order s; the condition $0 \leq j \leq s+2$ is necessary because at order s we work with a polynomial of degree $s+2$. With a moment's thought the reader will realize that the sequence α_s is monotonically non-increasing and decreases to zero for $s \to \infty$; it cannot vanish in view of the non-resonance condition.

Let us rewrite for convenience the relevant formulæ of the algorithm, including the information that $H_2 = H_3 = \cdots = 0$. The Lie transform is written as

$$T_\chi = \sum_{s\geq 0} E_s \ , \quad E_0 = 1 \ , \quad E_s = \sum_{j=1}^{s} \frac{j}{s} L_{\chi_j} E_{s-j} \ . \tag{7.47}$$

The homological equation at every order is

$$Z_s - \partial_\omega \chi_s = \Psi_s \ , \quad s = 1,\ldots,r \ , \tag{7.48}$$

where

$$\begin{aligned} \Psi_1 &= H_1 \ , \\ \Psi_s &= -\frac{s-1}{s} L_{\chi_{s-1}} H_1 - \sum_{j=1}^{s-1} \frac{j}{s} E_{s-j} Z_j \quad \text{for } 2 \leq s \leq r \ . \end{aligned} \tag{7.49}$$

Here we may recall the scheme of quantitative estimates described in Section 5.4.2, focusing on the divisors. Without performing a detailed calculation we can make the following observations:

(i) the solution of the homological equation makes a copy of part of Ψ_s into Z_s, and adds a divisor $\langle (j-k),\omega\rangle$ to the coefficients of Ψ_s which are put in χ_s;

(ii) the Poisson bracket between any two functions, say f and g, multiplies together the divisors which appear in f and g.

The simplest thing to do is to follow the process of accumulation of small divisors in the functions χ, Z and Ψ. It may help to follow the triangular diagram (7.7). With a little patience, and using the information in (i)–(ii),, the reader will be able to construct the following scheme, where the worst accumulation of divisors in every function is accounted for using the sequence α_s.

$$
\begin{array}{llll}
\Psi_1 \sim 1 & \Rightarrow & Z_1 \sim 1\,, & \chi_1 \sim \dfrac{1}{\alpha_1} \\
\Psi_2 \sim \dfrac{1}{\alpha_1} & \Rightarrow & Z_2 \sim \dfrac{1}{\alpha_1}\,, & \chi_2 \sim \dfrac{1}{\alpha_1\alpha_2} \\
\Psi_3 \sim \dfrac{1}{\alpha_1\alpha_2} & \Rightarrow & Z_3 \sim \dfrac{1}{\alpha_1\alpha_2}\,, & \chi_3 \sim \dfrac{1}{\alpha_1\alpha_2\alpha_3} \\
\vdots & & \vdots & \vdots \\
\Psi_r \sim \dfrac{1}{\alpha_1\cdots\alpha_{r-1}} & \Rightarrow & Z_r \sim \dfrac{1}{\alpha_1\cdots\alpha_{r-1}}\,, & \chi_r \sim \dfrac{1}{\alpha_1\alpha_2\cdots\alpha_r}\,.
\end{array}
$$

The scheme resumes, in short, the same result that we have found in more rigorous terms for the direct construction of first integrals, in Section 5.4, Lemma 5.19. A rigorous scheme leads to the same conclusion, indeed: see, for example, [80]. However, a closer examination may suggest that the argument presented here is too naive and simplistic.

In order to have deeper insight, let us consider a second sequence $\{\beta_s\}_{s\geq 0}$ defined as

$$
\beta_0 = 1\,, \quad \beta_s = \min_{|k|=s} \bigl|\langle k, \omega\rangle\bigr|\,. \tag{7.50}
$$

That is, β_s is the worst divisor associated to a Fourier mode $e^{i\langle k,q\rangle}$ with $|k| = s$. Such a divisor is expected to show up at order at least s (order $s-2$ for homogeneous polynomials) in the construction of the normal form, and it is reasonable to expect that it will actually appear, unless there are cancellations at the edge of a miracle. The sequence β_s exhibits a wild behaviour: an example is shown in Fig. 7.4, where the frequencies are $\omega_1 = 1$, $\omega_2 = \sqrt{2}/2$ and $\omega_3 = \sqrt{3}/2$. The jumps in the sequence α_s, also reported in the figure, occur when a new divisor, smaller than all the previous ones, appears in the sequence β_s.

With a little more attention we may look for particular paths of accumulation of divisors which appear to be dominant. Here we show how divergence of series may occur, generically; the argument is illustrated assuming that the frequencies are non-resonant, but it applies also to the resonant case. Using formulæ (7.49) and (7.47), extract from Ψ_s the contributions

$$
\Psi_s = E_s Z_1 + \cdots = L_{\chi_{s-1}} Z_1 + \cdots = \{Z_1, \chi_{s-1}\} + \cdots\,.
$$

The simplest way is to use action-angle variables. Then $Z_1(p)$ is independent of the angles, while we have $\chi_{s-1} = \sum_k c_k(p) e^{i\langle k,q\rangle}$; therefore, the Poisson bracket $\{Z_1, \chi_s\}$ acts diagonally on the Fourier expansion of χ_{s-1}; that is, we get $\{Z_1, c_k(p) e^{i\langle k,q\rangle}\} = c'_k(p) e^{i\langle k,q\rangle}$ where only the coefficient $c'_k(p)$ changes. Now, the solution of the homological equations divides $c'_k(p)$ by *the same*

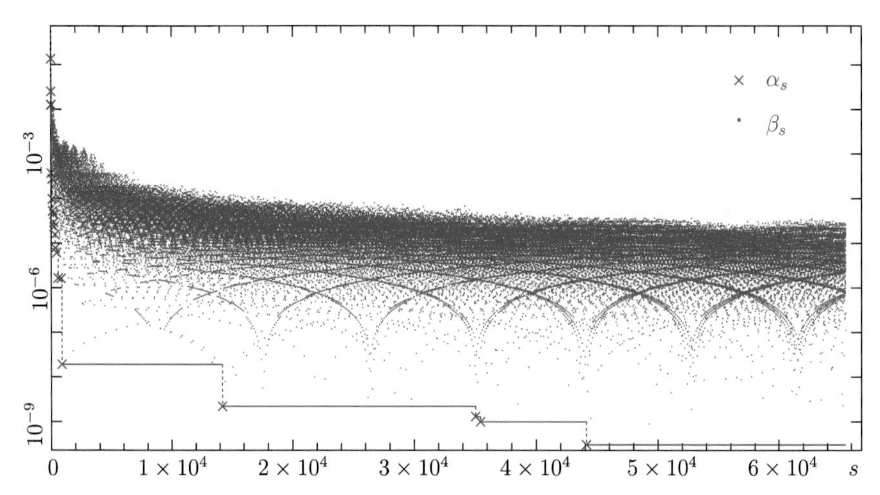

Figure 7.4 The sequences $\{\alpha_s\}_{s\geq 0}$ and $\{\beta_s\}_{s\geq 0}$, in semilog scale, up to s = 65,536. Here, $n = 3$ and the frequencies are $\omega_1 = 1$, $\omega_2 = \sqrt{2}/2$ and $\omega_3 = \sqrt{3}/2$. The horizontal lines highlight the step behaviour of the sequence α_s.

divisor $\langle k, \omega\rangle$ which was already in $c_k(p)$. Therefore, in the recurrent process we can extract the sequence of terms

$$\chi_1 \sim \frac{1}{\alpha_1} \to \{Z_1, \chi_1\} \to \chi_2 \sim \frac{1}{\alpha_1^2} \to \{Z_1, \chi_2\} \to \chi_3 \sim \frac{1}{\alpha_1^3} \to \ldots$$

with a coefficient which gains a divisor α_1 at every step. Since a series $\sum_s \varrho^s/\alpha_1^s$ has convergence radius α_1, we can hardly expect that the generating function will be convergent in a disk with radius bigger than α_1.

The obvious objection here is that in the case of an elliptic equilibrium we have $Z_1 = 0$. But the whole argument applies by replacing Z_1 with Z_2, or with the first Z_r which is non-zero: there is little to be changed.

The bad news is that in the process of construction of the normal form, at order $s+2$ (or maybe a bit later) a coefficient with the divisor $\beta_s < \alpha_{s-1}$ is expected to appear; we have called it α_s. By the same process, the same divisor will repeat at every next order, so that the convergence radius cannot be expected to be bigger than α_s. Since $\alpha_s \to 0$ for $s \to \infty$, the convergence radius of the series must vanish.

This argument is not a rigorous proof, of course; however, a numerical confirmation may be found in [200] for the simpler case of a map, and in [51][52] for a Hamiltonian case. There are exceptions, of course.

Example 7.17: *Essentially isochronous systems.* Suppose that the normal form $Z(p) = H_0(p) + Z_2(p) + Z_4(p) + \ldots$, with $H_0 = \langle \omega, p\rangle$, turns out to be a function of H_0 only; this is a very exceptional case, but it may occur. Then, writing $Z = Z(H_0)$ the canonical equations (in action-angle variables) take the form

$$\dot{p}_j = 0 \ , \quad \dot{q}_j = \frac{dZ}{dH_0}\omega_j \ , \quad j = 1, \ldots, n \ .$$

The system is essentially isochronous in the sense that the frequencies do depend on the actions only by a factor depending on the initial value of the actions themselves, so that they are merely rescaled without introducing any effect due to resonance. In this case the normal form turns out to be convergent, a fact which emerges with evidence if one replaces the Lie transform algorithm with a composition of Lie series (it remains concealed if one uses the Lie transform).[6] The reason is definitely not trivial: the Lie derivative $L_{\chi_j} Z_s$ cancels the last divisor generated by the solution of the homological equation, thus affecting the accumulation of small divisors and making it increase only geometriccaly: there are no factorials. A detailed proof may be found in [157]. *E.D.*

The resonant case may still raise some doubts; so let us consider it, shortly. Let $r = \dim \mathscr{M}_\omega$. Then we should redefine the sequence $\{\alpha_s\}_{s>0}$ as

$$\alpha_s = \min_{\substack{0<|k|\leq s+2 \\ k\in\mathbb{Z}^n\setminus\mathscr{M}_\omega}} \bigl|\langle k, \omega\rangle\bigr| \ .$$

If $r < n-1$, then the sequence is still monotonically decreasing to zero; therefore nothing changes in our considerations. Doubts are raised in case $r = n-1$, namely of complete resonance, for in this case the sequence α_s does not decrease to zero. It is true that $\langle k, \omega\rangle = 0$ for many values of k, but the corresponding terms go into the normal form Z and do not generate a zero divisor in χ.

The point is that, having focused our attention on small divisors, we have forgotten that the Poisson bracket $\{Z_1, \chi_s\}$ contains derivatives. Working with homogeneous polynomials, a derivative of χ_s produces a factor s which accumulates into a factorial $s!$ as in the estimates for the direct construction of a first integral. If we are using action-angle variables, the same factor comes from Cauchy estimates on derivatives with respect to p, due to the game of the restrictions on domains. This represents a second obstacle against the convergence of the normal form, which is hardly removed.

---

[6] A reader who finds the latter claim a very mysterious one is definitely right. The reason will perhaps become understandable after having read the proof of Kolmogorov's theorem in classical style, in Section 8.3.

8
Persistence of Invariant Tori

This chapter is devoted to the celebrated theorem of Andrey Nikolaevich Kolmogorov on persistence of invariant tori in near-to-integrable Hamiltonian systems. The relevance of the theorem for Classical and Celestial Mechanics was emphasized by V. I. Arnold, who wrote in [3]:

> *One of the most remarkable of A.N. Kolmogorov's mathematical achievements is his work on classical mechanics of 1954. A simple and novel idea, the combination of very classical and essentially modern methods, the solution of a 200 year-old problem, a clear geometric picture and a breadth of outlook – these are the merits of the work. Its deficiency has been that complete proofs have never been published.*

Kolmogorov announced his theorem at the International Congress of Mathematicians held in Amsterdam in 1954. Then he published a sketch of the proof in the short note [128]. The note contains in synthetic form the formal scheme of an iterative procedure and a few essential hints on the proof of convergence. Kolmogorov also professed a series of lectures in Moscow, where he gave the complete proof, as witnessed by some Russian mathematicians (see [38], ch. 11), but it seems that the text of the lectures has never been published (at least, not in Western countries). A proof of a related theorem for maps was later published by Jürgen Kurt Moser [174]. Two papers with proofs for Hamiltonian systems which generalize and extend the ideas of Kolmogorov have been published by Vladimir Igorevich Arnold [3][4]. The work of Kolmogorov, Arnold and Moser marked the beginning of a wide research field nowadays known as *KAM theory*. A lot of papers have been published on the subject: an exhaustive list would fill several pages. Different methods of proof have been devised by several authors, including also extensions of Moser's work on area-preserving maps and to reversible dynamical systems [176][201]; in particular, much work has been devoted to weakening the analytical assumptions, often at the cost of an increase of technicalities. On the other hand, proofs based on the original paper of Kolmogorov are rare; examples may be found, for example, in [13], [14], [16], [117]. For a thorough account on the development of KAM theory the reader may see the encyclopedic book [11]. For a tutorial text on KAM theory the reader may see [156].

8.1 The Work of Kolmogorov

In view of the great relevance of the work of Kolmogorov and of its impact on the progress of our knowledge about non-integrable systems, this section is devoted to a short exposition of the main ideas. The exposition closely follows the scheme that can be found in the short note of Kolmogorov [128]. It goes without saying that the exposition in this section reflects the personal experience of the author.

8.1.1 Statement of the Theorem

The working framework is still the general problem of dynamics as stated by Poincaré (see Chapter 4). With a negligible change of notation, let us write the Hamiltonian as

$$H(p,q) = h(p) + \varepsilon f(p,q,\varepsilon) \ , \tag{8.1}$$

where $p \in \mathscr{G} \subset \mathbb{R}^n$ are action variables in an open set $\mathscr{G}$, and $q \in \mathbb{T}^n$ are angle variables. The Hamiltonian is assumed to be analytic in all variables.

Kolmogorov's theorem claims that invariant tori *characterized by strongly non-resonant frequencies* (e.g., Diophantine frequencies in the sense of Section A.2.1) continue to exist for ε small enough and still carry quasiperiodic orbits.

Theorem 8.1: *Consider the Hamiltonian (8.1); assume:*

(i) $h(p)$ is non-degenerate, that is,

$$\det\left(\frac{\partial^2 h}{\partial p_j \partial p_k}\right) \neq 0 \ ;$$

(ii) $h(p)$ possesses an invariant torus p^* with frequencies ω satisfying the Diophantine condition

$$\bigl|\langle k,\omega\rangle\bigr| \geq \frac{\gamma}{|k|^\tau} \ , \quad k \neq 0 \ ,$$

for some $\gamma > 0$ and $\tau > n-1$.

Then there exists a positive ε^* such that the following holds true: for $\varepsilon < \varepsilon^*$ there exists a perturbed invariant torus carrying quasiperiodic motions with *frequencies ω, which is a small deformation and translation of the unperturbed one.*

The Diophantine condition of strong non-resonance is the one exploited by Kolmogorov. It may be replaced by a weaker condition, as we shall see later.

8.1.2 The Normal Form of Kolmogorov

The first idea is that there is a very simple Hamiltonian which brings immediately to evidence the existence of an invariant torus. Precisely, let the Hamiltonian be

$$H(q,p) = \langle \omega, p \rangle + R(q,p) \ , \quad R(q,p) = \mathcal{O}(p^2) \ , \tag{8.2}$$

where $\omega \in \mathbb{R}^n$ and $R(q,p)$ is at least quadratic in the actions p. Hamilton's equations read

$$\dot{q}_j = \omega_j + \frac{\partial R}{\partial p_j} \ , \quad \dot{p}_j = -\frac{\partial R}{\partial q_j} \ , \quad j = 1, \ldots, n \ .$$

Let now the initial point be $p = 0$ with arbitrary q. Since $\frac{\partial R}{\partial p_j} = \mathcal{O}(p)$ and $\frac{\partial R}{\partial q_j} = \mathcal{O}(p^2)$ we can immediately conclude that the torus $p = 0$ is invariant for the flow and carries a Kronecker flow with frequencies ω. Thus, the suggestion of Kolmogorov may be expressed as:

> *to reduce a Hamiltonian such as (8.1) to the normal form (8.2) in a neighbourhood of a strongly non-resonant unperturbed torus.*

Let us agree that the Hamiltonian (8.2) is in the *normal form of Kolmogorov*. The major problem, actually the long-standing one, is to prove the convergence of the normalization process.

8.1.3 Sketch of the Formal Scheme

The construction of the normal form is a clever revisitation of the method of the Poincaré–Birkhoff normal form which we have seen in Section 7.3.2. Here are a few points which will be expanded in detail later.

Let us consider the Hamiltonian

$$H(p,q) = h(p) + f(p,q)$$

with $h(p)$ non-degenerate and $f(p,q)$ small. The reader may want to add a parameter ε in front of $f(p,q)$, as in (8.1), if this helps in tracking the smallness of some terms. However, it will soon be realized at some point that in Kolmogorov's scheme one must get rid of a perturbation parameter.

Let us select a point p^* such that the corresponding frequencies $\omega(p^*)$ are non-resonant. With a translation $p' = p - p^*$ the Hamiltonian takes the form, omitting primes and writing $\omega \in \mathbb{R}^n$ in place of $\omega(p^*)$,

$$H(p,q) = \eta + \langle \omega, p \rangle + \frac{1}{2}\bigl\langle \mathsf{C}(q)p, p \bigr\rangle + A(q) + \bigl\langle B(q), p \bigr\rangle + \mathcal{O}(p^3), \tag{8.3}$$

where terms of degrees 0, 1 and 2 in p have been separated introducing the function $A(q)$, the vector function $B(q)$ and the symmetric matrix $\mathsf{C}(q)$. These terms are calculated as

$$\begin{aligned} A(q) &= f(p^*, q)\ , \quad B_j(q) = \frac{\partial f}{\partial p_j}(p^*, q)\ , \\ \mathsf{C}_{jk}(q) &= \frac{\partial^2 h}{\partial p_j \partial p_k}(p^*) + \frac{\partial^2 f}{\partial p_j \partial p_k}(p^*, q)\ . \end{aligned} \tag{8.4}$$

Here $A(q)$ and $B(q)$ are of the same order as f, while $\mathsf{C}(q)$ includes a small correction to the quadratic part of $h(p)$. The constant η may be ignored.

The suggestion of Kolmogorov is to kill the unwanted terms $A(q)$ and $\bigl\langle B(q), p \bigr\rangle$ through a sequence of canonical transformations with generating functions of the form[1]

$$S(P,q) = \langle \xi, q \rangle + X(q) + \langle Y(q), P \rangle\ , \quad (\xi \in \mathbb{R}^n)\ .$$

Without entering now into full detail, what we can imagine is that we may construct an algorithm quite similar to that of Section 7.3.2; we could indeed, but the reader, for the time being, should exercise some patience and wait a little (or, better, read the note of Kolmogorov [128]). For the moment, suffice it to say that the function $X(q)$ and the vector function $Y(q)$ are determined by solving an homological equation; hence they contain small divisors $\langle k, \omega \rangle$ with fixed frequencies. Therefore, as we have seen, we are able to determine the generating function so that it is analytic. The major problem, once more, is the accumulation of small divisors through the normalization process which may prevent the convergence of the iterations. Thus, we fall back on the problem discussed in Section 7.4.1.

8.1.4 The Method of Fast Convergence

Here comes the second new idea introduced by Kolmogorov: *to dominate the action of the small divisors through a rapidly convergent scheme*, named by Kolmogorov himself *the generalized Newton's method*, with a reference to [126]. In very rough heuristic terms we can describe the method as follows.

---

[1] Here, as in the short note of Kolmogorov, we use the formalism of generating functions in mixed variables. Later the whole algorithm will be restated using Lie series.

Do not use expansions in a parameter, ε say; rather, use ε only as a factor indicating that the perturbation $\varepsilon f(p, q)$ is small (e.g., in some norm). Collect all contributions independent of and linear in p in a single pair of functions $A(q)$ and $\langle B(q), p\rangle$, which are small, say of size ε; terms of higher degree in p are collected in the remainder $\mathcal{O}(p^3)$. The transformation produces a new Hamiltonian of the same form as (8.3), but with new functions $A(q)$ and $\langle B(q), p\rangle$ reduced to a smaller size, roughly ε^2. Again, do not expand in ε, but iterate the procedure after having collected all contributions in a few functions. Starting with functions of size ε and forgetting for a moment the contribution of small divisors, the procedure (in very rough terms) reduces step by step the size of the unwanted terms to ε^2, ε^4, ε^8, ...; that is, they decrease quadratically, as in Newton's method. Such a strong decrease compensates the growth of the coefficients of the expansion due to accumulation of small divisors, eventually assuring the convergence of the procedure. The latter heuristic argument was commonly used in the past, and often it has been synthetized in the terms "quadratic method," "quadratic convergence," "Newton method," "superconvergence" and so on. With a little more attention we realize that this is a naive way of presenting the issue. The accumulation of divisors causes some coefficients to grow very large: this is the mechanism explained in Section 7.4.1 for the Poincaré–Birkhoff normal form. Hence most of the power of the quadratic convergence scheme must be spent in order to contrast this growth. At the end of the proof, in Section 8.2.6, we shall see that we may even manage it so that the actual decrease of the size of unwanted terms may be as slow as an inverse power of the number of iterations; anyway, enough to assure convergence of the whole procedure. As a matter of fact, the details of the proof make use of a clever scheme which was not commonly known in Kolmogorov's time; such details are not reported in Kolmogorov's short note, and were published only ten years later by Arnold [3]; these facts may justify the widespread belief that Kolmogorov did not provide the complete proof.

It is an indisputable fact that the fast convergence provided by Kolmogorov's method has represented a breakthrough in solving a long-established problem. However, it may be natural to ask whether a proof based on some variant of the classical methods of expansion of Chapter 7 may be constructed.[2] It can, indeed: to the best of the author's knowledge

[2] Quoting Rüssman [195]: "It has often been said that the rapid convergence of the Newton iteration is necessary for compensating the influence of small divisors. But a deeper analysis shows that this is not true. (...) Historically, the Newton method was surely necessary to establish the main theorems of the KAM-theory. But for clarifying the structure of the small divisor problems the Newton method is not useful because it compensates not only the influence of small divisors, but also many bad estimates veiling the structure of the problems."

the first proof of existence of invariant tori which does not make use of the fast convergence assured by the quadratic method was published by Helmut Rüssman [196]. Later proofs have been published by Ugo Locatelli and the author [82][83][88].

It would be an inexcusable sin to neglect the original, crystal-clear scheme of Kolmogorov's proof. On the other hand, a classical approach based on composition of Lie series has the double advantage of clarifying the mechanism of accumulation of small divisors and of being constructive since it implements the method of splitting the Hamiltonian into trigonometric polynomials. Therefore, this chapter provides two different proofs; the reader is free to make a choice between them.

8.2 The Proof According to the Scheme of Kolmogorov

Suppose that we have performed the process of preparation of the Hamiltonian as in Section 8.1.3. Let us consider the Hamiltonian (8.3) with the function $A(q)$, the vector function $B(q)$ and the matrix $C(p,q)$ defined as by (8.4). However, we neglect the part $\mathscr{O}(p^3)$, so that the Hamiltonian is quadratic in the actions. This simplifies several technical estimates; however, all the critical points are taken into account: the general case requires longer calculations, but no really new idea. The Hamiltonian is thus written as

$$H(q,p) = \langle \omega, p \rangle + A(q) + \langle B(q), p \rangle + \frac{1}{2}\langle \mathsf{C}(q)p, p \rangle \ . \tag{8.5}$$

8.2.1 *Analytic Setting*

The discussion now is going to be technical; we start with some definitions.

(a) For real vectors $x \in \mathbb{R}^n$ we shall use the norm

$$|x| = \sum_{j=1}^{n} |x_j| \ .$$

(b) For a real analytic function $f(q)$, where $q \in \mathbb{T}^n$ are the angle variables, we shall use the weighted Fourier norm

$$\|f\|_\sigma = \sum_{k\in\mathbb{Z}^n} |f_k| e^{|k|\sigma} \ ,$$

where f_k are the (complex) Fourier coefficients, and σ is a positive constant. Since $f(q)$ is assumed to be analytic, there is σ such that this series defining the norm converges (see Section 6.7.1). The symbol $\|\cdot\|_{\varrho,\sigma}$ will denote the weighted Fourier norm of a function in a domain of type $\mathscr{D}_{\varrho,\sigma} = \Delta_\varrho(0) \times \mathbb{T}^n_\sigma$, as introduced in Section 6.7.

(c) For a vector-valued analytic function $w(q) = (w_1(q), \ldots, w_n(q))$, we shall use the norm

$$\|w\|_\sigma = \sum_{j=1}^{n} \|w_j\|_\sigma \ .$$

The following lemma collects some useful properties of the norms just introduced. (Labeling of the items continues that of the preceding definitions.)

Lemma 8.2: *Consider the domain $\Delta_\varrho(0) \times \mathbb{T}^n_\sigma$ where $\Delta_\varrho(p_0)$ is a polydisk with arbitrary radius $\varrho > 0$ and centre p_0. Let $w(q)$ and $v(q)$ be analytic vector functions and $\mathsf{C}(q)$ be a $n \times n$ matrix with elements $C_{jk}(q)$ which are analytic functions. The following properties hold true.*

(d) For the vector function $w(q)$ one has

$$\|\langle w, p\rangle\|_{\varrho,\sigma} \le \varrho \|w\|_\sigma \ .$$

(e) If $\|\langle w, p\rangle\|_{\varrho,\sigma} \le D\varrho$, with some positive D, then $\|w\|_\sigma \le D$.

(f) If $\|\langle \mathsf{C}(q)p, p\rangle\|_{d,\sigma} \le D\varrho^2$, then $\|\mathsf{C}_{j,k}\|_\sigma \le D$.

(g) We have $\|\langle w, v\rangle\|_\sigma \le \|w\|_\sigma \|v\|_\sigma$.

(h) For a function $f(q)$ one has $\|\overline{f}\|_\sigma \le \|f\|_\sigma$ and $\|f - \overline{f}\|_\sigma \le \|f\|_\sigma$, where $\overline{f}$ is the average of $f(q)$ over the angles q.

Proof. (d) We exploit the linearity in $p \in \Delta_\varrho$ and apply the definition of Fourier norm. Expanding $w_j(q)p_j$ in Fourier series with coefficients $w_{j,k}\, p_j$, we evaluate

$$\|w_j p_j\|_{\varrho,\sigma} = \sum_{k\in\mathbb{Z}^n} |w_{j,k} p_j|_\varrho \, e^{|k|\sigma} = \varrho \|w_j\|_\sigma \ .$$

By definition (c) we conclude $\|\langle w(q), p\rangle\|_{\varrho,\sigma} = \varrho \sum_j \|w_j p_j\|_{\varrho,\sigma}$, as claimed.

(e) We obviously have

$$w_j(q) = \frac{\partial \langle w(q), p\rangle}{\partial p_j} = \frac{\partial}{\partial p_j} w_j(q) p_j \ .$$

We use Cauchy's inequalities in order to estimate the derivatives at the centre of the polydisk $\Delta_\varrho(0)$, keeping σ unchanged. We exploit again the Fourier expansion of $w_j(q)$ with coefficients $w_{j,k} \in \mathbb{C}$. Using definition (c) we have

$$\begin{aligned}
\|w\|_\sigma &= \sum_j \|w_j\|_\sigma \le \frac{1}{\varrho} \sum_j \|w_j p_j\|_{\varrho,\sigma} \\
&= \frac{1}{\varrho} \sum_j \sum_k |w_{j,k} p_j|_\varrho \, e^{|k|\sigma} = \frac{1}{\varrho} \sum_k e^{|k|\sigma} \Bigl(\sum_j |w_{j,k} p_j|_\varrho \Bigr) \\
&= \frac{1}{\varrho} \sum_k e^{|k|\sigma} \Bigl| \sum_j w_{j,k} p_j \Bigr|_\varrho = \frac{1}{\varrho} \|\langle w(q), p\rangle\|_{\varrho,\sigma} \le D \ .
\end{aligned}$$

Here a comment is in order, concerning the seemingly incorrect equality at the beginning of the last line. We say indeed that $\sum_j |w_{j,k}\, p_j|_\varrho = \bigl|\sum_j w_{j,k}\, p_j\bigr|_\varrho$, which in this case is true, as we check as follows. The quantities to be added together vary independently of each other. Let us pick any j,k; since $|p_j| \le \varrho$, the maximum of $|w_{j,k}p_j|$ is reached at $p_j = \varrho w^*_{j,k}/|w_{j,k}|$, where we have $|w_{j,k}p_j| = w_{j,k}p_j = \varrho|w_{j,k}|$. Hence we conclude

$$\Bigl|\sum_j w_{j,k}\, p_j\Bigr|_\varrho = \sum_j \varrho|w_j| = \sum_j |w_{j,k}\, p_j|_\varrho \ ,$$

as claimed.

(f) We obviously have

$$C_{i,j}(q) = \frac{\partial^2}{\partial p_i \partial p_j}\frac{1}{2}\langle \mathsf{C}(q)p,p\rangle \ .$$

Using the general form of Cauchy's estimate (6.43) in Section 6.4.3 with $d=1$ the claim readily follows.

The proof of (g) requires only elementary considerations which are left to the reader. The proof of (h) is a straightforward consequence of the definition of Fourier norm, because the left side of the inequalities requires only a partial sum of terms in the right side. *Q.E.D.*

Now we may write the statement of Theorem 8.1 in a more detailed form.

Proposition 8.3: *Consider the Hamiltonian (8.5) and assume:*

(i) The function $A(q)$ and the vector function $B(q)$ are analytic in the angles q; moreover, there are positive constants σ and ε such that

$$\max\bigl(\|A\|_\sigma, \|B\|_\sigma\bigr) \le \varepsilon \ .$$

(ii) The matrix $\mathsf{C}(q)$ is analytic in the angles q; moreover, there is a positive constant $m \le 1$ such that

$$m|x| \le |\overline{\mathsf{C}}x| \quad \text{for all } x \in \mathbb{R}^n \ .$$

(iii) For a vector-valued function $w(q)$ with bounded norm $\|w\|_\sigma$, one has

$$\|\mathsf{C}w\|_\sigma \le m^{-1}\|w\|_\sigma \ ;$$

(iv) the frequencies ω satisfy the Diophantine condition

$$|\langle k,\omega\rangle| \ge \gamma|k|^{-\tau} \quad \text{for } 0 \ne k \in \mathbb{Z}^n$$

with some constants $\gamma > 0$ and $\tau > n-1$.

Then there exists a positive constant ε^* depending on n, τ, γ and σ such that the following holds true: if $\varepsilon < \varepsilon^*$, then there exists a canonical transformation $(q,p) = \mathscr{C}(q',p')$ satisfying

$$(8.6) \qquad |p_j - p'_j| < d^{\tau+1}\varrho \ , \quad |q_j - q'_j| < d^{\tau+1}\sigma \ , \quad j = 1,\ldots,n$$

for all $(q', p') \in \mathbb{T}^n_{(1-3d)\sigma} \times \Delta_{(1-3d)\varrho}(0)$ *which gives the Hamiltonian the Kolmogorov normal form (8.2).*

Exercise 8.4: Prove that hypotheses (ii) and (iii) are consequences of the hypothesis of non-degeneration of $h(p)$ in the Hamiltonian (8.1). *A.E.L.*

The proof of Proposition 8.3 is worked out in the following Sections 8.2.2 to 8.2.6. The whole proof is divided into three main parts. First, the formal algorithm is developed, thus establishing the scheme of operations which constitute a single normalization step. Next, the quantitative analytic estimates for a single step are obtained; this involves also a lemma on small divisors. Finally, the convergence of the infinite sequence of iterations is proved. Along the whole proof constant reference is made to the analytical estimates of Chapter 6 on Lie series.

8.2.2 Formal Algorithm

The algorithm is based on composition of Lie series; hence it consists in iterating infinitely many times a single normalization step. All formulæ needed to perform a single step are collected here. They implement the method sketched in Section 8.1.3; the sole difference with respect to Kolmogorov's paper is the use of Lie series to perform canonical transformations. The algorithm will be justified immediately after this statement.

To the Hamiltonian in the form (8.5) we apply two near-to-identity canonical transformations with generating functions

$$\chi_1(q) = X(q) + \langle \xi, q \rangle \ , \quad \chi_2(q,p) = \langle Y(q), p \rangle \ . \tag{8.7}$$

The function $X(q)$, the real vector ξ and the vector function $Y(q)$ are determined by the equations

$$\partial_\omega X = A \ , \tag{8.8}$$

$$\overline{\mathsf{C}}\xi = \overline{B - \mathsf{C}\frac{\partial X}{\partial q}} \ , \tag{8.9}$$

$$\partial_\omega \langle Y, p \rangle = \left\langle B - \mathsf{C}\left(\frac{\partial X}{\partial q} + \xi\right), p \right\rangle \ . \tag{8.10}$$

The transformed Hamiltonian is computed as

$$H'(q,p) = \exp\bigl(L_{\chi_2}\bigr) \circ \exp\bigl(L_{\chi_1}\bigr) H(q,p) \tag{8.11}$$

and has again the form (8.5) with a function $A'(q)$, a vector function $B'(q)$ and a matrix $\mathsf{C}'(q)$ given by

$$(8.12)\quad A' = \exp(L_{\langle Y,p\rangle})\hat{A}\ ,$$

$$(8.13)\quad \langle B',p\rangle = \sum_{j\ge 1}\frac{j}{(j+1)!}L^{j}_{\langle Y,p\rangle}\langle \hat{B},p\rangle\ ,$$

$$(8.14)\quad \langle \mathsf{C}'p,p\rangle = \langle \mathsf{C}p,p\rangle + \sum_{j\ge 1}\frac{1}{j!}L^{j}_{\langle Y,p\rangle}\langle \mathsf{C}p,p\rangle\ ,$$

$$(8.15)\quad \hat{A} = \frac{1}{2}\left\langle \mathsf{C}\left(\frac{\partial X}{\partial q}+\xi\right), \left(\frac{\partial X}{\partial q}+\xi\right)\right\rangle - \left\langle B, \left(\frac{\partial X}{\partial q}+\xi\right)\right\rangle\ ,$$

$$(8.16)\quad \hat{B} = B - \mathsf{C}\left(\frac{\partial X}{\partial q}+\xi\right)\ .$$

The canonical transformation is explicitly written as

$$(8.17)\quad \begin{aligned} q &= \exp\bigl(L_{\langle Y,p'\rangle}\bigr)q'\ ,\\ p &= \exp\bigl(L_{\langle Y,p'\rangle}\bigr)\circ\exp\bigl(L_{X+\langle \xi,q'\rangle}\bigr)p'\ . \end{aligned}$$

Note that $\exp\bigl(L_{X+\langle \xi,q'\rangle}\bigr)q' = q'$, which justifies the first line.

The formal algorithm is justified by playing a little with the Lie series expansions. Here is the complete calculation.

Let us write explicitly the first transformation with generating function $\chi_1(q) = X(q) + \langle \xi, q\rangle$, considering the function $X(q)$ and the real vector ξ as unknowns. Denoting by $\hat{H}$ the transformed Hamiltonian, we get[3]

$$\begin{aligned} \hat{H} = \exp\bigl(L_{\chi_1}\bigr)H = \langle \omega,p\rangle &+ \frac{1}{2}\langle \mathsf{C}p,p\rangle \\ &+ A + L_X\langle \omega,p\rangle + L_{\langle \xi,q\rangle}\langle \omega,p\rangle \\ &+ \langle B,p\rangle + \frac{1}{2}L_X\langle \mathsf{C}p,p\rangle + \frac{1}{2}L_{\langle \xi,q\rangle}\langle \mathsf{C}p,p\rangle \\ &+ \frac{1}{4}L^{2}_{\chi_1}\langle \mathsf{C}p,p\rangle + L_{\chi_1}\langle B,p\rangle\ . \end{aligned}$$

The equation has been arranged by aligning on the second line terms independent of p, in the third line terms linear in p and in the last line terms of higher order to be kept in the transformed Hamiltonian.

Recalling that we have defined $\partial_\omega = L_{\langle \omega,p\rangle}$, we kill the unwanted term A by determining X through the equation $A + L_X\langle \omega,p\rangle = A - \partial_\omega X = 0$, namely (8.8). Since a constant in the Hamiltonian function is irrelevant, we

[3] Notice that, due to χ_1 being independent of the actions p, most terms in the Lie series do actually vanish. The easy remark is that the Poisson bracket with χ_1 decreases the degree in p by one, so that, for example, $L_{\chi_1}A = 0$.

can always assume that $A(q)$ has zero average, that is, $\overline{A} = 0$; therefore, a formal solution of equation (8.8) exists (see Section 8.2.4). In view of this, and letting ξ be still undetermined, we can write the transformed Hamiltonian $\hat{H}$ as

$$\hat{H}(q,p) = \langle \omega, p \rangle + \hat{A}(q) + \langle \hat{B}(q), p \rangle + \frac{1}{2} \langle \mathsf{C}(q)p, p \rangle \ , \tag{8.18}$$

where $\hat{A}$ and $\hat{B}$ are given by (8.15) and (8.16). The expressions for $\hat{A}$ and $\hat{B}$ are easily determined by collecting all terms independent of p and linear in p, respectively, in the expression for $\hat{H}$, and calculating the Poisson brackets. The irrelevant constant $L_{\langle \xi, q \rangle} \langle \omega, p \rangle = \langle \xi, \omega \rangle$ has been omitted.

We now perform the second transformation with generating function $\chi_2(q,p) = \langle Y(q), p \rangle$, and write the transformed Hamiltonian H' as

$$\begin{aligned} H' &= \exp\bigl(L_{\langle Y,p \rangle}\bigr) \hat{H} \\ &= \langle \omega, p \rangle + \exp\bigl(L_{\langle Y,p \rangle}\bigr) \hat{A} \\ &\quad + \langle \hat{B}, p \rangle + L_{\langle Y,p \rangle} \langle \omega, p \rangle + \sum_{j \geq 2} \frac{1}{j!} L^j_{\langle Y,p \rangle} \langle \omega, p \rangle + \sum_{j \geq 1} \frac{1}{j!} L^j_{\langle Y,p \rangle} \langle \hat{B}, p \rangle \\ &\quad + \frac{1}{2} \langle \mathsf{C}p, p \rangle + \frac{1}{2} \sum_{j \geq 1} \frac{1}{j!} L^j_{\langle Y,p \rangle} \langle \mathsf{C}p, p \rangle \ . \end{aligned} \tag{8.19}$$

We kill the unwanted term $\langle \hat{B}, p \rangle$ by imposing

$$\langle \hat{B}, p \rangle + L_{\langle Y,p \rangle} \langle \omega, p \rangle = 0 \ . \tag{8.20}$$

Using (8.16), this gives the equation

$$\left\langle B - \mathsf{C}\left(\frac{\partial X}{\partial q} + \xi \right), p \right\rangle - \partial_\omega \langle Y, p \rangle = 0 \ ,$$

namely (8.10). However, the latter equation can be solved for Y only if the average of the known term is zero. Therefore, we kill the unwanted average part by determining the still-unknown vector ξ from the equation

$$\overline{B - \mathsf{C}\frac{\partial X}{\partial q}} - \overline{\mathsf{C}}\xi = 0 \ ,$$

namely (8.9). This admits a solution because $\overline{\mathsf{C}}$ is non-degenerate. Hence the homological equation (8.20) can be solved, too.

We come now to determining the explicit form of the transformed Hamiltonian H', which is tantamount to justifying (8.12)–(8.14). Since $\exp(L_{\langle Y,p \rangle})\hat{A}$ is the sole term independent of the actions p appearing in (8.19), then (8.12) is obviously verified. Replacing (8.20) in (8.19), we find

$$\begin{aligned}\langle B', p\rangle &= \sum_{j\geq 2}\frac{1}{j!}L^j_{\langle Y,p\rangle}\langle\omega,p\rangle + \sum_{j\geq 1}\frac{1}{j!}L^j_{\langle Y,p\rangle}\langle\hat{B},p\rangle \\ &= \sum_{j\geq 1}\frac{1}{(j+1)!}L^j_{\langle Y,p\rangle}\Bigl(L_{\langle Y,p\rangle}\langle\omega,p\rangle + (j+1)\langle\hat{B},p\rangle\Bigr) \ ;\end{aligned}$$

using again (8.20), we get (8.13). Finally, (8.14) is nothing but the last line of (8.19). This concludes the calculation of the transformed Hamiltonian.

Having defined the generating functions, the canonical transformation of variables is defined, too, and takes the form (8.17). This concludes the justification of the formal algorithm.

8.2.3 Iterative Lemma

The aim of this section is to translate the algorithm of Section 8.2.2 into a scheme of estimates for the norms of all functions involved.

Lemma 8.5: *Let $H(q,p)$ be of the form (8.5), and assume that the hypotheses (i)–(iv) of Proposition 8.3 are satisfied; we repeat the hypotheses here for the reader's convenience:*

(i) there are positive constants σ and ε such that

$$\max\bigl(\|A\|_\sigma, \|B\|_\sigma\bigr) \leq \varepsilon \ ;$$

(ii) there is a positive constant $m \leq 1$ such that

$$m|x| \leq |\overline{\mathsf{C}}x| \quad \text{for all } x\in\mathbb{R}^n \ ;$$

(iii) for a vector-valued function $w(q)$ with bounded norm $\|w\|_\sigma$, one has

$$\|\mathsf{C}w\|_\sigma \leq m^{-1}\|w\|_\sigma \ ;$$

(iv) the frequencies ω satisfy the Diophantine condition

$$|\langle k,\omega\rangle| \geq \gamma|k|^{-\tau} \quad \text{for } 0\neq k\in\mathbb{Z}^n \ .$$

Let $d\leq 1/6$ and σ_* be positive constants satisfying

$$(1-3d)\sigma \geq \sigma_* \ .$$

Then there exists a positive constant $\Lambda = \Lambda(n,\tau,\gamma,\sigma_*)$ such that the following holds true: if

$$\frac{\Lambda\varepsilon}{m^6 d^{3\tau+4}} \leq 1, \tag{8.21}$$

then there exists a canonical transformation $(q,p) = \mathscr{C}(q',p')$ satisfying

$$\begin{aligned}|p'_j - p_j| &< \frac{\Lambda\varepsilon}{m^6 d^{3\tau+4}}d^{\tau+1}\varrho < d^{\tau+1}\varrho \ , \\ |q'_j - q_j| &< \frac{\Lambda\varepsilon}{m^6 d^{3\tau+4}}d^{\tau+1}\sigma_* < d^{\tau+1}\sigma \ , \quad j=1\ldots,n \ ,\end{aligned} \tag{8.22}$$

for all $(q', p') \in \mathbb{T}^n_{(1-3d)\sigma} \times \Delta_{(1-3d)\varrho}(0)$, *which brings the Hamiltonian to the form (8.5) with the same* ω *and with new functions* A', B' *and* C' *satisfying the hypotheses (i)–(iii) with new positive constants* ε', σ' *and* m' *given by*

$$\begin{aligned} \varepsilon' &= \frac{\Lambda}{m^6 d^{3\tau+4}}\,\varepsilon^2\ , \\ \sigma' &= (1-3d)\,\sigma\ , \\ m' &= (1-d^{\tau+1})\,m\ . \end{aligned} \tag{8.23}$$

The proof is deferred to Section 8.2.5. We shall need also the following technical estimate.

Lemma 8.6: *Under hypothesis (iii), for a vector function* $g(p)$ *with bounded norm* $|g|_\varrho$ *in the domain* Δ_ϱ, *we have*

$$\|\mathsf{C}g\|_{\varrho,\sigma} \le m^{-1}|g|_\varrho\ . \tag{8.24}$$

Proof. The inequality (iii) obviously applies to a constant vector $v \in \mathbb{C}^n$; that is, we have

$$\|\mathsf{C}v\|_\sigma \le m^{-1}|v|$$

with the weighted Fourier norm calculated as

$$\|\mathsf{C}v\|_\sigma = \sum_{k\in\mathbb{Z}^n}\Bigl|\sum_l C_{j,l,k}v_l\Bigr|\,e^{|k|\sigma}\ , \tag{8.25}$$

where $C_{j,l,k}$ are the Fourier coefficients of the function $C_{j,l}(q)$. For the function $g(p)$ we have

$$\mathsf{C}g = \sum_{k\in\mathbb{Z}^n}\sum_{l=1}^{n} C_{j,l,k}g_l(p)e^{i\langle k,q\rangle}\ .$$

Hence the norm of $\mathsf{C}g$ is evaluated as

$$\|\mathsf{C}g\|_{\varrho,\sigma} = \sum_{k\in\mathbb{Z}^n}\Bigl|\sum_l C_{j,l,k}g_l\Bigr|_\varrho e^{|k|\sigma} \le \sum_{k\in\mathbb{Z}^n}\Bigl|\sum_l C_{j,l,k}\,|g_l|_\varrho\Bigr|e^{|k|\sigma}\ .$$

In view of (8.25) the latter expression coincides with $\|\mathsf{C}v\|_\sigma$ with the constant vector $v = \bigl(|g_1|_\varrho, \ldots, |g_n|_\varrho\bigr)$; therefore, (8.24) follows. *Q.E.D.*

8.2.4 Lemma on Small Divisors

A major role in the proof of the iterative lemma is played by the control of the effect of the small divisors in the solutions of Eqs. (8.8) and (8.10) for the generating functions. The problem of solving the homological equation has been already discussed in Section 4.5.1, in connection with non-existence of first integrals. However, here there is a nice difference: the frequencies are constant and non-resonant.

Let us state the problem as follows. Given a known function $\psi(p,q)$ with zero average, namely $\overline{\psi}=0$, find φ such that $\partial_\omega\varphi=\psi$. The actions p here are just parameters. The procedure is quite standard. Expand in Fourier series

$$\psi(p,q)=\sum_{0\neq k\in\mathbb{Z}^n}\psi_k(p)\exp\bigl(i\langle k,q\rangle\bigr)\ ,\quad \varphi(p,q)=\sum_{k\in\mathbb{Z}^n}c_k(p)\exp\bigl(i\langle k,q\rangle\bigr)\ ,$$

with $\psi_k(p)$ known and $c_k(p)$ to be found. Calculate

$$\partial_\omega\varphi=i\sum_k\langle k,\omega\rangle c_k(p)\exp\bigl(i\langle k,q\rangle\bigr)\ .$$

Therefore, assuming that the frequencies ω are non-resonant, we get the formal solution with coefficients

$$c_k(p)=-i\frac{\psi_k(p)}{\langle k,\omega\rangle}\ .$$

We could add to φ an arbitrary function $\overline{\varphi}(p)$ with zero average, but we shall not need to exploit this freedom.

In order to proceed, we need quantitative estimates on the solution $\varphi(p,q)$. The following lemma makes quantitative the argument of Poincaré set out in Section 7.3.2.

Lemma 8.7: *Let $\psi(p,q)$ be a zero average function, $\overline{\psi}=0$, with bounded norm $\|\psi\|_{(1-\delta)(\varrho,\sigma)}$ for some non-negative $\delta<1$, and let ω be Diophantine. Let φ be the unique zero average solution of the equation $\partial_\omega\varphi=\psi$. Then for every positive $d<1-\delta$ one has*

$$\|\varphi\|_{(1-\delta-d)(\varrho,\sigma)}\le\frac{1}{\gamma}\left(\frac{\tau}{ed\sigma}\right)^{\tau}\|\psi\|_{(1-\delta)(\varrho,\sigma)}\ ,\tag{8.26}$$

$$\left\|\frac{\partial\varphi}{\partial q}\right\|_{(1-\delta-d)(\varrho,\sigma)}\le\frac{1}{\gamma}\left(\frac{\tau+1}{ed\sigma}\right)^{\tau+1}\|\psi\|_{(1-\delta)(\varrho,\sigma)}\ .\tag{8.27}$$

Proof. Recall the elementary inequality[4]

$$x^\alpha e^{-\beta x}\le\left(\frac{\alpha}{e\beta}\right)^\alpha\tag{8.28}$$

for all positive α, β and x. Recall also that the solution φ is

$$\varphi=-i\sum_k\frac{\psi_k(p)}{\langle k,\omega\rangle}e^{i\langle k,q\rangle}\ ,$$

[4] The right member is the maximum over x of the function on the left.

where $\psi_k(p)$ are the known coefficients of the Fourier expansion of ψ. Using the definition of the norm, evaluate

$$\begin{aligned}\|\varphi\|_{(1-\delta-d)(\varrho,\sigma)} &\le \frac{1}{\gamma}\sum_k |k|^\tau |\psi_k|_\varrho e^{|k|(1-\delta-d)\sigma}\\ &\le \frac{1}{\gamma}\left(\frac{\tau}{ed\sigma}\right)^\tau \sum_k |\psi_k|_\varrho e^{|k|(1-\delta)\sigma}\ ,\end{aligned}$$

where use has been made of $|k|^\tau e^{-|k|d\sigma} \le \bigl(\tau/(ed\sigma)\bigr)^\tau$ in view of the inequality (8.28). The first inequality then follows from the definition of the norm of ψ. As to the second inequality, remark that we have

$$\frac{\partial\varphi}{\partial q_j} = \sum_k \frac{k_j\psi_k(p)}{\langle k,\omega\rangle} e^{i\langle k,q\rangle}\ .$$

In view of $|k| = |k_1| + \cdots + |k_n|$, estimate

$$\begin{aligned}\left\|\frac{\partial\varphi}{\partial q}\right\|_{(1-\delta-d)(\varrho,\sigma)} &\le \sum_j \frac{1}{\gamma}\sum_k |k_j|\,|k|^\tau |\psi_k|_\varrho e^{|k|(1-\delta-d)\sigma}\\ &\le \frac{1}{\gamma}\sum_k |k|^{\tau+1}|\psi_k|_\varrho e^{|k|(1-\delta-d)\sigma}\ .\end{aligned}$$

Then (8.27) follows by using again (8.28). *Q.E.D.*

8.2.5 Proof of the Iterative Lemma

The proof of Lemma 8.5 is a matter of a patient but straightforward application of the estimates given by Lemma 8.7 and the estimates for Poisson brackets and Lie series, Section 6.7

Let us start with the estimates for the generating functions. By Lemma 8.7 the solution $X(q)$ of (8.8), using hypothesis (i), satisfies

$$\left\|\frac{\partial X}{\partial q}\right\|_{(1-d)\sigma} \le \frac{K_1}{d^{\tau+1}}\varepsilon\ ,\quad K_1 = \frac{1}{\gamma}\left(\frac{\tau+1}{e\sigma_*}\right)^{\tau+1}\ . \tag{8.29}$$

Using hypotheses (i) and (iii), in (8.9) we have (recall that $m, d < 1$)

$$\left\|B - \mathsf{C}\frac{\partial X}{\partial q}\right\|_{(1-d)\sigma} \le \frac{K_1+1}{md^{\tau+1}}\varepsilon\ ; \tag{8.30}$$

therefore, using also hypothesis (ii), from (8.9) we get

$$m|\xi| \le |\overline{\mathsf{C}}\xi| \le \frac{K_1+1}{md^{\tau+1}}\varepsilon\ ,$$

and so also

$$|\xi| \le \frac{K_1+1}{m^2 d^{\tau+1}}\varepsilon \ . \tag{8.31}$$

By this and (8.30), we estimate $\hat{B}$ as given by (8.16). Recalling that $m \le 1$ and using again hypothesis (iii), we get

$$\|\hat{B}\|_{(1-d)\sigma} \le \frac{2(K_1+1)}{m^3 d^{\tau+1}}\varepsilon \ . \tag{8.32}$$

Furthermore, by solving (8.10) we get for the components of the vector functions $\hat{B}$ and Y,

$$\hat{B}_l = \sum_k \hat{b}_{l,k} e^{i\langle k,q\rangle} \ , \quad Y_l = \sum_k y_{l,k} e^{i\langle k,\omega\rangle} \quad \text{with} \quad y_{l,k} = \frac{\hat{b}_{l,k}}{i\langle k,\omega\rangle}$$

from which we get

$$\|Y\|_{(1-2d)\sigma} \le \frac{K_2}{m^3 d^{2\tau+1}}\varepsilon \ , \quad K_2 = \frac{2(K_1+1)}{\gamma}\left(\frac{\tau}{e\sigma_*}\right)^{\tau} \ . \tag{8.33}$$

This concludes the estimate of the generating functions.

We come now to the transformed Hamiltonian. From (8.15), by hypothesis (iii) and using (8.30) and (8.31) together with $d \le 1/6$, we get

$$\|\hat{A}\|_{(1-d)\sigma} \le \frac{(2K_1+1)^2}{m^5 d^{2\tau+2}}\varepsilon^2 \ .$$

On the other hand, by Lemma 8.2(d) and by (8.33), the norm of $\langle Y,p\rangle$ is estimated by

$$\|\langle Y,p\rangle\|_{(1-2d)(\varrho,\sigma)} \le \frac{K_2(1-2d)\varrho}{m^3 d^{2\tau+1}}\varepsilon \ . \tag{8.34}$$

Putting the two latter estimates in (8.12), namely $A' = \exp\bigl(\langle Y,p\rangle\bigr)\hat{A}$, we apply Lemma 6.18, formula (6.72). Thus, we assume the convergence condition

$$\frac{4eK_2}{m^3 d^{2\tau+3}\sigma_*}\varepsilon \le 1 \tag{8.35}$$

and get the estimate

$$\|A'\|_{(1-3d)\sigma} \le \frac{K_3}{m^5 d^{2\tau+2}}\varepsilon^2 \ , \quad K_3 = \frac{(e^2+1)(2K_1+1)^2}{e^2} \ . \tag{8.36}$$

The estimate of $\langle B',p\rangle$, given by (8.13), is found by using a similar argument. Indeed, it is enough to remark that one has

$$\begin{aligned}\|\langle B',p\rangle\|_{(1-3d)(\varrho,\sigma)} &\le \sum_{j\ge1}\frac{j}{(j+1)!}\big\|L^j_{\langle Y,p\rangle}\langle\hat B,p\rangle\big\|_{(1-3d)(\varrho,\sigma)}\\ &\le \sum_{j\ge1}\frac{1}{j!}\big\|L^j_{\langle Y,p\rangle}\langle\hat B,p\rangle\big\|_{(1-3d)(\varrho,\sigma)}\ ,\end{aligned}$$

which is, in fact, the estimate of $\big\|\exp\big(L_{\langle Y,p\rangle}\big)\langle\hat B,p\rangle - \langle\hat B,p\rangle\big\|_{(1-3d)(\varrho,\sigma)}$. Hence we apply again Lemma 6.18, formula (6.74) for the remainder of order one. Using the estimate of $\hat B$ in (8.32), we get

$$\|\langle B',p\rangle\|_{(1-3d)(\varrho,\sigma)} \le \frac{K_4(1-3d)\varrho}{m^6 d^{3\tau+4}}\varepsilon^2\ ,\quad K_4=\frac{8(K_1+1)K_2}{e\sigma_*}\ . \tag{8.37}$$

Hence, from Lemma 8.2(e) we get

$$\|B'\|_{(1-3d)(\varrho,\sigma)} \le \frac{K_4}{m^6 d^{3\tau+4}}\varepsilon^2\ . \tag{8.38}$$

We finally come to the estimate for C' as given by (8.14). Applying Lemma 8.6 to the real vector p and in view of $|p|_\varrho=\varrho$, we have

$$\|\langle\mathsf{C}p,p\rangle\|_{\varrho,\sigma}\le\frac{\varrho^2}{m}\ .$$

We apply Lemma 6.18 once more and get

$$\big\|\langle(\mathsf{C}'-\mathsf{C})p,p\rangle\big\|_{(1-3d)(\varrho,\sigma)} \le \frac{K_5(1-3d)^2\varrho^2}{m^4 d^{2\tau+3}}\varepsilon\ ,\quad K_5=\frac{4K_2}{e\sigma_*}\ . \tag{8.39}$$

Hence, by Lemma 8.2(f), we have

$$\|\mathsf{C}'_{j,k}-\mathsf{C}_{j,k}\|_{(1-3d)\sigma}\le\frac{K_5}{m^4 d^{2\tau+3}}\varepsilon\ .$$

This allows us to estimate

$$\big|\overline{\mathsf{C}'}x\big| \ge \big|\overline{\mathsf{C}}x\big| - \big|(\overline{\mathsf{C}'-\mathsf{C}})x\big| \ge \left(m-\frac{K_5}{m^4 d^{2\tau+3}}\varepsilon\right)|x|\ , \tag{8.40}$$

provided ε satisfies the condition

$$\tilde m = m-\frac{K_5}{m^4 d^{2\tau+3}}\varepsilon>0\ . \tag{8.41}$$

On the other hand, for a vector-valued function $w(q)$, we have

$$\begin{aligned}\|\mathsf{C}'w\|_{(1-3d)\sigma} &\le \|\mathsf{C}w\|_{(1-3d)\sigma}+\|(\mathsf{C}'-\mathsf{C})w\|_{(1-3d)\sigma}\\ &\le \left(\frac{1}{m}+\frac{K_5}{m^4 d^{2\tau+3}}\varepsilon\right)\|w\|_{(1-3d)\sigma} < \frac{1}{\tilde m}\|w\|_{(1-3d)\sigma}\ ,\end{aligned} \tag{8.42}$$

in view of (8.41) and of the elementary inequality $a^{-1}+b<(a-b)^{-1}$ for $0<b<a<1$. This concludes the estimates for the transformed Hamiltonian.

It remains to estimate the canonical transformation (8.17). Here, too, we proceed in two steps, first transforming with $\exp(L_{X+\langle\xi,q\rangle})$ and then transforming with $\exp(L_{\langle Y,p\rangle})$. The first transformation is explicitly written as[5]

$$q = \hat{q}\ , \quad p = \hat{p} - \xi - \frac{\partial X}{\partial q}\bigg|_{q=\hat{q},p=\hat{p}}\ . \tag{8.43}$$

The second transformation is

$$\hat{q} = \exp\bigl(L_{\langle Y,p\rangle}\bigr)q\Big|_{q=q'}\ , \qquad \hat{p} = \exp\bigl(L_{\langle Y,p\rangle}\bigr)p\Big|_{q=q',p=p'}\ . \tag{8.44}$$

The estimate is performed by using the fact that the sup norm is bounded by the weighted Fourier norm. By (8.29) and (8.31) the first transformation is estimated by

$$\bigl|p_j - \hat{p}_j\bigr|_{(1-2d)(\varrho,\sigma)} \le \frac{2K_1+1}{m^2 d^{\tau+1}}\varepsilon. \tag{8.45}$$

For the second transformation we show that

$$\begin{aligned} \bigl|q'_j - \hat{q}_j\bigr|_{(1-3d)(\varrho,\sigma)} &\le \frac{(e^2+1)K_2}{e^2 m^3 d^{2\tau+1}}\,\varepsilon\ , \\ \bigl|p'_j - \hat{p}_j\bigr|_{(1-3d)(\varrho,\sigma)} &\le \frac{4K_2\varrho}{e m^3 d^{2\tau+3}\sigma}\,\varepsilon\ . \end{aligned} \tag{8.46}$$

The first inequality can be verified by calculating

$$|q'_j - \hat{q}_j| \le \sum_{s\ge1} \frac{1}{s!}\bigl\|L^{s-1}_{\langle Y,p\rangle}Y_j\bigr\|_{(1-3d)(\varrho,\sigma)} \le \sum_{s\ge0}\frac{1}{s!}\bigl\|L^{s}_{\langle Y,p\rangle}Y_j\bigr\|_{(1-3d)(\varrho,\sigma)}\ ,$$

that is, using the estimate of a Lie series applied to Y_j. Hence, by (6.72) of Lemma 6.18 together with (8.33) and the convergence condition (8.35), the desired inequality follows. As to the second inequality, it follows immediately by putting $f = p_j$, and so $\|p_j\|_{(1-3d)(\varrho,\sigma)} = (1-3d)\varrho$, into (6.74) of Lemma 6.18 and from (8.34).

Thus, collecting (8.43), (8.44) and (8.46), the composition of the two canonical transformations is estimated as

$$\begin{aligned} \bigl|q'_j - q_j\bigr|_{(1-3d)(\varrho,\sigma)} &\le \frac{(e^2+1)K_2}{e^2 m^3 d^{2\tau+1}}\,\varepsilon\ , \\ \bigl|p'_j - p_j\bigr|_{(1-3d)(\varrho,\sigma)} &\le \left(\frac{2K_1+1}{m^2 d^{\tau+1}} + \frac{4K_2\varrho}{e m^3 d^{2\tau+3}\sigma}\right)\varepsilon\ . \end{aligned} \tag{8.47}$$

[5] Since the generating function is independent of p, the coordinate q is unchanged, and the series for $\exp(L_{X+\langle\xi,q\rangle})$ reduces to the first term only.

Now we should collect all the estimates of this section in order to conclude the proof of the iterative lemma. We define Λ as

$$\Lambda = \max\left(\frac{2(2K_1+1)}{\varrho}, \frac{8eK_2}{\sigma_*}, K_3, K_4, K_5 \right) , \tag{8.48}$$

with K_1, K_2, K_3, K_4 and K_5 given by (8.29), (8.33), (8.36), (8.37) and (8.39). It is seen that Λ depends only on n, τ, γ and σ_*, as claimed. By the way, the definition of Λ shows that the convergence condition (8.35) is satisfied in view of hypothesis (8.21). By (8.36) and (8.38) one has $\max\bigl(\|A'\|_{\sigma'}, \|B'\|_{\sigma'}\bigr) \le \varepsilon'$, with ε' given by the first of (8.23). Coming to the bounds for C', we use the convergence condition (8.21) and check that in (8.41) we have

$$\frac{K_5\varepsilon}{m^4 d^{2\tau+3}} \le \frac{\Lambda\varepsilon}{m^4 d^{2\tau+3}} \le m^2 d^{\tau+1} .$$

Hence (8.41) and (8.42) are satisfied with $\tilde{m}$ replaced by m' as given by (8.23), in view of $m' \le \tilde{m}$; this gives the bounds for C'. In order to check the estimates for the canonical transformation, let us first check that by (8.48) we have

$$\frac{2K_1+1}{m^2 d^{2\tau+1}} \le \frac{\Lambda}{m^6 d^{3\tau+4}} d^{\tau+3} , \qquad \frac{K_2}{m^3 d^{2\tau+3}} \le \frac{\Lambda}{4em^6 d^{3\tau+4}} d^{\tau+1} .$$

Putting this into (8.47) and recalling that $d \le 1/6$, we readily get

$$|p'_j - p_j|_{(1-3d)(\varrho,\sigma)} < \frac{\Lambda}{m^6 d^{3\tau+4}} d^{\tau+1}\varrho, \; |q'_j - q_j|_{(1-3d)(\varrho,\sigma)} < \frac{\Lambda}{m^6 d^{3\tau+4}} d^{\tau+1}\sigma_* ,$$

which in view of condition (8.21) gives (8.22). This concludes the proof of Lemma 8.5.

8.2.6 Conclusion of the Proof of Proposition 8.3

By repeated application of the iterative lemma, we construct an infinite sequence $\{\hat{\mathscr{C}}^{(k)}\}_{k\ge1}$ of canonical transformations of the form

$$\bigl(q^{(k-1)}, p^{(k-1)}\bigr) = \hat{\mathscr{C}}^{(k)}\bigl(q^{(k)}, p^{(k)}\bigr)$$

(the upper index labeling the coordinates at the kth step). This produces a sequence $\{H^{(k)}\}_{k\ge0}$ of Hamiltonians, where $H^{(0)} = H$ is the original one, satisfying

$$\begin{aligned} &\max\bigl(\|A^{(k)}\|_{\sigma_k}, \|B^{(k)}\|_{\sigma_k}\bigr) \le \varepsilon_k ,\\ &\bigl|\overline{\mathsf{C}}^{(k)} v\bigr| \ge m_k|v| \quad \text{for all } v\in\mathbb{R}^n ,\\ &\bigl\|\mathsf{C}^{(k)} w\bigr\|_{\sigma_k} \le \frac{1}{m_k}\|w\|_{\sigma_k} \end{aligned} \tag{8.49}$$

for all vector-valued functions $w(q)$, with sequences $\{\varepsilon_k\}_{k\geq 0}$, $\{\sigma_k\}_{k\geq 0}$ and $\{m_k\}_{k\geq 0}$ defined by $\varepsilon_0 = \varepsilon$, $\sigma_0 = \sigma$, $m_0 = m$ and

$$\varepsilon_k = \frac{\Lambda}{m_{k-1}^6 d_k^{3\tau+4}} \varepsilon_{k-1}^2 \ , \tag{8.50}$$

$$\sigma_k = (1 - 3d_k)\sigma_{k-1} \ , \tag{8.51}$$

$$m_k = (1 - d_k^{\tau+1}) m_{k-1} \ . \tag{8.52}$$

These sequences depend on the arbitrary sequence $\{d_k\}_{k\geq 1}$. In turn, the latter sequence must be chosen so that for every $k \geq 1$ the conditions $d_k \leq 1/6$ and

$$\frac{\Lambda \varepsilon_{k-1}}{m_{k-1}^6 d_k^{3\tau+4}} \leq 1 \ , \tag{8.53}$$

$$(1 - 3d_k)\sigma_{k-1} \geq \sigma_* > 0 \ , \tag{8.54}$$

$$(1 - d_k^{\tau+1}) m_{k-1} \geq m_* > 0 \tag{8.55}$$

are satisfied, with some constants σ_* and m_*. The canonical transformations satisfy

$$\begin{aligned} |p^{(k)} - p^{(k-1)}| &\leq \frac{\Lambda \varrho}{m_{k-1}^3 d_k^{2\tau+3}} \varepsilon_{k-1} < d_k^{\tau+1} \varrho \ , \\ |q^{(k)} - q^{(k-1)}| &\leq \frac{\Lambda \sigma_{k-1}}{m_{k-1}^3 d_k^{2\tau+1}} \varepsilon_{k-1} < d_k^{\tau+1} \sigma_{k-1} \ . \end{aligned} \tag{8.56}$$

The problem now is to show that the sequence $\{d_k\}_{k\geq 1}$ can be determined so that (8.53), (8.54) and (8.55) are satisfied for every $k \geq 1$, the sequence $\{\varepsilon_k\}_{k\geq 0}$ converges to zero, and moreover that the sequence of canonical transformations converges to an analytic canonical transformation which gives the Hamiltonian a normal form of Kolmogorov.

The suggestion is to look at (8.50) as a relation which can be used in order to determine d_k if ε_{k-1} and ε_k are known, since

$$d_k = \left(\frac{\Lambda \varepsilon_{k-1}^2}{m_{k-1}^6 \varepsilon_k} \right)^{1/(3\tau+4)} \ . \tag{8.57}$$

Therefore, we can regard $\{\varepsilon_k\}_{k\geq 0}$ as being the arbitrary sequence. Let us make the choice[6]

$$\varepsilon_k = \frac{\varepsilon_0}{(k+1)^{2(3\tau+4)}} \ , \tag{8.58}$$

---

[6] This choice is rather arbitrary and may be replaced by any other condition which assures that $\varepsilon_k \to 0$ for $k \to +\infty$, and $\sum_k d_k \leq 1/6$. For example, we may well make the choice $\varepsilon_k \sim C^{-k}$ with some constant $C > 1$. Most authors prefer to keep the fast convergence by asking, for example, $\varepsilon_k = \varepsilon_{k-1}^{3/2}$.

and suppose for a moment that the condition $m_k \geq m_* > 0$ is satisfied. Then we get

$$d_k < \frac{4}{k^2} \left(\frac{\Lambda \varepsilon_0}{m_*^6} \right)^{1/(3\tau+4)} ,$$

and so also[7]

$$\sum_{k \geq 1} d_k < \frac{2\pi^2}{3} \left(\frac{\Lambda \varepsilon_0}{m_*^6} \right)^{1/(3\tau+4)} . \tag{8.59}$$

Here we use the condition that $\varepsilon \equiv \varepsilon_0$ should be small enough, making the condition quantitative by asking

$$4\pi^2 \left(\frac{\Lambda \varepsilon_0}{m_*^6} \right)^{1/(3\tau+4)} \leq 1 . \tag{8.60}$$

This immediately gives

$$\sum_{k \geq 1} d_k < \frac{1}{6} , \tag{8.61}$$

which implies in particular $d_k < 1/6$ for all $k \geq 1$. We prove now that (8.54) and (8.55) are satisfied. Starting with (8.54), write

$$\ln \prod_{k \geq 1} (1 - 3d_k) = \sum_{k \geq 1} \ln(1 - 3d_k) .$$

Using the elementary inequality

$$0 \geq \ln(1 - x) \geq -2x \ln 2 \quad \text{for } 0 \leq x \leq 1/2 ,$$

evaluate

$$0 \geq \sum_{k \geq 1} \ln(1 - 3d_k) \geq -6 \ln 2 \sum_{k \geq 1} d_k \geq -\ln 2 .$$

Therefore, $\prod_{k \geq 1} (1 - 3d_k) \geq 1/2$, so that (8.54) is fulfilled with, for example,

$$\sigma_* = \frac{\sigma}{2} . \tag{8.62}$$

By the same argument, in view of $d_k^{\tau+1} < d_k$, we evaluate

$$\prod_{k \geq 1} (1 - d_k) \geq 1/2^{1/3} , \tag{8.63}$$

so that (8.55) is fulfilled with, for example,

$$m_* = \frac{m_0}{2^{1/3}} . \tag{8.64}$$

---

[7] Recall that $\sum_{k \geq 1} k^{-2} = \pi^2/6$.

As to condition (8.53), it is clearly satisfied by the choice (8.58) of the sequence ε_k. In fact, by comparing (8.53) with (8.50), one immediately realizes that it follows from the condition $\varepsilon_k \le \varepsilon_{k-1}$, which is clearly fulfilled because of (8.58). Thus, we are left only with the condition (8.60) on the smallness of ε_0.

It remains to prove that the canonical transformation is well defined on some domain. To this end, having fixed a positive initial value of the analyticity radius in the actions $\varrho_0 = \varrho$, consider the sequence of domains $\{\mathbb{T}^n_{\sigma_k} \times \Delta_{\varrho_k}(0)\}_{k\ge 0}$, with σ_k as in (8.51) and $\varrho_k = (1-3d_k)\varrho_{k-1}$. Then the canonical transformation

$$\begin{aligned} \hat{\mathscr{C}}^{(k)}\colon \mathbb{T}^n_{\sigma_k} \times \Delta_{\varrho_k}(0) &\;\to\; \mathbb{T}^n_{\sigma_{k-1}} \times \Delta_{\varrho_{k-1}}(0)\,, \\ (q^{(k)}, p^{(k)}) &\;\mapsto\; (q^{(k-1)}, p^{(k-1)}) = \mathscr{C}^{(k)}(q^{(k)}, p^{(k)}) \end{aligned}$$

is analytic. Therefore, by composition, the transformation

$$\mathscr{C}^{(k)}\colon \mathbb{T}^n_{\sigma_k} \times \Delta_{\varrho_k}(0) \to \mathbb{T}^n_{\sigma_0} \times \Delta_{\varrho_0}(0)$$

defined as

$$\mathscr{C}^{(k)} = \hat{\mathscr{C}}^{(k)} \circ \cdots \circ \hat{\mathscr{C}}^{(1)}$$

is canonical and analytic. On the other hand, by (8.56) we have

$$\bigl|q^{(k)} - q^{(0)}\bigr| < \bigl|q^{(k)} - q^{(k-1)}\bigr| + \cdots + \bigl|q^{(1)} - q^{(0)}\bigr| < \sigma \sum_{j=1}^{k} d_j^{\tau+1}\,,$$

$$\bigl|p^{(k)} - p^{(0)}\bigr| < \bigl|p^{(k)} - p^{(k-1)}\bigr| + \cdots + \bigl|p^{(1)} - p^{(0)}\bigr| < \varrho \sum_{j=1}^{k} d_j^{\tau+1}\,.$$

Since $\sum_{j\ge 1} d_j$ is convergent, the sequence $\mathscr{C}^k$ converges absolutely to a transformation

$$\mathscr{C}^{(\infty)}\colon \mathbb{T}^n_{\sigma_*} \times \Delta_{\varrho_*}(0) \to \mathbb{T}^n_{\sigma_0} \times \Delta_{\varrho_0}(0)$$

with, for example, $\varrho_* = \varrho_0/2$ in view of (8.63). The absolute convergence implies the uniform convergence in any compact subset of $\mathbb{T}^n_{\sigma_*} \times \Delta_{\varrho_*}(0)$, and so, by Weierstrass's theorem, $\mathscr{C}^{(\infty)}$ is also analytic. Finally, denoting by $(q^{(\infty)}, p^{(\infty)})$ the canonical coordinates in $\mathbb{T}^n_{\sigma_*} \times \Delta_{\varrho_*}(0)$, and combining (8.56) with (8.50), (8.53), (8.58) and (8.59), we have

$$\begin{aligned} \bigl|q_j^{(\infty)} - q_j^{(0)}\bigr| &< \frac{4\pi^2}{3}\left(\frac{\Lambda\varepsilon_0}{m_*^6}\right)^{1/(3\tau+4)} \sigma_* < d^{\tau+1}\varrho\,, \\ \bigl|p_j^{(\infty)} - p_j^{(0)}\bigr| &< \frac{2\pi^2}{3}\left(\frac{\Lambda\varepsilon_0}{m_*^6}\right)^{1/(3\tau+4)} \varrho_* < d^{\tau+1}\sigma\,, \quad j = 1,\ldots,n\,, \end{aligned} \tag{8.65}$$

where (8.60) and the definitions $\varrho_* = \varrho/2$ and $\sigma_* = \sigma/2$ have been used. Therefore, the transformation reduces to identity for $\varepsilon = 0$. By the properties of the Lie series transformation, one also has $H^{(k)} = H^{(0)} \circ \mathscr{C}^{(k)}$, so that the sequence $H^{(k)}$ converges to an analytic function $H^{(\infty)}$ which by construction is in normal form. This concludes the proof of Theorem 8.1.

In view of many discussions which have been raised in the literature, it may be interesting to add some additional remarks. The proof provides explicit estimates for the constant ε^* that in the statement of Proposition 8.3 is claimed to exist. In fact, we have

$$\varepsilon^* = m_*^6 / [\Lambda (4\pi^2)^{3\tau+4}] ,$$

where m_* is given in (8.64), $\sigma_* = \sigma/2$, $\varrho_* = \varrho/2$ and Λ can be explicitly calculated using the definitions (8.48), (8.29), (8.33), (8.36), (8.37) and (8.39). A point which is often mentioned, in particular for its interest for systems of many bodies (e.g., for nonlinear chains), is that $\varepsilon^* \to 0$ for $n \to \infty$, and even quite fast. Indeed the dependence of Λ on n is hidden in $\tau > n-1$. The worst case occurs for the constant $K_4 \sim (\tau+1)^{3(\tau+1)}$, according to (8.37), thus suggesting that $\varepsilon^* \sim n^{-3n}$, indeed. A further point concerns the size of the deformation of invariant tori. This is estimated in formula (8.65). Recalling the definition (8.57) of d_k, we see that the deformation is roughly evaluated as $\mathcal{O}(\varepsilon^{1/3})$. It should be stressed, however, that the estimates obtained here are very far from optimal.

8.3 A Proof in Classical Style

The aim of this section is to present an alternate proof of Kolmogorov's theorem, different from the previous one, which uses a classical scheme of expansion.[8] The main problem, however, is the same: to control the accumulation of small divisors. The key point is that the dramatic accumulation of divisors described in Section 7.4 does not occur here.

The reader will note that both the formal scheme and the quantitative estimates appear to be more complex here than in the original Kolmogorov's scheme. This is the price to be paid in order to bring into evidence the role of small divisors. Moreover, it is worth recalling once more that the scheme here is constructive.

In order to reduce the technical (and boring) troubles to a minimum, we further simplify the initial Hamiltonian by considering a system of coupled rotators as described by the Hamiltonian $H(p,q) = H_0(p) + \varepsilon H_1(p,q)$, where

$$(8.66) \qquad H_0(p) = \frac{1}{2}\sum_{j=1}^{n} p_j^2 \ , \quad H_1(p,q) = \sum_{k\in\mathbb{Z}^n} c_k(p) e^{i\langle k,q\rangle} \ .$$

[8] The proof given here is a complete exposition of the short scheme in [99].

The perturbation $H_1(p,q)$ is assumed to be analytic in a complex domain $\mathbb{T}^n_\sigma$ for some $\sigma > 0$, and the coefficients $c_k(p)$ are assumed to be polynomials of degree at most 2. As in the proof using a quadratic scheme, our restrictions are intended to reduce the technicalities of the proof (there remain enough, anyway), but all the essential difficulties are taken into account. Some hints on how to deal with the general case are provided later, in Section 8.4. The statement can be anticipated in short as follows.

Consider the Hamiltonian (8.66) on a domain $\mathscr{D}_{\varrho,\sigma} = \Delta_\varrho \times \mathbb{T}^n_\sigma$ with some $\varrho > 0$. Let $p_* \in \Delta_\varrho$ be such that the corresponding unperturbed frequencies $\omega(p_*) = p_*$ satisfy a suitable condition of strong non-resonance. If in a neighbourhood of p_* the perturbation H_1 satisfies $\|H_1\|_{\varrho,\sigma} < \varepsilon_*$ small enough, then there exists a perturbed invariant torus carrying quasiperiodic motions with frequencies $\omega(p^*)$, which is a small deformation and translation of the unperturbed one.

The statement is very similar to that of Theorem 8.1, except that the condition of non-resonance is left undecided until now. The proof extends over the rest of the present Section 8.3. The proof itself will provide a natural condition of non-resonance weaker than the Diophantine one: see Section 8.3.6, condition $\boldsymbol{\tau}$, formula (8.82). In order to make this part self-consistent, the proof is worked out with no reference to the previous sections, so that a reader who wants to follow it does not need to know details about Kolmogorov's scheme.

As in Kolmogorov's method, the proof requires an infinite sequence of canonical transformations which gives the Hamiltonian a normal form. The difference, as we have already said at the beginning of the chapter, is that the algorithm does not exploit the fast convergence of the quadratic method. Nevertheless, it is shown that the accumulation of small divisors can be kept under control. To this end it is essential that at every step the generating functions should be trigonometric polynomials of finite order. More precisely, the generating functions at the rth step must be a trigonometric polynomial of degree rK with $K > 0$ fixed since the beginning. Therefore, we pick $K > 0$ with some criterion and expand the perturbation $H_1(p,q)$ in trigonometric polynomials, as explained in Section 7.3.3. The analytical estimates which justify the expansion are given in Section 7.3.4. We also call to the reader's attention the short discussion concerning the choice of K at the end of Section 7.3.3.

8.3.1 The Formal Constructive Algorithm

According to the procedure outlined in Section 8.1, and considering the Hamiltonian (8.66) expanded in trigonometric polynomials with a fixed parameter K, we select a non-resonant unperturbed torus p^*. By translating the origin of the action variables to p^*, we write the Hamiltonian as[9]

$$(8.67)\qquad H(q,p) = \langle \omega, p\rangle + \frac{1}{2}\langle p,p\rangle + \sum_{s\geq 1} \varepsilon^s \bigl[A_s(q) + B_s(p,q) + C_s(p,q)\bigr].$$

This is rather obvious, because H_1 is a polynomial of degree 2. The aim is to construct an infinite sequence $H^{(0)}(p,q)$, $H^{(1)}(p,q)$, $H^{(2)}(p,q)$, ... of Hamiltonians, with $H^{(0)}$ coinciding with H in (8.67), which after r steps of normalization turn out to be written in the general form

$$(8.68)\qquad \begin{aligned} H^{(r)} &= \langle \omega, p\rangle + \sum_{s=0}^{r} \varepsilon^s h_s(p,q) \\ &\quad + \sum_{s>r} \varepsilon^s \left[A_s^{(r)}(q) + B_s^{(r)}(p,q) + C_s^{(r)}(p,q)\right] , \end{aligned}$$

where $h_0(p) = \frac{1}{2}\langle p,p\rangle$. The Hamiltonian $H^{(r)}(p,q)$ is in Kolmogorov's normal form up to order r, in the formal sense. The following properties will be preserved along the whole procedure:

(i) $h_1(p,q),\ldots,h_r(p,q)$ are quadratic in p, so that they are in normal form and are trigonometric polynomials of degree sK in q; they do not change after step r.
(ii) $A_s^{(r)}(q)$ is independent of p.
(iii) $B_s^{(r)}(p,q)$ is linear in p.
(iv) $C_s^{(r)}(p,q)$ is a quadratic polynomial in p.
(v) $A_s^{(r)}(q)$, $B_s^{(r)}(p,q)$ and $C_s^{(r)}(p,q)$ are trigonometric polynomials of degree sK.

The properties may appear to be quite strange, but are precisely the characteristics which allow us to control the impact of small divisors on convergence. In particular, property (iv) will be preserved in view of Lemmas 7.11 and 7.12.

The normalization process is worked out with a recasting of the method of Kolmogorov. Assuming that $r-1$ steps have been performed, so that the Hamiltonian $H^{(r-1)}(p,q)$ has the desired form (8.68) with $r-1$ in place of r, we construct in sequence the new Hamiltonians

$$\hat{H}^{(r)} = \exp\bigl(\varepsilon^r L_{\chi_{r,1}}\bigr) H^{(r-1)} , \quad H^{(r)} = \exp\bigl(\varepsilon^r L_{\chi_{r,2}}\bigr) \hat{H}^{(r)} ,$$

the first one being an intermediate Hamiltonian, and the second one being in normal form up to order r. At every step r we apply a first canonical transformation with generating function $\chi_{r,1}(q) = X_r(q) + \langle \xi_r, q\rangle$, followed

[9] Here we found it convenient to avoid the notations $\langle B(q),p\rangle$ and $\langle \mathsf{C}(q)p,p\rangle$ of Section 8.2.

by a second transformation with generating function $\chi_{r,2}(p,q) = \langle Y^{(r)}(q), p\rangle$. This completely agrees with the algorithm of Kolmogorov. Now comes the difference: at every step we refrain from adding up all terms in the expansion of the perturbation; we keep the expansion in trigonometric polynomials of degree sK. The explicit algorithm for a single step is summarized in Table 8.2, which contains the equations for the generating functions, and Tables 8.3 and 8.4, which provide the transformation in explicit form. The tables are arranged in three successive pages and include also the quantitative estimates; this is in order to collect the whole scheme, both formal and quantitative, in a form which can be seen with no need to jump over many pages.

A reader who has developed some expertise in perturbation methods will probably be able to check the formal part of the algorithm himself. However, some help may be welcome; hence, a detailed (close to pedantic) explanation is included here. The quantitative part will be developed in the next sections, always making reference to the tables.

We calculate the Hamiltonian $\hat{H}^{(r)}$ by applying $\exp(L_{\chi_{r,1}})$ to every term of $H^{(r-1)}$; the formulæ are collected in the first half of Table 8.3. It is convenient to follow the good old adage *divide et impera* and separate the calculation into three parts, considering first terms of order ε^0 and ε^r in $H^{(r-1)}$, next the functions $h_1 \ldots, h_{r-1}$ and finally terms of order ε^s with $s > r$. The Lie triangle for the part which involves terms of order ε^0 and ε^r reduces to the following:

$$
\begin{array}{llllll}
\varepsilon^0 : & \langle\omega,p\rangle & h_0 & & & \\
\varepsilon^r : & L_{\chi_{r,1}}\langle\omega,p\rangle & L_{\chi_{r,1}}h_0 & A_r^{(r-1)} & B_r^{(r-1)} & C_r^{(r-1)} \\
\varepsilon^{2r} : & 0 & \frac{1}{2}L^2_{\chi_{r,1}}h_0 & 0 & L_{\chi_{r,1}}B_r^{(r-1)} & L_{\chi_{r,1}}C_r^{(r-1)} \\
\varepsilon^{3r} : & 0 & 0 & 0 & 0 & \frac{1}{2}L^2_{\chi_{r,1}}C_r^{(r-1)}.
\end{array}
$$

Since the generating function is of order ε^r, the rows of the triangle proceed by powers of ε^r. The second line contains all terms which are involved in the process of removing the unwanted functions.

Extracting from the second line the two contributions independent of p, we get the part that we want to clear, namely

$$
L_{\chi_{r,1}}\langle\omega,p\rangle + A_r^{(r-1)} = -\partial_\omega X_r + A_r^{(r-1)} - \partial_\omega\langle\xi,q\rangle\ .
$$

Ignoring the last term $\partial_\omega\langle\xi,q\rangle$, which is an irrelevant constant, we get the first homological equation in Table 8.2 which determines the function $X_r(q)$. Forgetting also the average of $A_r^{(r-1)}(q)$, which is a constant, the equation is solved as explained in Section 8.2.4.

Table 8.2 Constructive algorithm for Kolmogorov's normal form. Generating functions.

- Equations for the generating functions
  - Expressions:

$$\chi_{r,1} = X_r + \langle \xi_r, q \rangle \ , \qquad \chi_{r,2} = \langle Y_r(q), p \rangle \ .$$

  - Equations:

$$\begin{aligned} \partial_\omega X_r &= A_r^{(r-1)} \ , \\ \langle \xi_r, p \rangle &= \overline{B_r^{(r-1)}} \ , \\ \partial_\omega \chi_{r,2} &= \hat{B}_r^{(r)} \ , \end{aligned}$$

with

$$\hat{B}_r^{(r)} = B_r^{(r-1)} - \overline{B_r^{(r-1)}} - \left\langle \frac{\partial X_r}{\partial q}, p \right\rangle \ .$$

- Quantitative estimates for the generating functions.
  - Set

$$\begin{aligned} &\Omega = \max_j |\omega_j| \ ; \\ &\alpha_0 = 1 \ , \quad \alpha_r = \frac{1}{\Omega} \min_{0<|k|\le rK} \bigl|\langle k, \omega \rangle\bigr| \ ; \\ &\delta_r = \frac{1}{2\pi^2} \cdot \frac{1}{r^2} \ ; \\ &d_0 = 0 \ , \quad d_r = 2(\delta_1 + \cdots + \delta_r) \ , \quad r > 0 \ . \end{aligned}$$

  - Estimate:

$$\begin{aligned} \|X_r\|_{1-d_{r-1}} &\le \frac{1}{\Omega \alpha_r} \|A_r^{(r-1)}\|_{1-d_{r-1}} \ , \\ \bigl|\xi_{r,j}\bigr| &\le \frac{2}{\varrho} \left\| \overline{B_r^{(r-1)}} \right\|_{1-d_{r-1}} , \\ \|\chi_{r,2}\|_{1-d_{r-1}-\delta_r} &\le \frac{1}{\Omega \alpha_r} \|\hat{B}_r^{(r)}\|_{1-d_{r-1}-\delta_r} . \end{aligned}$$

Extracting from the second line the contributions linear in p, we get

$$\tfrac{1}{2} L_{\chi_{r,1}} \langle p, p \rangle + B_r^{(r-1)} = -\left\langle \frac{\partial X_r}{\partial q}, p \right\rangle - \langle \xi_r, p \rangle + B_r^{(r-1)} \ .$$

Here we remark that the average of $B_r^{(r-1)}(q)$ might be non-zero, so that it would introduce an unwanted correction of the frequencies ω in $\hat{H}^{(r)}$. We avoid it by determining ξ_r from the second equation in Table 8.2, which is a

Table 8.3. Kolmogorov's normal form: intermediate Hamiltonian

---

- Intermediate Hamiltonian $\hat{H}^{(r)} = \exp\bigl(L_{\chi_{r,1}}\bigr)H^{(r-1)}$:

$$\hat{A}_r^{(r)} = 0\ , \quad \hat{B}_r^{(r)} = B_r^{(r-1)} - \overline{B_r^{(r-1)}} + L_{X_r}h_0\ , \quad \overline{\hat{B}_r^{(r)}} = 0\ .$$

For $s > r$,

$$\hat{A}_s^{(r)} = \begin{cases} A_s^{(r-1)}\ , & r < s < 2r\ ; \\ \frac{1}{2}L_{\chi_{r,1}}^2 h_{s-2r} + L_{\chi_{r,1}}B_{s-r}^{(r-1)} + A_s^{(r-1)}\ , & 2r \le s < 3r\ ; \\ \frac{1}{2}L_{\chi_{r,1}}^2 C_{s-2r}^{(r-1)} + L_{\chi_{r,1}}B_{s-r}^{(r-1)} + A_s^{(r-1)}\ , & s \ge 3r\ . \end{cases}$$

$$\hat{B}_s^{(r)} = \begin{cases} L_{\chi_{r,1}}h_{s-r} + B_s^{(r-1)}\ , & r < s < 2r\ ; \\ L_{\chi_{r,1}}C_{s-r}^{(r-1)} + B_s^{(r-1)}\ , & s \ge 2r\ . \end{cases}$$

- Quantitative estimates for the intermediate Hamiltonian. Set

$$G_{r,1} = \frac{2e}{\varrho\sigma}\biggl(\frac{\|A_r^{(r-1)}\|_{1-d_{r-1}}}{e\Omega} + \frac{e\alpha_r\delta_r\sigma}{\varrho}\bigl\|\overline{B_r^{(r-1)}}\bigr\|_{1-d_{r-1}}\biggr).$$

Get

$$\|\hat{B}_r^{(r)}\|_{1-d_{r-1}-\delta_r} \le \|B_r^{(r-1)}\|_{1-d_{r-1}} + \frac{2\|A_r^{(r-1)}\|_{1-d_{r-1}}}{\Omega\alpha_r\delta_r^2\varrho\sigma}\|h^{(0)}\|_1\ .$$

For $r < s < 2r$, $2r \le s < 3r$ and $s \ge 3r$, respectively, get

$$\|\hat{A}_s^{(r)}\|_{1-d_{r-1}-\delta_r} \le$$
$$\begin{cases} \|A_s^{(r-1)}\|_{1-d_{r-1}}\ ; \\ \Bigl(\frac{G_{r,1}}{\alpha_r\delta_r^2}\Bigr)^2\|h_{s-2r}\|_{1-d_{s-2r}} + \frac{G_{r,1}}{\alpha_r\delta_r^2}\|B_{s-r}^{(r-1)}\|_{1-d_{r-1}} + \|A_s^{(r-1)}\|_{1-d_{r-1}}\ ; \\ \Bigl(\frac{G_{r,1}}{\alpha_r\delta_r^2}\Bigr)^2\|C_{s-2r}^{(r-1)}\|_{1-d_{s-2r}} + \frac{G_{r,1}}{\alpha_r\delta_r^2}\|B_{s-r}^{(r-1)}\|_{1-d_{r-1}} + \|A_s^{(r-1)}\|_{1-d_{r-1}}\ . \end{cases}$$

For $r < s < 2r$ and $s \ge 2r$, respectively, get

$$\|\hat{B}_s^{(r)}\|_{1-d_{r-1}-\delta_r} \le \begin{cases} \frac{G_{r,1}}{\alpha_r\delta_r^2}\|h_{s-r}\|_{1-d_{s-r}} + \|B_s^{(r-1)}\|_{1-d_{r-1}}\ ; \\ \frac{G_{r,1}}{\alpha_r\delta_r^2}\|C_{s-r}^{(r-1)}\|_{1-d_{s-r}} + \|B_s^{(r-1)}\|_{1-d_{r-1}}\ . \end{cases}$$

Table 8.4. Kolmogorov's normal form: new Hamiltonian

- Transformed Hamiltonian $H^{(r)} = \exp\bigl(L_{\chi_{r,2}}\hat{H}^{(r)}\bigr)$.
  Set $k = \lfloor s/r \rfloor$, $m = s\ (\mathrm{mod}\, r)$, $s = kr + m$;

$$h_r = L_{\chi_{r,2}} h_0 + C_r^{(r-1)} \ .$$

$$A_s^{(r)} = \sum_{j=0}^{k-1} \frac{1}{j!} L^j_{\chi_{r,2}} \hat{A}^{(r)}_{s-jr} \ , \quad s > r \ .$$

$$B_s^{(r)} = \begin{cases} \dfrac{k-1}{k!} L^{k-1}_{\chi_{r,2}} \hat{B}_r^{(r)} + \displaystyle\sum_{j=0}^{k-2} \frac{1}{j!} L^j_{\chi_{r,2}} \hat{B}^{(r)}_{s-jr} \ , & k \geq 2 \ , \ m = 0 \ ; \\ \displaystyle\sum_{j=0}^{k-1} \frac{1}{j!} L^j_{\chi_{r,2}} \hat{B}^{(r)}_{s-jr} \ , & k \geq 1 \ , \ m \neq 0 \ . \end{cases}$$

$$C_s^{(r)} = \frac{1}{k!} L^k_{\chi_{r,2}} h_m + \sum_{j=0}^{k-1} \frac{1}{j!} L^j_{\chi_{r,2}} C^{(r)}_{s-jr} \ , \qquad s > r \ .$$

- Quantitative estimates for the new Hamiltonian. Set

$$G_{r,2} = \frac{2e}{\Omega \varrho \sigma} \|\hat{B}_r^{(r)}\|_{(1-d_{r-1}-\delta_r)} \ .$$

For $s > r$ get

$$\|h_r\|_{1-d_r} \leq \frac{G_{r,2}}{\alpha_r \delta_r^2} \|h_0\|_1 + \|C_r^{(r)}\|_{1-d_{r-1}} \ ;$$

$$\|A_s^{(r)}\|_{1-d_r} \leq \sum_{j=0}^{k-1} \left(\frac{G_{r,2}}{\alpha_r \delta_r^2}\right)^j \|\hat{A}^{(r)}_{s-jr}\|_{1-d_{r-1}-\delta_r} \ ;$$

$$\|B_s^{(r)}\|_{1-d_r} \leq \sum_{j=0}^{k-1} \left(\frac{G_{r,2}}{\alpha_r \delta_r^2}\right)^j \|\hat{B}^{(r)}_{s-jr}\|_{1-d_{r-1}-\delta_r} \ ;$$

$$\|\overline{B_s^{(r)}}\|_{1-d_r} \leq \delta_r \sum_{j=0}^{k-1} \left(\frac{G_{r,2}}{\alpha_r \delta_r^2}\right)^j \|\hat{B}^{(r)}_{s-jr}\|_{1-d_{r-1}-\delta_r} \ ;$$

$$\|C_s^{(r)}\|_{1-d_r} \leq \left(\frac{G_{r,2}}{\alpha_r \delta_r^2}\right)^k \|h_{s-kr}\|_{1-d_{s-kr}} + \sum_{j=0}^{k-1} \left(\frac{G_{r,2}}{\alpha_r \delta_r^2}\right)^j \|\hat{C}^{(r)}_{s-jr}\|_{1-d_{r-1}-\delta_r} \ .$$

trivial one in view of $h_0 = \frac{1}{2}\langle p, p\rangle$. Having removed the average, what is left is the function $\hat{B}_r^{(r)}$ in Table 8.3.
The remaining term $C_r^{(r-1)}$ in the second line is quadratic in p, and is left unchanged. Coming to the third and fourth lines, the term $L_{\chi_{r,1}} C_r^{(r-1)}$, of order ε^{2r} and linear in p, is included in $\hat{B}_{2r}^{(r)}$. The terms $\frac{1}{2} L_{\chi_{r,1}}^2 h_0$ and $L_{\chi_{r,1}} B_r^{(r-1)}$, of order ε^{2r}, and $\frac{1}{2} L_{\chi_{r,1}}^2 C_r^{(r-1)}$, of order ε^{3r}, are independent of p, and are included in $\hat{A}_{2r}^{(r)}$ and $\hat{A}_{3r}^{(r)}$, respectively. The triangle does not include anything else, due to $\chi_{r,1}$ being independent of p. No term of degree higher than 2 in p is generated.

For $h_1, \ldots, h_{r-1}$, which are all quadratic functions, we have

$$\varepsilon^s \exp L_{\varepsilon^r \chi_{r,1}} h_s = \varepsilon^s h_s + \varepsilon^{r+s} L_{\chi_{r,1}} h_s + \frac{\varepsilon^{2r+s}}{2} L_{\chi_{r,1}}^2 h_s \ ,$$

all the rest of the series being zero. The new term $L_{\chi_{r,1}} h_s$ is linear in p and is included in $\hat{B}_{r+s}^{(r)}$ (renamed as $\hat{B}_s^{(r)}$ with a translation of the lower index) in Table 8.3. Similarly, $\frac{1}{2} L_{\chi_{r,1}}^2 h_s$ is independent of p and is included in $\hat{A}_{r+s}^{(r)}$.

Finally, for the functions $A_s^{(r-1)}$, $B_s^{(r-1)}$ and $C_s^{(r-1)}$ of higher order $s > r$ we get a small triangle, namely

$$\begin{array}{lccc}
\varepsilon^s : & A_s^{(r-1)} & B_s^{(r-1)} & C_s^{(r-1)} \\
\varepsilon^{r+s} : & 0 & L_{\chi_{r,1}} B_s^{(r-1)} & L_{\chi_{r,1}} C_s^{(r-1)} \\
\varepsilon^{2r+s} : & 0 & 0 & \frac{1}{2} L_{\chi_{r,1}}^2 C_s^{(r-1)} .
\end{array}$$

The triangle contains all terms which are generated, which are at most quadratic in p, as expected. All unchanged functions of the first line are left in place. The functions $L_{\chi_{r,1}} B_s^{(r-1)}$ and $\frac{1}{2} L_{\chi_{r,1}}^2 C_s^{(r-1)}$ are independent of p and are included in $\hat{A}_{r+s}^{(r)}$ and $\hat{A}_{2r+s}^{(r)}$, respectively. Finally, $L_{\chi_{r,1}} C_s^{(r-1)}$ is linear in p and is included in $\hat{B}_{r+s}^{(r)}$. The functions $C_s^{(r-1)}$ for $s > r$ remain unchanged in this step.

Having performed the transformation with $\chi_{r,1}$, we have an intermediate Hamiltonian,

$$\begin{aligned}
\hat{H}^{(r)} = \omega \cdot p + \sum_{s=0}^{r-1} \varepsilon^s h_s(p, q) + \hat{B}_r^{(r)} + C_r^{(r-1)} \\
+ \sum_{s>r} \varepsilon^s \left[\hat{A}_s^{(r)}(q) + \hat{B}_s^{(r)}(p, q) + C_s^{(r)}(p, q) \right] \ ,
\end{aligned}$$

where all functions are defined in Table 8.3. Remark that there is no term $\hat{A}_r^{(r)}$, which has been cleared.

We come now to calculating the Hamiltonian $H^{(r)} = \exp L_{\chi_{r,2}} \hat{H}^{(r)}$; the relevant formulæ are collected in the first half of Table 8.4. We split again the calculation into three parts. First considering terms of order ε^0 and ε^r, we get the Lie triangle:

$$
\begin{array}{lllll}
\varepsilon^0 : & \langle\omega, p\rangle & h_0 & & \\
\varepsilon^r : & L_{\chi_{r,2}}\langle\omega, p\rangle & L_{\chi_{r,2}} h_0 & \hat{B}_r^{(r)} & C_r^{(r-1)} \\
\varepsilon^{2r} : & \frac{1}{2} L_{\chi_{r,2}}^2 \langle\omega, p\rangle & \frac{1}{2} L_{\chi_{r,2}}^2 h_0 & L_{\chi_{r,2}} \hat{B}_r^{(r)} & L_{\chi_{r,2}} C_r^{(r-1)} \\
\varepsilon^{3r} : & \frac{1}{3!} L_{\chi_{r,2}}^3 \langle\omega, p\rangle & \frac{1}{3!} L_{\chi_{r,2}}^3 h_0 & \frac{1}{2} L_{\chi_{r,2}}^2 \hat{B}_r^{(r)} & \frac{1}{2} L_{\chi_{r,2}}^2 C_r^{(r-1)} \\
\varepsilon^{4r} : & \frac{1}{4!} L_{\chi_{r,2}}^4 \langle\omega, p\rangle & \frac{1}{4!} L_{\chi_{r,2}}^4 h_0 & \frac{1}{3!} L_{\chi_{r,2}}^3 \hat{B}_r^{(r)} & \frac{1}{3!} L_{\chi_{r,2}}^3 C_r^{(r-1)} \\
 & \vdots & \vdots & \vdots & \vdots
\end{array}
$$

The triangle here is infinite, because $\chi_{r,2}$ is linear in p and so it does not change the degree in p of a function.

Extracting the linear terms in p from the second line and forcing them to zero, we get

$$
L_{\chi_{r,2}}\langle\omega, p\rangle + \hat{B}_r^{(r)} = -\partial_\omega \chi_{r,2} + \hat{B}_r^{(r)} = 0 \ ,
$$

which is the third equation for the generating functions in Table 8.2. In order to show that it may be solved we need a few more considerations. Recall that $\chi_{r,2} = \langle Y_r(q), p\rangle$. On the other hand, since $\hat{B}_r^{(r)}$ is linear in p, we may write it as, for example, $\hat{B}_r^{(r)} = \langle\Psi(q), p\rangle$. Then the equation to be solved splits into the n equations $\partial_\omega Y_{r,j}(q) = \Psi_j(q)$, which can all be solved because the right members have null average (it has been removed by appropriately determining ξ_r). Having determined $\chi_{r,2}$, we can construct all the rest of the triangle and move every element to the appropriate function in Table 8.4. Only the functions $B_{kr}^{(r)}$ (the first line in the right for $B_s^{(r)}$) deserve some more explanation. For $k > 1$ we observe that in view of the homological equation we have

$$
\tfrac{1}{k!} L_{\chi_{r,2}}^k \langle\omega, p\rangle = -\tfrac{1}{k!} L_{\chi_{r,2}}^{k-1} \partial_\omega \chi_{r,2} = -\tfrac{1}{k!} L_{\chi_{r,2}}^{k-1} \hat{B}_r^{(r)} \ .
$$

Thus, we calculate

$$
\tfrac{1}{k!} L_{\chi_{r,2}}^k \langle\omega, p\rangle + \tfrac{1}{(k-1)!} L_{\chi_{r,2}}^{k-1} \hat{B}_r^{(r)} = \left(\tfrac{1}{(k-1)!} - \tfrac{1}{k!}\right) L_{\chi_{r,2}}^{k-1} \hat{B}_r^{(r)} = \tfrac{k-1}{k!} L_{\chi_{r,2}}^{k-1} \hat{B}_r^{(r)} \ .
$$

The result is included in $B_{kr}^{(r)}$ (relabeled as $B_s^{(r)}$) in Table 8.4.

For the functions $h_1, \ldots, h_r$, we have

$$
\varepsilon^m \exp(\varepsilon^r L_{\chi_2}) h_m = \sum_{k \geq 0} \frac{\varepsilon^{kr+m}}{k!} L_{\chi_{r,2}}^k h_m \ , \quad m = 0, \ldots, r-1 \ .
$$

For $k = 0$, the right member is only h_m, which is left unchanged. For $k > 0$, we include every term in the sum in the corresponding function $C^{(r)}_{kr+m}$. Finally, for $s > r$ the functions $\hat{A}^{(r)}_s$, $\hat{B}^{(r)}_s$ and $C^{(r-1)}_s$ are transformed as

$$\varepsilon^s \exp\bigl(\varepsilon^r L_{\chi_2}\bigr)\hat{A}^{(r)}_s = \varepsilon^s \sum_{k\geq 0} \frac{\varepsilon^r}{k!} L^k_{\chi_{r,2}} \hat{A}^{(r)}_s \ ,$$

$$\varepsilon^s \exp\bigl(\varepsilon^r L_{\chi_2}\bigr)\hat{B}^{(r)}_s = \varepsilon^s \sum_{k\geq 0} \frac{\varepsilon^r}{k!} L^k_{\chi_{r,2}} \hat{B}^{(r)}_s \ ,$$

$$\varepsilon^s \exp\bigl(\varepsilon^r L_{\chi_2}\bigr)C^{(r-1)}_s = \varepsilon^s \sum_{k\geq 0} \frac{\varepsilon^r}{k!} L^k_{\chi_{r,2}} C^{(r-1)}_s \ .$$

Every term in the sums should be moved to the appropriate function in Table 8.4, according to the power of ε.

It remains to show that also the property (v) is satisfied, namely that every function of order ε^s is a trigonometric polynomial of degree sK. The generating functions X_r and $\chi_{r,2}$ have degree rK because they are solutions of homological equations with known members of degree rK. On the other hand, if a function f_l has degree lK, then $L_{\chi_{r,j}} f_l$ clearly has degree $(r + l)K$. Just apply this rule to every Lie derivative in the algorithm. Thus, the justification of the formal algorithm in Tables 8.2, 8.3 and 8.4 is complete.

8.3.2 Quantitative Estimates

In view of the form of the Hamiltonian, which is a polynomial in the actions p and a trigonometric polynomial in the angles, we shall consider all functions as analytic in a domain $\mathscr{D}_{\varrho,\sigma} = \Delta_\varrho \times \mathbb{T}^n_\sigma$, where the choice of ϱ is rather arbitrary, while σ is given in advance. Since we are interested in a local expansion of the Hamiltonian (8.5) around the unperturbed torus $p = 0$, we pick some value $\varrho < 1$ and keep it constant along the whole proof. Our choice is intended to make it easier to adapt the proof to more general situations, as will be sketched later in Section 8.4.

In view of our purposes, it is convenient to use the weighted Fourier norm, for which the estimates of Poisson brackets are provided in Section 6.4.3, skipping the obvious fact that the estimate of the derivatives of a polynomial of degree two does not require the sophisticated tool of Cauchy estimates.[10]

[10] Using Cauchy estimates makes things easier if one considers the general case of a Hamiltonian which is a power series in the actions p: most of the estimates in the present section are easily transported to this case. A proof that fully avoids Cauchy estimates should take into account that a derivative, both for polynomials and trigonometric polynomials, lowers an exponent in face of the function. This may be handled but requires some further adaptation of the scheme of proof, which we want to avoid.

We simplify a little the notation by writing the norm as, for example, $\|\cdot\|_{1-d}$ in place of $\|\cdot\|_{(1-d)(\varrho,\sigma)}$. In particular we write $\|\cdot\|_1$ in place of $\|\cdot\|_{\varrho,\sigma}$.

The strategy of the proof is organized as follows:

(i) Translate the formal algorithm into a sequence of estimates for the norms of all functions involved.

(ii) Isolate the problem of accumulation of small divisors and show that they obey some strict, perhaps surprising rules which make their contribution to grow not faster than geometrically with the order (no factorials).

(iii) Prove that the norms of the generating functions satisfy the convergence condition of Proposition 6.12.

All relevant estimates concerning point (i) are collected in Tables 8.2, 8.3 and 8.4, together with the formal expressions. Sections 8.3.3 and 8.3.4 provide some details. Most of the estimates are straightforward, since they represent a translation of the expressions of the formal algorithm into the corresponding inequalities for the norms using the generalized Cauchy estimates of Section 6.7. An exception is represented by the function $\langle \xi, q \rangle$ with $\xi \in \mathbb{R}^n$, which is part of the generating functions χ_1, but is not trigonometric. This requires some detailed treatment.

Point (ii) requires a separate analysis which is worked out in Sections 8.3.5 to 8.3.7, as in an *intermezzo*. Our analysis leads to the formulation of a condition of strong non-resonance named *condition* $\boldsymbol{\tau}$, formula (8.82) at the end of Section 8.3.6. A formal proposition concerning the convergence of the Kolmogorov normal form is stated at the end of Section 8.3.7.

Finally, we proceed with point (iii). Sections 8.3.9 to 8.3.11 are devoted to the details of the proof.

8.3.3 Estimates for the Generating Functions

The formal algorithm is based on the operations of solving the homological equations and of calculating Lie derivatives. For easy reference the equations are collected in Table 8.2, together with the corresponding estimates. The present section provides the details of the calculation.

In order to use the generalized Cauchy estimates of Section 6.7 we should make a choice for the restrictions of domains. We need an infinite sequence of restrictions $d_1 < d_2 < d_3 < \ldots$ tending to a limit $d < 1$. We shall actually impose the stronger condition $d \leq 1/6$. To this end, let us introduce the sequence $\{\delta_r\}_{r>0}$ defined as

$$\delta_r = \frac{1}{2\pi^2} \cdot \frac{1}{r^2}\ ; \quad \sum_{r>0} \delta_r = \frac{1}{12}\ . \tag{8.69}$$

Then set

$$d_0 = 0\ ,\quad d_r = 2(\delta_1 + \cdots + \delta_r)\ ,\quad r > 0\ , \tag{8.70}$$

which satisfies our request. Our aim is to find recurrent estimates for $\hat{H}^{(r)}$ in the domain $\mathscr{D}_{(1-d_{r-1}-\delta_r)(\varrho,\sigma)}$ and for $H^{(r)}$ in the domain $\mathscr{D}_{(1-dr)(\varrho,\sigma)}$.

Let us begin with the estimate of the solution of a homological equation. Having given a non-resonant frequency vector $\omega \in \mathbb{R}^n$, we introduce the real, non-increasing sequence $\{\alpha_r\}_{r\geq 0}$ defined as

$$\alpha_0 = 1\ ,\quad \alpha_r = \frac{1}{\Omega}\min_{0<|k|\leq rK}\bigl|\langle k,\omega\rangle\bigr|\ ,\quad \Omega = \min_j |\omega_j|\ . \tag{8.71}$$

That is, $\Omega\alpha_r$ is the smallest divisor which may appear in the solution of the homological equation for the generating functions $\chi_{r,1}$ and $\chi_{r,2}$ at step r of the normalization process. The choice of the normalization factor Ω is made so that we have $\alpha_r \leq 1$ for all $r > 0$; the definition always includes the case $|k| = 1$ for which we have exactly $\min\bigl|\langle k,\omega\rangle\bigr| = \Omega$). The property $\alpha_r \leq 1$ will be needed in the estimates for the small divisors. If the frequencies are non-resonant, then the sequence α_r does not vanish, but has zero limit for $r \to \infty$.

Lemma 8.8: *Let $\psi_s(p,q)$ be a trigonometric polynomial of degree sK, bounded in a domain $\mathscr{D}_{\varrho,\sigma}$. Let also $\omega \in \mathbb{R}^n$ be a non-resonant vector. Then the homological equation $\partial_\omega\chi_s = \psi_s$ possesses a solution with zero average satisfying*

$$\|\chi\|_{\varrho,\sigma} \leq \frac{\|\psi\|_{\varrho,\sigma}}{\Omega\alpha_s} \tag{8.72}$$

with α_s defined by (8.71) .

Proof. The solution is constructed as explained in Section 7.3.1: if $\psi_{s,k}(p)$ are the coefficients of $\psi_s(p,q)$, then the coefficients $c_k(p)$ of the Fourier expansion of χ and the corresponding estimates are

$$c_k(p) = -i\frac{\psi_{s,k}(p)}{\langle k,\omega\rangle}\ ,\quad |c_k|_\varrho \leq \frac{|\psi_{s,k}|_\varrho}{\Omega\alpha_s}\ .$$

Then we apply definition (6.64) of the weighted Fourier norm. *Q.E.D.*

The estimates for X_r and $\chi_{r,2}$ in Table 8.2 are a straightforward application of Lemma 8.8: just add the appropriate upper and lower labels.

The estimate of $|\xi_r|$ requires some more detailed considerations. In view of the special form[11] of $h_0(p) = \frac{1}{2}\langle p,p\rangle$, the equation for ξ_r is $\langle\xi,p\rangle = \overline{B_r^{(r-1)}}$, a

[11] This special form has no particular meaning. Choosing $h_0(p) = \frac{1}{2}\langle \mathsf{C}p,p\rangle$, with C a constant non-degenerate real matrix, as in the proof using the quadratic scheme of Kolmogorov, just adds a couple of constants which propagate through the estimates, with no substantial modification of the scheme.

very simple form. Therefore, by Cauchy's estimate on the centre of the disk $\Delta_{(1-d)\varrho}$ and in view of the inequality $d < 1/2$, we get

$$(8.73) \qquad |\xi_r| \le \frac{1}{(1-d)\varrho} \big\| \overline{B_r^{(r-1)}} \big\|_{1-d} \le \frac{2}{\varrho} \big\| \overline{B_r^{(r-1)}} \big\|_{1-d} \,.$$

The following lemma allows us to estimate the action of $L_{\langle \xi, q\rangle}$.

Lemma 8.9: *Let $\xi \in \mathbb{R}^n$. If $f(p,q)$ is analytic and bounded in $\mathscr{D}_1$, then*

$$\big\| L_{\langle \xi, q\rangle} f \big\|_{(1-d)} \le \frac{1}{d\varrho} |\xi| \, \|f\|_1 \ , \qquad |\xi| = |\xi_1| + \cdots + |\xi_n| \,.$$

Proof. We calculate

$$L_{\langle \xi, q\rangle} f = -\sum_{l=1}^{n} \xi_l \frac{\partial f}{\partial p_l} \,.$$

In view of the estimate (6.64) of Lemma 6.8, we get

$$\big\| L_{\langle \xi, q\rangle} f \big\|_{(1-d)} \le \frac{\|f\|_1}{d\varrho} \sum_{l=1}^{n} |\xi_l| \ ,$$

which immediately gives the desired inequality. *Q.E.D.*

8.3.4 The General Scheme of Estimates

We are now ready to construct a recurrent scheme of estimates, which is collected in Table 8.3 and in Table 8.4.

Let us come to the estimates for the intermediate Hamiltonian $\hat{H}^r$ collected in Table 8.3. The estimate of the Lie derivative L_{X_r} is evaluated using the generalized Cauchy estimates for the weighted Fourier norm, Section 6.7.2, Lemma 6.15, formula (6.66). The Lie derivative $L_{\langle \xi_r, q\rangle}$ will be estimated as in Lemma 8.9. If we know the norm $\|f\|_{1-d_{r-1}}$ of a function $f(p,q)$, then a straightforward calculation gives

$$(8.74) \qquad \begin{aligned} &\big\| L_{\chi_{r,1}} f \big\|_{1-d_{r-1}-\delta_r} \\ &\quad \le \frac{2\|X_r\|_{1-d_{r-1}}}{e\delta_r^2 \varrho\sigma} \|f\|_{1-d_{r-1}} + \frac{|\xi_r|}{\delta_r \varrho} \|f\|_{1-d_{r-1}} \\ &\quad \le \frac{2}{e} \left(\frac{\|A_r^{(r-1)}\|_{1-d_{r-1}}}{\Omega\alpha_r} + \frac{e\delta_r\sigma}{\varrho} \big\| \overline{B_r^{(r-1)}} \big\|_{1-d_{r-1}} \right) \frac{\|f\|_{1-d_{r-1}}}{\delta_r^2 \varrho\sigma} \,. \end{aligned}$$

The quantity in parentheses plays the role of $\|\chi_{r,1}\|$ in the estimate of the Poisson bracket. We should use Lemma 6.17 in order to estimate the multiple Lie derivative $\big\| L^s_{\chi_{r,1}} f \big\|_{1-d_{r-1}-\delta_r}$. It is convenient to highlight the divisor

$\alpha_r\delta_r^2$ which represents the actual trouble in the proof of convergence.[12] Hence we introduce the quantity $G_{r,1}$ as defined in Table 8.4 and write

$$\left\| L^s_{\chi_{r,1}} f \right\|_{1-d_{r-1}-\delta_r} \leq \frac{s!}{e^2}\left(\frac{G_{r,1}}{\alpha_r\delta_r^2}\right)^s \|f\|_{1-d_{r-1}} . \tag{8.75}$$

For $s > r$ all estimates of the functions $\hat{A}_s^{(r)}$ and $\hat{B}_s^{(r)}$ which enter the expansion of $\hat{H}^{(r)}$ are nothing but an application of (8.75). It is enough to replace the expressions of the functions in Table 8.3 with the corresponding estimates (8.75) for the norm (we skip the unessential factor $1/e^2$). All functions $C_s^{(r-1)}$ remain unchanged: just keep their norms.

The estimate of $\hat{B}_r^{(r)}$ in Table 8.3 must be treated separately. We proceed as in the estimate (8.74) without the term containing ξ_r, thus getting the estimate reported in Table 8.3. Only the norm $\left\|\overline{B_r^{(r-1)}}\right\|_{1-d_{r-1}}$ which appears in the quantity $G_{r,1}$ remains not fully explicited until now; indeed, it depends on the estimates for $\overline{B_s^{(r)}}$ in Table 8.4, which we are going to make explicit.

Let us come to the new Hamiltonian $H^{(r)}$; the estimates are collected in Table 8.4. The Lie derivative $L_{\chi_{r,2}}$ is evaluated using the generalized Cauchy estimates for the weighted Fourier norm, Section 6.7.2, Lemma 6.17 with the norm of χ_2 estimated in Table 8.2. Here, too, it is convenient to highlight the divisors $\alpha_r\delta_r^2$ and collect all the other factors in the quantity $G_{r,2}$. The estimates of h_r and of all functions $A_s^{(r)}$ and $C_s^{(r)}$ do not require any further comment: it is enough to replace the expressions at the beginning of Table 8.4 with the corresponding estimates for the norms. The functions $h_1, \ldots, h_{r-1}$ and their norms remain unchanged.

The estimate of $\overline{B_s^{(r)}}$ needs some care, as anticipated.[13] For $r = 0$, no problem arises, since $B_s^{(0)}$ is a known term in the initial Hamiltonian, and so is its average. For $r > 0$, proceeding by recurrence, we should take into consideration the expressions of $\hat{B}_s^{(r)}$ in Table 8.3 and of $B_s^{(r)}$ in Table 8.4 that we recall:

---

[12] The reader will note that the role of the small divisors is played not only by the quantities α_r which depend on the frequencies, but also by the restrictions δ_r of the domain which are requested by Cauchy estimates of Lie derivatives. By comparison, the reader may observe that in the case of formal expansion of first integrals of Chapter 5 the sources of divergence are both the divisors α_r and the factors coming from the exponents of polynomials, due to derivatives.

[13] It may appear natural to estimate $\left\|\overline{B_r^{(r-1)}}\right\|_{1-d} \leq \left\|B_r^{(r-1)}\right\|_{1-d}$; this is true, indeed, but not convenient since it adds a small divisor δ_r which badly affects the scheme of accumulation of divisors in the Sections 8.3.5 and 8.3.6.

$$(8.76)\qquad \hat{B}_s^{(r)} = \begin{cases} L_{\chi_{r,1}} h_{s-r} + B_s^{(r-1)}\,, & r < s < 2r\,;\\ L_{\chi_{r,1}} C_{s-r}^{(r-1)} + B_s^{(r-1)}\,, & s \geq 2r\,; \end{cases}$$

and, recalling that $k = \lfloor s/r \rfloor$ and $m = s \bmod r$,

$$(8.77)\qquad B_s^{(r)} = \begin{cases} \frac{k-1}{k!} L_{\chi_{r,2}}^{k-1} \hat{B}_r^{(r)} + \sum_{j=0}^{k-2} \frac{1}{j!} L_{\chi_{r,2}}^{j} \hat{B}_{s-jr}^{(r)}\,, & k \geq 2\,,\ m = 0\,;\\ \sum_{j=0}^{k-1} \frac{1}{j!} L_{\chi_{r,2}}^{j} \hat{B}_{s-jr}^{(r)}\,, & k \geq 1\,,\ m \neq 0\,. \end{cases}$$

The two expressions for $\hat{B}$ in (8.76) are much the same, the irrelevant difference being in the symbols h_{s-r} and $C_{s-r}^{(r-1)}$. Hence let us work only on the second one. The estimate of $\hat{B}_s^{(r-1)}$ is known from the previous step. Recalling that $\chi_{r,1} = X_r + \langle \xi_r, q \rangle$, in view of Corollary 6.16 we get

$$\left\| \overline{L_{X_r} C_{s-r}^{(r-1)}} \right\|_{1-d_{r-1}-\delta_r} \leq \frac{1}{e\delta_r \varrho\sigma} \|X_r\|_{1-d_{r-1}} \left\| C_{s-r}^{(r-1)} \right\|_{1-d_{r-1}}\,.$$

Similarly, using Lemma 8.9, we get

$$\left\| \overline{L_{\langle \xi, q\rangle} C_{s-r}^{(r-1)}} \right\|_{1-d_{r-1}-\delta_r} \leq \frac{1}{\delta_r \varrho} |\xi_r|\, \|C_{s-r}^{(r-1)}\|_{1-d_{r-1}}\,.$$

Adding the two expressions together and using the estimates for $\|X_r\|_{1-d_{r-1}}$ and $|\xi_r|$ in Table 8.2, we may write

$$\left\| \overline{L_{\chi_{r,1}} C_{s-r}^{(r-1)}} \right\|_{1-d_{r-1}-\delta_r} \leq \frac{G_{r,1}}{\alpha_r \delta_r} \left\| C_{s-r}^{(r-1)} \right\|_{1-d_{r-1}}\,,$$

with $G_{r,1}$ as in Table 8.3. Therefore, the average $\overline{\hat{B}_s^{(r)}}$ is estimated by an expression similar to, but not coinciding with, the estimate of $\hat{B}_s^{(r)}$ at the end of Table 8.3; for $r < s < 2r$ and $s \geq 2r$, respectively, we get indeed

$$\left\| \overline{\hat{B}_s^{(r)}} \right\|_{1-d_{r-1}-\delta_r} \leq \begin{cases} \frac{G_{r,1}}{\alpha_r \delta_r} \|h_{s-r}\|_{1-d_{s-r}} + \left\| \overline{B_s^{(r-1)}} \right\|_{1-d_{r-1}}\,;\\ \frac{G_{r,1}}{\alpha_r \delta_r} \|C_{s-r}^{(r-1)}\|_{1-d_{s-r}} + \left\| \overline{B_s^{(r-1)}} \right\|_{1-d_{r-1}}\,. \end{cases}$$

The fact to be remarked here is that there is a single divisor,[14] δ_r, instead of δ_r^2. We show that the same happens also for the average of $B_s^{(r)}$.

[14] The somewhat mysterious role of the single divisor is seen by considering the expression for $G_{r,1}$ in Table 8.3. As remarked in note 13, a naive estimate would produce a divisor δ_r^2; moreover, the general estimate of Lie derivatives would add a further divisor δ_r^2. This would result in an excess of factors δ_r which breaks the control of the accumulation of small divisors in the next section.

The two expressions in (8.77) are estimated the same way in view of the obvious inequality $\frac{k-1}{k!} < \frac{1}{(k-1)!}$. Separating the term $j = 0$ in the second expression for $B_s^{(r)}$, we write

$$(8.78) \qquad B_s^{(r)} = \hat{B}_s^{(r)} + L_{\chi_{r,2}}\Psi \ , \quad \Psi = \sum_{j=0}^{s} \frac{1}{j!} L_{\chi_{r,2}}^{j-1} \hat{B}_{s-jr}^{(r)} \ .$$

The average of $\hat{B}_s^{(r)}$ has already been estimated. Using Corollary 6.16, we get

$$(8.79) \qquad \left\| \overline{L_{\chi_{r,2}}\Psi} \right\|_{1-d_r-\delta_r} \leq \frac{1}{e\delta_r \varrho\sigma} \|\chi_{r,2}\|_{1-d_r-\delta_r} \|\Psi\|_{1-d_r} \ .$$

Therefore, we find again an expression similar to the general estimate, but with a single divisor, δ_r, instead of δ_r^2; thus, we may write the estimate of $\|\overline{B_s^{(r)}}\|_{1-d_r}$ as in Table 8.4. This completes the general scheme of estimates; from now on we may refer to Tables 8.2, 8.3 and 8.4.

8.3.5 The Accumulation of Small Divisors

This section contains some heuristic considerations which suggest how to isolate and control the action of small divisors. The fact to be noted and exploited is that all estimates collected in Tables 8.3 and 8.4 exhibit a common structure: they are sums of different contributions obtained by multiplying a factor $\frac{G_{r,1}}{\alpha_r \delta_r^2}$ or $\frac{G_{r,2}}{\alpha_r \delta_r^2}$ (or a power of it) by the known norm of some function. Therefore, we are led to consider the quantities $\beta_r = \alpha_r \delta_r^2$ as the small divisors to be put under observation. They form a non-increasing sequence with $\beta_1 \leq 1$; it will be convenient also to set $\beta_0 = 1$.

The accumulation of small divisors can be analyzed by exploiting the remark that we have just made. Consider any function h_s, $A_s^{(r)}$, $B_s^{(r)}$ and $C_s^{(r)}$ in the expansion of $H^{(r)}$. Paying a little attention to the recurrent scheme of estimates of Tables 8.3 and 8.4, the reader will get convinced that the norm of every function is estimated by a sum of terms with general form $c(\beta_{j_1} \cdots \beta_{j_m})^{-1}$, where c is a positive number and m is the number of divisors which have been generated through the scheme of estimates. As an example, let us see how to construct the collection of terms to be added up in the estimate for $\|\hat{A}_s^{(r)}\|$ (the domain is not relevant here). We multiply each term in $\|h_s\|$ by $\left(\frac{G_{r,1}}{\beta_r}\right)^2$; next we multiply every term in $\|B_{s-r}^{(r-1)}\|$ by

The single divisor here and in the estimate of Lemma 8.9 removes one of the factors δ_r^2, thus making the two addends in $G_{r,1}$ contain the same divisors (see also note 15).

$\frac{G_{r,1}}{\beta_r}$, taking the collection of terms in $\|A_s^{(r-1)}\|$ as is; and finally we make the union of the three collections so constructed without applying any algebraic simplification.

The heuristic argument just illustrated suggests to concentrate our attention on the products $\beta_{j_1} \cdots \beta_{j_m}$ which accumulate as denominators. More interesting, the mechanism of accumulation has nothing to do with the actual value of the divisors β: the sole information that we really need concerns the *indices* $j_1, \ldots, j_m$: if we know the indices, then we know which factors β are there.

8.3.6 The Kindness of Small Divisors

Let us open a long parenthesis. Forget for a moment the technicalities of Kolmogorov's theorem and focus attention on the propagation of lists of indices in the simplest case.

For $s \geq 1$, we call *list of indices* I_s a collection $\{j_1, \ldots, j_{s-1}\}$ of non-negative integers, with length $s-1 \geq 0$. The empty list $I_1 = \{\}$ of length 0 is allowed, as well as repeated indices. The index 0 is allowed, too, and will be used in order to pad a short list to the desired length, when needed. The lists of indices provide a full characterization of the products of small divisors: to the list $\{j_1, \ldots, j_{s-1}\}$ we associate the product $\beta_{j_1} \cdots \beta_{j_{s-1}}$. Adding any number of zeros to a list of indices is harmless, for we have set $\beta_0 = 1$.

On the set of lists of indices we introduce a partial ordering as follows. Let I, I' be lists with the same length s. We say that $I \triangleleft I'$ in case there is a permutation of the indices such that the relations $j_1 \leq j'_1, \ldots, j_s \leq j'_s$ hold true. If the lists have different lengths, then we pad the shorter one with zeros, and apply the criterion. The comparison is made easy by ordering all elements of every list (e.g., in increasing order) and comparing element by element. The order is clearly partial: for example, the criterion does not apply to the lists $\{1, 3\}$ and $\{2, 2\}$. This will be harmless, however.

A central role will be played by special lists of indices which we denote by I_s^*, with $s \geq 0$. We define

$$I_1^* = \{\}\ , \quad I_s^* = \left\{ \left\lfloor \frac{s}{s} \right\rfloor, \left\lfloor \frac{s}{s-1} \right\rfloor, \ldots, \left\lfloor \frac{s}{2} \right\rfloor \right\}\ , \quad s > 1\ . \tag{8.80}$$

In Table 8.5 we give examples of the special lists just defined.

Lemma 8.10: *For the sets of indices $I_s^* = \{j_1, \ldots, j_s\}$ the following statements hold true:*

(i) the maximal index is $j_{\max} = \left\lfloor \frac{s}{2} \right\rfloor$;

(ii) for every $k \in \{1, \ldots, j_{\max}\}$ the index k appears exactly $\left\lfloor \frac{s}{k} \right\rfloor - \left\lfloor \frac{s}{k+1} \right\rfloor$ times;

Table 8.5 The special lists I_s^* for $0 \le s \le 16$.

s	I_s^*
1	{}
2	{1}
3	{1, 1}
4	{1, 1, 2}
5	{1, 1, 1, 2}
6	{1, 1, 1, 2, 3}
7	{1, 1, 1, 1, 2, 3}
8	{1, 1, 1, 1, 2, 2, 4}
9	{1, 1, 1, 1, 1, 2, 3, 4}
10	{1, 1, 1, 1, 1, 2, 2, 3, 5}
11	{1, 1, 1, 1, 1, 1, 2, 2, 3, 5}
12	{1, 1, 1, 1, 1, 1, 2, 2, 3, 4, 6}
13	{1, 1, 1, 1, 1, 1, 1, 2, 2, 3, 4, 6}
14	{1, 1, 1, 1, 1, 1, 1, 2, 2, 2, 3, 4, 7}
15	{1, 1, 1, 1, 1, 1, 1, 1, 2, 2, 3, 3, 5, 7}
16	{1, 1, 1, 1, 1, 1, 1, 1, 2, 2, 2, 3, 4, 5, 8}

(iii) for $0 < r \le s$ we have

$$\bigl(\{r\} \cup I_r^* \cup I_s^*\bigr) \lhd I_{r+s}^* \ .$$

Proof. The claim (i) is a trivial consequence of the definition.
(ii) For each fixed value of $s > 0$ and $1 \le k \le \lfloor s/2 \rfloor$, we should determine the cardinality of the set $J_{k,s} = \{j \in \mathbb{N} : 2 \le j \le s\,,\ \lfloor s/j \rfloor = k\}$. For this purpose, we use the elementary inequalities

$$\left\lfloor \frac{s}{\lfloor s/k \rfloor} \right\rfloor \ge k \quad \text{and} \quad \left\lfloor \frac{s}{\lfloor s/k \rfloor + 1} \right\rfloor < k \ .$$

For any index $j > 0$ we have $j \in J_{k,s}$ *if and only if* $j \ge \lfloor s/(k+1) \rfloor + 1$ *and* $j \le \lfloor s/k \rfloor$; this is easily seen by rewriting the relations just given with $k+1$ in place of k. Therefore $\#J_{k,s} = \left\lfloor \frac{s}{k} \right\rfloor - \left\lfloor \frac{s}{k+1} \right\rfloor$, as claimed.
(iii) Since $r \le s$, the definition in (8.80) implies that neither $\{r\} \cup I_r^* \cup I_s^*$ nor I_{r+s}^* can include any index exceeding $\lfloor (r+s)/2 \rfloor$. Let us define some finite sequences of non-negative integers as follows:

$$\begin{aligned} R_k &= \#\{j \in I_r^* \ :\ j \le k\}\ , & S_k &= \#\{j \in I_s^* \ :\ j \le k\}\ , \\ M_k &= \#\{j \in \{r\} \cup I_r^* \cup I_s^* \ :\ j \le k\}\ , & N_k &= \#\{j \in I_{r+s}^* \ :\ j \le k\}\ , \end{aligned}$$

where $1 \le k \le \lfloor (r+s)/2 \rfloor$. When $k < r$, the property (ii) of the present lemma allows us to write

$$R_k = r - \left\lfloor \frac{r}{k+1} \right\rfloor\ , \quad S_k = s - \left\lfloor \frac{s}{k+1} \right\rfloor\ , \quad N_k = r + s - \left\lfloor \frac{r+s}{k+1} \right\rfloor\ ;$$

using the elementary estimate $\lfloor x \rfloor + \lfloor y \rfloor \leq \lfloor x + y \rfloor$, from these equations it follows that $M_k \geq N_k$ for $1 \leq k < r$. In the remaining cases, that is, when $r \leq k \leq \lfloor (r+s)/2 \rfloor$, we have that

$$R_k = r - 1 \ , \quad S_k = s - \left\lfloor \frac{s}{k+1} \right\rfloor \ , \quad N_k = r + s - \left\lfloor \frac{r+s}{k+1} \right\rfloor \ ;$$

therefore, $M_k = 1 + R_k + S_k \geq N_k$. Since we have just shown that $M_k \geq N_k$ $\forall\, 1 \leq k \leq \lfloor (r+s)/2 \rfloor$, it is now an easy matter to complete the proof. Let us first reorder both the set of indices $\{r\} \cup I_r^* \cup I_s^*$ and I_{r+s}^* in increasing order; moreover, let us recall that $\#(\{r\} \cup I_r^* \cup I_s^*) = \#I_{r+s}^* = r + s - 1$, because of the definition in (8.80). Thus, since $M_1 \geq N_1$, every element equal to 1 in $\{r\} \cup I_r^* \cup I_s^*$ has a corresponding index in I_{r+s}^*, the value of which is at least 1. Analogously, since $M_2 \geq N_2$, every index 2 in $\{r\} \cup I_r^* \cup I_s^*$ has a corresponding index in I_{r+s}^* which is at least 2, and so on up to $k = \lfloor (r+s)/2 \rfloor$. We conclude that $\{r\} \cup I_r^* \cup I_s^* \lhd I_{r+s}^*$. *Q.E.D.*

Lemma 8.11: *Let I_r and I_s be lists of length $r-1$ and $s-1$, respectively, with $1 \leq r \leq s$ (pad them with zeros, if needed). If $I_r \lhd I_r^*$ and $I_s \lhd I_s^*$, then we have*

$$r \cup I_r \cup I_s \lhd I_{r+s}^* \ .$$

The proof is elementary: use Lemma 8.10.

We come now to exploit the relation between lists of indices and products of small divisors. Let $1 = \alpha_0 \geq \alpha_2 \geq \alpha_3 \geq \ldots$ be a non-increasing arbitrary sequence of positive numbers; the sequence $\{\alpha_r\}_{r \geq 0}$ of small divisors defined by (8.71) satisfies this condition (and has also the additional property that it tends to zero). To a list I we associate the quantity $Q(I) = \prod_{j \in I} \frac{1}{\alpha_j}$. The following property is obvious:

$$\text{if} \quad I \lhd I', \quad \text{then} \quad Q(I) < Q(I') \ .$$

Let us consider the special sequence Q_s^* defined as

$$Q_s^* = \prod_{j \in I_s^*} \frac{1}{\alpha_j} \ , \qquad s \geq 1 \ . \tag{8.81}$$

We look for a sufficient condition assuring that the sequence Q_s^* grows no faster than geometrically. In view of Lemma 8.10 we may evaluate

$$\ln Q_s^* = \ln \prod_{j \in I_s^*} \frac{1}{\alpha_j} \leq -\sum_{k=1}^{s} \left(\left\lfloor \frac{s}{k} \right\rfloor - \left\lfloor \frac{s}{k+1} \right\rfloor \right) \ln \alpha_k \leq -s \sum_{k \geq 1} \frac{\ln \alpha_k}{k(k+1)} \ .$$

We are thus led to introduce

Condition τ: *The sequence $\{\alpha_r\}_{r \geq 0}$ satisfies*

$$-\sum_{r \geq 1} \frac{\ln \alpha_r}{r(r+1)} = \Gamma < \infty \ . \tag{8.82}$$

If so, then we also have the estimate $Q_s^* < e^{s\Gamma}$, that is, it grows no faster than geometrically.

Taking the sequence of small divisors associated to the frequencies $\omega \in \mathbb{R}^n$ as in (8.71), *condition* $\boldsymbol{\tau}$ *will provide the strong non-resonance condition for the validity of Kolmogorov's theorem.*

8.3.7 Small Divisors in the Algorithm of Kolmogorov

Let us say that a *function* f *owns a list of indices* $I = \{j_1, \ldots, j_k\}$ *if its estimate contains a term with the divisor* $\beta_{j_1} \cdots \beta_{j_k}$. The following considerations are immediate.

(i) If ψ_r (trigonometric polynomial of degree r) owns a list I, then the solution χ_r of the homological equation $\partial_\omega \chi_r = \psi_r$ adds a divisor α_r; the Lie derivative L_{χ_r} generates a divisor δ_r^2, thus producing an actual divisor $\beta_r = \alpha_r \delta_r^2$. We say that χ_r owns the list $\{r\} \cup I$, the union meaning concatenation of lists (repeated indices are kept in the list).

(ii) If f_s (of order s) owns a list I' and χ_r is as at point (i), then $L_{\chi_r} f_s$ owns the list $\{r\} \cup I \cup I'$.

(iii) For the multiple Lie derivative, the following scheme applies:

$$\begin{array}{rll} f_s & \text{owns} & I' \; ; \\ L_{\chi_r} f & \text{owns} & I_1 = \{r\} \cup I \cup I' \; ; \\ L_{\chi_r}^2 f & \text{owns} & I_2 = \{r\} \cup I \cup I_1 \; ; \\ L_{\chi_r}^3 f & \text{owns} & I_3 = \{r\} \cup I \cup I_2 \; ; \end{array}$$

and so on, so that we may proceed by recurrence.

Hence points (i) and (ii) contain all the relevant information about the mechanism of accumulation of divisors.

Now we may go back to the estimates for the normal form of Kolmogorov and try to identify the worst possible product of divisors in every coefficient of every function. To this end, in view of the discussion of the previous Section 8.3.6, we focus attention on the indices and look for two informations, namely

(i) the *number* of divisors β_j;

(ii) a *selection rule* specifying which lists of coefficients may occur.

Definition 8.12: *For all integers* $r \geq 0$ *and* $s > 0$, *we introduce the collection of lists*

$$\mathscr{I}_{r,s} = \bigl\{ I = \{j_1, \ldots, j_{s-1}\} \colon 0 \leq j_m \leq \min\{r, \lfloor s/2 \rfloor\} \, , \; I \lhd I_s^* \bigr\} \, . \tag{8.83}$$

Table 8.6 The number of indices, the selection rule and the upper bound $T_{r,s}$ (defined by (8.86)) for the function h_r for $r \geq 0$ and for the functions $A_s^{(r)}$, $B_s^{(r)}$, $\overline{B_s^{(r)}}$ and $C_s^{(r)}$ for $1 \leq r < s$.

function	#of indices	selection rule	bounded by
H_0	0	$\{\}$	$T_{0,0}$;
$A_s^{(r)}$	$2s-2$	$\mathscr{I}_{r,s} \cup \mathscr{I}_{r,s}$	$T_{r,s}^2$;
$\alpha_r \delta_r \overline{B_s^{(r)}}$	$2s-2$	$\mathscr{I}_{r,s} \cup \mathscr{I}_{r,s}$	$T_{r,s}^2$;
$B_s^{(r)}$	$2s-1$	$\{r\} \cup \mathscr{I}_{r,s} \cup \mathscr{I}_{r,s}$	$\frac{1}{\beta_r} T_{r,s}^2$;
h_r, $C_s^{(r)}$	$2s$	$\{r\} \cup \{r\} \cup \mathscr{I}_{r,s} \cup \mathscr{I}_{r,s}$	$\frac{1}{\beta_r^2} T_{r,s}^2$;

Lemma 8.13: *The following properties hold true: for $r' < r$ we have*

$$\mathscr{I}_{r',s} \subset \mathscr{I}_{r,s} \qquad \text{for} \quad r' < r \; ; \tag{8.84}$$

for $0 \leq r \leq s$ we have

$$\{r\} \cup \mathscr{I}_{r-1,r} \cup \mathscr{I}_{r,s} \subset \mathscr{I}_{r,r+s} \; . \tag{8.85}$$

The first property is obvious. For the second, check that the number of indices in the two members is the same; the maximum value of the indices is respected; the selection rule is respected in view of property (iii) of Lemma 8.10.

Our goal now is to identify the list of divisors owned by every function, with particular care for the generating functions.

Lemma 8.14: *Every function in the sequence $H^{(r)}$ of Hamiltonians owns only lists of indices as specified in Table 8.6.*

Proof. For $r = 0$, there are no divisors, therefore every function in $H^{(0)}$ owns an empty list, namely $\{\}$; by padding every list with an appropriate number of zeros, the table is correct, because $\mathscr{I}_{0,s}$ is a list of $s-1$ zeros. For $r > 0$, we may proceed by induction, supposing that the table applies up to step $r-1$. We follow the path of the formal algorithm of Tables 8.2, 8.3 and 8.4, taking into account that new lists of indices are generated by recurrence as described in Section 8.3.5. The part concerning the generating functions gives[15] (the verb "owns" is used as a shortening of the expression "owns only lists in the set")

$$\begin{aligned}
A_r^{(r-1)} &\quad \text{owns} \quad \mathscr{I}_{r-1,r} \cup \mathscr{I}_{r-1,r}\ ;\\
\alpha_{r-1}\delta_{r-1}\overline{B_r^{(r-1)}} &\quad \text{owns} \quad \mathscr{I}_{r-1,r} \cup \mathscr{I}_{r-1,r}\ ;\\
G_{r,1} &\quad \text{owns} \quad \mathscr{I}_{r-1,r} \cup \mathscr{I}_{r-1,r}\ ;\\
\chi_{r,1} &\quad \text{owns} \quad \{r\} \cup \mathscr{I}_{r-1,r} \cup \mathscr{I}_{r-1,r}\ ;\\
\hat{B}_1^{(1)},\, G_{r,2} &\quad \text{owns} \quad \{r\} \cup \mathscr{I}_{r-1,r} \cup \mathscr{I}_{r-1,r}\ ;\\
\chi_{r,2} &\quad \text{owns} \quad \{r\} \cup \{r\} \cup \mathscr{I}_{r-1,r} \cup \mathscr{I}_{r-1,r}\ .
\end{aligned}$$

Using the information on $\chi_{r,1}$ and selecting the worst case among all terms in the sum in Table 8.3, we get (multiplication by 2 means the union of identical lists)

$$\begin{aligned}
&\hat{A}_s^{(r)} \text{ owns } 2 \times \bigl(\{r\} \cup \mathscr{I}_{r-1,r}\bigr) \cup \mathscr{I}_{r,s-2r+1} \subset \mathscr{I}_{r,s}\ ;\\
&\hat{B}_s^{(r)} \text{ owns } \{r\} \cup \mathscr{I}_{r-1,r} \cup \mathscr{I}_{r-1,s-r} \subset \{r\} \cup \mathscr{I}_{r,s}\ .
\end{aligned}$$

The inclusion relations follow from Lemma 8.13.
Coming to Table 8.4, use the information on $\chi_{r,2}$; for all allowed values of j in the sums, get

$$\begin{aligned}
&h_r \text{ owns } \{r\} \cup \{r\} \cup \mathscr{I}_{r-1,r} \cup \mathscr{I}_{r-1,r} \subset \{r\} \cup \{r\} \cup \mathscr{I}_{r,r}\ ;\\
&A_s^{(r)} \text{ owns } 2j \times \bigl(\{r\} \cup \mathscr{I}_{r-1,r}\bigr) \cup \mathscr{I}_{r,s-jr} \subset \{r\} \cup \mathscr{I}_{r,s}\ ;\\
&B_s^{(r)} \text{ owns } 2j \times \bigl(\{r\} \cup \mathscr{I}_{r-1,r}\bigr) \cup \{r\} \cup \mathscr{I}_{r,s-jr} \subset \{r\} \cup \{r\} \cup \mathscr{I}_{r,r}\ ;\\
&C_s^{(r)} \text{ owns } 2j \times \bigl(\{r\} \cup \mathscr{I}_{r-1,r}\bigr) \cup \{r\} \cup \{r\} \cup \mathscr{I}_{r,s-jr} \subset \{r\} \cup \{r\} \cup \mathscr{I}_{r,r}\ .
\end{aligned}$$

Here, too, the inclusion relations follow from Lemma 8.13 and hold true for every term in the sums. This completes the induction. *Q.E.D.*

We associate to the collections of lists $\mathscr{I}_{r,s}$ introduced in Definition 8.12 the double-indexed sequence of positive numbers

$$T_{r,s} = \prod_{j \in \mathscr{I}_{r,s}} \beta_j^{-1}\ , \quad r \ge 0\ ,\ s > 0\ . \tag{8.86}$$

---

[15] Here the claims concerning the functions χ should be understood as follows: after solving the homological equations and making a Lie derivative L_χ, a divisor $\alpha_r\delta_r^2$ is added, as indicated by $\{r\}$. A little care should be exerted with the function $\overline{B_r^{r-1}}$ in $\chi_{r,1}$. At first sight it seems that there is a divisor δ_r^2 in excess which breaks the scheme. However, one of the divisors δ_r disappears as explained in Section 8.3.4 and reported in Table 8.3; the second one disappears in view of Lemma 8.9 (see also notes 13 and 14).

From Lemma 8.13 we immediately get the inequalities

$$\text{(8.87)}\qquad \begin{aligned} &T_{r',s} \le T_{r,s} &&\text{for } 0 \le r' < r\ ;\\ &\frac{1}{\beta_r} T_{r-1,r} T_{r,s} \le T_{r,r+s} &&\text{for } 0 \le r \le s. \end{aligned}$$

The bounds for the products of small divisors in Table 8.6 follow.

8.3.8 A Detailed Statement

Having settled the scheme for the control of small divisors, including the precise statement of condition $\boldsymbol{\tau}$, we can give a formal statement; the proof is completed in the next two sections.

Proposition 8.15: *Consider the Hamiltonian*

$$H(q,p) = \langle \omega, p\rangle + h_0(p) + \sum_{s\ge 1} \varepsilon^s \bigl[A_s(q) + B_s(p,q) + C_s(p,q)\bigr]\ ,$$

with $h_0(p) = \frac{1}{2}\langle p,p\rangle$, on a neighbourhood $\mathscr{D}_{\varrho,\sigma} = \Delta_\varrho \times \mathbb{T}^n$ of a non-resonant unperturbed torus $p = 0$, with $0 < \varrho < 1$ and $\sigma > 0$. Assume:

(i) The functions $A_s(q)$, $B_s(p,q)$ and $C_s(p,q)$ are trigonometric polynomials of degree sK for some fixed $K > 0$; moreover, $A_s(q)$ is independent of p, $B_s(q)$ is linear in p and $C_s(p,q)$ is quadratic in p.

(ii) All quantities $\|A_s\|_{\varrho,\sigma}$, $\|B_s\|_{\varrho,\sigma}$ and $\|C_s\|_{\varrho,\sigma}$ are bounded.

(iii) The frequencies ω satisfy condition $\boldsymbol{\tau}$ as in (8.82) with a constant Γ.

Let

$$\text{(8.88)}\qquad \begin{aligned} E &= \max_{s>0}\bigl(\|H_0\|_1,\ \|A_s^{(0)}\|_1,\ \|B_s^{(0)}\|_1,\ \|C_s^{(0)}\|_1\bigr)\ ,\\ M &= \max\left(1,\ \frac{eE}{\varrho\sigma}(1+2E\sigma)\right)\ . \end{aligned}$$

Then there exists a positive constant ε^* such that the following holds true: if $\varepsilon < \varepsilon^*$, then there exists a canonical transformation $(q,p) = \mathscr{C}(q',p')$ satisfying

$$\text{(8.89)}\qquad |p_j - p'_j| < d^{\tau+1}\varrho\ ,\quad |q_j - q'_j| < d^{\tau+1}\sigma\ ,\quad j = 1,\ldots,n$$

for all $(q',p') \in \mathbb{T}^n_{(1-3d)\sigma} \times \Delta_{(1-3d)\varrho}(0)$ which gives the Hamiltonian the Kolmogorov normal form (8.2). Moreover, the normal form is expanded as

$$H(q,p) = \langle \omega, p\rangle + h_0(p) + \sum_{s\ge 1} \varepsilon^s C_s(p,q)$$

with $\|C_s\|_{(1-d)} \le b^{s-1}F$ with some positive constants b and F.

8.3.9 Recurrent Estimates

Let us look again at Tables 8.3 and 8.4, and take into account the control of divisors which we have found in the previous section. Our goal is now to obtain a manageable estimate of the generating sequences. We shall prove that the norms are bounded geometrically. It is worthwhile to recall also that we have actually made an expansion in a small parameter ε, so that a geometric growth of the generating sequences $\chi_{r,1}$ and $\chi_{r,2}$ as, for example, M^r will only mean that the threshold ε^* for the validity of the theorem is $\varepsilon^* \sim M^{-1}$.

We go back to considering the Hamiltonian (8.5), namely

$$H(q,p) = \langle \omega, p\rangle + h_0(p) + \sum_{s\geq 1} \varepsilon^s \bigl[A_s(q) + B_s(p,q) + C_s(p,q)\bigr] \ ,$$

with $h_0(p) = \frac{1}{2}\langle p,p\rangle$, on a neighbourhood $\mathscr{D}_{\varrho,\sigma} = \Delta_\varrho \times \mathbb{T}^n$ of a non-resonant unperturbed torus $p = 0$; we may assume $\varrho \leq 1$. Possibly with a rescaling of ε we may also assume that all quantities $\|A_s\|_1$, $\|B_s\|_1$ and $\|C_s\|_1$ are bounded.

Lemma 8.16: *Let the Hamiltonian be as in Proposition 8.15 and let the constants E and M be defined as in (8.88). Let the double-indexed sequences $\nu_{r,s}$ and $\hat{\nu}_{r,s}$ be defined as*

$$
\begin{aligned}
&\nu_{0,s} = 1 && \text{for } s \geq 0 \ ,\\
&\hat{\nu}_{r,s} = \sum_{j=0}^{\min(\lfloor s/r\rfloor,2)} \nu_{r-1,r}^{j}\,\nu_{r-1,s-jr} && \text{for } r \geq 1\ ,\ s \geq 0\ ,\\
&\nu_{r,s} = \sum_{j=0}^{\lfloor s/r\rfloor} \nu_{r-1,r}^{j}\,\hat{\nu}_{r,s-jr} && \text{for } r \geq 1\ ,\ s \geq 0\ .
\end{aligned}
\tag{8.90}
$$

Then for $1 \leq r < s$ the following estimates hold true

$$
\begin{aligned}
\|A_s^{(r)}\|_{1-d_r} &\leq \nu_{r,s} M^{s-2} T_{r,s}^2 E \ ,\\
\|B_s^{(r)}\|_{1-d_r} &\leq \nu_{r,s} M^{s-1} \frac{1}{\beta_r} T_{r,s}^2 E \ ,\\
\|h_r\|_{1-d_r} &\leq \nu_{r-1,r} M^{r} \frac{1}{\beta_r^2} T_{r,r}^2 E \ ,\\
\|C_s^{(r)}\|_{1-d_r} &\leq \nu_{r,s} M^{s} \frac{1}{\beta_r^2} T_{r,s}^2 E \ ,\\
G_{r,1} &\leq \nu_{r-1,r} T_{r-1,r}^2 M^{r-1} \ ,\\
G_{r,2} &\leq \nu_{r-1,r} M^{r} \frac{1}{\beta_r} T_{r-1,r}^2 \ ,
\end{aligned}
$$

with d_r, δ_r defined by (8.69) and (8.70) and with $T_{r,s}$ defined by (8.86).

Proof. The estimates are clearly true for $r = 0$ in view of $T_{0,s} = 1$ and $\nu_{0,s} = 1$. For $r \geq 1$ we use induction. Assuming the claim be true up to $r - 1$, we have

$$\alpha_r \delta_r \left\| \overline{B_r^{(r-1)}} \right\|_{1-d_{r-1}} \leq \alpha_r \delta_r^2 \varrho \, \nu_{r-1,r} T_{r-1,r}^2 M^{r-2} E^2 \ .$$

Thus, we get

$$G_{r,1} \leq \nu_{r-1,r} T_{r-1,r}^2 M^{r-2} \frac{eE}{\varrho\sigma}(1 + 2E\sigma) \leq \nu_{r-1,r} T_{r-1,r}^2 M^{r-1} \ .$$

Now we estimate the intermediate Hamiltonian $\hat{H}^{(r)}$. For $\hat{A}_s^{(r)}$, remarking that there is no relevant difference between the estimates of h_r and of $C_s^{(r-1)}$, we get

$$\begin{aligned} \|\hat{A}_s^{(r)}\|_{1-d_{r-1}-\delta_r} &\leq \nu_{r-1,r}^2 \left(\frac{1}{\beta_r} T_{r-1,r}^2 M^{r-1} \right)^2 \times \nu_{r-1,s-2r} T_{r-1,s-2r}^2 M^{s-2r} E \\ &\quad + \nu_{r-1,r} \frac{1}{\beta_r} T_{r-1,r}^2 M^{r-1} \times \nu_{r-1,s-r} T_{r-1,s-r}^2 M^{s-r-1} E \\ &\quad + \nu_{r-1,s} T_{r-1,s} M^{s-2} E \\ &\leq \hat{\nu}_{r,s} T_{r,s} M^{s-2} E \ , \end{aligned}$$

where the definition (8.90) of $\hat{\nu}_{r,s}$ has been used. Next we have

$$\begin{aligned} \|\hat{B}_s^{(r)}\|_{1-d_{r-1}-\delta_r} &\leq \frac{1}{\beta_r} T_{r-1,r}^2 M^{r-1} \nu_{r-1,r} \times \frac{1}{\beta_r} \nu_{r-1,s-r} T_{r-1,s-r}^2 M^{s-r} E \\ &\quad + \nu_{r-1,s} T_{r-1,s}^2 M^{s-1} E \\ &\leq \frac{1}{\beta_r} T_{r-1,s}^2 M^{s-1} E \hat{\nu}_{r,s} \ . \end{aligned}$$

Coming to the transformed Hamiltonian $H^{(r)}$, we first have

$$G_{r,2} \leq \hat{\nu}_{r-1,r} \frac{1}{\beta_r} T_{r-1,r}^2 M^r \leq \nu_{r-1,r} \frac{1}{\beta_r} T_{r-1,r}^2 M^r \ .$$

Then we estimate the functions. For h_r we get

$$\begin{aligned} \|h_r\|_{1-d_r} &\leq \frac{1}{\beta_r^2} T_{r-1,s}^2 \hat{\nu}_{r-1,r} M^r E + \nu_{r-1,r} T_{r-1,s}^2 M^r E \\ &\leq \frac{1}{\beta_r^2} \nu_{r,r} T_{r-1,s}^2 M^r E. \end{aligned}$$

For $A_s^{(r)}$, recalling that $k = \lfloor s/r \rfloor$, we get

$$\begin{aligned}\|A_s^{(r)}\|_{1-d_r} &\le \sum_{j=0}^{k-1} \left(\frac{1}{\beta_r} T_{r-1,s}^2 \hat{\nu}_{r,s} M^r \right)^j \times \hat{\nu}_{r,s} T_{r,s-jr} M^{s-jr-2} E \\ &\le \nu_{r,s} T_{r,s} M^{s-2} E ,\end{aligned}$$

where the definition of $\nu_{r,s}$ has been used. For $B_s^{(r)}$, still recalling $k = \lfloor s/r \rfloor$, we get

$$\begin{aligned}\|B_s^{(r)}\|_{1-d_r} &\le \sum_{j=0}^{k-1} \left(\frac{1}{\beta_r} T_{r-1,s}^2 \hat{\nu}_{r,s} M^r \right)^j \times \frac{1}{\beta_r} \hat{\nu}_{r,s} T_{r,s-jr} M^{s-jr-2} E \\ &\le \frac{1}{\beta_r} \nu_{r,s} T_{r,s} M^{s-1} E .\end{aligned}$$

For $C_s^{(r)}$, recalling once more that $k = \lfloor s/r \rfloor$, we get

$$\begin{aligned}\|C_s^{(r)}\|_{1-d_r} \le& \left(\frac{1}{\beta_r} T_{r-1,r}^2 \hat{\nu}_{r,s} M^r \right)^k \times T_{r,s-kr} M^{s-kr} E \\ &+ \sum_{j=0}^{k-1} \left(\frac{1}{\beta_r} T_{r-1,r}^2 \hat{\nu}_{r,s} M^r \right)^j \times \frac{1}{\beta_r^2} \hat{\nu}_{r,s-jr} T_{r,s-jr} M^{s-jr} E \\ \le& \frac{1}{\beta_r^2} \nu_{r,s} T_{r,s} M^s E .\end{aligned}$$

Q.E.D.

8.3.10 Completion of the Proof

In order to complete the proof of Proposition 8.15 we must show that the generating functions $\chi_{r,1}$ and $\chi_{r,2}$ satisfy the convergence condition of Proposition 6.12. To this end, let us look at the scheme of estimates of Tables 8.2, 8.3 and 8.4, from which we see that we may use the estimate

$$\|\chi_{r,1}\|_{1-d_{r-1}} \le \frac{G_{r,1}}{\alpha_r} , \quad \|\chi_{r,2}\|_{1-d_{r-1}} \le \frac{G_{r,2}}{\alpha_r} .$$

Therefore, from the estimates of Lemma 8.16, we have

$$\begin{aligned}\|\chi_{r,1}\|_{1-d_{r-1}} &\le \frac{1}{\alpha_r} \nu_{r-1,r} T_{r-1,r}^2 M^{r-1} , \\ \|\chi_{r,2}\|_{1-d_{r-1}} &\le \frac{1}{\alpha_r \beta_r} T_{r-1,r}^2 \nu_{r-1,r} M^r .\end{aligned} \tag{8.91}$$

Thus, the problem is to show that the quantities $\nu_{r,s}$ and $T_{r,s}$ can be bounded geometrically.[16] This would be enough because we should not forget that the generating functions are actually $\varepsilon^r \chi_{r,1}$ and $\varepsilon^r \chi_{r,2}$.

Lemma 8.17: *Let the sequence $\{\alpha_r\}_{r\geq 1}$, introduced by (8.71), satisfy condition $\boldsymbol{\tau}$ and the sequence $\{\delta_r\}_{r\geq 1}$ be defined as in (8.70). Then, the sequence $\{T_{r,s}\}_{r\geq 0\,,\,s\geq 0}$ defined by (8.86) is bounded by*

$$T_{r,s} \leq \frac{1}{\alpha_s \delta_s^2} T_{r,s} \leq \left(2^{15} e^{\Gamma}\right)^s \qquad \text{for } r \geq 1\,,\ s \geq 1\,.$$

Proof. Since $\alpha_s \delta_s^2 < 1$ (see (8.71) and (8.70)), it is enough to prove the second part of the inequality stated in the lemma, that is, $T_{r,s}/(\alpha_s \delta_s^2) \leq A^s e^{s\Gamma}$ for $r \geq 1\,,\ s \geq 1$. Starting from Definition (8.86), using properties (i) and (ii) of Lemma 8.10, the selection rule in (8.83) and the decreasing character of the sequence $\{\alpha_s \delta_s^2\}_{s\geq 1}$, we get

$$\frac{T_{r,s}}{\alpha_s \delta_s^2} = \frac{1}{\alpha_s \delta_s^2} \max_{I\in \mathscr{J}_{s,s}} \prod_{j\in I\,,\,j\geq 1} \frac{1}{\alpha_j \delta_j^2} \leq \prod_{j\in \{s\}\cup I_s^*\,,\,j\geq 1} \frac{1}{\alpha_j \delta_j^2}\ .$$

Starting from this estimate, we have

$$\begin{aligned}
\log \frac{T_{r,s}}{\alpha_s \delta_s^2} &\leq -\log(\alpha_s \delta_s^2) - \sum_{k=1}^{\lfloor s/2 \rfloor} \left(\left\lfloor \frac{s}{k} \right\rfloor - \left\lfloor \frac{s}{k+1} \right\rfloor \right) \log(\alpha_k \delta_k^2) \\
&\leq -\sum_{k=1}^{s} \left(\left\lfloor \frac{s}{k} \right\rfloor - \left\lfloor \frac{s}{k+1} \right\rfloor \right) (\log \alpha_k + 2\log \delta_k) \qquad (8.92)\\
&\leq -s \sum_{k\geq 1} \frac{\log \alpha_k + 2\log \delta_k}{k(k+1)} = s\left(\Gamma - \sum_{k\geq 1} \frac{2\log \delta_k}{k(k+1)} \right) ,
\end{aligned}$$

where we used properties (i) and (ii) of Lemma 8.10, the fact that the sequence $\{\alpha_s \delta_s^2\}_{s\geq 1}$ is decreasing and the condition $\boldsymbol{\tau}$ in (8.82). A numerical evaluation is done as follows. Recalling the definition of δ_r we get

---

[16] In the discussion that follows, coherently with the attitude of the present notes, there are also explicit estimates of the constants. However, one should keep in mind that also the evaluation of the constants is made with estimates which are typically far from being optimal. The suggestion is only that actual values can be found, and possibly (certainly) be improved with respect to the ones given here.

$$
\begin{aligned}
-\sum_{k\geq 1} \frac{2\log \delta_k}{k(k+1)} &= 2\sum_{k\geq 1} \frac{\log \frac{8\pi^2}{3} + 2\log k}{k(k+1)} \\
&< 2\log\frac{8\pi^2}{3} + 4\left(\frac{\log 2}{6} + \int_2^{\infty} \frac{\log x\,dx}{x^2}\right) \\
&< 15\log 2\ ,
\end{aligned}
$$

where the known relation $\sum_{k\geq 1} 1/[k(k+1)] = 1$ is used. Putting this estimate into (8.92), we conclude the proof. *Q.E.D.*

It remains to verify that the sequence $\nu_{r-1,r}$ is bounded geometrically. This requires some boring calculations, worked out in Section 8.3.11 where we prove the inequality $\nu_{r-1,r} < 4^{r-1}$.

Finally, we have shown that all terms in the estimates (8.91) for the generating functions are bounded geometrically. Therefore, setting $d = 2\sum_r \delta_r$ as in (8.70), there exists a positive ε^* such that for $\varepsilon < \varepsilon^*$, we have

$$
\frac{1}{d^2\varrho\sigma}\sum_r \varepsilon^r \|\chi_{r,1}\|_{1-d} < \frac{1}{2e}\ , \quad \frac{1}{d^2\varrho\sigma}\sum_r \varepsilon^r \|\chi_{r,2}\|_{1-d} < \frac{1}{2e}\ ,
$$

namely the convergence condition of Proposition 6.12. Having chosen $d = 1/6$, the canonical transformation is analytic in a domain, for example, $\mathscr{D}_{\varrho/2,\sigma/2}$, and the sequence of Hamiltonians $H^{(r)}$ converges to an analytic Hamiltonian,

$$
H^{(\infty)} = \langle \omega, p\rangle + \mathcal{O}(p^2),
$$

in Kolmogorov normal form. This concludes the proof that there exists a perturbed invariant torus carrying quasiperiodic motions with frequencies ω, which is close to the unperturbed one.

8.3.11 Estimate of the Sequence ν

With a minor change, let us replace the sequence with the majorant one (let $k = \lfloor s/r \rfloor$):

$$
\begin{aligned}
&\nu_{0,s} = 1 && \text{for } s \geq 0\ , \\
&\hat\nu_{r,s} = \sum_{j=0}^{k} \nu_{r-1,r}^{j}\,\nu_{r-1,s-jr} && \text{for } r\geq 1\ ,\ s\geq 0\ , \\
&\nu_{r,s} = \sum_{j=0}^{k} \nu_{r-1,r}^{j}\,\hat\nu_{r,s-jr} && \text{for } r\geq 1\ ,\ s\geq 0\ .
\end{aligned}
\tag{8.93}
$$

Note that we are interested in an estimate of the norms of the generating functions $\chi_{r,1}$ and $\chi_{r,2}$; therefore, it is enough to have an estimate for the elements $\nu_{r-1,r}$.

Two remarks are in order here. The first one is that $\nu_{r,s} \geq \hat{\nu}_{r,s}$, for by extracting the term $j = 0$ in the sum for $\nu_{r,s}$ we have $\nu_{r,s} = \hat{\nu}_{r,s} + \ldots$, the dots standing for a non-negative quantity. The second is that for $r > s$ we have $k = \lfloor s/r \rfloor = 0$; hence the sums reduce to the sole term $j = 0$. Thus, we get

$$\nu_{r,s} = \hat{\nu}_{r,s} = \nu_{r-1,s} = \cdots = \nu_{s,s} \; . \tag{8.94}$$

For $r \leq s$ we may simplify the sequences by calculating

$$\begin{aligned}
\hat{\nu}_{r,s} &= \nu_{r-1,s} + \nu_{r-1,r} \sum_{m=0}^{k-1} \nu_{r-1,r}^{m} \nu_{r-1,(s-r)-mr} = \nu_{r-1,s} + \nu_{r-1,r} \hat{\nu}_{r,s-r} \; , \\
\nu_{r,s} &= \hat{\nu}_{r-1,s} + \nu_{r-1,r} \sum_{m=0}^{k-1} \nu_{r-1,r}^{m} \hat{\nu}_{r,(s-r)-mr} = \hat{\nu}_{r-1,s} + \nu_{r-1,r} \nu_{r,s-r} \; .
\end{aligned}$$

As a matter of fact, we may discard the first formula defining $\hat{\nu}_{r,s}$, for from the second formula we get

$$\nu_{r,s} \leq \nu_{r-1,s} + \nu_{r-1,r} \nu_{r,s-r} \; .$$

Using the latter formula together with (8.94), we find a majorant of the sequence $\nu_{r,s}$, namely $\nu_{r,s} \leq \eta_{r,s}$, by setting

$$\begin{aligned}
&\eta_{0,0} = 0 \; , \quad \eta_{0,s} = 1 && \text{for} \quad s \geq 0 \; , \\
&\eta_{r,r} = \eta_{r,r-1} = \cdots = \eta_{r,0} && \text{for} \quad r \geq 1 \; , \\
&\eta_{r,s} = \eta_{r-1,s} + \eta_{r-1,r} \eta_{r,s-r} && \text{for} \quad r \leq s \; .
\end{aligned} \tag{8.95}$$

We show now that we have $\eta_{r-1,r} \leq \lambda_r$, where $\{\lambda_r\}_{r \geq 1}$ is *Catalan's sequence*, defined as

$$\lambda_1 = 1 \; , \quad \lambda_r = \sum_{j-1}^{r-1} \lambda_j \lambda_{r-j} \quad \text{for} \quad r > 1 \; . \tag{8.96}$$

To this end, let us apply $r - 1$ times the definition (8.95), so that we get in sequence

$$\begin{aligned}
\eta_{r-1,r} &= \eta_{r-2,r} + \eta_{r-2,r-1} \eta_{r-1,1} \; , \\
\eta_{r-2,r} &= \eta_{r-3,r} + \eta_{r-3,r-2} \eta_{r-2,2} \; , \\
&\cdots\cdots\cdots\cdots\cdots \\
\eta_{2,r} &= \eta_{1,r} + \eta_{1,2} \eta_{2,r-2} \; , \\
\eta_{1,r} &= \eta_{0,r} + \eta_{0,1} \eta_{1,r-1} \; .
\end{aligned}$$

Then we get

$$
\begin{aligned}
\eta_{r-1,r} &= \eta_{0,1}\eta_{1,r-1} + \eta_{1,2}\eta_{2,r-2} + \cdots + \eta_{r-2,r-1}\eta_{r-1,1} \\
&= \sum_{j=1}^{r-1} \eta_{j-1,j}\eta_{j,r-j} \ .
\end{aligned} \tag{8.97}
$$

Now, from (8.95) we also have, trivially, $\eta_{r-1,s} \leq \eta_{r,s}$. If $j < r-j$, then we have the chain of inequalities

$$\eta_{j,r-j} < \eta_{j+1,r-j} < \cdots < \eta_{r-j-1,r-j} \ ;$$

on the other hand, if $j \geq r-j$, then, always from (8.86), we have the chain of equalities

$$\eta_{j,r-j} = \eta_{j-1,r-j} = \cdots = \eta_{r-j-1,r-j} \ .$$

Thus, we may replace (8.97) with

$$\eta_{r-1,r} \leq \sum_{j=1}^{r-1} \eta_{j-1,j}\eta_{r-j-1,r-j} \ .$$

We conclude that $\nu_{r-1,r} \leq \eta_{r-1,r} \leq \lambda_r$.

The last step is to show that the Catalan's sequence is bounded geometrically. It is known that

$$\lambda_r = \frac{2^{r-1}(2r-3)!!}{r!} \leq \frac{4^{r-1}}{r} \ , \tag{8.98}$$

where the usual notation of the *semifactorial* $(2n+1)!! = 1 \cdot 3 \cdot \cdots \cdot (2n+1)$ has been used. Let us see how the formula can be proved, using the method of generating function due to Gauss. Let the function $g(z)$ be defined as $g(z) = \sum_{r \geq 1} \lambda_r z^r$, so that $\lambda_r = g^{(r)}(0)/r!$. Then it is necessary to check that the recurrent definition (8.96) is equivalent to the equation $g = z + g^2$. By repeated differentiation of the latter equation, one readily finds

$$g' = \frac{1}{1-2g} \ , \ \ldots \ , \ g^{(r)} = \frac{2^{r-1}(2r-3)!!}{(1-2g)^{2r-1}}$$

(check by induction). From this, (8.98) follows. The inequality is just a rough estimate: we do not need to find a better one for our purposes.

8.4 Concluding Remarks

We close this chapter with a couple of comments. The first one concerns the extension of the proof to the general case of an analytic Hamiltonian, not just a polynomial one. The second concerns the method of Lindstedt.

8.4.1 A General Scheme

The scheme of proof of the previous section may be extended to a Hamiltonian in the form $H(p,q) = H_0(p) + \varepsilon H_1(p,q)$ of the general problem of dynamics. There are three differences to be taken into account, namely:

(i) In general, the quadratic part of the unperturbed Hamiltonian will not have the particularly simple form, $H_0 = \frac{1}{2}\langle p,p\rangle$, of a system of unperturbed rotators.

(ii) In the neighbourhood of an unperturbed torus the Hamiltonian must be expanded in power series of the actions, convergent in a polydisk $\Delta_{\varrho'}$ for some $\varrho' > 0$.

(iii) The power expansion in ε of the perturbation $H_1(p,q)$ has coefficients which are not restricted to be trigonometric polynomials: the Fourier expansion may actually contain infinite terms, but it must be convergent in a complexified torus $\mathbb{T}^n_{\sigma'}$ for some positive σ'.

The difference at point (i) is not a really serious problem. In general, we have $h_0 = \frac{1}{2}\langle \mathsf{C}p,p\rangle$ with a real symmetric matrix C, and we add the hypothesis that C is non-degenerate.[17]

The difference at point (ii) is not really harmful. The immediate consequence is that the arbitrary parameter ϱ must satisfy $0 < \varrho < \varrho'$, so that the norms of the functions are bounded, as requested by all estimates. The complexity of the algorithm will increase, because the polynomial degree also must be taken into account. The first property to be exploited is that the Lie derivative $L_{\chi_{r,1}}$ lowers by one the degree of any polynomial in p. Therefore, the scheme of Table 8.3 becomes definitely more complex. For example, if $f_s(p,q)$ is a homogeneous polynomial of degree s, then

$$\exp\bigl(L_{\chi_{r,1}}\bigr) f_s = f_s + L_{\chi_{r,1}} f_s + \frac{1}{2} L^2_{\chi_{r,1}} f_s + \cdots + \frac{1}{s!} L^s_{\chi_{r,1}} f_s$$

is a sum of homogeneous polynomials in p of decreasing degree from s to 0. This requires a major rearrangement of Table 8.3. The derivative $L_{\chi_{r,2}}$ keeps the degree of a polynomial; therefore, Table 8.4 should be modified only in order to take into account that there are polynomials of any degree.[18]

The difference at point (iii) may be puzzling, because the infinite Fourier expansion of the perturbation destroys the control on the accumulation of

[17] For a study concerning a particular class of Hamiltonian with a degenerate quadratic part, see [71], [72] and [162].

[18] As an exercise, it may be useful to consider the case of a Hamiltonian which is a polynomial of degree 3 in p. It takes only a moment to realize that the normalization algorithm never increases the polynomial degree. This may suggest how to deal with the general case of a Hamiltonian which is an infinite power series.

small divisors based on lists. However, the problem has been already discussed in Section 7.3.3, where also the remark has been made that a constructive algorithm for an infinite Fourier expansion can hardly be proposed. The way out is to choose an integer parameter $K > 0$ and expand the Hamiltonian in trigonometric polynomials, as, for example, in (7.38).

An attentive reader will perhaps raise two objections, namely: (i) there is no minimal degree of the trigonometric polynomial at every order, because it will be lost during the normalization process; (ii) the expansion in powers of ε is lost, because terms with low power of ε but high $|k|$ will be moved at higher orders. The answer to the first question is direct: the only thing we really need is that the maximal degree sK is preserved at every normalization step; this remains true. The answer to the second question is to relax the requests concerning the expansion in ε and let the lower indices, s say, denote the order. The smallness of every term will be assured by its norm.

There is, however, a point which deserves to be pointed out once more, recalling the discussion at the end of Section 7.3.3. The parameter K should be chosen independent of ε; typically, one can set $K \sim 1/\sigma$. The mechanism of control of small divisors will prove to be effective with no changes.

We can thus conclude with a formal statement of a theorem which applies in a general case. Recall that to the frequencies $\omega \in \mathbb{R}^n$ we associate the monotonically non-increasing sequence $\{\alpha_r\}_{r\geq 0}$ defined as

$$\alpha_0 = 1\ , \quad \alpha_r = \frac{1}{\Omega} \min_{0<|k|\leq rK} \bigl|\langle k, \omega\rangle\bigr|\ , \quad \Omega = \max_j |\omega_j|\ .$$

Theorem 8.18: *Consider the Hamiltonian*

$$H(p,q) = H_0(p) + \varepsilon H_1(p,q,\varepsilon)$$

analytic in $(p,q) \in \mathscr{G}_\varrho \times \mathbb{T}^n_\sigma$ and analytic in ε for $|\varepsilon|$ small. Assume:

(i) $H_0(p)$ is non-degenerate, that is,

$$\det\left(\frac{\partial^2 H_0}{\partial p_j \partial p_k}\right) \neq 0\ ;$$

(ii) $H_0(p)$ possesses an invariant torus p^* with frequencies ω satisfying condition $\boldsymbol{\tau}$, namely

$$-\sum_{r\geq 1} \frac{\ln \alpha_r}{r(r+1)} = \Gamma < \infty\ .$$

Then there exists a positive ε^* such that the following holds true: for $\varepsilon < \varepsilon^*$, there exists a perturbed invariant torus carrying quasiperiodic motions with frequencies ω, which is a small deformation and translation of the unperturbed one.

Adapting the proof in classical style proposed here to the general statement just presented may be tedious and boring but does not involve new concepts. A proof for the general case which exploits the suggestions made here may be found in [82]; but (in order not to make the exercise too easy) the non-resonance condition is formulated there in a slightly different form.

8.4.2 A Comment on Lindstedt's Series

We come now to a short comment concerning the method of Lindstedt series [155], published in 1883 and thoroughly discussed by Poincaré (see [188], tome II, ch. VIII and ch. XIII). The model considered by Lindstedt is a non-linear oscillator with a time-periodic perturbation. The method goes back to the classical series expansion of the solution in the form introduced by Lagrange, but Lindstedt exploits a crucial idea: to look for a single solution with a given frequency and find the corresponding amplitude.

The basis of Lindstedt's idea may be traced back to Lagrange (see [134], § 42–46). The problem raised is the following. One looks for a solution by a series of trigonometric functions of time t – what we call now Fourier series or quasiperiodic motion. Unfortunately, a straightforward calculation produces the so-called *secular terms*,[19] namely terms which contain powers of t. These terms are produced precisely by resonances and correspond to small divisors that we have seen until now. Lagrange shows how these terms may be removed by suitably modifying the construction, but the method of Lindstedt appears definitely more direct. Lindstedt shows that solutions which contain only trigonometric functions of t may be constructed as *formal* expansions.

Poincaré extends the method of Lindstedt and shows that it is fully consistent by using the indirect approach of constructing a normal form. Then he proposes to prove that Lindstedt's series are divergent, but here he fails to reach a definite conclusion. He closes chapter XIII by writing (n_1, n_2 are the frequencies that we denote by ω_1, ω_2):

> *... les séries ne pourraient-elles pas, par example, converger quand ... le rapport n_1/n_2 soit incommensurable, et que son carré soit au contraire commensurable (ou quand le rapport n_1/n_2 est assujetti à une autre condition analogue à celle que je viens d'énoncer un peu au hasard)?*
>
> *Les raisonnements de ce chapitre ne me permettent pas d'affirmer que ce fait ne se présentera pas. Tout ce qu'il m'est permis de dire, c'est qu'il est fort invraisemblable.*

---

[19] The name *secular terms* was introduced by Kepler referring to slow changes in the planetary orbits which become visible over the arc of centuries. These terms have represented a major trouble for the dynamics of the planetary system.

It is clear that the idea of Lindstedt of looking for solutions with a given frequency is the same as in the approach of Kolmogorov. On the other hand, the ratio n_1/n_2 suggested by Poincaré corresponds precisely to Diophantine frequencies. So, *why should Lindstedt's series not converge?*

The question has been awakened after more than 70 years by Moser [175]: he showed that Kolmogorov's theorem provides an *indirect* proof that Lindstedt's series should converge. However, the classical method of expansion produces trigonometric series with coefficients which grow as factorials – which we have seen happen for the formal series for first integrals. Moser's conclusion is that the convergence of Lindstedt series cannot be proved using a majorant method. Therefore, he proposes the conjecture that there must be *cancellations* due to algebraic compensation of diverging terms.

The existence of such cancellations has been established by Lars Håkan Eliasson in a report of the University of Stockholm in 1988 which remained unpublished till 1996 [65]; however, he did not find an explicit form of the cancellations. The question was eventually settled by Giovanni Gallavotti [71] in 1994; using a clever representation of Lindstedt's series by trees, he could discover the critical cancellations, thus concluding for convergence.

The reader may legitimately ask why the classical scheme proposed in Section 8.3 does not even mention the word "cancellations." This depends on the different algorithms used. The point to be understood is that different algorithms produce the coefficient of a given mode in the Fourier expansion as a sum of many ones; among these coefficients there are diverging ones in Lindstedt's method, while there are none in the algorithm of composition of Lie series. The final series coincide only after algebraic resummation of all coefficients of every mode. Such bizarre behaviour can be revealed by recourse to an explicit construction using algebraic manipulations. An example may be found in [87].

A full discussion of Lindstedt's method and of the works mentioned would be interesting; but the number of pages would increase too much for a textbook like the present one. The reader asking for a quick summary of the method may want to see [86].

9
Long Time Stability

This chapter is devoted to recent results concerning stability in the general problem of dynamics. A short excursus over the concept of stability in a wider sense seems appropriate, on account of the thorough investigations made by Lyapounov [166].

The entry level is concerned with stability of equilibrium points, which are the simplest case of particular but interesting orbits. Further developments concern stability of more complex particular solutions, such as periodic orbits (actually one-dimensional invariant tori) or quasiperiodic motions on invariant tori of dimension between 2 and the number n of degrees of freedom. A further aspect is stability in a wider sense: if an integrable system (in the Liouville–Arnold sense) is subject to a small perturbation, a natural question is whether the *actions* of the system, which are constants along the unperturbed orbits, remain close to their initial value when the system is perturbed. Particular attention is paid to the actions, because it is clearly hard to introduce a concept of stability for the angles – unless they lose their character of being angles, that is, to evolve monotonically in time.

9.1 Overview on the Concept of Stability

Let us begin with an informal illustration of the problem, referring to the two cases discussed in the present notes, namely the stability of an elliptic equilibrium and the general problem of dynamics as enunciated by Poincaré.

The traditional quest is for perpetual stability. For example, an equilibrium point P is said to be perpetually stable in case, for every neighbourhood U of P, there exists a second neighbourhood V satisfying the property that every orbit with initial point $P(0) \in V$ satisfies $P(t) \in U$ for all $t \in \mathbb{R}$. It is well known that this is a very strong property which excludes even most equilibria of linear systems. Lyapounov introduced some useful *variazioni*, such as the distinction of stability either in the past or in the future or the concept of asymptotic stability. However, the problem of perpetual stability of an equilibrium remains a fundamental one for conservative (Hamiltonian) systems.

The stability of an elliptic equilibrium has been considered in Chapter 5. The unperturbed Hamiltonian $H_0(x,y) = \sum_j \omega_j^2(y_j^2 + x_j^2)/2$ represents a system of harmonic oscillators, and the perturbation is expanded in power series as $H_1(x,y) + H_2(x,y) + \ldots$, with $H_s(x,y)$ a homogeneous polynomial of degree $s+2$. The perturbation parameter here is the energy E of the system. If all frequencies have the same sign, then the Hamiltonian itself plays the role of a Lyapounov function, since it has a minimum (or a maximum) at the equilibrium: for values of E close to that of the equilibrium, the energy manifold $H(x,y) = E$ has a connected component which encloses the equilibrium, and the invariance of energy bounds the orbits inside this component. However, if the frequencies have different signs, then the energy manifold is no more connected. Thus, the problem is a major one and can be dealt with, for example, by trying to control the behaviour of the actions $p_j = (y_j^2 + x_j^2)/2$ or of some other appropriate quantity, as we did in Chapter 5.

The general problem of dynamics is concerned with an unperturbed Hamiltonian $H_0(p)$, where p are the action variables, subject to a perturbation $\varepsilon f(p,q,\varepsilon)$, where q are angles and ε is a small parameter measuring the size of the perturbation. If we are looking for perpetual stability, then the strongest result is represented by the theorem of Kolmogorov, which was the subject of Chapter 8. As we have seen, the claim is that if the perturbation is small enough, then *most* invariant tori of the unperturbed system survive, being only deformed. For these orbits the actions remain forever close to their initial value. However, "most" is to be intended in the sense of measure (or probability). From the topological viewpoint, the situation looks puzzling, because the complement of the set of invariant tori is open and dense, and for more than two degrees of freedom it is also connected. Therefore, the existence of orbits which explore all the complements of invariant tori cannot be excluded. For such orbits the actions would change considerably during the evolution of the system. This phenomenon has been first described in a particular (and somehow artificial) model by Arnold in [6], and since then it has been named *Arnold diffusion*. However, a proof of the existence of diffusion as a generic fact has required long work and has been achieved only recently (see, e.g., [163], [164], [124], [125], [24]).

What we have just said can be resumed in crude (and simplistic) words as follows: perpetual stability in the mathematical sense is a concept basically extraneous to Physics. A different approach consists in looking for stability for a time which is finite but significantly long.

Let us take the most direct approach by stating the problem, in rough terms, as follows: *to prove that the actions p of the system are bounded by $|p(t) - p(0)| \simeq \varepsilon^b$ (with $b > 0$) for a large time $|t| < T(\varepsilon)$.*

Formulated this way, the question may appear deprived of meaning. For everybody who has some knowledge of the theory of differential equation will immediately remark that this appears to be just a property of continuity of the solutions with respect to initial data or to parameters. The relevant point, indeed, concerns the meaning of "large $T(\varepsilon)$."

The question may be meaningful if we take into consideration some characteristic of the dynamical system which is (more or less accurately) described by our equations. In this case "large" should be interpreted as *large with respect to some characteristic time of the physical system, or comparable with the lifetime of it.* For instance, if one is concerned with the planetary system, then the question may be reformulated as follows: to prove that the semimajor axes, the eccentricities and the inclinations of the planetary orbits will remain close to the present values for a time comparable with the age of the solar system. If one considers a galaxy, then we may refer to the age of the universe. Another example is represented by particle accelerators: the request is that the particles do not escape and hit the walls during the lifetime of the experiment (typically a few days).[1] Thus, the desired stability time $T(\varepsilon)$ may be quite reasonably defined if one is concerned with a specific physical system.

From a mathematical viewpoint the adjective "large" is more difficult to explain, since there is no typical lifetime directly associated to a differential equation. Hence, in order to make the word "stability" meaningful it is essential to consider the actual dependence of T on ε. In this sense the property of continuity with respect to initial data does not help too much. For instance, if we consider the trivial example of the differential equation $\dot{x} = x$, one will immediately see that if $x(0) = x_0 > 0$ is the initial point, then we have $x(t) > 2x_0$ for $t > T = \ln 2$ no matter how small x_0 is; hence T may hardly be considered to be "large," since it does not even increase when one puts the initial point closer and closer to equilibrium. Thus, we should at least make the question more significant by adding a minimal request, namely that $T(\varepsilon)$ should increase to infinity when ε decreases to zero.

---

[1] It may be interesting to note that the number of revolutions of the particles in an accelerator in a few days is roughly of the same order of magnitude as the number of revolutions of Jupiter around the Sun over a time interval comparable with the estimated age of the universe, which means that the mathematical difficulty is essentially the same for both problems. For a galaxy, things are quite different: the stars may perform a few hundred revolutions during a time as long as the age of the universe, which means that a galaxy does not really need to be very stable in order to exist.

We may also raise a further, more significant question: how does $T(\varepsilon)$ go to infinity? In the light of our current knowledge we may give different answers.

i. $T(\varepsilon) \simeq 1/\varepsilon$: this is the theory of *adiabatic invariants*, which has been widely used in Astronomy and Physics, and is essentially based on the *averaging methods* introduced by Lagrange.
ii. $T(\varepsilon) \simeq 1/\varepsilon^r$ with $r > 1$: this is the theory of *complete stability*, developed by Birkhoff for the elliptic equilibria of Hamiltonian systems (see [27], cap. IV, §2 and §4).
iii. $T(\varepsilon) \simeq \exp(1/\varepsilon^a)$ with $0 < a \leq 1$: this is the theory of *exponential stability*, anticipated by Moser [172] and Littlewood [153][154] and fully stated by Nekhoroshev [179][180].
iv. $T(\varepsilon) \simeq \exp\bigl(\exp(1/\varepsilon^a)\bigr)$ with $0 < a \leq 1$: this is the theory of *superexponential stability*, first developed by Alessandro Morbidelli and the author [170].
v. $T(\varepsilon) = \infty$: this is the *perpetual stability* investigated by Lyapounov; it is the dream of most mathematicians indeed, and is the case often considered in textbooks.

9.1.1 A Short Historical Note

Long-time stability for the general problem of dynamics may be seen as a complementary result to Kolmogorov's theorem. As a matter of fact Kolmogorov's theorem is strongly related to the quest for perpetual stability: as already remarked, the invariant tori are parameterized by action variables which remain constant for an infinite time, so that the unperturbed actions undergo a small change, typically $\mathcal{O}(\sqrt{\varepsilon})$. However, the concept of stability applies only in a probabilistic sense, because the subset of invariant tori has relative measure of order $1 - \mathcal{O}(\sqrt{\varepsilon})$ in phase space. Long-time stability means that we are interested in a property which applies to *all* initial conditions in an open set. The finite (but long) time is the price to be paid in order to have a control on the effect of diffusion: there is no way to keep it perpetually bounded, but one tries to bound it for a time long enough.

Adiabatic invariance is a very classical problem, which will not be dealt with in the present notes.[2] For a nice introduction see, for example, [10] or [159].

---

[2] By using the averaging method, Lagrange was able to give the first long-time stability result for the solar system [135]. He proved that the semimajor axes remain invariant for the first-order approximation in the masses of the planets. Later he used a similar approach in order to determine the evolution of the inclinations (which remain small) and the precession of the nodes [136]. The

Exponential stability is a quite strong property, first proposed by Moser [172] and Littlewood [153] [154]. The case of an elliptic equilibrium discussed in Section 5.2 constitutes a very simple example; Littlewood considered precisely this case, having in mind the triangular equilibria of the restricted problem of three bodies. Initially the concept of exponential stability was soon superseded by the newborn theory of preservation of invariant tori characterized by strongly non-resonant frequencies. Several years later a general formulation of the theory of exponential stability was developed by Nekhoroshev ([179] and [180]).

In crude words, the theorem may be seen as a remarkable generalization of the stability result illustrated in Chapter 5, Section 5.2, while dealing with the dynamics in the neighbourhood of an elliptic equilibrium. The mathematical difficulty is considerably bigger because the unperturbed Hamiltonian $H_0(p)$ is not isochronous (the frequencies do depend on the actions). The main achievement of Nekhoroshev is the combination of quantitative analytical methods of classical perturbation theory with a kind of geography of resonances in the action space. This matter will be discussed in detail in this chapter. The proof of Nekhoroshev's theorem given here is based on the papers by Benettin et al. [17], by Zehnder and the author [81], and [90]; it completes the short exposition in [99]. These papers were deeply inspired by the original work of Nekhoroshev, with some simplifying hypotheses. The second paper considers also a potential depending generically (i.e., non-periodically but still analytically) on time. Other proofs may be found in [20], [193] and [58].

Superexponential stability is an enhancement of exponential stability which will be discussed in Chapter 10.

9.2 The Theorem of Nekhoroshev

It is now time to go back to the questions raised in Chapter 7, in particular in Section 7.3. The framework is still the general problem of dynamics as stated by Poincaré (see Chapter 4), namely a canonical system with Hamiltonian

$$H(p,q) = H_0(p) + \varepsilon f(p,q,\varepsilon), \tag{9.1}$$

where $(p,q) \in \mathscr{G} \times \mathbb{T}^n$ are action-angle variables in the Arnold–Jost sense and the perturbation $f(p,q)$ is assumed to be analytic in the canonical variables and can be expanded in power series of ε in a neighbourhood of $\varepsilon = 0$.

---

extension to the eccentricities and perihelia was soon made by Laplace [145] (although the Laplace's note was published before Lagrange's). A more complete version was later published by Lagrange with the aim of completing his "proof" of stability of the solar system [137] [138]. For a brief historical introduction see, for example, [148].

This chapter answers the question raised in Section 7.3.1; that is, in some sense, how to overcome the negative impact of Poincaré's theorem on non-integrability.

9.2.1 Statement of the Theorem

Let us consider the Hamiltonian (7.27), which we may recast as

$$H(p,q) = H_0(p) + \varepsilon f(p,q,\varepsilon) \ , \tag{9.2}$$

where p, q are action-angle variables in a domain $\mathscr{G} \times \mathbb{T}^n$, with $\mathscr{G} \subset \mathbb{R}^n$ open. The perturbation $f(p,q,\varepsilon)$ collects all terms in the ε expansion $\varepsilon H_1(p,q) + \varepsilon^2 H_2(p,q) + \ldots$ in (7.27), which is legitimate because the expansion is assumed to be convergent for small ε. We denote $\omega(p)$ and $C(p)$ the frequencies and the Hessian matrix of the unperturbed Hamiltonian, namely

$$\omega_j(p) = \frac{\partial H_0}{\partial p_j} \ , \quad C_{jk}(p) = \frac{\partial^2 H_0}{\partial p_j \partial p_k} \ , \quad j,k = 1,\ldots,n \ . \tag{9.3}$$

The Hamiltonian will be characterized by five parameters, namely ϱ, σ (analyticity parameters), ε (perturbation parameter) and two further positive parameters m and M characterizing the unperturbed part of the Hamiltonian. On the Hamiltonian (9.2) we make the following hypotheses:

(i) Both H_0 and f are assumed to be analytic bounded functions on the complex extension $\mathscr{D}_{\varrho,2\sigma} = \mathscr{G}_\varrho \times \mathbb{T}^n_{2\sigma}$ of the real domain $\mathscr{G} \times \mathbb{T}^n$; moreover, $f(p,q)$ is assumed to be analytic in the parameter ε.

(ii) The perturbation f is assumed to be bounded, that is,

$$|f|_{\varrho,2\sigma} < \infty \ , \tag{9.4}$$

where $|f|_{\varrho,2\sigma}$ denotes the supremum norm over the domain $\mathscr{D}_{\varrho,2\sigma}$, namely $|f|_{\varrho,2\sigma} = \sup_{(p,q)\in\mathscr{D}_{\varrho,2\sigma}} |f(p,q)|$; thus, $\varepsilon |f|_{\varrho,2\sigma}$ is a true estimate of the size of the perturbation.

(iii) For all p in $\mathscr{G}_\varrho$ the unperturbed Hamiltonian $H_0(p)$ is assumed to satisfy[3]

$$\|C(p)v\| \le M\|v\| \ , \quad |\langle C(p)v, v\rangle| \ge m\|v\|^2 \quad \text{for all } v \in \mathbb{R}^n \tag{9.5}$$

with positive constants $m \le M$.

The theorem of Nekhoroshev is stated as follows.

Theorem 9.1: *Let $H = H_0(p) + \varepsilon f(p,q)$ satisfy the hypotheses (i), (ii) and (iii). Then there exist positive constants μ_* and T_* depending on ϱ, σ,*

[3] The assumption $m \le M$ is not restrictive in view of the trivial inequality $|\langle C(p)v, v\rangle| \le \|C(p)v\| \, \|v\| \le M\|v\|^2$ which follows from the first equation of (9.5).

$|f|_{\varrho,2\sigma}$, m, M *and the number* n *of degrees of freedom such that the following statement holds true: if*

$$3^4\mu_*\varepsilon < 1,$$

then for every orbit $p(t), q(t)$ *satisfying* $p(0) \in \mathscr{G}$, *one has the estimate*

$$\operatorname{dist}\bigl(p(t) - p(0)\bigr) \le (\mu_*\varepsilon)^{1/4}\varrho$$

for all times t *satisfying*

$$|t| \le \frac{T_*}{\varepsilon}\exp\left[\left(\frac{1}{\mu_*\varepsilon}\right)^{1/2a}\right],$$

where $a = n^2 + n$.

Explicit estimates for the values of the constants μ_* and T_* will be given in the proof. However, one should not consider these estimated values as optimal ones, of course. The reader should rather pay attention to the dependence of the constants on some relevant parameters of the Hamiltonian. For example, it is interesting to investigate the dependence on the number n of degrees of freedom. In this respect, it must be noted that the value $a = n^2 + n$ is far from being optimal, at least if one makes the assumption of convexity for the unperturbed Hamiltonian $H_0(p)$. An attempt to optimize the dependence of a on n was first made in [160] and [161], where it is found that $a = 2n$, with a suggestion that this should be the best value.

9.2.2 An Overview of the Method

The starting point is the expansion of the Hamiltonian in trigonometric polynomials as explained in Section 7.3.3, completed with the analytical estimates of Section 7.3.4. As we did for Kolmogorov's theorem, such an expansion allows us to work out the normalization process, taking care at every order of a finite number of resonances. This exhibits some similarity with the procedure used in the case of fixed frequencies, as we have done formally in Section 7.1.1 for an elliptic equilibrium and in Section 7.3.2 for an isochronous system.

The problem here is that we want to investigate also the case of a generic function $H_0(p)$, so that the frequencies $\omega(p) = \frac{\partial H_0}{\partial p}$ of the unperturbed system exhibit a true dependence on p. To this end we must exploit the idea which has been suggested in the example of Section 4.6.2, namely that the normal form may be constructed in suitable local domains. The process is illustrated in panel **a** of Fig. 9.1, where we limit $|k| \le K$. At the first step of normalization we must separate small strips in the action domain where a divisor $\bigl|\langle k, \omega(p)\rangle\bigr|$ is small; this is the case of a single resonance. The intersections between two such strips correspond to double resonances, and here we should separate a small region where two divisors with different k

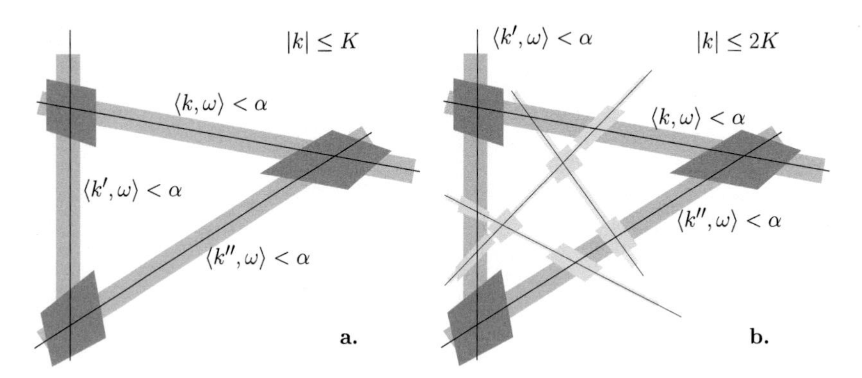

Figure 9.1 The structure of resonant subsets of the action space for increasing order of resonance.

are small; for reasons which will emerge from the proof, the width of such a region must be larger than the single resonance's one. The process continues as the dimension increases.

In each resonant region the construction of the normal form must take into account the resonances inside the region; therefore, the construction of normal form has only a *local* character. This does not contradict the theorem of non-integrability of Poincaré, because we must take care of a finite number of resonances, which do not form a dense set.

Proceeding with the second normalization step, we must separate more regions with resonances of order $|k| \le 2K$, as in panel **b** of Fig. 9.1, and for higher orders the regions to be separated increase, but if we stop normalization at order r, then we must separate only resonances with $|k| \le rK$, which are not dense.

Thus, we are confronted with two problems, namely:

(i) To construct a suitable normal form truncated at finite order r in every region, taking into account the resonances inside the region considered.

(ii) To cover the action space with a family of domains where the construction of the normal form may be consistently performed.

The two questions are answered separately. Item (i) is handled in the so-called *analytic part*; essentially the construction of the normal form as in Chapter 7. However, here the formal construction must be completed by adding quantitative estimates. Item (ii) is the innovative contribution of Nekhoroshev and constitutes the so-called *geometric part* of the proof. The two parts are discussed separately in the next two sections.

9.3 Analytic Part

The analytic part exploits the formal construction of Section 7.3, using in particular the splitting of the Hamiltonian in trigonometric polynomials as in Section 7.3.3. For the reader's convenience we summarize the main points.

9.3.1 *Formal Scheme*

For a non-negative integer K, we denote by $\mathscr{F}_K$ the distinguished class of functions which do not contain Fourier harmonics with $|k| > K$; this is Definition 7.10, as we did in Section 7.3.3. Such a splitting is preserved (in a formal sense) by the algorithm of construction of the normal form, as stated by Lemmas 7.11, 7.12 and 7.13. Thus, we may re-expand the Hamiltonian (9.2) without taking care of the parameter ε, namely as

$$H(p,q) = H_0(p) + H_1(p,q) + H_2(p,q) + \ldots\ , \qquad H_s(p,q) \in \mathscr{F}_{sK}\ . \tag{9.6}$$

The geometrically decreasing size of terms in the expansion is assured by Lemma 7.15 and Corollary 7.16; we have indeed

$$\|H_s\|_{\varrho,\sigma} \leq h^{s-1}F\ , \quad s \geq 1\ , \tag{9.7}$$

with the constants h and F defined by

$$h = e^{-K\sigma/2}\ , \quad F = \varepsilon\left(\frac{1+e^{-\sigma/2}}{1-e^{-\sigma/2}}\right)^n |f|_{\varrho,2\sigma}\ . \tag{9.8}$$

Let us come to the construction of the normal form. To this end we need two definitions.

(i) A subgroup $\mathscr{M} \in \mathbb{Z}^n$ is said to be a *resonance module* if it satisfies[4] $\mathrm{span}(\mathscr{M}) \cap \mathbb{Z}^n = \mathscr{M}$.

(ii) A function $Z(p,q)$ is said to be in *normal form with respect to the resonance module* $\mathscr{M}$ in case its Fourier expansion has the form

$$Z(p,q) = \sum_{k\in\mathscr{M}} z_k(p)\exp(i\langle k,q\rangle)\ ,$$

namely, if it contains only harmonics belonging to $\mathscr{M}$ (resonant harmonics).

---

[4] Here, both $\mathscr{M}$ and $\mathbb{Z}^n$ are considered as subsets of $\mathbb{R}^n$, and $\mathrm{span}(\mathscr{M})$ is the linear subspace in $\mathbb{R}^n$ spanned by $\mathscr{M}$. In order to understand the definition, remark that having given a subgroup $\mathscr{M} \subset \mathbb{Z}^n$ one can always extract a proper subset $\mathscr{M}' \subset \mathscr{M}$ of the same dimension as $\mathscr{M}$, which is still a subgroup: see Appendix A, Section A.1. The apparently strange condition $\mathrm{span}(\mathscr{M}) \cap \mathbb{Z}^n = \mathscr{M}$ excludes such a module $\mathscr{M}'$, which is clearly unsuitable for our purposes.

The normal form will be constructed using the Lie transform method, namely the algorithm of Section 7.1.1, truncated at an arbitrary order r. Let us recall it for convenience. We determine a generating sequence $\chi^{(r)} = \{\chi_1, \ldots, \chi_r\}$ and the normal form $Z^{(r)} = Z_0 + \cdots + Z_r$ by solving the equation

$$T_{\chi^{(r)}} Z^{(r)} = H_0 + \cdots + H_r + \mathscr{R}^{(r+1)} \ , \tag{9.9}$$

where $\mathscr{R}^{(r+1)}$ is a remainder not normalized. We also recall the definition of the Lie transform,

$$T_\chi = \sum_{s\geq 0} E_s \ , \quad E_0 = 1 \ , \quad E_s = \sum_{j=1}^{s} \frac{j}{s} L_{\chi_j} E_{s-j} \ . \tag{9.10}$$

The construction requires solving at every order the homological equation

$$Z_s - \partial_\omega \chi_s = \Psi_s \ , \quad s = 1, \ldots, r \ , \tag{9.11}$$

where

$$\begin{aligned} \Psi_1 &= H_1 \ , \\ \Psi_s &= H_s - \sum_{j=1}^{s-1} \frac{j}{s} \bigl(L_{\chi_j} H_{s-j} + E_{s-j} Z_j \bigr) \quad \text{for } 2 \leq s \leq r \ . \end{aligned} \tag{9.12}$$

The solution of the homological equation is found by exploiting the expansion of Ψ_s in Fourier series as

$$\Psi_s(p,q) = \sum_{k\in\mathbb{Z}^n} \psi_k(p) \exp(i\langle k, q\rangle) \ ,$$

with known coefficients $\psi_k(p)$, and considering the same expansion for χ_s and Z_s, with unknown coefficients $c_k(p)$ and $z_k(p)$, respectively. Then equation (9.11) splits into the system of equations for the coefficients

$$z_k(p) - i\bigl\langle k \cdot \omega(p) \bigr\rangle\, c_k(p) = \psi_k(p) \ , \quad \omega(p) = \frac{\partial H_0}{\partial p} \ . \tag{9.13}$$

Having chosen a resonance module $\mathscr{M}$, the solution is found in the form

$$\begin{aligned} Z_s &= \sum_{k\in\mathscr{M}} \psi_k(p) \exp(i\langle k, q\rangle) \ , \\ \chi_s &= i \sum_{k\notin\mathscr{M}} \frac{\psi_k(p)}{\bigl\langle k \cdot \omega(p) \bigr\rangle} \exp(i\langle k, q\rangle) \ . \end{aligned} \tag{9.14}$$

Vanishing denominators are avoided by associating to $\mathscr{M}$ a suitable local domain: this is the crucial point which will be completely solved at the end of the geometric part of the proof.

9.3.2 An Explicit Expression for the Remainder

The algorithm of the previous section shows how to construct the normal form $Z^{(r)}$. However, the canonical transformation to normal form also produces a remainder $\mathscr{R}^{(r+1)}$ of order $r+1$. An explicit expression which will be useful for the analytical estimates may be found as follows.

Recalling (9.9), and since the operator T_χ is invertible, we may construct a function $\mathscr{R}^{(r+1)}$ starting with terms of order $r+1$ and satisfying

$$T_{\chi^{(r)}}\big(Z^{(r)} + \mathscr{R}^{(r+1)}\big) = H \ . \tag{9.15}$$

Indeed, this equation says nothing new with respect to (9.9), but it is useful because it gives immediately

$$\mathscr{R}^{(r+1)} = T^{-1}_{\chi^{(r)}} H - Z^{(r)}. \tag{9.16}$$

Using the expression (6.22) for T_χ^{-1}, we get

$$\mathscr{R}^{(r+1)} = \sum_{s>r}\Bigg(\sum_{j=0}^{s-1} D_j H_{s-j} - \sum_{j=1}^{s} \frac{j}{s} D_{s-j}(\Psi_j - Z_j)\Bigg) \ . \tag{9.17}$$

This is found by writing the remainder as $\mathscr{R}^{(r+1)} = \sum_{s>r}\sum_{j=0}^{s} D_j H_{s-j}$, and by using

$$D_s H_0 = -\sum_{j=1}^{s} \frac{j}{s} D_{s-j} L_{\chi_j} H_0 = -\sum_{j=1}^{s} \frac{j}{s} D_{s-j} L_{\chi_j}(\Psi_j - Z_j) \ ,$$

in view of (9.11).

9.3.3 Truncated First Integrals

Having constructed the generating sequence $\chi^{(r)}$ and the normal form $Z^{(r)}$, it is an easy matter to construct truncated first integrals. Indeed, it is enough to use the property of the Lie transform of preserving the Poisson brackets. For let $\Phi_0(p,q)$ be a function which is in involution with $Z^{(r)}$, that is, $\{\Phi_0, Z^{(r)}\} = 0$. Then the function $T_\chi \Phi_0$ is an approximated first integral for the Hamiltonian H in the sense that the Poisson bracket $\{T_\chi\Phi_0, H\}$ is of order $r+1$. This is easily seen by computing

$$\{T_\chi\Phi_0, H\} = \{T_\chi\Phi_0, T_\chi\big(Z^{(r)} + \mathscr{R}^{(r+1)}\big)\} = T_\chi\{\Phi_0, \mathscr{R}^{(r+1)}\} \ . \tag{9.18}$$

Thus, any function Φ_0 which is in involution with $Z^{(r)}$ generates an approximate first integral $\Phi = T_\chi\Phi_0$. The choice of Φ_0 is a trivial matter. Recall that $Z^{(r)}$ is in normal form, that is, its Fourier expansion contains only Fourier harmonics $\exp\big(i\langle k, q\rangle\big)$ with $k \in \mathscr{M}$, a resonance module. It follows that any function

$$\Phi_0 = \langle \lambda, p\rangle \ , \quad 0 \neq \lambda \perp \mathscr{M} \tag{9.19}$$

is a first integral. For, writing $Z^{(r)} = \sum_{k\in\mathscr{M}} z_k(p)\exp\bigl(i\langle k,q\rangle\bigr)$, calculate

$$\{\Phi_0, Z^{(r)}\} = -i\sum_{k\in\mathscr{M}} \langle k,\lambda\rangle z_k(p)\exp\bigl(i\langle k,q\rangle\bigr) = 0$$

in view of $\lambda \perp \mathscr{M}$. We conclude that there exist $n - \dim\mathscr{M}$ first integrals for $Z^{(r)}$ which are independent and in involution: just choose $n-\dim\mathscr{M}$ independent real vectors $\lambda \perp \mathscr{M}$. The reader will easily check that, in general, there are no further independent first integrals for $Z^{(r)}$, unless the latter happens to have a very special form.

9.3.4 Quantitative Estimates on the Generating Sequence

It is now time to complete the formal scheme by adding quantitative estimates.

Definition 9.2: *A subset $\mathcal{V} \subset \mathbb{R}^n$ is said to be a non-resonance domain of type $(\mathscr{M},\alpha,\delta,N)$ in case*

$$|\langle k,\omega(p)\rangle| > \alpha \text{ for all } p \in \mathcal{V}_\delta\ ,\quad k \in \mathbb{Z}^n \setminus \mathscr{M} \text{ and } |k| \le N\ .$$

Here, α and δ are positive parameters, N is a positive integer, $\mathscr{M}$ is a resonance module and $\mathcal{V}_\delta$ is the complex extension of the real domain $\mathcal{V}$, as described in Section 7.2.2. The non-resonance domain is the subset of the action space where the Hamiltonian can be given a normal form with respect to the given resonance module. Indeed, the solution (9.14) of the homological equation avoids vanishing divisors due to the limitation of the local domain.

Let us now come to quantitative estimates. We restrict the Hamiltonian function to the domain $\mathcal{V}_\delta \times \mathbb{T}^n_\sigma$ and look for estimates for the new functions $\chi_1,\ldots,\chi_r$ and $Z_1,\ldots,Z_r$ in a restricted domain $\mathcal{V}_{(1-d)\delta} \times \mathbb{T}^n_{(1-d)\sigma}$.

First of all, note that if $\|\Psi_s\|_{(1-d')(\delta,\sigma)}$ is known for some $d' < 1$, then the definition of the norm together with the form of the solution of (9.11) and the definition of non-resonance domain immediately give

$$\begin{aligned}\|\chi_s\|_{(1-d')(\delta,\sigma)} &\le \frac{1}{\alpha}\|\Psi_s\|_{(1-d')(\delta,\sigma)}\ ,\\ \|Z_s\|_{(1-d')(\delta,\sigma)} &\le \|\Psi_s\|_{(1-d')(\delta,\sigma)}\ .\end{aligned} \tag{9.20}$$

The quantitative estimates for the truncated generating sequence are collected in the following.

Lemma 9.3: *On a non-resonance domain $\mathcal{V}$ of type $(\mathscr{M},\alpha,\delta,N)$ consider the Hamiltonian (9.6) and let $\|H_s\|_{\delta,\sigma} \le h^{s-1}F$ for some $F > 0$ and $h \ge 0$. Then for any positive $d < 1$, the truncated generating sequence $\chi^{(r)} = \{\chi_1,\ldots,\chi_r\}$, which gives the Hamiltonian the normal form (9.9), satisfies*

$$\|\Psi_s\|_{(1-d)(\delta,\sigma)} \le \frac{b^{s-1}}{s}F\ ,\quad \|\chi_s\|_{(1-d)(\delta,\sigma)} \le \frac{b^{s-1}}{s}G \tag{9.21}$$

with

$$b = \frac{16(r-1)F}{e\alpha d^2\delta\sigma} + 4h \ , \quad G = \frac{F}{\alpha} \ . \tag{9.22}$$

Proof. Noting that the definition (9.12) of Ψ_s involves Poisson brackets, and so a restriction of the domains in the estimate, fix the final restriction $d < 1$ of the domain, and define

$$d_s = \sqrt{\frac{s-1}{r-1}}\, d \ , \quad 1 \le s \le r \ . \tag{9.23}$$

Look now for an estimate of $\|\Psi_s\|_{(1-d_s)(\delta,\sigma)}$, from which the estimate for $\|\chi_s\|_{(1-d_s)(\delta,\sigma)}$ is obtained in view of (9.20). To this end, look for two sequences $\{\eta_s\}_{1\le s\le r}$ and $\{\tilde{\vartheta}_{s,j}\}_{0\le s\le r\,,\,1\le j\le r}$ such that

$$\|\Psi_s\|_{(1-d_s)(\delta,\sigma)} \le \eta_s F \ , \quad \|E_s Z_j\|_{(1-d_{s+j})(\delta,\sigma)} \le \tilde{\vartheta}_{s,j} F \ . \tag{9.24}$$

In view of $\Psi_1 = H_1$ we may choose $\eta_1 = 1$, and in view of $E_0 Z_j = Z_j$ and of $\|Z_j\|_{(1-d_j)(\delta,\sigma)} \le \|\Psi_j\|_{(1-d_j)(\delta,\sigma)}$ we can set $\tilde{\vartheta}_{0,j} = \eta_j$. Using the second of (9.12), for $s \ge 2$ we get

$$\|\Psi_s\|_{(1-d_s)(\delta,\sigma)} \le h^{s-1}F + \sum_{j=1}^{s-1} \frac{j}{s} \left(\frac{2\eta_j F h^{s-j-1}}{e d_s (d_s - d_j)\delta\sigma\alpha} + \tilde{\vartheta}_{s-j,j} \right) F \ .$$

On the other hand, by the definition of the operator E_s, for $s \ge 1$ we have

$$\|E_s Z_j\|_{(1-d_{s+j})(\delta,\sigma)} \le \sum_{l=1}^{s} \frac{l}{s} \cdot \frac{2F^2 \eta_l \tilde{\vartheta}_{s-l,j}}{e(d_{s+j} - d_l)(d_{s+j} - d_{s+j-l})\delta\sigma\alpha} \ .$$

In view of (9.23) and of the elementary inequality[5]

$$\left(\sqrt{s-1} - \sqrt{j-1}\right)\left(\sqrt{s-1} - \sqrt{s-j-1}\right) \ge \frac{1}{2} \quad \text{for } 1 \le j \le s-1 \, ,$$

[5] For $s = 2$, we have only $j = 1$, and the inequality is trivially true. For $s > 2$, let $x = (j-1)/(s-2)$ and $a = \sqrt{(s-1)/(s-2)}$, so that the expression to be minimized is $(s-2)\varphi(x)$, with $\varphi(x) = (a - \sqrt{x})(a - \sqrt{1-x})$, for $0 \le x \le 1$ and $a > 1$. With a further change of variable, set $x = \sin^2\vartheta$ with $0 \le \vartheta \le \pi/2$, and get $\varphi(\vartheta) = (a - \sin\vartheta)(a - \cos\vartheta) = a^2 - a(\sin\vartheta + \cos\vartheta) + \sin\vartheta\cos\vartheta$. By differentiation, get $\varphi'(\vartheta) = -a(\cos\vartheta - \sin\vartheta) + \cos^2\vartheta - \sin^2\vartheta = -(\cos\vartheta - \sin\vartheta)(a - \cos\vartheta - \sin\vartheta)$. For $\cos\vartheta - \sin\vartheta = 0$ with $0 \le \vartheta \le \pi/2$, the function $\varphi(\vartheta)$ has a maximum, which we forget. For $\cos\vartheta + \sin\vartheta = a$, and so also $\sin\vartheta\cos\vartheta = (a^2-1)/2$, the function $\varphi(\vartheta)$ has a minimum with value $(a^2-1)/2$. At that point we have $(s-2)\varphi(x) = 1/2$.

we get

$$\frac{1}{(d_{s+j}-d_l)(d_{s+j}-d_{s+j-l})} = \frac{r-1}{\big(\sqrt{s+j-1}-\sqrt{l-1}\big)\big(\sqrt{s+j-1}-\sqrt{s+j-l-1}\big)d^2} \le \frac{2(r-1)}{d^2}$$

and, since $1 \le j < s$ in estimate for $\|\Psi_s\|_{(1-d_s)(\delta,\sigma)}$, we get

$$\frac{1}{d_s(d_s-d_j)} = \frac{r-1}{\sqrt{s-1}\big(\sqrt{s-1}-\sqrt{j-1}\big)d^2} \le \frac{2(r-1)}{d^2}\ .$$

Thus, the sequences $\{\eta_s\}$ and $\{\tilde{\vartheta}_{s,j}\}$ may be defined as

$$\begin{aligned} \eta_s &= h^{s-1} + \frac{C_r}{s}\sum_{j=1}^{s-1} j\eta_j h^{s-j-1} + \frac{1}{s}\sum_{j=1}^{s-1} j\tilde{\vartheta}_{s-j,j} \\ \tilde{\vartheta}_{s,j} &= \frac{C_r}{s}\sum_{l=1}^{s} l\eta_l\tilde{\vartheta}_{s-l,j}\ , \end{aligned} \tag{9.25}$$

with

$$C_r = \frac{4(r-1)F}{e\alpha d^2\delta\sigma}\ . \tag{9.26}$$

We are thus led to investigate the behaviour of the sequences (9.25), with initial values $\eta_1 = 1$ and $\tilde{\vartheta}_{0,j} = \eta_j$. As a matter of fact, all we need is an estimate of the sequence η_s, because this gives the estimate for the generating function.

The second sequence is immediately seen to give $\tilde{\vartheta}_{s,j} = \eta_j\tilde{\vartheta}_{s,1}$; for this is true for $s = 0$ and is easily checked by induction for $s > 0$. Hence, we may set $\vartheta_s = \tilde{\vartheta}_{s,1}$ and consider the double sequence

$$\begin{aligned} \eta_s &= h^{s-1} + \frac{C_r}{s}\sum_{j=1}^{s-1} j\eta_j h^{s-j-1} + \frac{1}{s}\sum_{j=1}^{s-1} j\eta_j\vartheta_{s-j} \\ \vartheta_s &= \frac{C_r}{s}\sum_{j=1}^{s} j\eta_j\vartheta_{s-j}\ , \end{aligned} \tag{9.27}$$

with initial values $\eta_1 = \vartheta_0 = 1$. Multiplying the first by C_r and subtracting it from the second, we get

$$\vartheta_s = C_r\eta_s - C_r h^{s-1} - \frac{C_r^2}{s}\sum_{j=1}^{s-1} j\eta_j h^{s-j-1}\ ,$$

so that ϑ_s is determined by $\eta_1, \ldots, \eta_{s-1}$ only. Replacing ϑ_{s-j} as given by the latter expression in the first of (9.27), we estimate[6]

$$\begin{aligned}
\eta_s &= h^{s-1} + \frac{C_r}{s}\sum_{j=1}^{s-1} j\eta_j\eta_{s-j} - \frac{C_r^2}{s}\sum_{j=1}^{s-1} j\eta_j \sum_{l=1}^{s-j-1} \frac{l}{s-j}\eta_l h^{s-j-l-1} \\
&< h^{s-1} + \frac{C_r}{s}\sum_{j=1}^{s-1} j\eta_j\eta_{s-j} \\
&= h^{s-1} + C_r\sum_{j=1}^{s-1} \eta_j\eta_{s-j}\ .
\end{aligned}$$

Here, we have removed the negative term in the first line in view of the fact that all contributions in the sum are positive and that, in view of (9.27), all quantities $\eta_1, \ldots, \eta_{s-1}$ are positive. The sequence η_s satisfies the inequality

$$\eta_s \le (C_r + h)^{s-1}\mu_s\ , \tag{9.28}$$

where the sequence $\{\mu_s\}_{s\ge 1}$ is defined as

$$\mu_1 = 1\ ,\quad \mu_s = \sum_{j=1}^{s-1} \mu_j\mu_{s-j}\ . \tag{9.29}$$

This is easily seen by induction.[7] The latter sequence is known as Catalan's sequence and gives (see the end of Section 8.3.11)

$$\mu_s = \frac{2^{s-1}(2s-3)!!}{s!} \le \frac{4^{s-1}}{s}\ , \tag{9.30}$$

where the notation $(2n+1)!! = 1\cdot 3\cdot \cdots \cdot (2n+1)$ has been used. Replacing the latter inequality in (9.28) and recalling (9.20) and (9.24), the claim follows. *Q.E.D.*

---

[6] Use $\sum_{j=1}^{s-1} j\eta_j\eta_{s-j} = \sum_{j=1}^{s-1}(s-j)\eta_j\eta_{s-j} = (s/2)\sum_{j=1}^{s-1}\eta_j\eta_{s-j}$.

[7] It is true for $s = 1$. For $s > 1$, evaluate

$$\begin{aligned}
\eta_s &< h^{s-1} + C_r(C_r+h)^{s-2}\textstyle\sum_{j=1}^{s-1}\mu_j\mu_{s-j} \\
&< (C_r+h)^{s-2}\Big[h + C_r\textstyle\sum_{j=1}^{s-1}\mu_j\mu_{s-j}\Big] \\
&< (C_r+h)^{s-1}\textstyle\sum_{j=1}^{s-1}\mu_j\mu_{s-j} = (C_r+h)^{s-1}\mu_s\ ,
\end{aligned}$$

where the induction hypothesis has been used in the first inequality.

9.3.5 Estimates for the Normal Form

We consider again the Hamiltonian (9.2) characterized by the parameters ϱ, σ and ε as stated at the beginning of Section 9.2.1, with the hypotheses (i) and (ii) formulated there. No use will be made here of the hypothesis (iii), which will play instead a relevant role in the geometric part. We actually consider the Hamiltonian on a restricted domain $\mathcal{V}_\delta \subset \mathscr{G}_\varrho$, where $\mathcal{V}$ is a non-resonance domain.

The results concerning the canonical transformation are collected here.

Proposition 9.4: *Consider the Hamiltonian (9.2); let $\mathcal{V}$ be a non-resonance domain of type $(\mathscr{M}, \alpha, \delta, rK)$, and assume*

$$\mu := \frac{r\varepsilon A}{\alpha\delta} + 4e^{-K\sigma/2} \le \frac{1}{2}\,, \quad A = \frac{2^{10}}{e\sigma}\left(\frac{1+e^{-\sigma/2}}{1-e^{-\sigma/2}}\right)^n |f|_{\varrho,2\sigma}\,. \tag{9.31}$$

Then:

(i) there exists a finite generating sequence $\chi^{(r)} = \{\chi_1, \ldots, \chi_r\}$, with $\chi_s \in \mathscr{F}_{sK}$ which transforms the Hamiltonian to

$$H^{(r)}(p,q) = H_0(p) + Z_1(p,q) + \cdots + Z_r(p,q) + \mathscr{R}^{(r+1)}(p,q)\,,$$

where $Z_1, \ldots, Z_r$ are in normal form with respect to the resonance module $\mathscr{M}$ and $Z_s \in \mathscr{F}_{sK}$;

(ii) the generating sequence defines an analytic canonical transformation on the domain $\mathcal{V}_{\frac{3}{4}\delta} \times \mathbb{T}^n$, with the properties

$$\mathcal{V}_{\frac{5}{8}\delta} \subset T_\chi \mathcal{V}_{\frac{3}{4}\delta} \subset \mathcal{V}_{\frac{7}{8}\delta}\,, \quad \mathcal{V}_{\frac{5}{8}\delta} \subset T_\chi^{-1} \mathcal{V}_{\frac{3}{4}\delta} \subset \mathcal{V}_{\frac{7}{8}\delta}\,,$$

and, moreover, in $\mathcal{V}_{\frac{3}{4}\delta}$ one has

$$\bigl|p - T_{\chi^{(r)}} p\bigr| \le \frac{\delta}{16}\,, \quad \bigl|p - T_{\chi^{(r)}}^{-1} p\bigr| \le \frac{\delta}{16}\,; \tag{9.32}$$

(iii) the remainder $\mathscr{R}^{(r+1)}$ is estimated by

$$\bigl\|\mathscr{R}^{(r+1)}\bigr\|_{\frac{3}{4}(\delta,\sigma)} \le B\varepsilon\mu^r\,, \quad B = \frac{e\sigma}{2^7}A\,.$$

Proof. The proof of (i) and (ii) is a rather straightforward application of the results of the previous sections in this chapter, together with the quantitative estimates on the Lie transform in Section 6.6.1. Split the perturbation f as we did in Section. 9.3.1, and use the estimates of Lemma 7.15. Then apply Proposition 6.13 and Lemma 6.14 with the constants b and G given by Lemma 9.3 and with $d = 1/8$. The value of μ comes from

$$\frac{2eG}{d^2\delta\sigma} + b = \frac{2eF}{\alpha d^2\delta\sigma} + \frac{16(r-1)F}{e\alpha d^2\delta\sigma} + 4h \le \frac{16rF}{e\alpha d^2\delta\sigma} + 4h =: \mu_d\,; \tag{9.33}$$

set $d = 1/8$ and substitute h and F as given by Lemma 7.15. The proof of (iii) requires a little more work. Recall the expression (9.17) of the remainder, and use the following estimates which follow from the form of the solution of the homological equation (9.11) and from (9.33):

$$\|D_j H_{s-j}\|_{(1-2d)(\delta,\sigma)} \leq \mu_d^j h^{s-j-1} F \leq \frac{\mu_d^{s-1} F}{4^{s-j-1}} ,$$

$$\|\Psi_j - Z_j\|_{(1-d)(\delta,\sigma)} \leq \|\Psi_j\|_{(1-d)(\delta,\sigma)} \leq \frac{\mu_d^{j-1} F}{j} ,$$

$$\|D_{s-j}(\Psi_j - Z_j)\|_{(1-2d)(\delta,\sigma)} \leq \frac{\mu_d^{s-1} F}{j} .$$

Then evaluate

$$\|\mathscr{R}^{(r+1)}\|_{(1-2d)(\delta,\sigma)} \leq F \sum_{s>r} \mu_d^{s-1} \biggl(\sum_{j=0}^{s-1} \frac{1}{4^{s-j-1}} + \sum_{j=1}^{s} \frac{1}{s} \biggr) \leq 3F \frac{\mu_d^r}{1-\mu_d} ,$$

and (iii) follows by setting $d = 1/8$ and using condition (9.31). *Q.E.D.*

9.3.6 Local Stability Lemma

We now use the existence of truncated first integrals for the Hamiltonian in normal form in order to investigate the dynamics inside a non-resonance domain $\mathcal{V}$ of type $(\mathscr{M}, \alpha, \delta, rK)$. In view of statement (i) of Proposition 9.4 and of the discussion in Section 9.3.3 we know that there are $n - \dim \mathscr{M}$ approximate first integrals of the form $\Phi = T_\chi \Phi_0$, with $\Phi_0 = \langle \lambda, p \rangle$, with $\lambda \perp \mathscr{M}$. We may assume that $|\lambda| = 1$, which is not a restriction.

Lemma 9.5: *(Local stability lemma.) With the same hypotheses as Proposition 9.4, the following statement holds true: if $\bigl(q(t), p(t)\bigr)$ is an orbit satisfying $p(t) \in \mathcal{V}$ for $t \in [\tau^-, \tau^+]$, with $\tau^- < 0 < \tau^+$, then we have*

$$\mathrm{dist}\bigl(p(t), \Pi_{\mathscr{M}}(p(0))\bigr) < \tfrac{1}{2}\delta$$

for all $t \in [\tau^-, \tau^+] \cap [-t^*, t^*]$, where

$$t^* = \frac{D\delta}{4\varepsilon\mu^r} , \quad D = \frac{2^4}{A} ,$$

where $\Pi_{\mathscr{M}}(p)$ is the plane through p generated by $\mathscr{M}$, namely,

$$\Pi_{\mathscr{M}}(p) = \{p' \in \mathbb{R}^n \ : \ p' - p \in \mathrm{span}(\mathscr{M})\} ,$$

and with A as in Proposition 9.4.

The plane $\Pi_{\mathscr{M}}\bigl(p(0)\bigr) = p(0) + \mathrm{span}(\mathscr{M})$ will be called *plane of fast drift*. Moreover, we call *deformation* the nonlinear contributions to the functions Φ, which cause the orbit to oscillate around the plane of fast drift. Finally, we say that the remainder generates a *noise* which may cause a slow motion of the orbit in a direction transversal to the plane of fast drift; we call this slow motion *diffusion*.

Proof. Let $\Phi = T_\chi \Phi_0 = \Phi_0 + \Phi_1 + \ldots + \Phi_r$ be an approximate first integral with $\Phi_0 = \langle \lambda, p \rangle$, where $\lambda \perp \mathscr{M}$ and $|\lambda| = 1$. Use the elementary estimate:

$$\left|\Phi_0(t) - \Phi_0(0)\right| \leq \underbrace{\left|\Phi_0(t) - \Phi(t)\right|}_{\text{deformation}} + \underbrace{\left|\Phi(t) - \Phi(0)\right|}_{\text{diffusion}} + \underbrace{\left|\Phi(0) - \Phi_0(0)\right|}_{\text{deformation}} .$$

In view of (9.32) the first and the third terms in the r.h.s. is overall bounded by $\delta/8$. In order to estimate the second term, we use

$$\left|\dot{\Phi}(p,q)\right| \leq \left\|T_\chi\{\Phi_0, \mathscr{R}^{(r+1)}\}\right\|_{\frac{1}{2}(\delta,\sigma)} \leq \frac{16B\varepsilon}{e\sigma}\mu^r \text{ for } (p,q) \in \mathcal{V}_{\delta/2} \times \mathbb{T}^n .$$

This follows from the expression (9.18) of $\dot{\Phi}$, the estimate of the remainder in Proposition 9.4, the estimate (6.65) of the derivative of a function with respect to the angles q together with $|\lambda| = 1$, and the estimate (6.59) for the transformation with[8] T_χ. Thus,

$$\left|\Phi(t) - \Phi(0)\right| \leq |t| \frac{1}{D}\varepsilon\mu^r ,$$

which is smaller than $\delta/4$ if $|t| \leq t^*$, as claimed. Thus, the distance of the orbit from the plane through $p(0)$ determined by Φ_0 of the form just given is bounded by $\delta/2$, whatever is λ. Hence, the same is true for the plane of fast drift $\Pi_{\mathscr{M}}(p(0))$, which by definition is the intersection of planes of the type just considered. *Q.E.D.*

The result of the previous lemma may look mysterious and disappointing: in crude words, it's hard to see where the result is. Thus, it may be convenient to take a short break and illustrate it in an informal way, with the help of Fig. 9.2.

Let us consider the Hamiltonian $Z^{(0)}$ in normal form, forgetting for a moment the remainder $\mathscr{R}^{(r+1)}$; that is, let us suppose that the first integrals Φ are true ones. Let $p_0 \in \mathcal{V}$ be the initial point of an orbit. The $n - \dim \mathscr{M}$ functions $\Phi_0 = \langle \lambda, p \rangle$ determine a plane $\Pi_{\mathscr{M}}(p_0) = p_0 + \operatorname{span}(\mathscr{M})$ through the point p_0 which is invariant for the orbit in the coordinates of the normal form. Taking into account the coordinate transformation, which we must

[8] The reader may need a hint on how to deal with the restriction of domains. Note that $\{\Phi_0, \mathscr{R}^{(r+1)}\} = -\sum_{l=1}^{n} \lambda_l \frac{\partial \mathscr{R}^{(r+1)}}{\partial q_l}$. Thus, making a restriction $\frac{1}{8}(\delta,\sigma)$ of the domain and recalling that $|\lambda| = 1$, we get $\|\{\Phi_0, \mathscr{R}^{(r+1)}\}\|_{\frac{5}{8}(\delta,\sigma)} \leq \frac{8B}{e\sigma}\varepsilon\mu^r$. Finally, use the general estimate (6.59) for the Lie transform of a function with a further restriction of the domain by $\frac{1}{8}(\delta,\sigma)$, so that the final estimate is valid in $\frac{1}{2}(\delta,\sigma)$. This gives the desired estimate.

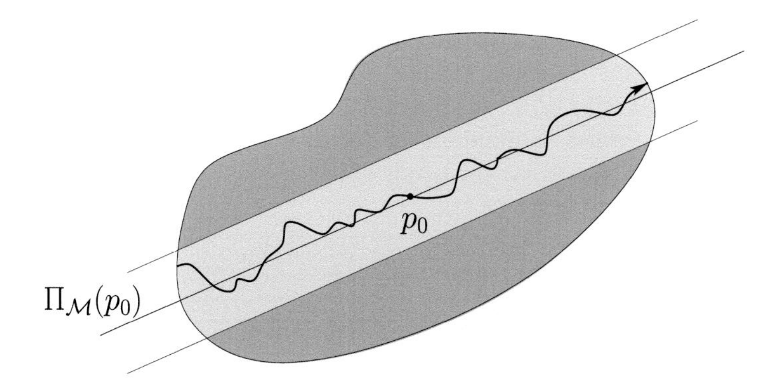

Figure 9.2 Illustrating the local stability estimate. The orbit starting at the point p remains in a small neighbourhood of the plane of fast drift, unless it leaves the local domain through a base.

apply then in the original coordinates, the plane is approximately invariant, due to the deformation. Therefore the orbit moves inside a strip of width $\delta/4$ around the plane: indeed, the deformation is bounded in view of the invariance of the functions Φ and of the estimate (9.32). However, this description ceases to apply if the orbit escapes from the domain $\mathcal{V}$ – an event which we cannot exclude a priori. For example, in the situation illustrated in Fig. 9.2 the orbit may well escape through the bases of the strip, namely, the intersections of the strip with the boundary of the domain $\mathcal{V}$.

Let us now take into account the effect of the remainder. In view of (9.18) the time derivative of the approximate first integral Φ is of the order of the remainder, that is, it is hopefully very small. This implies that the orbit may present a very slow motion in an arbitrary direction, including the direction transversal to the plane of fast drift, which is superimposed to the deformation. The relevant point is that if we consider a strip of width, for example, $\delta/2$ around the plane of fast drift, then the orbit must travel a distance $\delta/4$ in addition to that allowed by the deformation, which takes a long time, anyway.

Thus, we know a lot about the orbit as long as it remains in $\mathcal{V}$. The natural question is: *what happens if the orbit leaves the non-resonance domain?*

9.4 Geometric Part

The aim of the geometric part is to construct a covering of the action domain $\mathscr{W}$ with a family of non-resonance domains which are suitable for the analytic construction of the normal form. Thus, the main goal is to identify the domains where unwanted resonances do not appear. However, this is not enough: we must also take into account the motion along the plane of fast drift due to the flow: it could happen that the planes of fast drift create paths which allow the orbit to diffuse through the action space. This is the phenomenon called *overlapping of resonances*, and we must avoid it; conditions (9.5), in particular the condition of convexity of the unperturbed Hamiltonian, will play a major role.

The plan is the following. Suppose we are given:

(i) the domain $\mathscr{W}$ of the action variables, which may be extended to a complex domain $\mathscr{W}_\delta \subset \mathscr{G}_\varrho$;

(ii) a positive integer N, to be identified with rK of the analytic part;

(iii) a real analytic mapping $\omega(p) = \frac{\partial h}{\partial p}$ satisfying on $\mathscr{W}_\delta$ the convexity and boundedness conditions (9.5).

We cover $\mathscr{W}$ with non-resonance domains of type $\mathscr{M}, \beta_s/2, \delta_s, N$, where $\mathscr{M}$ is a resonance module, and the parameters β_s and δ_s are determined as functions of δ, N, $s = \dim \mathscr{M}$, the convexity and boundedness parameters m and M in (9.5) and the number n of degrees of freedom.

9.4.1 Geography of Resonances

Let us start with a series of definitions. The reader who is not familiar with this matter may find it useful to look at the discussion with examples and figures which is included after the definitions.

(i) N-moduli *and resonance parameters.* Let $\mathscr{M} \subset \mathbb{Z}^n$ be a resonance module, and let the subset $\mathscr{M}_N$ of $\mathscr{M}$ be defined as

$$\mathscr{M}_N = \left\{k \in \mathscr{M} \;:\; |k| \le N\right\} .$$

$\mathscr{M}$ will be said to be an N-module in case $\mathscr{M}_N$ contains $s = \dim \mathscr{M}$ independent vectors.

To every N-module of dimension s we shall associate a positive resonance parameter β_s with $\beta_0 < \beta_1 \ldots < \beta_n$.

(ii) Resonant manifold. To a resonant N-module $\mathscr{M}$ we associate the resonant manifold $\Sigma_{\mathscr{M}}$ defined as the set of points in $\mathscr{W}$ which are exactly resonant with $\mathscr{M}$. Formally,

$$\Sigma_{\mathscr{M}} = \left\{p \in \mathscr{W} \colon \langle k, \omega(p)\rangle = 0 \;\forall k \in \mathscr{M}\right\} .$$

(iii) Resonant zone. The resonant zone of multiplicity s associated to the resonance N-module $\mathscr{M}$, with $\dim_{\mathscr{M}} = s$, is defined as

$$\mathscr{Z}_{\mathscr{M}} = \left\{p \in \mathscr{W} \colon \bigl|\langle k, \omega(p)\rangle\bigr| < \beta_s \text{ for } s \text{ independent } k \in \mathscr{M}\right\} .$$

(iv) Resonant region of multiplicity s. The union of all resonant zones of multiplicity s forms the resonant region $\mathscr{Z}_s^*$. For $s = n + 1$ we define the corresponding resonant region as being empty. Formally,

$$\mathscr{Z}_s^* = \bigcup_{\dim \mathscr{M}=s} \mathscr{Z}_{\mathscr{M}} \quad \text{for} \quad 0 \le s \le n \; ; \qquad \mathscr{Z}_{n+1}^* = \emptyset \; .$$

Note that, by definition (iii), $\mathscr{Z}_0^* = \mathscr{W}$.

(v) Resonant block. We subtract from a resonant zone of multiplicity s everything belonging to the resonant region of multiplicity $s+1$; this defines the resonant block $\mathscr{B}_{\mathscr{M}}$. Formally,

$$\mathscr{B}_{\mathscr{M}} = \left(\mathscr{Z}_{\mathscr{M}} \setminus \mathscr{Z}_{s+1}^* \right) \; .$$

(vi) Resonant plane. To every $p \in \mathscr{B}_{\mathscr{M}}$ we associate the s-dimensional resonant plane through p defined as

$$\Pi_{\mathscr{M}}(p) = p + \mathrm{span}(\mathscr{M}) \; .$$

(vii) Extended plane. A strip of width $2\delta_s$ around the resonant plane $\Pi_{\mathscr{M}}(p)$ is the extended resonant plane. Formally,

$$\Pi_{\mathscr{M},\delta_s}(p) = \left\{ p' \in \mathbb{R}^n \; : \; \mathrm{dist}\big(p', \Pi_{\mathscr{M}}(p)\big) < \delta_s \right\} \; .$$

(viii) Cylinder and its basis. We take the intersection of an extended resonant plane through p with the resonant zone that p belongs to, and take the connected part of this set which contains p; this defines the cylinder $\mathscr{C}_{\mathscr{M},\delta_s}(p)$. Formally, we construct

$$\mathscr{C}_{\mathscr{M},\delta_s}(p) = \Pi_{\mathscr{M},\delta_s}(p) \cap \mathscr{Z}_{\mathscr{M}}$$

and select the connected component as stated earlier. The basis of the cylinder is the intersection of the cylinder with the boundary of the resonant zone, that is, $\partial \mathscr{Z}_{\mathscr{M}} \cap \Pi_{\mathscr{M},\delta_s}(p)$.

(ix) Extended block. The union of all cylinders generated by points of the same block defines the extended block. Formally,

$$\mathscr{B}_{\mathscr{M},\delta_s} = \bigcup_{p \in \mathscr{B}_{\mathscr{M}}} \mathscr{C}_{\mathscr{M},\delta_s}(p) \; .$$

Moreover, we will denote with $\left(\mathscr{B}_{\mathscr{M},\delta_s}\right)_\delta$ the further δ extension in the complex of the set $\mathscr{B}_{\mathscr{M},\delta_s}$.

At this point it is perhaps necessary to take a deep breath and allow ourselves an instant of pause, so that we may reconsider in more detail the chain of definitions just presented. We illustrate the construction in general terms; the reader may want to consider the case of a system of two degrees of freedom with unperturbed Hamiltonian $H_0(p) = (p_1^2 + p_2^2)/2$ (see, e.g., [90]); in this case $\mathscr{W}$ may be chosen to be an elliptic neighbourhood of the origin of the action space $\mathbb{R}^2$. This may be a useful exercise for a first approach.

The definition of N-moduli is clearly related to the construction of the normal form in the analytic part: a Fourier cutoff on the perturbation is

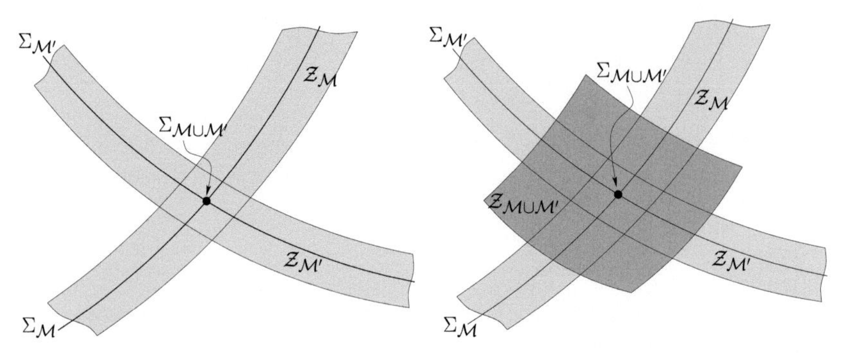

Figure 9.3 The construction of resonant zones. The module $\{0\}$, namely of multiplicity zero, generates a resonant manifold $\Sigma_{\{0\}}$; it coincides with its resonant zone $\mathscr{Z}_{\{0\}}$, which in turn coincides with the whole domain $\mathscr{W}$. The (different) modules $\mathscr{M}$ and $\mathscr{M}'$ of multiplicity one generate the resonant manifolds $\Sigma_{\mathscr{M}}$ and $\Sigma_{\mathscr{M}'}$, actually two different curves in the two-dimensional figure. The corresponding resonant zones are the strips $\mathscr{Z}_{\mathscr{M}}$ and $\mathscr{Z}_{\mathscr{M}'}$ around the manifolds (left panel). The intersection $\Sigma_{\mathscr{M}}\cap\Sigma_{\mathscr{M}'}$ is the resonant manifold $\Sigma_{\mathscr{M}\cup\mathscr{M}'}$ of multiplicity 2 which degenerates into a point in the two-dimensional figure. The corresponding resonant zone $\mathscr{Z}_{\mathscr{M}\cup\mathscr{M}'}$ is the curvilinear rectangle around $\Sigma_{\mathscr{M}\cup\mathscr{M}'}$ (right panel).

essential in order to overcome the negative result of Poincaré's theorem. The resonance parameters at this point look like guys playing a mysterious role. Let's skip discussing them for the moment: they will show up during the construction.

Definitions (i) to (v) establish the bases for construction of non-resonance domains.

The resonant manifolds defined in (i) form the skeleton of the geography of resonances. On these manifolds the resonances are exact.

The resonant zones defined in (ii) are the parts of the domain where some expression $\langle k,\omega\rangle$ with $|k|\le N$ becomes small (Fig. 9.3); we may say that they are strips around resonant manifolds where the normal form must be a resonant one in order to avoid divisors which are too small. To every zone we associate a multiplicity, namely the dimension of the corresponding module $\mathscr{M}$. The positive information is that, having fixed a module, we know that in the corresponding zone the resonances in $\mathscr{M}$ are there; however, we cannot exclude the existence of further ones which do not belong to the module. For example, in the left panel of Fig. 9.3 this happens in the intersection between the zones $\mathscr{Z}_{\mathscr{M}}$ and $\mathscr{Z}_{\mathscr{M}'}$. The reader may notice that in the right panel of Fig. 9.3 the resonant zone $\mathscr{Z}_{\mathscr{M}\cup\mathscr{M}'}$ is larger than the zones of less multiplicity; this will be explained later.

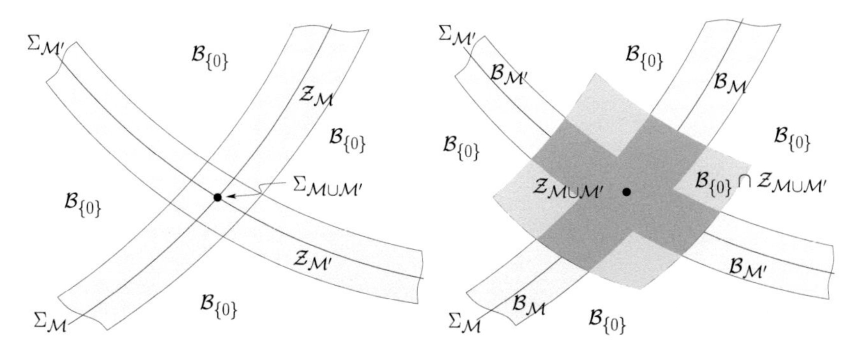

Figure 9.4 The construction of resonant regions and blocks. The resonant region $\mathscr{Z}_0^*$ of multiplicity zero is the whole domain $\mathscr{W}$. The resonant region $\mathscr{Z}_1^*$ of multiplicity one is the cross-shaped area in the left panel, namely the union of the two zones of multiplicity one in the left panel of Fig. 9.3. The resonant region $\mathscr{Z}_2^*$ in the right panel of the figure coincides with the resonant zone in the right panel of Fig. 9.3. This is an artifact of the plane representation together with our choice to represent only two manifolds with one intersection; we expect that the reader will have no difficulty in figuring out what happens in higher dimensions with many resonant manifolds. The block $\mathscr{B}_{\{0\}}$ is the complement of the cross-shaped area in the left panel. The blocks $\mathscr{B}_{\mathscr{M}}$ and $\mathscr{B}_{\mathscr{M}'}$ are obtained by taking out the rectangle $\mathscr{Z}_{\mathscr{M}\cup\mathscr{M}'}$ from the zones $\mathscr{Z}_{\mathscr{M}}$ and $\mathscr{Z}_{\mathscr{M}'}$ in the right panel. The block $\mathscr{B}_{\mathscr{M}\cup\mathscr{M}'}$ in our case coincides with $\mathscr{Z}_{\mathscr{M}\cup\mathscr{M}'}$.

The resonant regions defined in (iii) are characterized as the parts of the whole domain where we know that there are at least s resonances (see Fig. 9.4). Again, this does not exclude more resonances The definition of the resonant region of multiplicity $n+1$ is just technical: it allows us to simplify the definition of blocks, in particular of the block of maximal multiplicity.

Definition (iv) takes care of eliminating unwanted resonances. A block of multiplicity s is constructed by picking a resonant zone of the same multiplicity and taking out everything which belongs to a resonant region of multiplicity $s+1$. Recall that a resonant zone has a resonance module associated with it; taking out the region of multiplicity $s+1$ removes the resonances which do not belong to the module. Therefore, the resonant blocks of multiplicity s turn out to be the domains where there are exactly s resonances, and we know which ones. The blocks are a *covering* of $\mathscr{W}$, not a partition. For instance, the blocks $\mathscr{B}_{\{0\}}$ and $\mathscr{B}_{\mathscr{M}\cup\mathscr{M}'}$ in our example have a non-empty intersection – which is harmless, as we shall see later. However, by construction, *there is no overlap between blocks of the same multiplicity:* this is the relevant property.

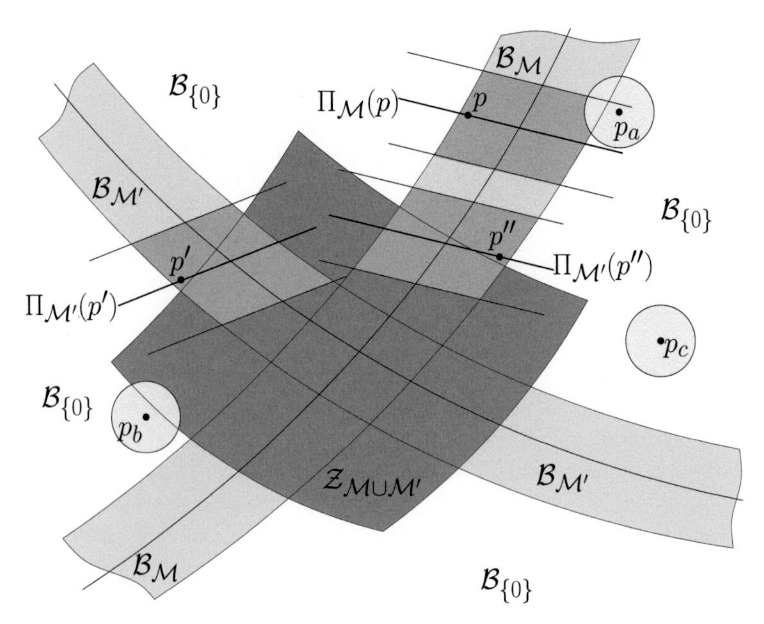

Figure 9.5 Illustrating the construction of the cylinders in the case of dimension 2. The points p_a, p_b and p_c belong to a resonant block of multiplicity 0; the corresponding cylinders are disks. The points p, p' and p'' belong to resonant blocks of multiplicity 1, the planes of fast drift are $\Pi_{\mathscr{M}}(p)$, $\Pi_{\mathscr{M}'}(p')$ and $\Pi_{\mathscr{M}'}(p'')$, respectively. The corresponding cylinders $\mathscr{C}_{\mathscr{M}(p)}$, $\mathscr{C}_{\mathscr{M}'(p')}$ and $\mathscr{C}_{\mathscr{M}'(p')}$ are the strips around the planes trimmed at the border of the resonant zone.

The structure of the blocks is an excellent basis for constructing the non-resonance domains which we need in the analytic part: we know exactly which resonances are inside the blocks. However, we cannot stop here: the original action coordinates which we are using are not the best ones. We know from the analytic part that the action coordinates of the normal form are better. We also know that the latter coordinates differ from the original ones by a small amount, which we have called *the deformation*. The new coordinates are better than the original ones, but they are not perfect: besides the deformation there is a *noise* in the dynamics, due to the remainder of the normalization procedure. The noise could cause a slow *diffusion*, which we should take into account. All this makes somewhat fuzzy the plain picture of the blocks; but this is unavoidable.

Definitions (vi)–(viii) continue the construction while paying some attention to the dynamics. Take any point p in a given resonant block associated to a resonance N-module $\mathscr{M}$, and try to imagine where the orbit through p may go. In view of the local stability, Lemma 9.5, the orbit remains close

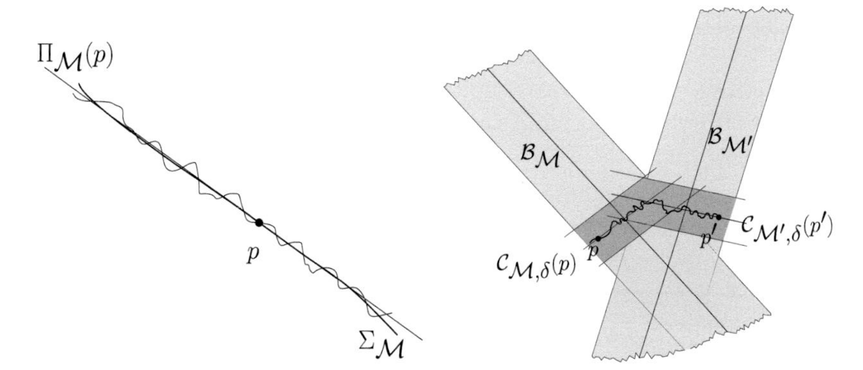

Figure 9.6 Ways to diffusion. Left: the fast drift plane follows the resonant manifold. Right: two different cylinders of the same multiplicity have a non-empty intersection. An orbit with initial point p may move in the cylinder $\mathscr{C}_{\mathscr{M},\delta}(p)$ until it reaches the cylinder $\mathscr{C}_{\mathscr{M}',\delta}(p')$. Then it moves in the latter cylinder, possibly getting out of its initial resonant block. This mechanism may create channels of diffusion for the orbits.

to a plane through p parallel to $\mathscr{M}$, but is subject to both deformation and noise. The plane is exactly the resonant plane of definition (vi), and the strip used there in order to control both the deformation and the noise is the extended plane of definition (vii). The construction of planes, extended planes and cylinders is illustrated in Fig. 9.5. For a point belonging to a block of multiplicity 0, such as the points p_a, p_b and p_c in the figure, there is actually no resonant plane (just because $\dim \mathscr{M} = 0$); the cylinder is indeed a disk. For p in a block $\mathscr{B}_{\mathscr{M}}$ of multiplicity 1, such as the points p, p' and p'' in the figure, the cylinder is a strip around the plane trimmed at the border of the corresponding zone $\mathscr{Z}_{\mathscr{M}}$. Note that both the planes and the cylinders may well extend over the border of the block: for instance, the cylinder $\mathscr{C}_{\mathscr{M},\delta}(p')$ in Fig. 9.5 overlaps the resonant zone of multiplicity 2. This is harmless, provided the cylinder still has an empty intersection with other resonant zones of the same multiplicity 1. This is a delicate point that deserves some further attention.

A first mechanism which generates diffusion is illustrated in a very rough manner in the left panel of Fig. 9.6. Suppose that at the point $p \in \Sigma_{\mathscr{M}}$ the resonant plane $\Pi_{\mathscr{M},\delta}(p)$ turns out to be tangent to the resonant manifold $\Sigma_{\mathscr{M}}$, and that $p \in \Sigma_{\mathscr{M}}$ and $\Pi_{\mathscr{M},\delta}(p)$ lie very close each other. Then the cylinder may extend very far inside the resonant zone, thus creating a diffusion channel where fast drift may occur. The resonance parameters provide no help in this case. We need instead a condition which assures that the resonant manifold $\Sigma_{\mathscr{M}}$ and the resonant plane $\Pi_{\mathscr{M},\delta}(p)$ are transversal, or at least that the tangency is of a finite order, so that they cannot lie too close to each other. Here the condition of convexity plays a main role: it assures the transversality between $\Sigma_{\mathscr{M}}$ and $\Pi_{\mathscr{M},\delta}(p)$. A weaker condition, namely *steepness*, which allows a finite-order tangency, has been used by Nekhoroshev. However, the proof turns out to be considerably simpler by using convexity.

The following easy example of a system which is neither convex nor steep has been proposed by Nekhoroshev himself. Consider the Hamiltonian $H(p,q) = (p_1^2 - p_2^2)/2 + \varepsilon \sin(q_1 - q_2)$. The orbit $q_1(t) = -\varepsilon^2 t/2$, $q_2(t) = -\varepsilon^2 t/2$, $p_1(t) = -\varepsilon t$ and $p_2(t) = \varepsilon t$ moves at constant speed along the resonant manifold $p_1 + p_2 = 0$, which coincides with the plane of fast drift at every point.

A second dangerous situation which must be avoided is illustrated in the right panel of fig. 9.6: the cylinder associated to $p \in \mathscr{B}_{\mathscr{M}}$ has a non-empty intersection with a resonant zone $\mathscr{Z}_{\mathscr{M}'}$ of the same multiplicity, with $\mathscr{M}' \neq \mathscr{M}$. The orbit through p may well enter the resonant zone $\mathscr{Z}_{\mathscr{M}'}$ – this is not prevented by Lemma 9.5. But in this case the orbit may get close to a different resonant plane associated to another point in the resonant zone $\mathscr{Z}_{\mathscr{M}'}$ – the point p' in the figure. Again, Lemma 9.5 does not forbid the orbit to follow the latter plane, thus getting out of the cylinder $\mathscr{C}_{\mathscr{M},\delta}(p)$. This process may possibly repeat, thus generating a sort of communication channel which allows the orbit to diffuse through different resonant regions of the same multiplicity but with different resonances, and to get very far from its initial point. This phenomenon was discovered by Contopoulos [48] and later by Chirikov [43], and has become known as *overlapping of resonances.* The choice of the parameters $\beta_0 < \beta_1 < \cdots < \beta_n$ plays a main role here. The increasing width of the resonant zones with the order of resonance creates a sort of safety border which allows the cylinders to enter the resonant zones of higher multiplicity, still being bounded far from the zones of the same multiplicity.

The final step is the construction of the extended block, illustrated in Fig. 9.7. The extension takes into account (i) the cylinders of fast drift which are created by the dynamics, (ii) the deformation due to the transformation of coordinates, and (iii) the slow diffusion of orbits due to the noise. The non-resonance domain required by the analytic part must include all these extensions, and a further complex extension which allows us to apply the analytic theory. This justifies the definition of extended block as the union of all the cylinders based on points of the same resonant block. The role of the safety region created by the resonance parameters is illustrated in the figure: an extended block of multiplicity s is not allowed to overlap a different resonant zone of the same multiplicity. This will act as a condition on both the convexity and the resonance parameters. A further complex extension will be necessary in order to allow the safety border required by application

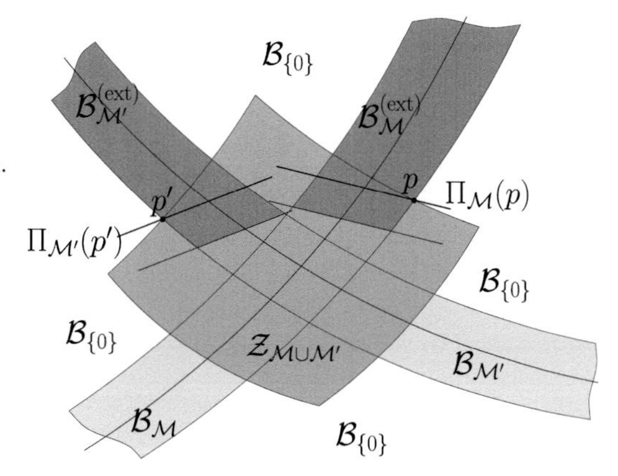

Figure 9.7 The extended block. The condition that extended blocks of the same multiplicity have empty intersection must be respected.

of Cauchy's estimates. The latter extension is not represented in the figure, but it should be easy to imagine it.

9.4.2 Three Technical Lemmas

The following lemmas are crucial in order both to estimate the diameter of the cylinders and to prove the property of non-overlapping of resonances.

The first lemma states a crucial property of the resonant blocks $\mathscr{B}_{\mathscr{M}}$. It shows that they are the likely candidates of the distinguished sets in $\mathbb{R}^n$ which meet the assumptions in Proposition 9.4.

Lemma 9.6: *Let $\mathscr{M}$ be an N-module of $\dim \mathscr{M} = s$. Then for every $p \in \mathscr{B}_{\mathscr{M}}$ the following non-resonance condition holds true:*

$$|k \cdot p| \geq \beta_{s+1} \quad \text{if} \quad k \notin \mathscr{M} \quad \text{and} \quad |k| \leq N \ .$$

Proof. By contradiction, assume that $|k \cdot p| < \beta_{s+1}$ for some $p \in \mathscr{B}_{\mathscr{M}}$ and $k \notin \mathscr{M}$ with $|k| \leq N$. Since $\mathscr{B}_{\mathscr{M}} \subset \mathscr{Z}_{\mathscr{M}}$ and since $\beta_s < \beta_{s+1}$, we conclude from the definition of $\mathscr{Z}_{\mathscr{M}}$ that there are $s+1$ independent vectors $k \in \mathscr{M}_N$ satisfying $|k \cdot p| < \beta_{s+1}$. Consequently, $p \in \mathscr{Z}_{\mathscr{M}'}$ for a resonant N-module $\mathscr{M}'$ with $\dim \mathscr{M}' = s+1$. Therefore, $p \in \mathscr{Z}_{\mathscr{M}'} \subset \mathscr{Z}^*_{s+1}$, and so $p \notin \mathscr{B}_{\mathscr{M}} = \mathscr{Z}_{\mathscr{M}} \setminus \mathscr{Z}^*_{s+1}$, contradicting $p \in \mathscr{B}_{\mathscr{M}}$. *Q.E.D.*

The second lemma has a geometric character. It contains an estimate of the length of a vector when something is known concerning its projections on a particular, non-orthonormal basis.

Lemma 9.7: *Let $1 \leq s \leq n$, and let $u^{(1)}, \ldots, u^{(s)}$ be linearly independent vectors in $\mathbb{R}^n$ satisfying $|u^{(j)}| \leq N$ for some positive N and for $1 \leq j \leq s$. Denote by $\mathrm{Vol}(u^{(1)}, \ldots, u^{(s)})$ the s-dimensional volume of the parallelepiped with sides $u^{(1)}, \ldots, u^{(s)}$; let, moreover, $w \in \mathrm{span}\big(u^{(1)}, \ldots, u^{(s)}\big)$ be any* *vector, and α a positive constant. Then the following statement holds true: if $|\langle w, u^{(j)}\rangle| \le \alpha$ for $1 \le j \le s$, then the Euclidean norm $\|w\|$ is bounded by*

$$\|w\| \le \frac{sN^{s-1}\alpha}{\mathrm{Vol}(u^{(1)},\ldots,u^{(s)})} \ .$$

Proof. For $s = 1$, the claim is trivially true. For $2 \le s \le n$, we proceed by induction. Let $V = \mathrm{span}(u^{(1)},\ldots,u^{(s)})$ and $V' = \mathrm{span}(u^{(1)},\ldots,u^{(s-1)})$ be the linear subspaces in $\mathbb{R}^n$ spanned by $u^{(1)},\ldots,u^{(s)}$ and by $u^{(1)},\ldots,u^{(s-1)}$, respectively. Introduce the decomposition

$$w = w' + w'' \quad , \ u^{(s)} = u' + u'' \ ,$$

where w' and u' belong to V', and w'' and u'' are orthogonal to V'. In view of $u^{(j)} \in V'$ for $1 \le j < s$ we have $\left|\langle w', u^{(j)}\rangle\right| = \left|\langle w, u^{(j)}\rangle\right| < \alpha$ for $1 \le j < s$. Thus, by the induction hypothesis we have also

$$\|w'\| \,\mathrm{Vol}(u^{(1)},\ldots,u^{(s-1)}) < (s-1)N^{s-2}\alpha \ , \tag{9.34}$$

so that we immediately get

$$\begin{aligned} \|w'\| \,\mathrm{Vol}(u^{(1)},\ldots,u^{(s)}) &= \|w'\| \,\mathrm{Vol}(u^{(1)},\ldots,u^{(s-1)})\|u''\| \\ &< (s-1)N^{s-2}\alpha\|u''\| \ . \end{aligned} \tag{9.35}$$

On the other hand, from

$$\left\|w''\right\| \left\|u''\right\| = \left|\langle w'', u''\rangle\right| = \left|\langle w, u^{(s)}\rangle - \langle w', u'\rangle\right| < \alpha + \left\|w'\right\| \left\|u'\right\| ,$$

we also obtain

$$\begin{aligned} \left\|w''\right\| \mathrm{Vol}(u^{(1)},\ldots,u^{(s)}) &= \left\|w''\right\| \left\|u''\right\| \mathrm{Vol}(u^{(1)},\ldots,u^{(s-1)}) \\ &< \left(N + (s-1)\|u'\|\right) N^{s-2}\alpha \ , \end{aligned} \tag{9.36}$$

where (9.34) and the inequality

$$\mathrm{Vol}(u^{(1)},\ldots,u^{(s-1)}) \le \left\|u'^{(1)}\right\| \ldots \left\|u'^{(s-1)}\right\| \le \left|u'^{(1)}\right| \ldots \left|u'^{(s-1)}\right| \le N^{s-1}$$

have been used. From (9.35) and (9.36), we conclude

$$\begin{aligned} \|w\| \,\mathrm{Vol}(u^{(1)},\ldots,u^{(s)}) &= \left(\|w'\|^2 + \|w''\|^2\right)^{1/2} \mathrm{Vol}(u^{(1)},\ldots,u^{(s)}) \\ &\le \left[(s-1)^2\|u''\|^2 + \left(N + (s-1)\|u'\|\right)^2\right]^{1/2} N^{s-2}\alpha \\ &= \left[(s-1)^2\|u^{(s)}\|^2 + N^2 + 2N(s-1)\|u'\|\right]^{1/2} N^{s-2}\alpha \\ &\le \left[(s-1)^2N^2 + N^2 + 2N^2(s-1)\right]^{1/2} N^{s-2}\alpha \\ &= sN^{s-1}\alpha \ , \end{aligned}$$

as claimed. *Q.E.D.*

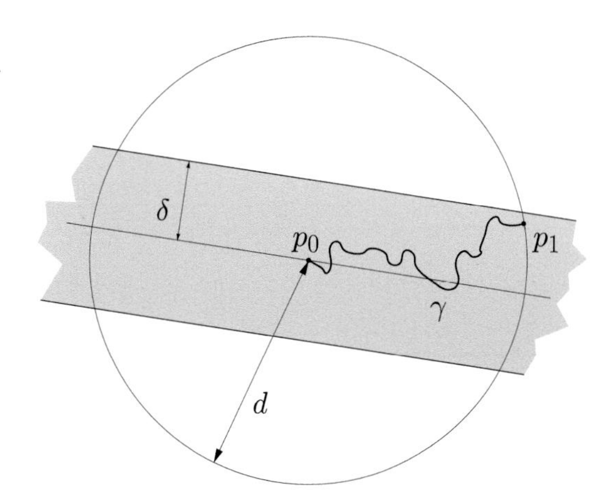

Figure 9.8 The geometric picture of Lemma 9.8. A ball of diameter d is intersected by an extended plane of width $\delta < d$. The curve γ connects the centre p_0 of the ball with a point p_1 belonging to both the extended plane and the sphere of radius d. Moreover, the curve γ lies entirely in the intersection of the closed ball with the extended plane.

The third lemma exploits the convexity property of the unperturbed Hamiltonian $H_0(p)$. Recall that the convexity hypothesis is assumed to hold in the domain $\mathscr{W}_\delta \subset \mathscr{G}_\varrho$, as stated at the beginning of Section 9.3. Looking at Fig. 9.8 may help the reader in understanding the statement of the lemma.

We need some local notations. Let Λ be an arbitrary s-dimensional subspace of $\mathbb{R}^n$, with $1 \le s < n$. The plane through $p \in \mathbb{R}^n$ parallel to Λ will denoted by $\Pi_\Lambda(p)$. The strip of width 2δ around the plane $\Pi_\Lambda(p)$ will be denoted by $\Pi_{\Lambda,\delta}(p)$. Formally,

$$\Pi_{\Lambda,\delta}(p) = \left\{p' \in \mathbb{R}^n \ : \ \operatorname{dist}(p', \Pi_\Lambda(p)) < \delta\right\} .$$

The open ball of radius $d > 0$ and centre p will be denoted by $B_d(p)$, and the corresponding sphere will be denoted by $\partial B_d(p)$. Finally, the projection of a vector $w \in \mathbb{R}^n$ on the subspace Λ will be denoted by w_Λ.

Lemma 9.8: *Let the unperturbed Hamiltonian $H_0(p)$ satisfy (9.5). Let Λ be a subspace of $\mathbb{R}^n$, with $1 \le \dim\Lambda < n$. Let $p_0 \in \mathscr{W}$ and $p_1 \in \partial B_d(p_0) \cap \Pi_{\Lambda,\delta}(p_0)$, and let $\gamma \subset B_d(p_0) \cap \Pi_{\Lambda,\delta}(p_0)$ be a continuous curve joining p_0 and p_1. Then the following statement holds true: if*

$$\delta \le \frac{md}{4M} \tag{9.37}$$

(so that $\delta \le d/4$), then there exists a point $\tilde{p} \in \gamma$ such that the projection $\omega_\Lambda(p)$ of $\omega(p)$ on Λ satisfies

$$\left\|\omega_\Lambda(\tilde{p})\right\| > \frac{md}{4} .$$

Proof. Let $\xi = p_1 - p_0$, and write $\xi = \xi_\Lambda + \xi'$, where $\xi_\Lambda \in \Lambda$ and $\xi' \perp \Lambda$.

Thus, $\|\xi\| = d$, and in view of $\|\xi'\| < \delta$, we have

$$\|\xi_\Lambda\| > d\sqrt{1 - \left(\frac{\delta}{d}\right)^2} \tag{9.38}$$

by Pythagoras' theorem. Let now the function $\varphi(p)$ be defined as

$$\varphi(p) = \frac{\langle \omega(p), \xi_\Lambda \rangle}{\|\xi_\Lambda\|} \ .$$

Consider the restriction of the function $\varphi(p)$ to the segment joining p_0 with p_1, namely $\varphi(p_0 + t\xi)$ with $t \in [0, 1]$. By the mean value theorem there exists $t' \in [0, 1]$ such that

$$\begin{aligned}\varphi(p_1) - \varphi(p_0) &= \left.\frac{d\varphi}{dt}\right|_{p_0 + t'\xi} = \frac{\langle C(p_0 + t'\xi)\xi, \xi_\Lambda \rangle}{\|\xi_\Lambda\|} \\ &= \frac{\langle C(p_0 + t'\xi)\xi_\Lambda, \xi_\Lambda \rangle}{\|\xi_\Lambda\|} + \frac{\langle C(p_0 + t'\xi)\xi', \xi_\Lambda \rangle}{\|\xi_\Lambda\|} \ .\end{aligned}$$

In view of (9.5) we estimate

$$|\varphi(p_1) - \varphi(p_0)| > m\|\xi_\Lambda\| - M\|\xi'\| \ ,$$

which holds true provided $M\|\xi'\| < m\|\xi_\Lambda\|$. This is assured by the condition (9.37), also recalling that $m \le M$. By a straightforward use of (9.38), we immediately get

$$\begin{aligned}|\varphi(p_1) - \varphi(p_0)| &> md\sqrt{1 - \left(\frac{\delta}{d}\right)^2} - M\delta \\ &\ge md\left[\sqrt{1 - \left(\frac{m}{4M}\right)^2} - \frac{1}{4}\right] > \frac{md}{2} \ .\end{aligned}$$

Consider now the restriction of the function $\varphi(p)$ to the curve γ. In view of the latter inequality there exists a point $\tilde{p} \in \gamma$ such that

$$|\varphi(\tilde{p})| > \frac{md}{4} \ .$$

The claim follows from the obvious inequality $\|\omega_\Lambda(p)\| \ge |\varphi(p)|$. *Q.E.D.*

9.4.3 Geometric Properties of the Geography of Resonances

This section contains the statement of the main properties which follow from the geography of resonances. All results are collected in the following.

Proposition 9.9: *Let N be a positive integer. For a positive δ, let the function $H_0(p)$ be real analytic in $\mathscr{W}_\delta$ and in this domain satisfy the hypothesis (9.5). Define the parameters $\beta_0 < \cdots < \beta_n$ and $\delta_0 < \cdots < \delta_n \le \delta/2$ as*

$$
\begin{aligned}
\beta_0 &= \left[\left(\frac{4M}{m}\right)^{n+1} (n+2)!\, N^{(n^2+n-2)/2} \right]^{-1} m\delta \ , \\
\beta_s &= \left(\frac{4M}{m}\right)^{s} (s+1)!\, N^{s(s-1)/2} \beta_0 \quad \text{for } 1 \le s \le n \ , \\
\delta_s &= \frac{\beta_s}{2NM} \quad \text{for } 0 \le s \le n \ .
\end{aligned}
\tag{9.39}
$$

Then the following statements hold true:

(i) *The blocks are a covering of $\mathscr{W}$; that is,*

$$\mathscr{W} \subset \bigcup_{\mathscr{M}} \mathscr{B}_{\mathscr{M}} \ ,$$

the union being extended to all N-moduli.

(ii) *For $0 \le s \le n$*

$$\mathscr{W} \setminus \mathscr{Z}^*_{s+1} \subset \bigcup_{0 \le \dim \mathscr{M} \le s} \mathscr{B}_{\mathscr{M}} \ .$$

(iii) *If $\mathscr{M}$ is an N-module with $\dim \mathscr{M} = s$, then for every $p \in \mathscr{B}_{\mathscr{M}}$, we have*

$$\operatorname{diam} \mathscr{C}_{\mathscr{M},\delta_s}(p) < d_s := \frac{4sN^{s-1}}{m} \beta_s < \delta \ .$$

(iv) *If $\mathscr{M}$ and $\mathscr{M}'$ are N-moduli with $\mathscr{M} \ne \mathscr{M}'$ and $\dim \mathscr{M} = \dim \mathscr{M}' = s$, then*

$$\overline{\mathscr{B}_{\mathscr{M},\delta_s}} \cap \mathscr{Z}_{\mathscr{M}'} = \emptyset \ ,$$

with the overline denoting the closure.

(v) *For an N-module $\mathscr{M}$ with $\dim \mathscr{M} = s$, the complex extension $\left(\mathscr{B}_{\mathscr{M},\delta_s}\right)_{\delta_s}$ of the extended block satisfies*

$$\left(\mathscr{B}_{\mathscr{M},\delta_s}\right)_{\delta_s} \cap \mathbb{R}^n \subset \mathscr{W}_\delta \ .$$

(vi) *For an N-module $\mathscr{M}$ with $\dim \mathscr{M} = s$, the extended block $\mathscr{B}_{\mathscr{M},\delta_s}$ is a non-resonance domain of type $(\mathscr{M}, \beta_s/2, \delta_s, N)$.*

Proof. The statements (i) and (ii) readily follow from the definition. Indeed, observe that

$$
\begin{aligned}
\bigcup_{\dim \mathscr{M} = s} \mathscr{B}_{\mathscr{M}} &= \bigcup_{\dim \mathscr{M} = s} \left[(\mathscr{Z}_{\mathscr{M}} \setminus \mathscr{Z}^*_{s+1}) \cap \mathscr{W} \right] \\
&= \left(\bigcup_{\dim \mathscr{M} = s} \mathscr{Z}_{\mathscr{M}} \right) \setminus \mathscr{Z}^*_{s+1} = \mathscr{Z}^*_s \setminus \mathscr{Z}^*_{s+1} \ .
\end{aligned}
$$

For $s = 0$, the statement (ii) is trivially true. For we have $\mathscr{Z}_0^* = \mathscr{W}$, and so also $\mathscr{W} \setminus \mathscr{Z}_1^* \subset \mathscr{Z}_0^* \setminus \mathscr{Z}_1^*$. For $s > 0$, proceed by induction assuming that the statement (ii) is true up to $s-1$; then, using the trivial equality $\mathscr{W} = \mathscr{W} \setminus \mathscr{Z}_s^* \cup \mathscr{Z}_s^*$, we have

$$\begin{aligned}\mathscr{W} \setminus \mathscr{Z}_{s+1}^* &= (\mathscr{W} \setminus \mathscr{Z}_s^*) \setminus \mathscr{Z}_{s+1}^* \cup \left(\mathscr{Z}_s^* \setminus \mathscr{Z}_{s+1}^*\right) \\ &\subset \Bigl(\bigcup_{\dim \mathscr{M} < s} \mathscr{B}_{\mathscr{M}}\Bigr) \setminus \mathscr{Z}_{s+1}^* \cup \Bigl(\bigcup_{\dim \mathscr{M} = s} \mathscr{B}_{\mathscr{M}}\Bigr) \\ &\subset \bigcup_{\dim \mathscr{M} \le s} \mathscr{B}_{\mathscr{M}} \ .\end{aligned}$$

This proves the statement (ii) for $s = 1, \ldots, n$. Moreover, the statement (i) follows from (ii) with $s = n$, since $\mathscr{Z}_{n+1}^* = \emptyset$.
The proof of (iii) is based on Lemmas 9.7 and 9.8, together with the definition of the sequences β_s and δ_s. We actually prove that $\mathscr{C}_{\mathscr{M},\delta_s}(p) \subset B_{d_s/2}(p)$, the open ball with centre p and radius $d_s/2$. In view of Definition (9.39) of the sequence β_s the inequality $d_s < \delta$ immediately follows, so that $B_{d_s/2}(p) \subset \mathscr{W}_\delta$. By contradiction, assume that there exists a point $p' \in \mathscr{C}_{\mathscr{M},\delta_s}(p) \setminus B_{d_s/2}(p)$. Let γ' be a curve entirely contained in $\mathscr{C}_{\mathscr{M},\delta_s}(p)$ and joining p with p'. Such a curve exists because $\mathscr{C}_{\mathscr{M},\delta_s}(p)$ is connected, by construction. Let $\tilde{p}$ be any point of γ', and consider the projection $\omega_{\mathscr{M}}(\tilde{p})$ of $\omega(\tilde{p})$ on the plane $\Pi_{\mathscr{M}}$. Since $\tilde{p} \in \mathscr{Z}_{\mathscr{M}}$ there are s independent vectors $k \in \mathscr{M}$ such that $\bigl|\langle k, \omega_{\mathscr{M}}(\tilde{p})\rangle\bigr| < \beta_s$, and so by Lemma 9.7 and by (9.39), we have

$$\bigl\|\omega_{\mathscr{M}}(\tilde{p})\bigr\| < sN^{s-1}\beta_s = \frac{md_s}{4} \ . \tag{9.40}$$

On the other hand, recall that $\gamma' \subset \Pi_{\mathscr{M},\delta_s}(p)$ by construction, and that $p' \notin B_{d_s/2}$ by assumption. Then there exists $p_1 \in \gamma' \cap \bigl(\partial B_{d_s/2}(p) \cap \Pi_{\mathscr{M}}(p)\bigr)$. Denoting by γ the part of γ' joining p to p_1 we apply Lemma 9.8 with p, $\Pi_{\mathscr{M}}$, $d_s/2$ and δ_s in place of p_0, Π_Λ, d and δ, respectively. The lemma states that there exists $\tilde{p} \in \gamma$ such that

$$\bigl\|\omega_{\mathscr{M}}(\tilde{p})\bigr\| > \frac{md_s}{4} \ ,$$

contradicting (9.40). This proves that $\mathscr{C}_{\mathscr{M},\delta_s}(p) \subset B_{d_s/2}(p)$, from which the estimate of the diameter in (iii) readily follows.
Coming to (iv), let us first check that if $p \in \mathscr{B}_{\mathscr{M},\delta_s}$, then

$$\bigl|\langle k, \omega(p)\rangle\bigr| \ge \beta_s \quad \text{for} \quad k \notin \mathscr{M} \quad \text{and} \quad |k| \le N \ . \tag{9.41}$$

For, in view of the definition of $\mathscr{B}_{\mathscr{M},\delta_s}$, there exists a $p' \in \mathscr{B}_{\mathscr{M}}$ such that $p \in \mathscr{C}_{\mathscr{M},\delta_s}(p')$. If k is chosen as in (9.41), then $|\langle k, \omega(p')\rangle| \ge \beta_{s+1}$. For assume by contradiction that the claim is false. Then there are s independent vectors $k \in \mathscr{M}_N$ such that $|\langle k, \omega(p')\rangle| < \beta_s < \beta_{s+1}$, and there is also a vector

$k' \notin \mathscr{M}$ with $|k'| \le N$ such that $|\langle k', \omega(p')\rangle| < \beta_{s+1}$. Therefore, there are $s+1$ independent integer vectors satisfying the latter condition. Thus, $p' \in \mathscr{Z}^*_{s+1}$, contradicting the hypothesis that $p' \in \mathscr{B}_{\mathscr{M}} = \mathscr{Z}_{\mathscr{M}} \setminus \mathscr{Z}^*_{s+1}$. Consider now the segment joining p with p'; it lies in $\mathscr{W}_\delta$ because $p \in \mathscr{C}_{\mathscr{M},\delta_s}(p') \subset B_{\delta/2}(p')$, as shown while proving (iii). By using

$$\omega(p) - \omega(p') = C(p'')(p - p')$$

with some p'' on the segment joining p with p', we have

$$\bigl|\langle k, \omega(p) - \omega(p')\rangle\bigr| \le |k| M \,\mathrm{dist}(p, p') \le \frac{NM}{2} \,\mathrm{diam}\, \mathscr{C}_{\mathscr{M},\delta_s} < \frac{4sMN^s}{m} \beta_s \ ,$$

where the estimate of $\mathrm{diam}\,\mathscr{C}_{\mathscr{M},\delta_s}$ in (iii) has been used. Therefore,

$$\bigl|\langle k, \omega(p)\rangle\bigr| \ge \bigl|\langle k, \omega(p')\rangle\bigr| - \bigl|\langle k, \omega(p) - \omega(p')\rangle\bigr| > \beta_{s+1} - \frac{4sMN^s}{m} \beta_s \ ,$$

which exceeds β_s in view of our choice (9.39) of the constants β_s. This proves (9.41). Assume now, by contradiction, that there is $\tilde{p} \in \overline{\mathscr{B}_{\mathscr{M},\delta_s}} \cap \mathscr{Z}_{\mathscr{M}'}$. Then, since $\tilde{p} \in \mathscr{Z}_{\mathscr{M}'}$ and $\mathscr{M}' \ne \mathscr{M}$, there exists $k \notin \mathscr{M}$ with $|k| \le N$ such that $|\langle k, \omega(\tilde{p})\rangle| < \beta_s$. On the other hand, in any neighbourhood of $\tilde{p}$ there is a point $p \in \mathscr{B}_{\mathscr{M},\delta_s}$ satisfying the same inequality, which contradicts (9.41). This proves (iv).
The claim (v) immediately follows from $\delta_s < \delta/2$ for $1 \le s \le n$.
In order to prove the claim (vi) let $p \in (\mathscr{B}_{\mathscr{M},\delta_s})_{\delta_s}$. By definition of complex extension there exists $p' \in \mathscr{B}_{\mathscr{M},\delta_s}$ such that $|p' - p| \le \delta_s$. Consequently, if $|k| \le N$,

$$\bigl|\langle k, \omega(p') - \omega(p)\rangle\bigr| \le MN\delta_s \le \frac{\beta_s}{2} \ ,$$

by our choice (9.39) of δ_s. Therefore, in view of (9.41), if $k \notin \mathscr{M}$, then

$$\bigl|\langle k, \omega(p)\rangle\bigr| \ge \bigl|\langle k, \omega(p')\rangle\bigr| - \bigl|\langle k, \omega(p) - \omega(p')\rangle\bigr| \ge \frac{\beta_s}{2} \ .$$

Thus, $|\langle k, \omega(p)\rangle| \ge \beta_s/2$ for all $p \in \bigl(\mathscr{B}_{\mathscr{M},\delta_s}\bigr)_{\delta_s}$ and all $k \notin \mathscr{M}$ with $|k| \le N$, which is the definition of non-resonance domain. This proves (vi), so that the proof of the proposition is complete. *Q.E.D.*

9.5 The Exponential Estimates

We now collect the results of the analytic and of the geometric part in order to investigate the global dynamics in the phase space. Using the geography of resonances, we prove that an orbit remains confined in a cylinder for a time whose length depends on some parameters. The final step is to make a good choice of the parameters so that the exponential estimate naturally comes out.

Let us recall the hypotheses on the Hamiltonian made in Theorem 9.1. In this section the domain $\mathscr{W}$ of the geometric part will be identified with the real part of $\mathscr{G}_\delta$, where $0 < \delta < \varrho/2$. Thus, the hypotheses of Theorem 9.1 are satisfied in $\mathscr{W}_\delta \subset \mathscr{G}_\varrho$. Let us also set $N = rK$, where r is the normalization order and K is the parameter controlling the splitting of the Hamiltonian. Finally, the non-resonance domain $\mathcal{V}$ of type $\mathscr{M}, \delta, \alpha, N$ of the analytic part will be replaced with an extended block $\mathscr{B}_{\mathscr{M},\delta_s}$, which, according to statement (vi) of Proposition 9.9, is a non-resonance domain of type $\mathscr{M}, \beta_s/2, \delta_s, rK$. The parameters β_s and δ_s are defined in Proposition 9.9. It should be recalled that they do depend on the parameters N and δ.

9.5.1 Global Estimates Depending on Parameters

Let us start with the definition of exit and entrance time for a domain in the action space.

Definition 9.10: *Let $\bigl(q(t), p(t)\bigr)$ be an orbit starting at $p(0) \in \operatorname{Int} \mathcal{V}$ for some set $\mathcal{V} \subset \mathscr{G}$. The exit resp. entrance time of this solution in $\mathcal{V}$ is defined as*

$$\tau^+ = \sup\{t > 0 \,:\, p(s) \in \operatorname{Int} \mathcal{V} \text{ for } \quad 0 \le s \le t\} \,,$$

respectively,

$$\tau^- = \inf\{t < 0 \,:\, p(s) \in \operatorname{Int} \mathcal{V} \text{ for } \quad t \le s \le 0\} \,.$$

Lemma 9.11: *Let r and K be positive integers, $N = rK$ and let $0 < \delta < \varrho/2$. Consider the N-module $\mathscr{M}$ of $\dim \mathscr{M} = s$ with the associate parameters β_s and δ_s. Let*

$$\begin{aligned} \mu_s &= \frac{2r\varepsilon A}{\beta_s \delta_s} + 4e^{-K\sigma/2} \,, \\ t_s^* &= \frac{D\delta_s}{4\varepsilon} \mu_s^{-r} \,, \end{aligned}$$

with A as in Proposition 9.4 and D as in Lemma 9.5. Assume $\mu_s \le 1/2$. Then for a solution $p(t)$with $p(0) \in \mathscr{B}_{\mathscr{M}}$ and with exit (resp. entrance) time τ^+ (resp. τ^-) in the cylinder $\mathscr{C}_{\mathscr{M},\delta_s}\bigl(p(0)\bigr)$ the following holds true: if $\tau^+ \le t_s^*$ (resp. if $\tau^- \ge -t_s^*$), then

$$p(\tau^+) \quad (\text{resp. } p(\tau^-)\,) \in \bigl(\partial \mathscr{Z}_{\mathscr{M}} \cap \Pi_{\mathscr{M},\delta_s}(p(0)\bigr) \,.$$

In other words, a solution with initial point $p(0) \in \mathscr{B}_{\mathscr{M}} \cap \mathscr{C}_{\mathscr{M},\delta_s}(p(0))$ can leave the cylinder $\mathscr{C}_{\mathscr{M},\delta_s}(p(0))$ within the time interval $|t| \le t_s^*$ only through its base (see Fig. 9.2).

Proof. By assumption, $p(0) \in \mathscr{B}_{\mathscr{M}} \cap \mathscr{C}_{\mathscr{M},\delta_s}(p(0))$; on the other hand, by definition, $\mathscr{B}_{\mathscr{M}} \subset \mathscr{B}_{\mathscr{M},\delta_s}$. In view of Proposition 9.9, statement (vi), the set $\mathscr{B}_{\mathscr{M},\delta_s}$ satisfies the assumptions on $\mathcal{V}$ in Proposition 9.4, with δ, however, replaced by δ_s and α replaced by $\beta_s/2$. Since $\mathscr{C}_{\mathscr{M},\delta_s}(p(0)) \subset \mathscr{B}_{\mathscr{M},\delta_s}$ we therefore conclude from Lemma 9.5 that if $\tau^+ \le t_s^*$, then $p(\tau^+) \in \operatorname{Int}\Pi_{\mathscr{M},\delta_s}$. Moreover, by definition of the exit time, $p(\tau^+) \in \partial\mathscr{C}_{\mathscr{M},\delta_s}(p(0))$, so that $p(\tau^+) \in \partial\mathscr{C}_{\mathscr{M},\delta_s}(p(0)) \cap \operatorname{Int}\Pi_{\mathscr{M},\delta_s}(p(0))$. Similarly, we get the same conclusion about τ^- and $p(\tau^-)$ if $\tau^- \ge t_s^*$. From the definition of the corresponding sets, one readily concludes that

$$\left(\partial\mathscr{C}_{\mathscr{M},\delta_s}(p(0)) \cap \operatorname{Int}\Pi_{\mathscr{M},\delta_s}(p(0))\right) \subset \left(\partial\mathscr{Z}_{\mathscr{M}} \cap \Pi_{\mathscr{M},\delta_s}(p(0))\right) ,$$

and this concludes the proof. *Q.E.D.*

The crucial step in proving Theorem 9.1 is the following lemma, which combines the analytical and the geometrical considerations. Recall that the geometric construction has been performed for the domain $\mathscr{W}$ which is the real part of $\mathscr{G}_\delta$.

Lemma 9.12: *With the parameters r, K, $N = rK$, $\delta > 0$ as in Lemma 9.11, and with the constants δ_0 and β_0 as given in Proposition 9.9, define*

$$\mu_0 = \frac{2r\varepsilon A}{\beta_0\delta_0} + 4e^{-K\sigma/2} ,$$
$$t_0^* = \frac{D\delta_0}{4\varepsilon}\mu_0^{-r} ,$$

with A as in Proposition 9.4 and D as in Lemma 9.5. Assume $\mu_0 \le 1/2$. Then the following statement holds true: if a solution satisfies $p(t) \in \mathscr{G}_\delta$ for $t \in [a,b]$, then

$$\operatorname{dist}(p(t), p(s)) < \delta$$

for all $t, s \in [a,b]$, provided $b - a < t_0^*$.

Proof. We shall prove the existence of a resonance module of $\dim\mathscr{M} = s$ for some s and of a point $p^* \in \mathscr{B}_{\mathscr{M}}$ such that

$$p(t) \in \mathscr{C}_{\mathscr{M},\delta_s}(p^*) \quad \text{for} \quad t \in [a,b] . \tag{9.42}$$

The lemma then follows in view of $\operatorname{diam}\mathscr{C}_{\mathscr{M},\delta_s}(p^*) < \delta$, which is the statement (iii) of Proposition 9.9. In order to prove (9.42), observe that due to the statement (i) in Proposition 9.9 there exists for every $t \in [a,b]$ an N-module $\mathscr{M}_t$ such that $p(t) \in \mathscr{B}_{\mathscr{M}_t}$, which has minimal dimension. Since the set of all N-moduli is finite, there exists among these N-moduli $\mathscr{M}_t$, $t \in [a,b]$, an N-module $\mathscr{M}$ such that $\dim\mathscr{M} = s$ is minimal. Then $p^* \equiv p(t_0) \in \mathscr{B}_{\mathscr{M}}$ for some $t_0 \in [a,b]$. Let now τ^+ be the exit time of the solution $p(t)$ starting at p^* contained in the cylinder $\mathscr{C}_{\mathscr{M},\delta_s}(p^*)$, and assume that $\tau^+ \le t_0^*$. We claim that

$$p(t_0 + \tau^+) \in \partial\mathscr{Z}_{\mathscr{M}} \cap \Pi_{\mathscr{M},\delta_s}(p^*) . \tag{9.43}$$

For, by definition, $\beta_0 \le \beta_s$ and $\delta_0 \le \delta_s$, so that $\mu_0 \ge \mu_s$ and $t_0^* \le t_s^*$. Therefore, the claim (9.43) follows immediately from the previous Lemma 9.11. Next we claim that

$$(9.44) \qquad b < t_0 + \tau^+ \ .$$

Indeed, assume by contradiction that $b \geq t_0 + \tau^+$. Then we conclude that $p(t_0 + \tau^+) \in \mathscr{G}_\delta$, and, by (9.43), that $p(t_0 + \tau^+) \notin \mathscr{Z}_{\mathscr{M}}$. Since, by Proposition 9.9, statement (iv), we have $\overline{\mathscr{B}_{\mathscr{M},\delta_s}} \cap \mathscr{Z}_{\mathscr{M}'} = \emptyset$ for every $\mathscr{M}' \neq \mathscr{M}$ satisfying $\dim \mathscr{M} = \dim \mathscr{M}' = s$, we find

$$p(t_0 + \tau^+) \notin \bigcup_{\dim \mathscr{M} = s} \mathscr{Z}_{\mathscr{M}} = \mathscr{Z}_s^* \ .$$

Therefore, $p(t_0 + \tau^+) \in \mathscr{G}_\delta \setminus \mathscr{Z}_s^*$. But, by Proposition 9.9, statement (ii),

$$\mathscr{G}_\delta \setminus \mathscr{Z}_s^* \subset \bigcup_{\dim \mathscr{M} = s-1} \mathscr{B}_{\mathscr{M}} \ .$$

Thus, $p(t_0 + \tau^*) \in \mathscr{B}_{\mathscr{M}^*}$ with $\dim \mathscr{M}^* \leq s - 1$. Since $t_0 + \tau^+ \in [a, b]$, this contradicts the definition of $\mathscr{M}$ which is assumed to have minimal dimension. Therefore, the claim $b < t_0 + \tau^+$ is proved. The same arguments show that $a > t_0 - \tau^-$, and we conclude that $p(t) \in \mathscr{C}_{\mathscr{M},\delta_s}(p^*)$ for all $t \in [a, b]$, as claimed in (9.42). This proves the lemma. *Q.E.D.*

Assume now $p(0) \in \mathscr{G}$, and pick as time interval $[a, b] = [\tau^-, \tau^+]$, where τ^+ and τ^- are the exit and entrance times of an orbit starting from $p(0)$ in the centre of an open ball of radius δ:

$$\begin{aligned} a &= \inf \{t < 0 \mid \mathrm{dist}(p(s), p(0)) < \delta \text{ for } t \leq s \leq 0\} \ , \\ b &= \sup\{t > 0 \mid \mathrm{dist}(p(s), p(0)) < \delta \text{ for } 0 \leq s \leq t\} \ . \end{aligned}$$

From Lemma 9.12 one concludes immediately

Proposition 9.13: *Let r, K, $N = rK$, $\varrho > 0$ be given, and let D, μ_0 and t_0^* be as in Lemma 9.12. Then for any positive $\delta < \varrho/3$, every solution with $p(0) \in \mathscr{G}$ satisfies*

$$\mathrm{dist}(p(t), p(0)) < \delta \quad \text{if} \quad |t| \leq t_0^* = \frac{D\delta_0}{4\varepsilon} \mu_0^{-r} \ ,$$

provided $\mu_0 \leq 1/2$.

9.5.2 Choice of the Parameters and Exponential Estimate

It remains to choose the parameters $K \geq 1$, $r \geq 1$ and $\delta > 0$ as functions of the parameters ϱ, σ, ε, m and M which characterize the Hamiltonian. The aim is to make a good choice, so that the stability time t_0^* is as large as possible. In the following, μ_0, t_0^* and A are as defined in Lemma 9.12, and β_0 and δ_0 are as defined in Proposition 9.9 with $N = rK$. We shall also assume $\mu_0 \leq 1/e$, which is stronger than the condition in Lemma 9.12, and $0 < \delta < \varrho/3$. This will complete the

Proof of Theorem 9.1. Let us first make a choice for the parameters r and K. Set

$$a = n^2 + n \ , \quad \delta_* = (n+2)! \left(\frac{4M}{m}\right)^{n+1} K^{a/2} \ , \tag{9.45}$$

so that we have

$$\beta_0 = \frac{rKm\delta}{\delta_* r^{a/2}} \ ,$$
$$\delta_0 = \frac{m\delta}{2M\delta_* r^{a/2}} \ ,$$
$$\mu_0 = \frac{\mu_*}{2e} \cdot \frac{r^a \varepsilon \varrho^2}{\delta^2} + 4e^{-K\sigma/2} \ ,$$

where

$$\mu_* = \frac{2^{13} M \delta_*^2}{\sigma \varrho^2 m^2 K} \left(\frac{1+e^{-\sigma/2}}{1-e^{-\sigma/2}}\right)^n |f|_{\varrho,2\sigma} \ .$$

Let us ask $\mu_0 \leq 1/e$, which holds true if

$$\mu_* \frac{r^a \varepsilon \varrho^2}{\delta^2} \leq 1 \ , \quad e^{-K\sigma/2} \leq \frac{1}{8e} \ .$$

Thus, it is natural to make the choice

$$r = \left(\frac{\delta^2}{\mu_* \varepsilon \varrho^2}\right)^{1/a} \ , \quad K = \left\lceil \frac{2(1+3\ln 2)}{\sigma} \right\rceil \ , \tag{9.46}$$

where $\lceil \cdot \rceil$ indicates rounding to the next integer. We have $K \geq 1$ by definition, while the condition $r \geq 1$ is satisfied provided

$$\frac{\mu_* \varepsilon \varrho^2}{\delta^2} \leq 1 \ . \tag{9.47}$$

With the values of r and K and using $\mu_0 \leq 1/e$, from Proposition 9.13 we get

$$t_0^* = \frac{T_*}{\varepsilon} \exp\left[\left(\frac{\delta^2}{\mu_* \varepsilon \varrho^2}\right)^{1/a}\right] \tag{9.48}$$

with a constant T_*.

It remains to make a choice for δ so that the condition (9.47) is satisfied. This leaves some freedom in choosing an actual value. For instance, we may set

$$\delta = (\mu_* \varepsilon)^{1/4} \varrho \ , \tag{9.49}$$

which also gives

$$\frac{\mu_* \varepsilon \varrho^2}{\delta^2} = (\mu_* \varepsilon)^{1/2} \ .$$

Thus, all parameters are well determined, provided the conditions $\delta < \varrho/3$ and (9.47) are fulfilled. This gives a condition of smallness of the perturbation, namely

$$\mu_* \varepsilon < \frac{1}{3^4} ,$$

that is, the condition in Theorem 9.1. The statement of the theorem then follows by Proposition 9.13 with the values of δ and t_0^* given by (9.48) and (9.49).

This concludes the proof of Nekhoroshev's theorem. *Q.E.D.*

9.6 An Alternative Proof by Lochak

A different scheme of proof has been proposed by Pierre Lochak [160] [161]. We resume the main ideas, though not providing a detailed proof which can be found in the papers just quoted. The beautiful idea in Lochak's work is to put the accent on the fact that resonances may exert a stabilizing effect on dynamics. This may appear a strange idea, since resonances are often considered as causes of instability. There are many examples of such a destructive action, which in some cases may produce dramatic effects. On the other hand, the stability of resonant states had been already investigated, for example, in [20] and [22].

The starting point is in a sense the opposite with respect to Kolmogorov's: Lochak proposes to focus attention on a completely resonant torus. All orbits on such a torus are periodic, and the frequencies, ω^* say, are all multiples of a given quantity. In particular, we have $|\langle k, \omega^* \rangle| > \alpha$ with some positive α for all $k \notin \mathcal{M}_{\omega^*}$, the resonance module corresponding to the completely resonant torus. This has a first relevant effect on what we have called the analytic part of Nekhoroshev's theorem. The part of the Hamiltonian in normal form depends on $n-1$ resonant angles $\langle k, q \rangle$, where q are the angle variables and $k \in \mathcal{M}_{\omega^*}$. With respect to the analytic part, as we have developed it in the proof, Lochak considers only resonant domains of multiplicity $n-1$; our analytic part may be given a slightly simpler form.

An additional technical remark is that the complete resonance simplifies the task of solving the homological equation: the periodic character of the unperturbed orbits allows us to solve the homological equation by just performing an average over a closed orbit.

The confinement of the orbit inside the resonant domain is obtained by exploiting the convexity of the unperturbed Hamiltonian h_0. The latter hypothesis may be replaced by the weaker condition of quasi-convexity, that is, convexity restricted to the invariant energy surface. There is no need of the property of non-overlapping of resonance; by the way, the latter property causes the worst dependence of the estimates on the number n of degrees of freedom, since it forces $a = n^2 + n$ in the statement of Theorem 9.1. Convexity or quasi-convexity, is essential, and leads to estimate $a = 2n$.

The more interesting novelty introduced by Lochak is the replacement of the geometric part of Nekhoroshev's proof (and of our proof in this chapter) with the theory of simultaneous approximation as developed by Dirichlet. *Let* $\omega \in \mathbb{R}^n$ *and* $Q > 1$ *be a real number. There exists an integer* q *with* $1 \leq q < Q$ *such that* $\mathrm{dist}(q\omega, \mathbb{Z}^n) \leq Q^{-1/n}$. For $\omega \in U(\omega^*) \subset \mathbb{R}^n$, a small neighbourhood of ω^*, then for $k \notin \mathscr{M}_{\omega^*}$ the quantity $|\langle k, \omega \rangle|$ does not vanish. The neighbourhood $U(\omega^*)$ may be mapped back to the action domain through the frequency map $\omega(p) = \frac{\partial h_0}{\partial 0}$ of the unperturbed Hamiltonian, thus giving a neighbourhood $V(p^*)$ of the action p^* of the resonant torus. The non-trivial part is to prove that for every perturbation ε there exists a finite set of completely resonant tori p^* such that the corresponding neighbourhoods $V(p^*)$ cover the action space, and that in every such neighbourhood the confinement assured by the analytic part of the proof is effective.

The conclusion is that *every orbit remains confined for an exponentially long time inside a completely resonant domain.* In this sense the resonances may be said to have a stabilizing effect. The dynamics inside the completely resonant domain may well be chaotic, but the orbit cannot escape.

10
Stability and Chaos

In the previous three chapters we have focused our attention on methods and results connected with an ordered, close-to-integrable behaviour. Only in Chapter 5 have we taken into consideration the chaotic phenomena which emerge already in apparently innocuous systems, such as the models of Contopoulos and of Hénon and Heiles; but our discussion was merely phenomenological, based only on numerical simulation of dynamics. It is time now to pay a little attention to the phenomenon of coexistence of ordered and chaotic behaviour.

The two main theorems of Kolmogorov and Nekhoroshev represent the strongest results which have been obtained in the past century. We have remarked that Kolmogorov's theorem, in the form discussed in Chapter 8, produces a wonderful but puzzling result: it claims that most orbits are quasi-periodic, as in an integrable system, but the complement of the set of invariant tori is dense, and in systems with more than two degrees of freedom it is also connected. The theorem says nothing about the fate of orbits in the complement. For a physical system (e.g., the planetary system) it is hard to decide whether or not the orbits lie on invariant tori.

Nekhoroshev's theorem seems to be in a better position, because it provides a somehow global picture: it puts strong limits on the possible diffusion of orbits in phase space, telling us that in a global sense it takes an exponentially long time. In this sense it is more interesting for Physics, because it provides information on orbits on open sets. On the other hand, one may observe that an overall explanation of the chaotic dynamics is out of the scope of Nekhoroshev's theory.

A natural question is whether we may figure out a somehow global picture of the typical behaviour of a nearly integrable system. In this respect, omitting a short account of the discovery of chaotic dynamics would be an unforgivable sin. The argument could fill not less than a whole book; however, it is worthwhile to describe at least the phenomenon of homoclinic points discovered by Poincaré (see [187] and the last chapter of [188], Tome III).

In this chapter the reader will not find a full answer to this question raised above: it is just a first, partial attempt. The chapter focuses on three main arguments:

(i) The superexponential stability in the neighbourhood of an invariant torus, let us say Kolmogorov's.
(ii) A description of the phenomenon of homoclinic and heteroclinic orbits discovered by Poincaré.
(iii) An enhancement of Nekhoroshev's theorem, pointing out its interrelationship with Kolmogorov's theorem, in the global version proved by Arnold.

In view of the general character of the discussion, the contents of the present chapter will be less technical in comparison to the previous chapters. Most proofs are only sketched, thus leaving to the reader the burden of completing them or to look for them in the references which are provided. It goes without saying that reading this chapter and completing the proofs requires a good familiarity with the contents of Chapters 7, 8 and 9.

10.1 The Neighbourhood of an Invariant Torus

We start by investigating the stability of orbits in the neighbourhood of an invariant, strongly non-resonant torus. We know about the normal form of the Hamiltonian either from Proposition 8.3, with the fast convergence method of Kolmogorov, or from Proposition 8.15, with a classical construction; the latter may be more suitable as a starting point. Therefore, we are dealing with a Hamiltonian,

$$H(p,q) = \langle \omega, p \rangle + R(p,q), \tag{10.1}$$

with the following properties:

(i) $R(p,q)$ is a real analytic function of $p \in \Delta_{\varrho}(p)$, a ball of radius $\varrho > 0$ centred on $p = 0$, and real analytic in $q \in \mathbb{T}^n_{\sigma}$, for some $\sigma > 0$;
(ii) we have $R(p,q) = \mathcal{O}(p^2)$; that is, it may be expanded in power series of p starting with terms of degree 2, with coefficients analytic in q.

It takes only a moment to reconnect the Hamiltonian (10.1) with the one considered in Section 7.3.2. The apparent trouble is that there is no small factor ε in front of the perturbation, for the function $R(p,q)$ includes the quadratic part of the unperturbed Hamiltonian (8.1) in the statement of Kolmogorov's theorem; the reader may doubt that $R(p,q)$ can be considered as a small perturbation. However, the relevant point is that we are considering a *local* neighbourhood of the invariant torus $p = 0$ for the Hamiltonian (10.1), since we restrict $p \in \Delta_{\varrho}$. Therefore, we may replace the original perturbation parameter ε with ϱ: a straightforward application of Cauchy's inequalities allows us to estimate $\|R\|_{\varrho,\sigma} \sim \mathcal{O}(\varrho^2)$ (e.g., in weighted Fourier norm), so that the ratio between $R(p,q)$ and the unperturbed part $\langle\omega,p\rangle$ is $\mathcal{O}(\varrho)$, which vanishes in the limit $\varrho\to 0$.

In Section 7.3.2 we have seen that the Hamiltonian (10.1) may be cast into Poincaré–Birkhoff normal form, though only in the formal sense. Here, using the theorems of Kolmogorov and of Nekhoroshev, we go beyond the formal level by completing the formal argument with quantitative estimates.

10.1.1 Poincaré–Birkhoff Normal Form

Here we should exploit the arguments of Section 7.3, by expanding $R(p,q)$ in trigonometric polynomials, with coefficients $\mathcal{O}(p^2)$. Thus, we may start with a Hamiltonian,

$$H(p,q) = \langle\omega,p\rangle + H_1(p,q) + H_2(p,q) + \ldots\,, \tag{10.2}$$

where $H_s(p,q)\in\mathscr{F}_{sK}$ is a polynomial of degree at least 2 and of finite degree in p, with coefficients which are polynomials of finite degree sK in q for some $K\geq 1$, as explained in Section 7.3.3.

Example 10.1: *The case of a Hamiltonian quadratic in the actions.* If we consider a Hamiltonian such as (8.66) and perform the Kolmogorov normalization in classical style as in Section 8.3, then we end up with a Hamiltonian in Kolmogorov normal form,

$$H(p,q) = \langle\omega,p\rangle + \frac{1}{2}p^2 + \sum_{s\geq 1} C_s(p,q)\ ,\quad C_s(p,q)\in\mathscr{F}_{sK}\ ,$$

which has already the desired form. *E.D.*

More generally, we may proceed as in Section 7.3.3 using both the powers of ϱ and the exponential decay of coefficients in order to split the Hamiltonian in the desired form.

Exercise 10.2: Show that by a suitable choice of the parameters σ and K we can estimate the Hamiltonian (10.2) as

$$\|H_s\|_{\varrho,\sigma}\leq h^{s-1}\varrho^2 E\ ,\quad s>0\ , \tag{10.3}$$

with, for example, $h=e^{-K\sigma/2}$ and some $E>0$. (Hint: see Lemma 7.15; the parameter ε is replaced by the radius ϱ of the action domain Δ_ϱ, and since the perturbation is $\mathcal{O}(p^2)$ there is a common factor ϱ^2). *A.E.L.*

The process of normalization may be performed using the algorithm of the Lie transform. The algorithm is fully expanded in Section 7.3.1, including also the solution of the homological equation. As explained in Section 7.4, the process must be stopped at some finite order, r say, thus producing a truncated normal form. The concept of complete stability of Birkhoff applies in this case (see Section 9.1), but we can do better by adding quantitative estimates.

The scheme of quantitative estimates for both the generating sequence and the normal form, including the remainder, is that of Lemma 9.3 and Proposition 9.4. To this end we should identify the non-resonance domain $\mathcal{V} \subset \mathbb{R}^n$ of type $(\mathscr{M}, \alpha, \delta, N)$, where we construct the normal form. We proceed as follows.

(i) The local non-resonance domain $\mathcal{V}$ is chosen to be the polydisk Δ_ϱ; hence we set $\delta = \varrho$. By the way, this removes the need of the geometric part of Nekhoroshev's theorem.

(ii) The frequencies ω of the invariant torus are non-resonant, according to Kolmogorov's theorem; hence the resonance module $\mathscr{M}$ is chosen to be $\mathscr{M} = \{0\}$.

(iii) Since at every step s of the normalization procedure we deal only with a finite number of Fourier components, bounded by $|k| \le sK$, we may use the sequence $\{\alpha_s\}$ of the smallest divisor defined as

$$\alpha_s = \min_{0<|k|\le sK} \bigl|\langle k, \omega\rangle\bigr| \ . \tag{10.4}$$

We may also want to replace this sequence with the Diophantine inequality, thus setting

$$\alpha_s = \frac{\gamma}{(sK)^\tau}$$

with some $\gamma > 0$ and $\tau > n-1$. Since we plan to construct the normal form up to the finite order $r > 0$, we may assign the parameter α the element α_{rK} of the sequence (10.4). With the further assumption that the frequencies are Diophantine, we take the definite choice

$$\alpha = \frac{\gamma}{(rK)^\tau} \ .$$

Having so settled the questions concerning the local non-resonance domain, and recalling the estimate (10.3) which results from Exercise 10.2, the quantitative estimates are a repetition of the analytic part of the proof of Nekhoroshev's theorem, Sections 9.3.4 and 9.3.5. The reader may easily check that it is only a matter of changing some symbols. Therefore we may restate the claim of Proposition 9.4 in a form adapted to the present context.

Proposition 10.3: *Consider the Hamiltonian (10.2) in the domain $\Delta_\varrho \times \mathbb{T}^n_\sigma$, and assume that the frequencies ω are Diophantine with constants $\gamma > 0$ and $\tau \ge n-1$, and that $\|H_s\|_{\varrho,\sigma} \le h^{s-1}\varrho^2 E$ for $s > 0$ with some $E > 0$. Then there are positive constants A and B such that the following statement holds true: if*

$$\mu := \frac{r^{\tau+1}\varrho K^\tau A}{\gamma} + 4e^{-K\sigma/2} \le \frac{1}{2}, \tag{10.5}$$

then:

(i) the Hamiltonian is given the Poincaré–Birkhoff normal form

$$H^{(r)}(p,q) = \langle \omega, p\rangle + Z_1(p) + \cdots + Z_r(p) + \mathscr{R}^{(r+1)}(p,q) \tag{10.6}$$

via a finite generating sequence $\chi^{(r)} = \{\chi_1, \ldots, \chi_r\}$, *with* $\chi_s \in \mathscr{F}_{sK}$*;*

(ii) the generating sequence defines an analytic canonical transformation on the domain $\Delta_{\frac{3}{4}\varrho} \times \mathbb{T}^n_{\frac{3}{4}\sigma}$, *with the properties*

$$\Delta_{\frac{5}{8}\varrho} \subset T_\chi \Delta_{\frac{3}{4}\varrho} \subset \Delta_{\frac{7}{8}\varrho}\,, \quad \Delta_{\frac{5}{8}\varrho} \subset T_\chi^{-1} \Delta_{\frac{3}{4}\varrho} \subset \Delta_{\frac{7}{8}\varrho}\,,$$

and, moreover, in $\Delta_{\frac{3}{4}\varrho}$ *one has*

$$\bigl|p - T_{\chi^{(r)}}p\bigr| \le \frac{\varrho}{16}\,, \quad \bigl|p - T_{\chi^{(r)}}^{-1}p\bigr| \le \frac{\varrho}{16}\,; \tag{10.7}$$

(iii) the remainder $\mathscr{R}^{(r+1)}$ *is estimated by*

$$\bigl\|\mathscr{R}^{(r+1)}\bigr\|_{\frac{3}{4}(\varrho,\sigma)} \le B\mu^r\,.$$

Again, the proof is a straightforward adaptation of that of Proposition 9.4; hence the details are left to the reader. It may be useful to add a couple of (perhaps pedantic) remarks. The smallness parameter ε is replaced by ϱ; since the perturbation is $\mathcal{O}(\varrho^2)$ the divisor ϱ resulting from Cauchy estimates is canceled, leaving the small factor ϱ at the numerator. Moreover, α is replaced by $\gamma/(rK)^\tau$.

The local stability Lemma 9.5 may now be applied to the Hamiltonian (10.6) in Poincaré–Birkhoff normal form. The conclusion in this case is meaningful: since $\mathscr{M} = \{0\}$ (non-resonance) there is no plane of fast drift through $p = 0$, so that an orbit remains confined in a neighbourhood Δ_ϱ (with some restriction of ϱ) for a time as long as $\sim 1/\mu^r$; this is the complete stability of Birkhoff, indeed. However we can do better, as we did in Section 5.4.4 for the case of an elliptic equilibrium.

10.1.2 Exponential Stability

We exploit the arbitrariness of both the radius ϱ and of the normalization order r in our procedure, choosing r as a function of ϱ so as to minimize the size of the remainder $\mathscr{R}^{(r+1)}(p,q)$. The calculation is a considerable simplification of the argument used in Section 9.5.2; let us work it out in detail. We replace condition (10.5) on μ with the slightly stronger one $\mu \le 1/e$. To this end, the easiest way is to ask

$$\frac{r^{\tau+1}\varrho K^\tau A}{\gamma} \le \frac{1}{2e}\,, \quad 4e^{-K\sigma/2} \le \frac{1}{2e}\,.$$

Hence, setting

$$\varrho_* = \frac{\gamma}{2eK^\tau A}\,, \tag{10.8}$$

we make the choice

$$(10.9) \qquad r = \left(\frac{\varrho_*}{\varrho}\right)^{1/(\tau+1)}, \quad K = \left\lceil \frac{2(1+3\ln 2)}{\sigma} \right\rceil ,$$

where $\lceil\cdot\rceil$ indicates rounding to the next integer. We have $K \geq 1$ by definition, while the condition $r \geq 1$ is satisfied, provided $\varrho < \varrho_*$. The latter is a condition on ϱ: essentially it means that our estimates hold true, provided the radius ϱ is small enough, so that ϱ_* can be seen as a threshold for the applicability of our estimates. In view of $\mu \leq 1/e$ we restate Proposition 10.3 in the following improved form.

Proposition 10.4: *With the same hypotheses as in Proposition 10.3, there exist positive constants ϱ_* and B such that the remainder $\mathscr{R}^{(r+1)}(p,q)$ is estimated in the local domain $\Delta_{\frac{3}{4}\varrho} \times \mathbb{T}^n_{\frac{3}{4}\sigma}$ as*

$$(10.10) \qquad \left\|\mathscr{R}^{(r+1)}\right\|_{\frac{3}{4}(\varrho,\sigma)} \leq B \exp\left[-\left(\frac{\varrho_*}{\varrho}\right)^{1/(\tau+1)}\right] .$$

The latter proposition allows us to conclude that a neighbourhood of an invariant torus of Kolmogorov is exponentially stable. In more explicit terms we have

Proposition 10.5: *With the same hypotheses as in Proposition 10.3, there exists a constant T_* such that for any initial point $p(0) \in \Delta_{\varrho/2}$, we have*

$$\left|p(t) - p(0)\right| < \frac{\varrho}{16} \quad \text{for} \quad |t| < T_* \exp\left[\left(\frac{\varrho_*}{\varrho}\right)^{1/(\tau+1)}\right] .$$

Proof. Let us denote by p the original actions and by $p' = T_{\chi^{(r)}}p$ the coordinates of the normal form, and use the triangle inequality

$$\left|p(t) - p(0)\right| \leq \left|p(0) - p'(0)\right| + \left|p'(0) - p'(t)\right| + \left|p'(t) - p(t)\right| .$$

We must recall that the canonical transformation induces a deformation of the actions which, according to Proposition 10.3, is estimated as

$$\left|p(0) - p'(0)\right| < \frac{\varrho}{16} , \quad \left|p'(t) - p(t)\right| < \frac{\varrho}{16} .$$

The diffusion speed due to the remainder is calculated as

$$\dot{p} = \left\{p, \mathscr{R}^{(r+1)}(p,q)\right\} = -\frac{\partial \mathscr{R}^{(r+1)}}{\partial q} ,$$

which in view of Proposition 10.3 is estimated in the domain $\Delta_{\frac{3}{4}\varrho}$ by

$$\dot{p} < \beta \exp\left[-\left(\frac{\varrho_*}{\varrho}\right)^{1/(\tau+1)}\right],$$

with some constant β. Therefore the diffusion is bounded by

$$\left|p'(t) - p'(0)\right| < |t|\,\beta \exp\left[-\left(\frac{\varrho_*}{\varrho}\right)^{1/(\tau+1)}\right] .$$

If we allow the diffusion to induce a change which does not exceed the deformation, still avoiding the orbit to leave the domain of definition of the Hamiltonian, we may ask it to be bounded by $\varrho/16$. This gives the exponential bound for $|t|$ in the statement with some constant T_*. *Q.E.D.*

10.1.3 Superexponential Stability

Proposition 10.5 claims that an invariant torus of Kolmogorov is exponentially stable. The latter property has often been named *stickiness* and has been pointed out in [183]. However, the result of exponential stability may still be considerably improved, thus introducing the concept of *superexponential stability*, first proposed in [170].

The argument consists in applying again Nekhoroshev's theorem in its complete form to the Hamiltonian (10.10) of Proposition 10.5. Comparing it with the Hamiltonian (9.2) in the statement of Nekhoroshev's theorem, Section 9.2.1, we set

$$H_0(p) = \langle\omega, p\rangle + Z_1(p) + \cdots + Z_r(p) \ , \quad \varepsilon f(p,q) = \mathscr{R}^{(r+1)}(p,q) \ ,$$

thus recovering the form of the Hamiltonian in Theorem 9.1. The crucial remark, however, is that in view of Proposition 10.4, we may set

$$\varepsilon = B \exp\left[-\left(\frac{\varrho_*}{\varrho}\right)^{1/(\tau+1)}\right] ,$$

namely an exponentially small quantity depending on ϱ. The choice of the domain is immediate: in view of Proposition 10.3 the Hamiltonian is analytic in the domain $\Delta_{3\varrho/4} \times \mathbb{T}^n_{3\sigma/4}$, which just requires an unessential restriction of ϱ and σ. Only the convexity hypothesis (iii) of Section 9.2.1 is not automatically assured: we must add it. To this end, we should replace the condition of non-degeneration

$$\det\left(\frac{\partial^2 h}{\partial p_j \partial p_k}\right) \neq 0 \ ;$$

in Kolmogorov's Theorem 8.1 with the stronger condition of convexity.

With the setting just given we complement Kolmogorov's theorem with the property of superexponential stability of an invariant torus.

Theorem 10.6: *Consider the Hamiltonian (8.1), namely*

$$H(p,q) = h(p) + \varepsilon f(p,q,\varepsilon) \ ; \tag{10.11}$$

assume:

(a) *$h(p)$ is convex, that is,*

$$\bigl|\langle \mathsf{C}(p)v, v\rangle\bigr| \ge m\|v\|^2 \quad \text{for all } v \in \mathbb{R}^n \ , \quad \mathsf{C}(p) = \det\left(\frac{\partial^2 h}{\partial p_j \partial p_k}\right) \ ;$$

(b) *$h(p)$ possesses an invariant torus p^* with frequencies ω satisfying the Diophantine condition*

$$\bigl|\langle k, \omega\rangle\bigr| \ge \frac{\gamma}{|k|^\tau} \ , \qquad k \ne 0 \ ,$$

for some $\gamma > 0$ and $\tau > n - 1$.

Then there exists positive ε^*, ϱ_*, a and T_* such that the following holds true: for $\varepsilon < \varepsilon^*$ and for every $\varrho < \varrho^*$, there is an analytic canonical transformation mapping $(p', q') \in \Delta_\varrho(0) \times \mathbb{T}^n$ to a neighbourhood $U(p^*) \times \mathbb{T}^n$ with the properties:

(i) *$p' = 0$ is an invariant torus carrying a quasi-periodic flow with frequencies ω^*;*

(ii) *the invariant torus is a small deformation of the unperturbed torus $p = p^*$;*

(iii) *for every initial datum $p'_0 \in \Delta_\varrho(0)$, one has*

$$|p'(t) - p'_0| \simeq \exp\bigl(-(\varrho^*/\varrho)^{1/(\tau+1)}\bigr)$$

for

$$|t| < T_* \exp\Bigl[\exp\bigl(\tfrac{1}{2a}(\varrho^*/\varrho)^{1/(\tau+1)}\bigr)\Bigr] \ ; \quad a = n^2 + n \ .$$

The claims (i) and (ii) are nothing but the statements of Kolmogorov's theorem; note that non-degeneration of $h(p)$ follows from the convexity condition. The map $(p', p') \to (p, q) \in U(p^*) \times \mathbb{T}^n$ which gives the Hamiltonian a Kolmogorov's normal form is a by-product of the proof of Kolmogorov's theorem. The claim (iii) follows from Proposition 10.5, applied to the Hamiltonian in Kolmogorov's normal form. If one refers to the original variables p, q of the Hamiltonian (10.11), then the exponentially small diffusion of point (iii) must be added to the deformation induced by the construction of Kolmogorv's normal form, the latter being the predominant effect anyway.

10.2 The Roots of Chaos and Diffusion

In this section we focus attention on the case of a non-autonomous periodically perturbed system with one degree of freedom; it is often said that such a system has $1 + \frac{1}{2}$ degrees of freedom. This is indeed the simplest case for which the relevant phenomena discovered by Poincaré do occur.[1] The Hamiltonian will be written as

$$H(x,y,t) = h(x,y) + \varepsilon f(x,t)\ ,\qquad f(t+T) = f(t)\quad \forall t\ .$$

Here (x,y) are the canonical coordinate and momentum on a two-dimensional phase space, ε is a small perturbation parameter and the perturbation $f(t)$ is periodic with period T. We shall also assume that $(x,y) = 0$ is an equilibrium for the canonical flow of $h(x,y)$, with non-vanishing real eigenvalues $\lambda, -\lambda$. Such an equilibrium is said to be *hyperbolic*, as opposed to the elliptic equilibrium which was the subject of the whole of Chapter 5.

The main tool used here is the Poincaré map ϕ^T over one period; that is, considering that $t \in \mathbb{T}$ in view of the periodicity of the perturbation, we are dealing with a one-to-one map which is the Poincaré section at $t = 0$. The map inherits the smoothness properties of the Hamiltonian flow and is area preserving.

10.2.1 Asymptotic Orbits

The objects to be put under observations are orbits which are *asymptotic* to an unstable periodic orbit. In our language these orbits are often characterized as *separatrices*, or more generally *stable/unstable manifolds*, but let us freely use also the language of the discoverer of such a surprising and interesting phenomenon. There are a lot of examples of such kinds of orbits in the figures of Section 5.3.3; following Poincaré, we concentrate on a single periodic orbit and on orbits which are asymptotic to it.

Definition 10.7: *The stable manifold $W^{(s)}$ and the unstable manifold $W^{(u)}$ of a hyperbolic fixed point O are defined, respectively, as*

$$\begin{aligned} W^{(s)} &= \{P : \phi^{kT} P \to_{t\to+\infty} O\}\ , \\ W^{(u)} &= \{P : \phi^{kT} P \to_{t\to-\infty} O\}\ . \end{aligned} \tag{10.12}$$

Therefore, the stable (resp. unstable) manifolds are sets of orbits which are all asymptotic to the equilibrium in the future (resp. past).

The example of a periodically forced pendulum is widely studied nowadays. The most common form of the Hamiltonian is

---

[1] It is necessary to remark that Poincaré discovered the phenomena described here while studying the dynamics of the problem of three bodies. Our simple models were introduced about seventy years later, starting around 1960 with the works of Contopolus and of Hénon and Heiles already quoted, when people unexpectedly became aware of the existence of chaotic orbits in apparently very simple systems.

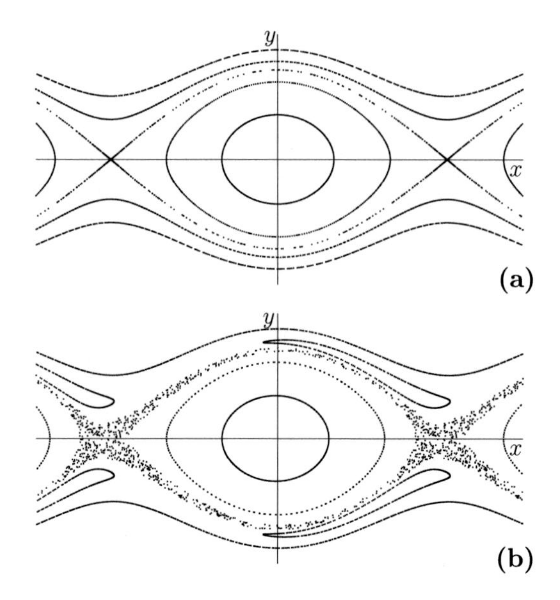

Figure 10.1 Poincaré section for the periodically forced pendulum.
(a) $\varepsilon = 0$, the free pendulum: the separatrices are smooth curves.
(b) $\varepsilon = 0.05$, the perturbed case: there is a chaotic zone around the separatrices.

(10.13) $$H_\varepsilon(x,y) = \frac{y^2}{2} - \cos x - \varepsilon\cos(x-t)\ , \quad x \in \mathbb{T}\ ,\ y \in \mathbb{R}\ .$$

The phenomenon to be investigated is illustrated in Fig. 10.1. We consider the Poincaré map at time $T = 2\pi$, the period of the forcing term; that is, we mark the sequence of points $P, \ldots \phi^T P, \ldots \phi^{2T} P, \ldots$; Poincaré called *théorie des conséquents* the study of these sequences.

If we set $\varepsilon = 0$, then the orbits of the map lie on the level curves of constant energy, $H_0(x,y) = E$; the forcing term controls only the period of the map, like in a stroboscope. The phase portrait of the Poincaré section is represented in panel **(a)** of Fig. 10.1; we are particularly interested in the orbits on the separatrix: Poincaré called them *orbites doublement asymptotiques*, because they are asymptotic to the upper unstable equilibrium of the pendulum both for $t \to -\infty$ and for $t \to +\infty$.

For $\varepsilon \neq 0$ (small) the phase portrait is represented in panel **(b)**. The remarkable fact is that around the separatrix of the unperturbed case we see a chaotic region: this is the phenomenon discovered by Poincaré, which we should understand.

We make a further simplification by considering a nonlinear system in the neighbourhood of a hyperbolic equilibrium, as described, for example, by the non-autonomous Hamiltonian

(10.14) $$H(x,y) = \frac{y^2 - x^2}{2} + \frac{x^3}{3} + \varepsilon x^2 \cos t\ , \quad (x,y) \in \mathbb{R}^n\ .$$

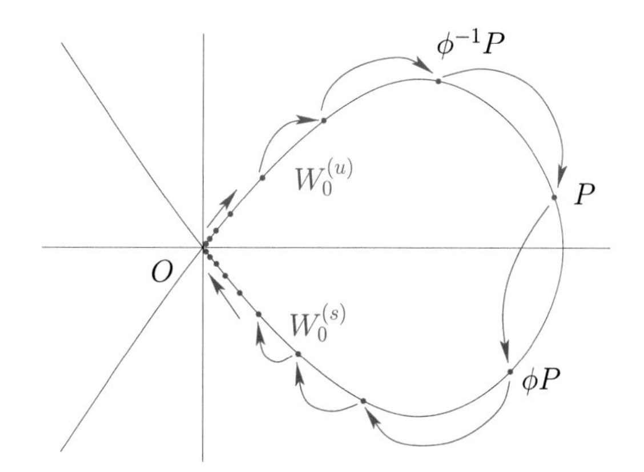

Figure 10.2 Poincaré section for the Hamiltonian (10.14), for $\varepsilon = 0$. The separatrices of the unstable equilibrium $x = y = 0$ form a closed loop in the right half-plane. The map $\phi^T P$ at time T with initial point P on the loop generates an orbit asymptotic to the unstable point for both $t \to -\infty$ and $t \to +\infty$.

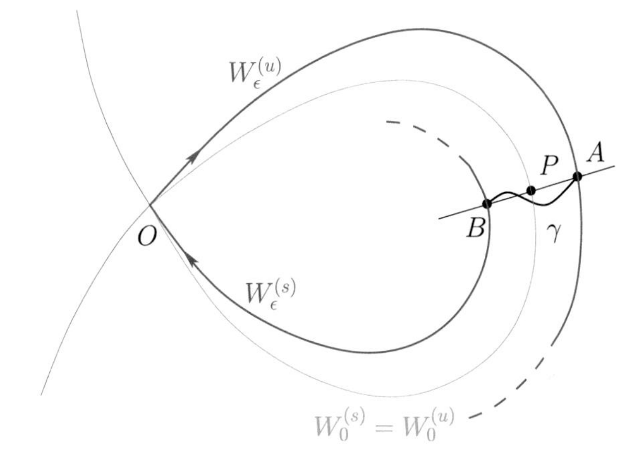

Figure 10.3 For $\varepsilon \neq 0$, the stable and unstable manifolds on the right of the fixed point need not coincide, a priori. For small ε, two points A, B on the unstable and the stable manifolds, respectively, and close to P, are joined by a curve γ with length tending to zero with ε.

For $\varepsilon = 0$ the phase portrait exhibits stable and unstable manifolds, as represented in Fig. 10.2.

We focus our attention on the loop on the right part of the figure, where the arcs of stable and unstable manifolds, marked by $W_0^{(s)}$ and $W_0^{(u)}$ respectively, do coincide, forming a loop (the manifolds extend also on the left part of the figure, where they are distinct). We consider again the Poincaré map ϕ^T at time $T = 2\pi$; for brevity, we denote it simply by ϕ. The map is area preserving, since the flow is a canonical one. Any orbit with initial point $P \in W_0^{(u)}$ generates a backward sequence $\phi^{-1}P$, $\phi^{-2}P$, $\phi^{-3}P$, ... asymptotic to the equilibrium in the past. On the other hand, we also have $P \in W_0^{(s)}$, so that the forward sequence ϕP, $\phi^2 P$, $\phi^3 P$, ... is still asymptotic to the equilibrium in the future, as represented in the figure. This is the kind of loop considered by Poincaré.

Let us now turn the perturbation on by making $\varepsilon \neq 0$ (small). The question is: *what about the fate of asymptotic orbits?* Poincaré had proved that, for $\varepsilon \neq 0$, asymptotic orbits continue to exist and are smooth; this result is known nowadays as the theorem of existence of stable and unstable manifolds. But there is a subtler question: *let the stable and unstable manifolds of a hyperbolic equilibrium coincide for $\varepsilon = 0$; do they continue to coincide for $\varepsilon \neq 0$, being only deformed?* The answer is by no means trivial.[2]

Without devoting much space to the latter question, we may recall a few facts. The Hamiltonian (10.14) may be cast in Poincaré–Birkhoff normal form; the canonical transformation is analytic (expressed as a convergent series) in a neighbourhood of the equilibrium. This implies that the Poincaré section at time $T = 2\pi$ for (10.14) has a hyperbolic equilibrium point with stable and unstable manifolds which may be locally represented via a convergent series expansion. The latter result was first proved by Cherry [41] and restated in a more general form by Moser [173]; for a proof using the method of normal form, see [89]. A spontaneous naive conclusion would be that the stable and unstable manifolds continue to coincide; but this is untrue. The point is that the expansion which produces the normal form provides also the expansions of the stable and unstable manifolds, but is convergent only in a neighbourhood of the equilibrium which does not include common points of the stable and unstable manifolds. In order to represent a longer part of each manifold a separate process of continuation is necessary, so that coincidence of manifolds does not follow. A thorough discussion, including explicit calculation of separatrices using computer algebra, may be found in [63]; detailed references to previous work on the subject may be found there. The authors show that the two manifolds actually do intersect transversally.

10.2.2 Stable and Unstable Manifolds

Let us follow the argument of Poincaré. We briefly recall some facts commonly known nowadays. We refer to Fig. 10.3.

The most known proofs of the so-called *stable manifold theorem* begin by proving a local result. In a local neighbourhood of the equilibrium O, the unperturbed stable (resp. unstable) manifold $W_0^{(s)}$ (resp. $W_0^{(u)}$) is slightly

---

[2] Poincaré raised the question in his paper on the problem of three bodies [187]. Initially he got convinced that the answer is in the positive, as he wrote in a first version of the paper which he had submitted for a prize banned on the occasion of the 60th birthday of King Oscar II of Sweden. After some objections raised by Lars Edvard Phragmén, a considerable part of the paper was completely rewritten by Poincaré, including the description of the phenomenon of homoclinic intersection described here. The history of the prize is widely narrated in [15]. A short report may be found in [214].

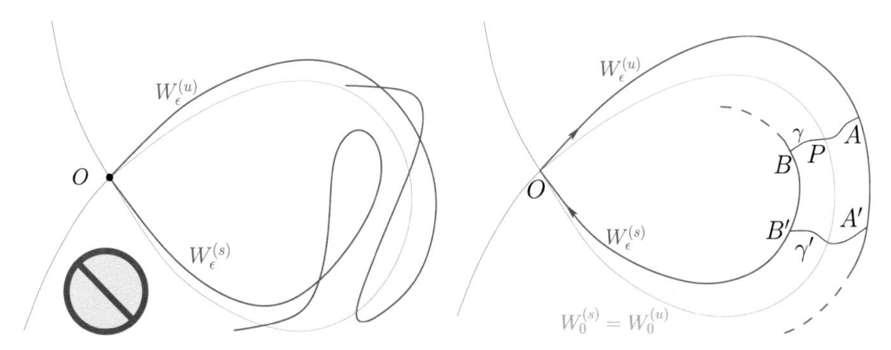

Figure 10.4 Left panel: a forbidden situation; the stable and unstable manifolds cannot have self-intersection points. Right panel: generally speaking, the stable and unstable manifolds may become separated, a possibility which we cannot exclude and should be further explored.

deformed into $W_\varepsilon^{(s)}$ ($W_\varepsilon^{(u)}$) which is tangent to $W_0^{(s)}$ ($W_0^{(u)}$) in O and is the graph of a function over $W_0^{(s)}$ ($W_0^{(u)}$); this provides a local slice which is prolonged by successive applications of the map ϕ^{-T}, thus proving separately the global existence of smooth manifolds $W_\varepsilon^{(s)}$ and $W_\varepsilon^{(u)}$.

By continuity, the manifold $W_\varepsilon^{(s)}$ ($W_\varepsilon^{(u)}$) initially follows closely the unperturbed manifold $W_0^{(s)}$ ($W_0^{(u)}$). We may state this property in a slightly more precise form by saying that for a given $P \in W_0^{(s)}$ there are points $A \in W_\varepsilon^{(u)}$ and $B \in W_\varepsilon^{(s)}$ such that $\operatorname{dist}(A, P) \sim 2 \operatorname{dist}(P, B)$ tends to zero with ε and can be connected by a continuous arc γ with length $\ell(\gamma)$ vanishing with ε. A very naive estimate may lead one to conclude that $\operatorname{dist}(A, P) \sim 2 \operatorname{dist}(P, B) = \mathcal{O}(\sqrt{\varepsilon})$, which is too pessimistic since we may find $\mathcal{O}(\varepsilon)$. However, such an improvement has no impact on the following; so let us avoid a long digression and use the rough estimate $\mathcal{O}(\sqrt{\varepsilon})$.

The manifold $W_\varepsilon^{(s)}$ ($W_\varepsilon^{(u)}$) has no double point; that is, *a self-intersection of the manifolds cannot occur* (left panel of Fig. 10.4). For, if $P \in W_\varepsilon^{(s)}$ is a double point, then for k large enough, $\phi^{-kT} P \in W_\varepsilon^{(s)}$ falls into a neighbourhood of O and is again a double point, contradicting the fact that $W_\varepsilon^{(s)}$ is locally the graph of a function.

10.2.3 Existence of a Homoclinic Orbit

Let us refer to the right panel of Fig. 10.4. Let $A \in W_\varepsilon^{(u)}$ and $B \in W_\varepsilon^{(s)}$ be joined by a continuous simple curve γ of length $\mathcal{O}(\sqrt{\varepsilon})$; two such points do exist, as we have already claimed. Let also $A' = \phi A$, $B' = \phi B$ and $\gamma' = \phi\gamma$ be the images of A, B and γ under the map ϕ. Let $AA' \subset W_\varepsilon^{(u)}$

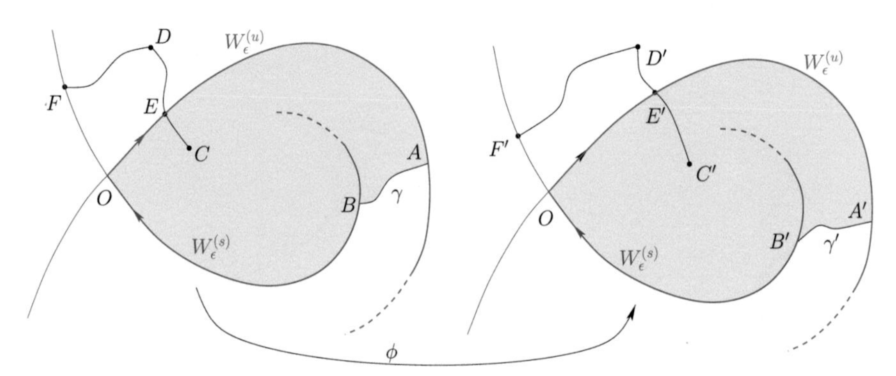

Figure 10.5 The arcs OA, γ and BO in the left panel delimit a curvilinear polygon which is mapped by ϕ onto the corresponding curvilinear polygon delimited by the arcs OA', γ' and $B'O$, with a one-to-one correspondence. Internal points of the first polygon are mapped to internal points of the image, and the same holds true for external points.

and $BB' \subset W_\varepsilon^{(s)}$ be the arcs between AA' and BB', respectively, on the corresponding manifolds. *We want to prove that* $AA' \cap BB' \neq \emptyset$.

We need a preliminary observation: the image of the closed curvilinear polygon $OABO$ is the curvilinear closed polygon $\phi(OABO) = OA'B'O$. That is, the image of an internal point $D \in OABO$ is an internal point $D' = \phi D \in OA'B'O$, and conversely for an external point; let us see a proof, with the help of Fig. 10.5.

The polygon $OABO$ is a simply connected region, by construction; hence $OA'B'O$ is simply connected, by continuity. Moreover, the image of $\partial(OABO)$ is obviously $\partial(OA'B'O)$, by construction.
Let now ψ be a simple curve connecting two arbitrary points; then $\psi' = \phi(\psi)$ is also a simple curve connecting the image of the two extreme points of ψ, by continuity of ϕ. Moreover, the map ϕ provides a one-to-one correspondence between the sets $\{\psi \cap \partial(OABO)\}$ and $\{\psi' \cap \partial(OA'B'O)\}$, for $G \in \psi \cap \partial(OABO)$ implies $\phi G \in \psi' \cap \partial(OA'B'O)$, and conversely.
Let $F \in W_\varepsilon^{(s)} \setminus OA$ (so that it lies on the arc of $W_\varepsilon^{(s)}$ on the left). Then we have $F' = \phi F \in W_\varepsilon^{(s)} \setminus OA'$, which implies $F' \notin OA'B'O$. We use now contradiction. Take an internal point $C \in OABO$ and an external point $D \notin OABA$. We claim that $C' \in OA'B'O$ remains an internal point, and $D' \notin OA'B'O$ remains external; for if ψ is a simple continuous arc joining either C with D and ψ' is an arc joining F with D, the number of intersections of either ψ or ψ' with $\partial(OABO)$ would change, which cannot occur.

After the preliminary observation, we come now to the proof that the arcs AA' and BB' must have an intersection at some point. Suppose for a

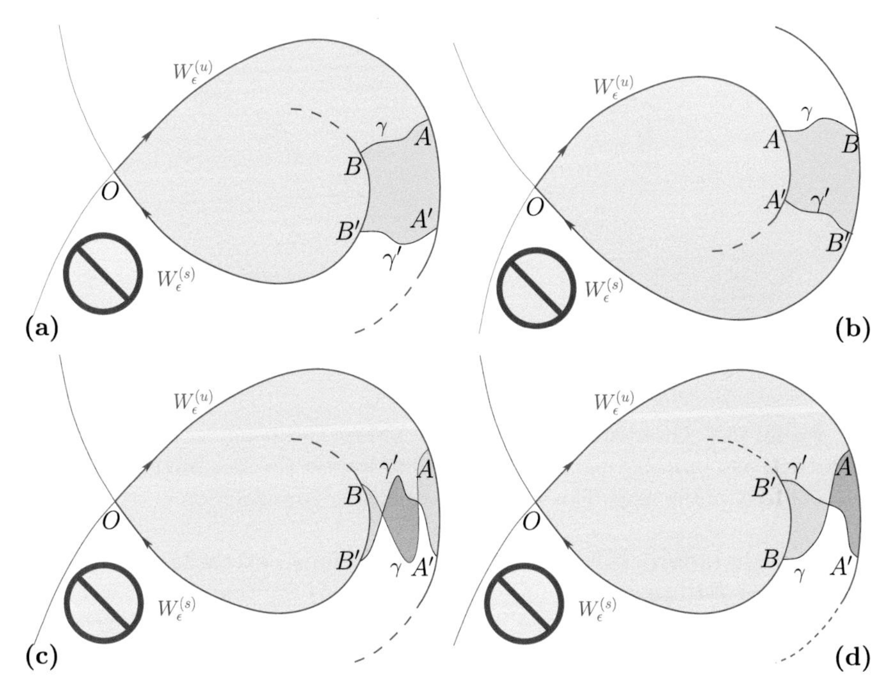

Figure 10.6 Four situations which do not involve intersection between the stable and unstable manifolds. However, all of them are forbidden: cases **(a)** and **(b)** violate the area-preserving property of the map; cases **(c)** and **(d)** contradict the fact that for $\varepsilon \to 0$ one has $\ell(\gamma) \sim \ell(\gamma') = \mathcal{O}(\sqrt{\varepsilon})$.

moment that there are no intersections; we prove that this cannot occur. We should consider two different cases, represented in Fig. 10.6.

(i) We have (panels **(a)** and **(b)**)

$$AA' \cap BB' = \emptyset\ , \quad \gamma \cap \gamma' = \emptyset\ .$$

(ii) We have (panels **(c)** and **(d)**)

$$AA' \cap BB' = \emptyset\ , \quad \gamma \cap \gamma' \neq \emptyset\ ,$$

with the points A, A', B, B' placed as in either panel **(c)** or **(d)**.

In case (i), consider the curvilinear polygon $OABO$, which is mapped into its image $\phi(OABO) = OA'B'O$; in view of the area-preserving property of the map ϕ we should have area$(OABO)$ = area$(OA'B'O)$, which is clearly false because the polygon $ABB'A'A$ either is added by the map, as in panel **(a)**, or it is subtracted, as in panel **(b)**. Hence this case cannot occur.

Case (ii), referring to panels **(c)** and **(d)**, seems acceptable, because the area added by the map (in light orange) may compensate an equivalent area which

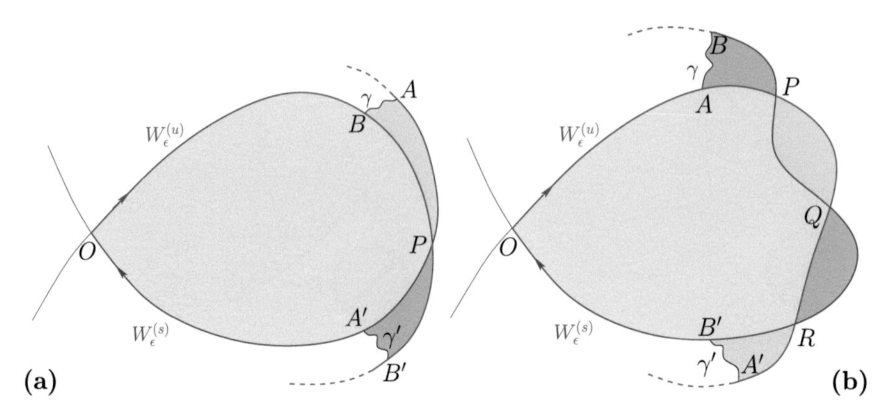

Figure 10.7 The existence of at least one intersection point between the stable and unstable manifolds allows us to respect the area-preserving property of the map. This case deserves further consideration.

is lost (in light brown). However, we observe that for $\varepsilon \to 0$ the length of both arcs AA' and BB' tends to a finite quantity; let us say that $\ell(AA') = \mathcal{O}(1)$ and $\ell(BB') = \mathcal{O}(1)$. On the other hand, we know that $\ell(\gamma') \simeq \ell(\gamma) = \mathcal{O}(\sqrt{\varepsilon})$, while in the cases in the figure at least one of $\ell(\gamma') = \mathcal{O}(1)$ and $\ell(\gamma) = \mathcal{O}(1)$ must occur. Hence none of these cases is allowed.

Having excluded these two cases, we must suppose that

$$AA' \cap BB' \neq \emptyset\ , \quad \gamma \cap \gamma' = \emptyset\ ,$$

as represented in Fig. 10.7, panel **(a)**. The preservation of area is respected, provided that the curvilinear triangles APB and $A'PB'$ have the same area; therefore, this is an acceptable case. There are, in fact, different possibilities:

(i) $AA' \cap BB'$ is a discrete set of points generated by transversal intersections: a single point as in panel **(a)** of Fig. 10.7, or possibly more than one as in panel (**(b)**);
(ii) $AA' \cap BB'$ is a set of arcs;
(iii) $W_\varepsilon^{(s)} \cap W_\varepsilon^{(s)}$ do coincide, as in the unperturbed case.

We conclude anyway that $W_\varepsilon^{(s)} \cap W_\varepsilon^{(u)} \neq \emptyset$.

The following conclusion is immediate:

Let $P \in W_\varepsilon^{(s)} \cap W_\varepsilon^{(u)}$, and let $\Omega(P) = \bigcup_{k\in\mathbb{Z}} \phi^k P$ be the orbit with initial point P. Then we have

$$\Omega(P) \subset W_\varepsilon^{(u)} \cap W_\varepsilon^{(s)}\ , \quad \phi^k P \underset{k\to+\infty}{\longrightarrow} O\ , \quad \phi^k P \underset{k\to-\infty}{\longrightarrow} O\ .$$

The latter claim is true because the stable and unstable manifolds are invariant under the map. We have thus shown that there is at least one doubly

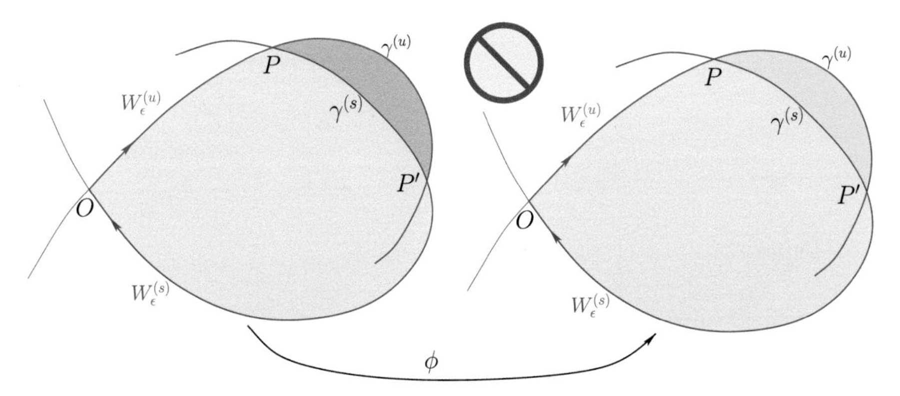

Figure 10.8 A further forbidden situation: if there is only one homoclinic orbit, then the area-preserving property is violated, again.

asymptotic orbit. Poincaré has named it *a homoclinic orbit*, and the points of the orbit are named *homoclinic points*.

10.2.4 Existence of a Second Homoclinic Orbit

Assume for simplicity that the case of panel **(a)** of Fig. 10.7 occurs, that is, $AB \cap A'B'$ contains only one intersection point.

By contradiction, let us suppose that there is only one homoclinic orbit, as represented in Fig. 10.8. Let P be a homoclinic point and $P' = \phi P$ be its image under the map. Denote by $\gamma^{(u)}$ the arc of $W_\varepsilon^{(u)}$ joining P and P', and by $\gamma^{(s)}$ the arc of $W_\varepsilon^{(s)}$ between P' and P, as in the left panel. Let us now apply the map, thus getting the figure of the right panel. Pay attention to the action of the map.

(i) The arc $OP \subset W_\varepsilon^{(u)}$ is mapped to $OP' \subset W_\varepsilon^{(u)}$; hence the new arc $\gamma^{(u)}$ is added to it.

(ii) The arc $OP \subset W_\varepsilon^{(s)}$ is mapped to $OP' \subset W_\varepsilon^{(s)}$; hence the arc $\gamma^{(s)}$ is subtracted from it.

(iii) By our assumption, $\gamma^{(u)} \cap \gamma^{(s)}$ contains only the extreme points P and P'; for if a third point $Q \in \left(\gamma^{(u)} \cap \gamma^{(s)} \setminus \{P, P'\}\right)$ exists, then $Q \notin \Omega(P)$ is a homoclinic point and $\Omega(Q) \neq \Omega(P)$ is a second homoclinic orbit.

(iv) The curvilinear polygon $OP\gamma^{(s)}P'O$ is mapped on $OP\gamma^{(u)}P'O$; therefore, it gains the area of the curvilinear polygon $P\gamma^{(s)}P'\gamma^{(u)}P$, as represented in the figure.

The latter claim contradicts the preservation of area by the map; hence the hypothesis that there is only one homoclinic orbit must be rejected.

There exists at least one homoclinic point $Q \notin \Omega(P)$, with a corresponding homoclinic orbit $\Omega(Q)$ distinct from $\Omega(P)$.

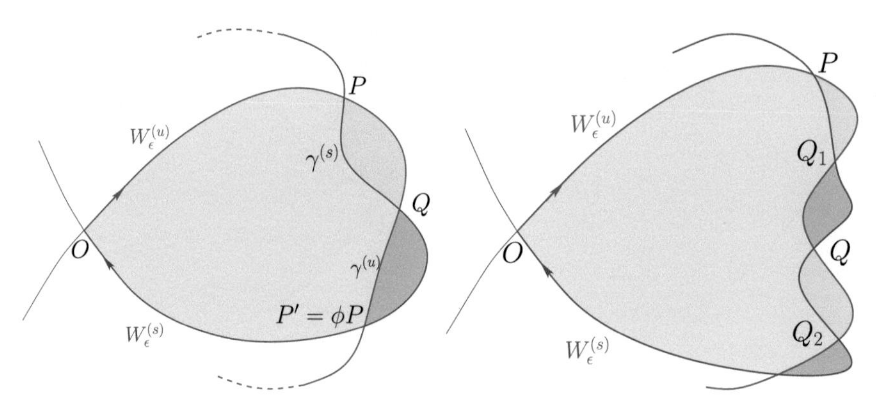

Figure 10.9 The area-preserving property forces the existence of at least a second homoclinic point.

It may well happen that $AA' \cap BB'$ contains more points, as, for example, points Q_1 and Q_2 in the right panel of Fig. 10.9; in this case every such point generates its own homoclinic orbit, distinct from the others.

10.2.5 Existence of Infinitely Many Homoclinic Orbits

We shall now follow the successive iterations of the map, illustrated in Figs. 10.10 and 10.11.

Let us start with the first iteration, illustrated in panel **(a)** of Fig. 10.10. We should pay particular attention to areas. Initially the arcs $OP \subset W_\varepsilon^{(u)}$ (blue) and $OP \subset W_\varepsilon^{(s)}$ (red) delimit the curvilinear polygon represented as the union of $\mathscr{N} \cup \mathscr{L}$, where $\mathscr{N}$ (in light blue) is the curvilinear polygon $OW_\varepsilon^{(u)}P\gamma^{(u)}Q\gamma^{(s)}P'W_\varepsilon^{(s)}O$, and the *lobe* $\mathscr{L}$ (in light orange) is delimited by the two arcs of $W_\varepsilon^{(u)}$ and $W_\varepsilon^{(s)}$ joining P with Q. With the first iteration the point P is mapped to $P' = \phi P$ and creates the new homoclinic point Q; the arc $OP \subset W_\varepsilon^{(u)}$ (blue) is prolonged by adding the arc $\gamma^{(u)}$, and the arc $\gamma^{(s)}$ is subtracted from $OP \subset W_\varepsilon^{(s)}$ (red). Thus, the lobe $\mathscr{L}$ is subtracted by the initial area and replaced with the addition of the lobe $\mathscr{M}$ (light brown), with the same area as $\mathscr{L}$; moreover, the lobe $\mathscr{L}$ is mapped to $\mathscr{L}' = \phi\mathscr{L}$ (light orange), represented in panel **(b)** of Fig. 10.10.

After the first iteration, the area delimited by the arcs $OP' \subset W_\varepsilon^{(u)}$ (blue) and $OP' \subset W_\varepsilon^{(s)}$ (red) is the union of the new region $\mathscr{N}$ (light blue) and of the lobe $\mathscr{L}'$ (light orange), while the lobe $\mathscr{L}$ is lost, as in panel **(b)**; thus, we are again in a situation similar to the initial one.

Now we apply the second iteration, mapping P' to $P'' = \phi P'$, subtracting the lobe $\mathscr{L}'$ and adding the lobe $\mathscr{M}'$, both with the same area. The point Q is mapped to $Q' = \phi Q$. This produces panel **(b)** of Fig. 10.10.

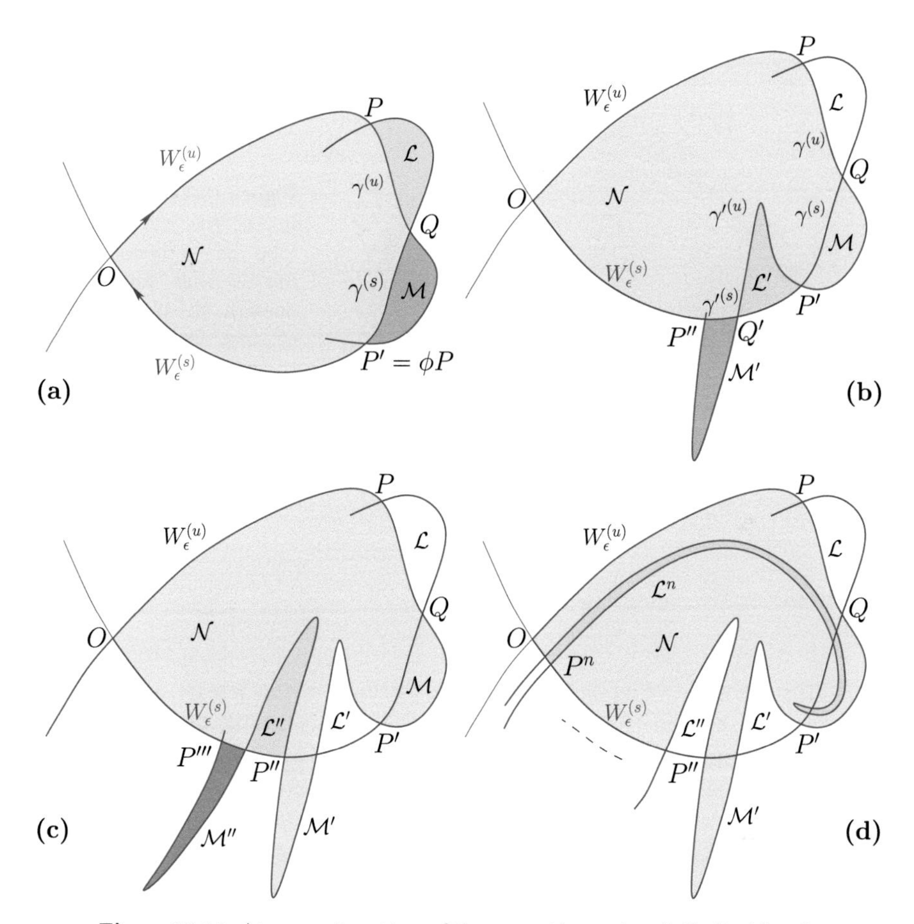

Figure 10.10 At every iteration of the map, the region delimited by the arcs $O\phi^k P \subset W^{(u)}$ and $O\phi^k P \subset W^{(s)}$ (light blue and orange) loses a lobe $\mathscr{L}$ and gains a lobe $\mathscr{M}$ with the same area, so that its shape is changed, but its area is preserved. For some iteration n the lobe $\mathscr{L}^n$ crosses again the stable manifold $W^{(s)}$, thus creating new homoclinic points independent from the previous ones.

The third iteration removes the lobe $\mathscr{L}''$ replacing it with the lobe $\mathscr{M}''$ with the same area, as illustrated in panel **(c)** of Fig. 10.10.

The process continues the same way with the successive iterations, but sooner or later something new shows up. Let us denote by $\mathscr{N}_0$ the initial region in panel **(a)** of Fig. 10.10. If we pay attention to the regions $\mathscr{N}$ at every step, we see that they keep the same area at every iteration (the lobes $\mathscr{L}$ and $\mathscr{M}$ have the same area), but the shape changes due to addition and subtraction of lobes – like an amoeba. But the lobes $\mathscr{L}$, $\mathscr{L}'$, $\mathscr{L}''$,... which

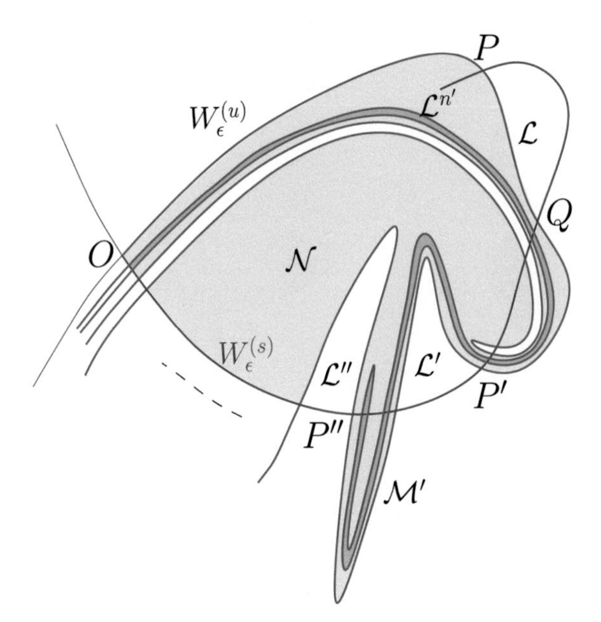

Figure 10.11 For increasing n, the width of the lobe $\mathscr{L}^{(n)}$ becomes so narrow that, for n large enough, the preservation of area forces the lobe to cross again the stable manifold; further homoclinic points are created. The process repeats ad infinitum.

are subtracted all have a positive area, and during the first few iterations they cover part of $\mathscr{N}_0$. On the other hand, all these lobes are disjoint, for $W_\varepsilon^{(u)}$ cannot have double points. Hence the iterates of $\mathscr{L}$ cannot all be contained in $\mathscr{N}_0$ (which has finite area); but they are always contained in $\mathscr{N}$ at the corresponding step. Consequently, at some iteration ϕ^n the lobe $\mathscr{L}^n$ must invade the lobe $\mathscr{M}$ of the first iteration, thus crossing the stable manifold $W_\varepsilon^{(s)}$, as in panel **(d)** of Fig. 10.10. This is not forbidden, because it is not a self-intersection of $W_\varepsilon^{(u)}$. The new intersections between the unstable and the stable manifolds are again homoclinic points which do not belong to the previous orbits $\Omega(P)$ or $\Omega(Q)$; hence they generate new, distinct homoclinic orbits.

The next iteration ϕ^{n+1} creates the next lobe $\phi\mathscr{L}^n$ which has the same area as $\mathscr{L}^n$, but cannot cross the unstable manifold $W_\varepsilon^{(u)}$; on the other hand, the distance between the points $P^{n+1} = \phi^{n+1}P$ and $Q^{n+1} = \phi^{n+1}Q$ must decrease to zero because they are approaching the limit point O. Therefore, the new lobe $\mathscr{L}^{n+1}$ must increase its length, while exploring the tight aisle between the previous lobe $\mathscr{L}^n$ and the manifold $W_\varepsilon^{(u)}$, which it is not allowed to cross. Therefore, it must cross again the border of the first lobe $\mathscr{M}$ along the stable manifold $W_\varepsilon^{(s)}$, as represented in Fig. 10.11. With the successive iterations the new generated lobes $\mathscr{L}$ must cross also the border of the next lobes $\mathscr{M}'$, $\mathscr{M}''$, ..., ad infinitum. Furthermore, after a sufficiently large number of iterations new intersection must be created, which do not belong to orbits already found. The successive intersections create more and more homoclinic points with corresponding homoclinic orbits.

If we consider the inverse map ϕ^{-1}, then the same process applies to the inverse images of P and of the lobe $\mathscr{L}$: the roles of the stable and unstable manifolds are merely exchanged. Thus, we conclude:

There are infinitely many homoclinic orbits which are distinct.

The extremely complicated behaviour of the stable and unstable manifolds has been vividly pointed out by Poincaré:

> *Que l'on cherche à se représenter la figure formée par ces deux courbes et leurs intersections en nombre infini dont chacune correspond à une solution doublement asymptotique, ces intersections forment une sorte de treillis, de tissu, de reseau à mailles infiniment serrées; chacune de ces deux courbes ne doit jamais se recouper elle–même, mais elle doit se replier sur elle même d'une manière très complexe pour venir recouper une infinité de fois toutes les mailles du reseau.*
>
> . . . .
>
> *On sera frappé de la complexité de cette figure, que je ne cherche même pas à tracer. Rien n'est plus propre à nous donner une idée de la complication du problème des trois corps et en général de tous les problèmes de Dynamique où il n'y a pas d'intégrale uniforme et où les séries de Bohlin sont divergentes.*

With a robust amount of presumption, we may want to exploit the power of our computers and try to plot the *homoclinic tangle* which Poincaré refused to draw. The initial part represented in Fig. 10.12 is a possible outcome.

10.2.6 Heteroclinic Orbits and Diffusion

The phenomenon of transversal intersections may well occur also between the stable and unstable manifolds of different hyperbolic points (or orbits in the case of a flow). The phenomenon is schematically represented in Fig. 10.13; the reader should not care too much that the lobes must cover the same area. Such intersection points have been named *heteroclinic* by Poincaré, and they generate doubly asymptotic orbits which connect different equilibria. It should be noted that two stable manifolds cannot have an intersection point (and the same holds true for two unstable manifolds); for the orbit corresponding to an intersection point should be asymptotic in the future (or the past) to two distinct periodic points, which cannot be.

Heteroclinic connections provide an effective mechanism for diffusion of orbits. Consider a small neighbourhood U of the point A. It contains at least one heteroclinic point which generates an orbit. The orbit is asymptotic in the future to the point B; therefore, after a number n large enough

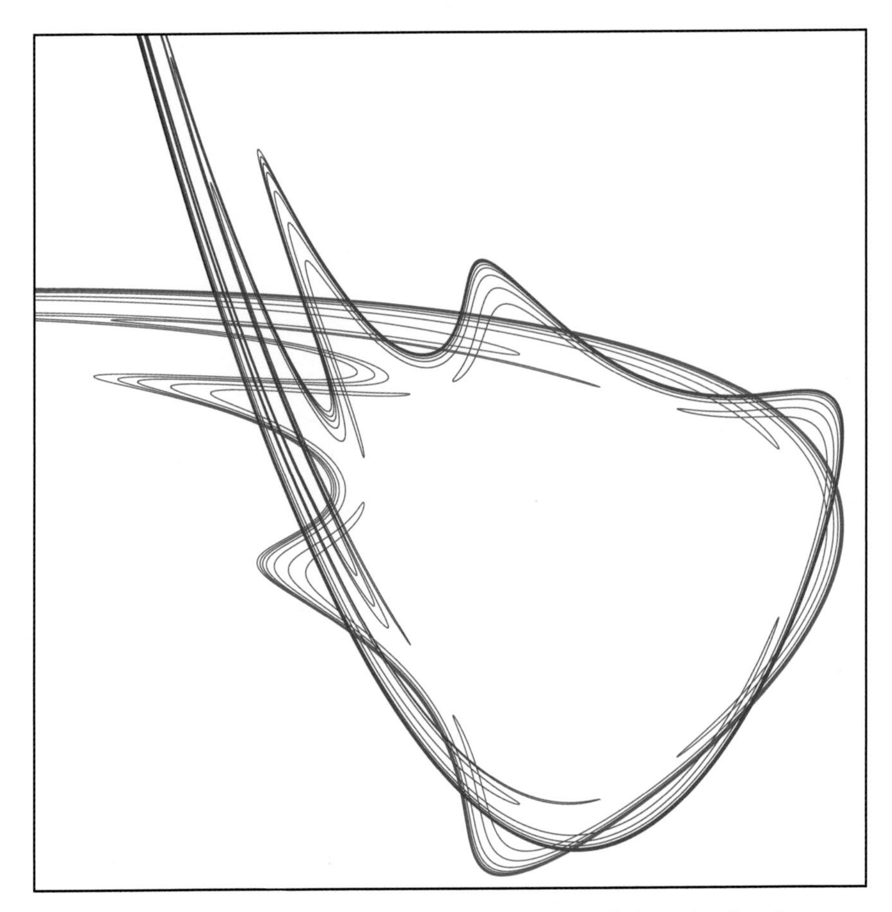

Figure 10.12 A pale image of the figure which Poincaré refused to draw, obtained by drawing the successive images of a very short segment of each manifold close to the unstable equilibrium. The initial length is 10^{-9}, and the map has been iterated 32 times.

of iterations the initial neighbourhood U is mapped into a neighbourhood $\phi^n U$ of B, by continuity. The neighbourhood $\phi^n U$ covers a slice of the unstable manifold of B, hence there is a point $Q \in U$ which is mapped onto a point $\phi^n Q$ belonging to the unstable manifold of B. The next iterates $\phi^{n+1}Q$, $\phi^{n+2}Q, \ldots$ move along the unstable manifold of B, dragging a small neighbourhood with them.

The phenomenon of diffusion becomes even more effective when there is a chain of heteroclinic connections; Fig. 10.14 provides a schematic representation. This happens if there is a sequence of hyperbolic equilibrium points $P_1, P_2, \ldots, P_n$, each of them with the corresponding stable mani-

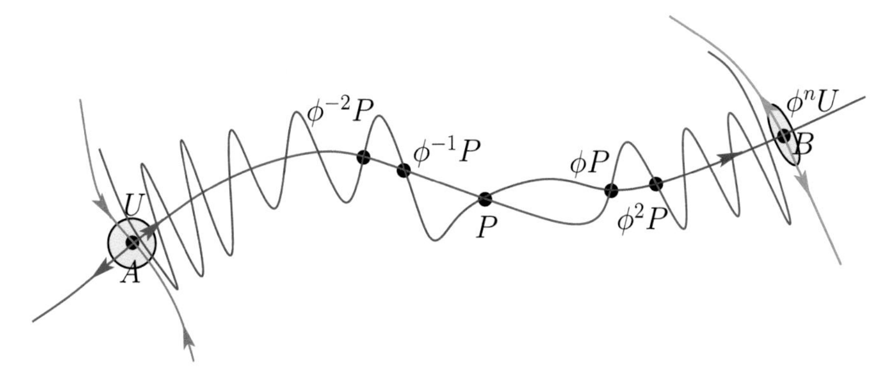

Figure 10.13 Schematic representation of a heteroclinic connection.

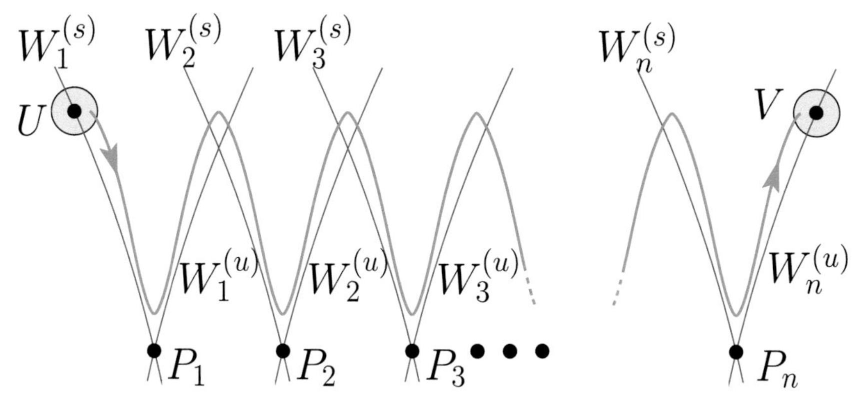

Figure 10.14 Schematic representation of a chain of heteroclinic connections and of a diffusion orbit which follows the chain.

folds $W_1^{(s)}$, $W_2^{(s)}, \ldots, W_n^{(s)}$ and unstable manifolds $W_1^{(u)}$, $W_2^{(u)}, \ldots, W_n^{(u)}$. If $W_1^{(u)} \cap W_2^{(s)} \neq \emptyset$, $W_2^{(u)} \cap W_3^{(s)} \neq \emptyset, \ldots, W_{n-1}^{(u)} \cap W_n^{(s)} \neq \emptyset$, then we can apply the preceding argument to the sequence of pairs P_1P_2, P_2P_3, $\ldots$ $P_{n-1}P_n$. Therefore, there is an orbit moving from a neighbourhood $U(P_1)$ which visits in sequence the neighbourhoods $U(P_2), \ldots, U(P_n)$. Similarly, an orbit with initial point in a neighbourhood U of $W_1^{(s)}$, as in Fig. 10.14, is allowed to follow the chain until it reaches a neighbourhood V of $W_n^{(u)}$.

The weak point of the latter argument is that we should *prove* that transverse heteroclinic connections do exist, generically. This problem has been raised by Poincaré. Proving that such orbits do exist generically for any perturbation, however small, has been revealed to be a major challenge.

10.3 An Example in Dimension 2: the Standard Map

Let us pause for a breath and see what happens in a concrete example of minimal dimension. A good choice is to consider the so-called *standard map*. The model was introduced by Boris Valerianovich Chirikov [42][43] in order to simulate the Poincaré section for a kicked pendulum; that is, a pendulum subject to periodic instantaneous impulses. The map writes

$$y' = y + \varepsilon \sin x \ , \quad x' = x + y' \ , \quad (x, y) \in \mathbb{T} \times \mathbb{R} \ . \tag{10.15}$$

As a matter of fact, the map turns out to be 2π-periodic also in the y variable (the momentum of the pendulum); hence it is usual to represent the phase portrait on a torus $\mathbb{T}^2$. The rich dynamical behaviour of the map makes it an excellent example which exhibits most of the interesting phenomena.

We recall some elementary facts. A point P is periodic with period N if $\phi^N P = P$; it is convenient to identify N as the minimal period. Thus, P is a fixed point for the map ϕ^N, so that a periodic orbit appears as a finite number of isolated points. The theory of normal form for linear systems in a neighbourhood of an equilibrium may be transported to the case of a map, with some *caveats*, by considering a linear map $\xi \to \mathsf{M}\xi$, where M is the Jacobian matrix of the map ϕ^N.

We are particularly interested in periodic orbits of an area-preserving two-dimensional map which are either *elliptic* or *hyperbolic*. In the elliptic case the eigenvalues are complex conjugated, being written as $e^{\pm i\omega}$, so that the linear approximation of the map is a rotation with angle ω. In the hyperbolic case the eigenvalues are real and are written as λ, λ^{-1} (we may assume $\lambda > 1$, thus excluding the exceptional case $\lambda = 1$), so that the eigenvectors define two invariant directions, respectively expanding and contracting.

A quasi-periodic orbit on an invariant torus appears in the diagram of the map as a curve densely filled by the section points. If the curve is parameterized by a real coordinate $\vartheta \in [0, 2\pi)$, then the frequency ω of the flow is replaced by the *rotation number* α, namely the average increment of ϑ induced by ϕ.

Our aim here is to use the standard map in order to illustrate the role of homoclinic and heteroclinic orbits as roots of chaotic behaviour, together with the role of invariant curves – the equivalent of Kolmogorov's invariant tori for the flow – as barriers which confine the orbits in separate regions.

10.3.1 Boxes into Boxes

An overall representation of the phase portrait of the map for increasing values of the perturbation parameter ε is provided in Fig. 10.15. Remark that for $\varepsilon = 0$ the map reduces to a horizontal translation $x \to x + y$,

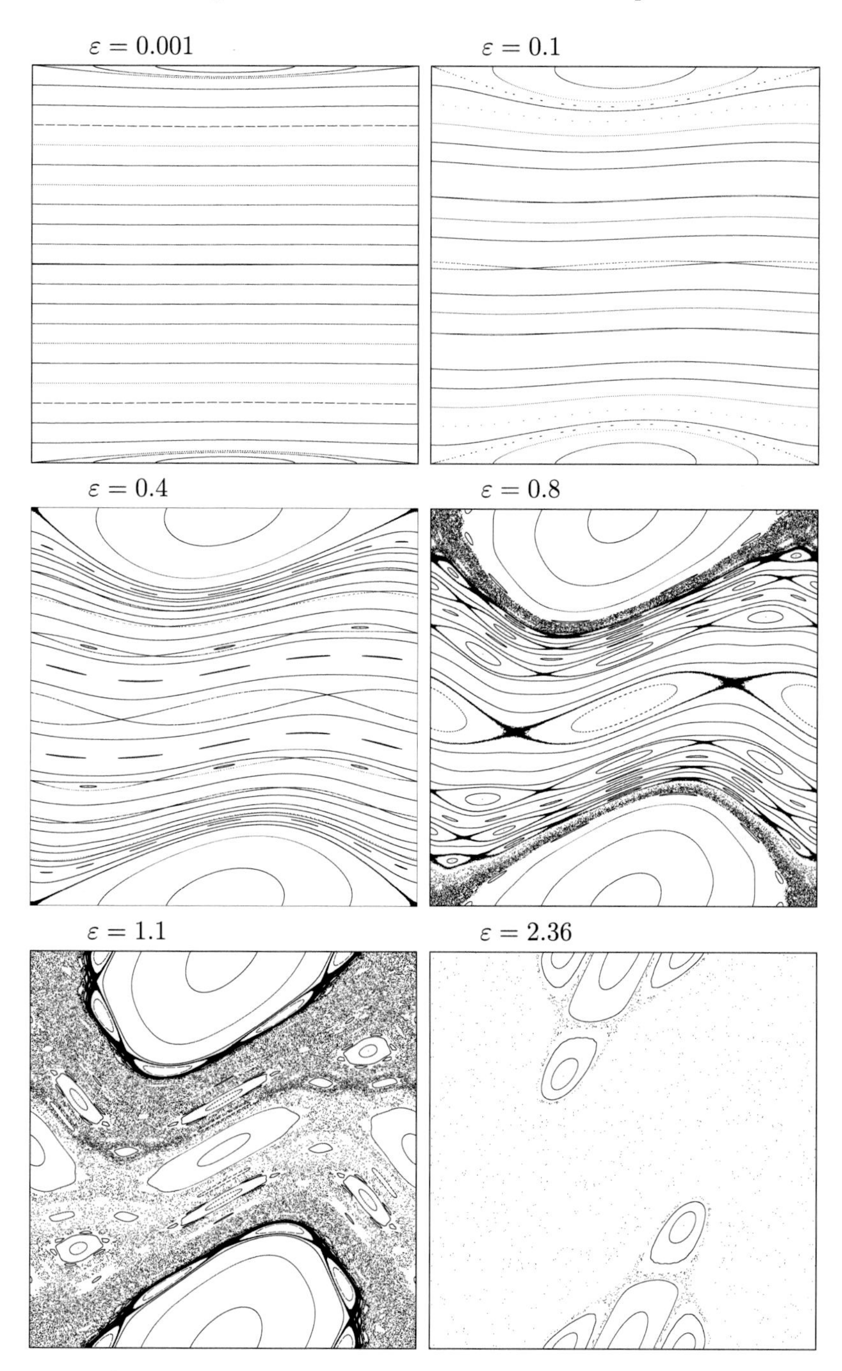

Figure 10.15 The phase portrait of the standard map for increasing perturbation parameter ε, represented on the torus $\mathbb{T}^2 = [0, 2\pi) \times [0, 2\pi)$.

with y constant. The orbit is periodic with period N if $y = k/N$ is a rational number. If y is irrational, then the orbit is quasi-periodic with rotation number y.

For a low value $\varepsilon = 0.001$, one observes that most of the invariant tori seem to be preserved. Kolmogorov's theorem assures indeed that most (in the sense of a measure) of the strongly non-resonant tori do persist. This is what we see in the upper left panel of Fig. 10.15, with an exception: the periodic orbit of period 1 which creates the islands and the separatrix in the lower and upper sides of the figure. An orbit of period 2 on the central line becomes visible if one draws the phase portrait by suitably enlarging the vertical scale.[3]

The reader will note that this claim looks puzzling, for according to Kolmogorov's theorem an invariant torus persists if its frequency – here its rotation number – is strongly irrational. Therefore, the unperturbed tori with rational rotation number are not expected to survive, in contrast with what we see in the figure.

The latter problem was already discussed by Poincaré, though without the support of Kolmogorov's theorem, of course [191]. It is remarkable, and worthy to be mentioned, that Poincaré begins his paper with this claim:

> *Je n'ai jamais présenté au public un travail aussi inachevé; je crois donc nécessaire d'expliquer en quelques mots les raisons qui m'ont determiné à le publier, et d'abord celles qui m'avaient engagé à l'entrependre.*

He explains that his efforts to investigate the appearance and the disappearance of periodic orbits of the problem of three bodies had led him to get convinced that the problem could be reduced to a property of geometric nature, which could be stated in a simple form. But he soon realized that the proof was unexpectedly difficult, and he concluded by saying:

> *Ma conviction qu'il est toujours vrai s'affermissait de jour en jour, mais je restais incapable de l'asseoir sur des fondements solides. Il semble que dans ces conditions, je devrais m'abstenir de toute publication tant que je n'aurai pas résolu la question; mais après les inutiles efforts que j'ai faits pendant de longs mois, il m'a paru que le plus sage était de laisser le problème mûrir, en m'en*

---

[3] We should not forget that every actual representation is limited by the resolution of the image: with our current technology the limit is the width of the pixel. We can enlarge some portions of the image, of course, but we should take care of the unavoidable numerical roundoff. We may also try to use multiprecision arithmetics, thus reducing the troubles due to roundoff and allowing ourselves to enlarge very small portions of the image. But if the size of the details to be visualized decreases exponentially with the inverse of ε, then even the multiprecision method will soon become useless.

reposant durant quelques années; cela serait très bien si j'étais sûr de pouvoir le reprendre un jour; mais à mon âge je ne puis en répondre. D'un autre côté, l'importance du sujet est trop grande (et je chercherai plus loin à la faire comprendre) et l'ensemble des résultats obtenus trop considérable déjà, pour que je me résigne à les laisser définitivement infructueux. Je puis espérer que les géomètres qui s'interesseront a ce problème et qui seront sans doute plus heureux que moi, pourront en tirer quelque parti et s'en servir pour trouver la voie dans laquelle is doivent se diriger. Je pense que ces considérations suffisent à me justifier.

Poincaré died a few months later, and the first proof of the theorem was published by Birkhoff [26]. The *last geometric theorem*, enunciated by Poincaré in [191], is resumed as follows.

Consider an annulus enclosed by two circumferences, and suppose that we are given a smooth one-to-one map of the annulus onto itself. Assume: (i) the circumferences are invariant for the map, and rotate in opposite directions; (ii) the map is area preserving. Then the map possesses two fixed points internal to the annulus.

In our case we may rephrase the theorem for our model by saying that between two invariant tori with different rotation numbers there exists a pair of periodic orbits, one of them being elliptic and the second one hyperbolic. Let us accept in addition that the separatrices of the hyperbolic points generate a homoclinic tangle which covers a small region.

The progressive destruction of invariant tori and the corresponding growth of the regions dominated by chaotic orbits is well evident in the sequence of Fig. 10.15. For $\varepsilon = 0.1$, the periodic orbit of period 2 is well visible, together with its separatrix. For $\varepsilon = 0.4$, more resonances show up, and moreover for $\varepsilon = 0.8$, a spot formed by chaotic orbits in the neighbourhood of the unstable orbit of period 2 becomes well visible on the scale of the figure. The reader will also remark that in all the four upper panels there are also invariant curves (tori) which extend from the left to the right margin and delimit the regions dominated by resonant islands and separatrices. For $\varepsilon = 1.1$, all such tori have disappeared, and for the highest value, $\varepsilon = 2.36$, only an island around the stable orbit of period 1 remains visible in the figure, together with four islands surrounding it and generated by a stable orbit of period 4.

The destruction of invariant tori is clearly due to the increase of the perturbation; such a phenomenon has been widely studied. For the standard map it turns out that the last torus to disappear has rotation number

$$\frac{\sqrt{5}-1}{2} = 0.618033988\ldots\ ,$$

the *golden number*.[4] The critical value ε_c at which the golden torus disap-

[4] The golden number $(\sqrt{5}-1)/2$ expands in continued fraction as $[1, 1, 1, \ldots]$,

pears has been determined numerically by John Morgan Greene [103], who found the value $\varepsilon_c = 0.971635\ldots$. More recently, analytical techniques combined with computer-assisted estimates have confirmed the value found by Greene: see, for example, [35][67].

The next interesting phenomenon is that around the stable periodic orbits there are islands of ordered orbits, surrounded by chaotic ones, which persist also when the last torus of the unperturbed map has been broken. We have already seen the same phenomenon around elliptic equilibrium points in Section 5.3. Let us, for a moment, restrict our attention to the local region inside an island, around one of the fixed points of a periodic orbit. If we introduce as local coordinates the distance from the centre of the island and an angle, then we realize that there is the same qualitative structure of the phase portrait of the standard map. That is, we can again introduce a rotation number which depends on the radial distance, and observe that up to some distance there are visually invariant tori. Here the distance plays the role of perturbation parameter, and the structure of invariant tori breaks when the distance increases. At the border we observe new islands generated by periodic orbits around the central one. The successive enlargements of Fig. 10.16 show that the same phenomenon repeats, on a smaller and smaller scale.

What is not evident in the figure is that pairs of stable/unstable orbits are created in any neighbourhood of the central orbit, at every scale, thus generating a new set of invariant tori and a new structure of homoclinic tangle at every level. The existence of homoclinic points in every neighbourhood of an elliptic fixed point was proved by Eduard Zehnder [215]. We do not see such a complicated structure in our figures because the size of islands and of chaotic regions is exceedingly small, but it is there. Thus, we are led to conclude that the phase portrait is, in fact, an intricate collection of *boxes into boxes* which reproduce essentially the same local qualitative behaviour, on smaller and smaller scales.

10.3.2 Homoclinic and Heteroclinic Tangles

We come now to illustrate the role of homoclinic and heteroclinic orbits. Let us first see that it is reasonable to expect that homoclinic tangles do show up in the neighbourhood of any hyperbolic orbit. Looking at the phase portraits of Fig. 10.15, we remark that for $\varepsilon = 0.4$ (left middle panel) we

---

which makes it considered to be the most irrational one. This may have motivated the naive belief that the torus with this frequency is the most robust one. It is true for the standard map, indeed, but not in general; see, for example, [35] pp. 106–107.

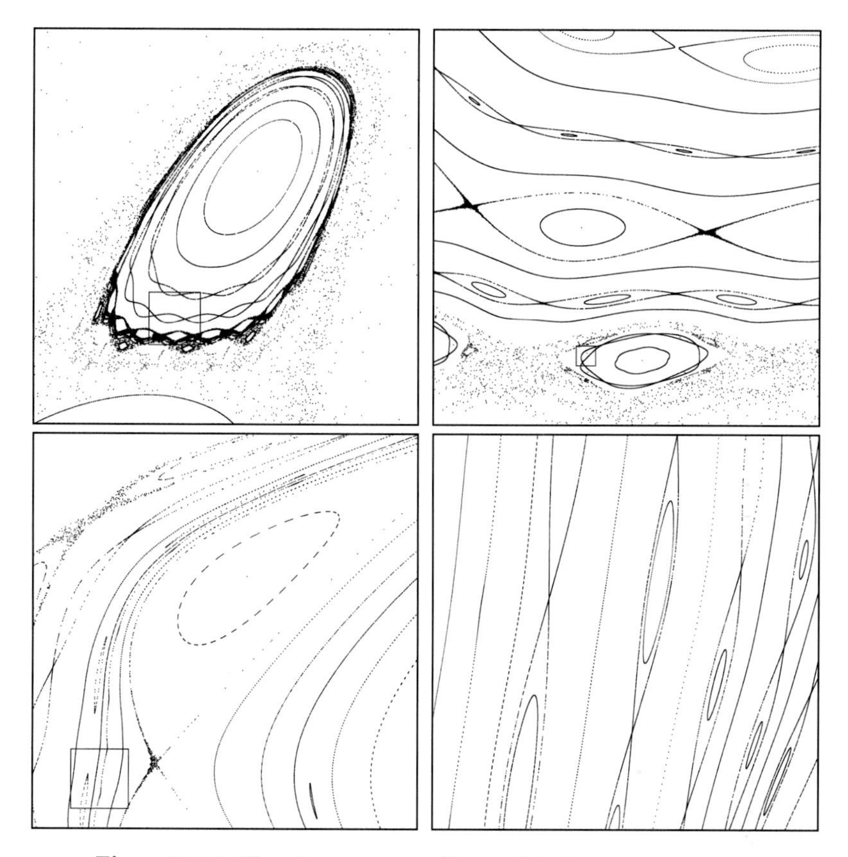

Figure 10.16 The phenomenon of boxes into boxes.

see two separatrices crossing at the periodic points, but we do not observe any homoclinic tangle on the scale of the figure; on the other hand, for $\varepsilon = 0.8$ (right middle panel) there are two well-visible spots around the periodic points, which we can interpret as homoclinic tangles which invade an observable region. What we can check numerically is that a homoclinic tangle exists in both cases, indeed. Fig. 10.17 provides an enlargement of two small rectangles around the periodic point on the left of the phase portraits. We see the same qualitative structure, but we should pay attention to the sides of the rectangle. For $\varepsilon = 0.4$, the sides are $(3 \times 10^{-3}, 6.5 \times 10^{-4})$; for $\varepsilon = 0.8$, the sizes are $(6.6 \times 10^{-1}, 3 \times 10^{-1})$. The conspicuous difference immediately suggests that trying to make the same phenomenon visible for lower values of ε such as, for example, $\varepsilon = 0.1$ will pose serious difficulties: there is no hope of seeing it without making a clever use of multiprecision arithmetics on our computers.

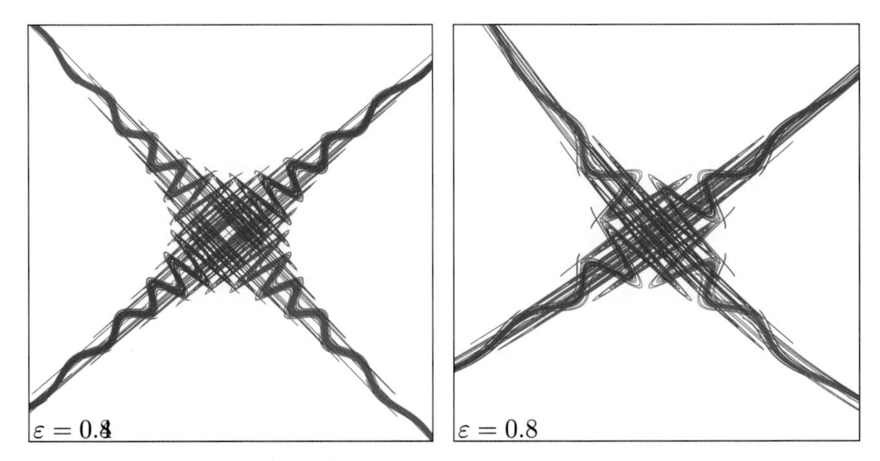

Figure 10.17 The homoclinic (heteroclinic) tangle of the separatices in the neighbourhood of the unstable orbit of period 2, for two different values of ε. The size of the rectangle is $(3 \times 10^{-3},\ 6.5 \times 10^{-4})$ for $\varepsilon = 0.4$ (left panel), and $(6.6 \times 10^{-1},\ 3 \times 10^{-1})$ for $\varepsilon = 0.8$ (right panel).

Let us now investigate the case of homoclinic tangles separated by invariant tori. In Fig. 10.18 the phase portrait for $\varepsilon = 0.8$ is represented together with the tangle formed by the stable and unstable manifols which belong to four different periodic orbits. Proceeding from the bottom up, we find periodic orbits of period 1, 3, 2, 3 and again 1. The latter orbit, on the top of the figure, coincides with the lower one if we take $\mathbb{T}^2$ as the phase space but should be considered as different if we prefer to take $\mathbb{T} \times \mathbb{R}$, as we shall do in the following. Every orbit has stable/unstable manifolds which are represented in different colours. In addition, the six curves in darker colour separate regions dominated by different periods: these curves, up to the resolution of the figure, are examples of invariant tori; there are many, indeed.

Some more orbits are represented in light colour, in order to allow the tangles to stand out. The reader will notice the ordered islands around the stable periodic orbits; the separatrices do not enter the islands, as expected. It is also interesting to look at chaotic orbits around the homoclinic/heteroclinic tangles: the reader will note that they extend a little around the homoclinic/heteroclinic tangles but do not cross the invariant tori which are effective barriers for the dynamics. It must be emphasized, however, that this is true only in a two-dimensional phase space, corresponding to a system of 2 degrees of freedom for the Hamiltonian flow on the invariant energy manifold; in higher dimensions, an invariant torus does not separate different regions of phase space.

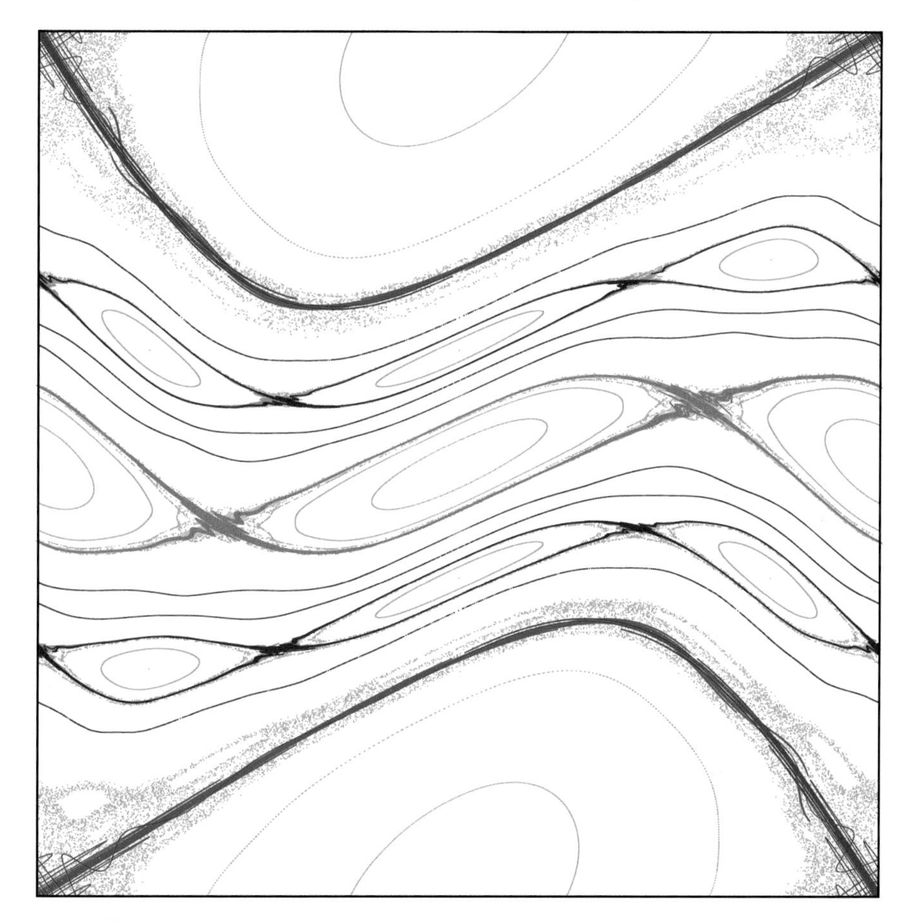

Figure 10.18 Phase portrait for the standard map at $\varepsilon = 0.9$. The figure includes the homoclinic/heteroclinic tangles generated by stable/unstable manifolds for the orbits of period 1, 2 and 3. Six orbits lying on invariant curves (Kolmogorov invariant tori for a flow) are put into evidence: they separate the phase space into invariant regions. The phase portrait including both ordered islands and chaotic orbits is also represented in light colour.

For a higher value of the perturbation parameter ε, the separating invariant tori have disappeared; therefore, the phenomenon of heteroclinic intersections may show up. This is illustrated in the Fig. 10.19. With some patience, the reader will see that there is a chain of heteroclinic connections: some intersection points are marked by circles labeled from bottom up by A (period 1 with 3), B (period 3 with 2), C (period 2 with 3) and D (period 3 with 1). The regions dominated by orbits of period 1–3–2–3–1 of Fig. 10.18

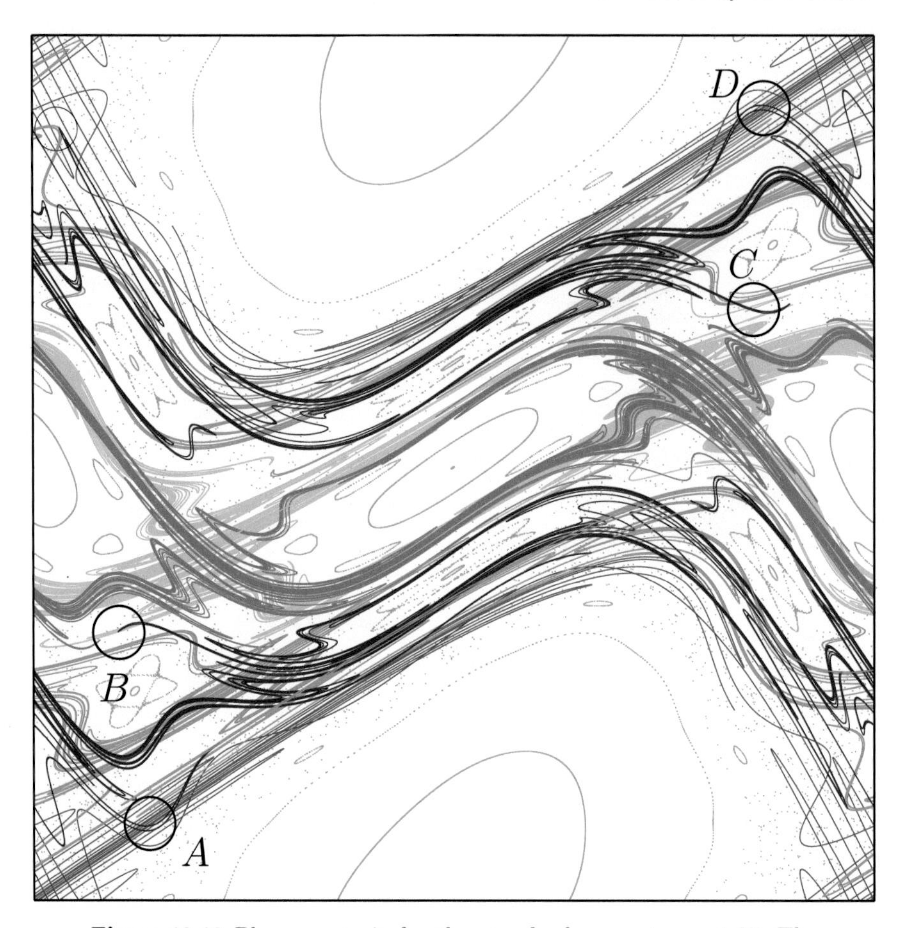

Figure 10.19 Phase portrait for the standard map at $\varepsilon = 1.25$. The figure is similar to Fig. 10.18, with two differences to be remarked. The invariant curves have been destroyed, so they do not appear any more. Moreover, heteroclinic intersection between the stable/unstable orbits of different period do appear, as pointed out with circles.

turn out to be connected, thus allowing a chaotic orbit to fill a considerable part of the phase space, except for the islands around stable periodic orbits.

The reader may also notice that heteroclinic intersections occur when several lobes have been created. The choice $\varepsilon = 1.25$ of the parameter is justified because a direct connection between the regions controlled by the widest resonance occurs with a reasonable number of lobes. Maybe some further explanation is useful. A further increase of the parameter makes the resonance region of period 3 disappear, and a direct connection between the regions of period 1 and 2 shows up with a limited number of lobes. This

Figure 10.20 Heteroclinic intersections for $\varepsilon = 1.6$ (left panel) and $\varepsilon = 2.6$ (right panel).

is observed in the left panel of Fig. 10.20, for $\varepsilon = 1.6$. For $\varepsilon = 2.6$, the lower and the upper regions of period 1, considered as distinct regions in $\mathbb{T} \times \mathbb{R}$, are directly connected. What happens if the value of ε is decreased can be imagined. When some invariant tori disappear, the separatrices of nearby resonances may create a chain of heteroclinic connections, thus creating channels between far resonant regions. Thus, diffusion orbits may well appear, but the diffusion speed is limited, because an orbit must travel along the whole chain.

Such a state of affairs is illustrated in Fig. 10.21. The figure should be compared with the previous Fig. 10.19, representing a chain of heteroclinic intersections which connects orbits of period 1-3-2-3-1. The value $\varepsilon = 1.25$ is the same. Starting with a set of 2^{20} points in a small circle of radius 10^{-10} (extremely small compared with the size of a pixel) around the point $(0, 0)$, namely the hyperbolic orbit of period 1, the various panels represent the sets of evolved points $\phi^n P$ for six different values of n, as in the labels of the different panels – note that what is represented is the state after n applications of the map, not the orbit up to iteration n. The phase space is taken to be $\mathbb{T} \times \mathbb{R}$; therefore, the upper border does not coincide with the lower one. For $n = 32$, we see that all points are distributed around the unstable manifold, represented in blue in Fig. 10.19. At $n = 64$, some orbits follow the separatrices of period 3, and for $n = 2^7$, 2^8 and 2^9 also the regions of period 2, 3 (the upper one) and 1 (on the top) have been reached successively by some orbits. The non-uniform density of points shows that traveling through different resonant regions along a chain of heteroclinic intersection may slow down the diffusion. After 2^{10} iterations the density

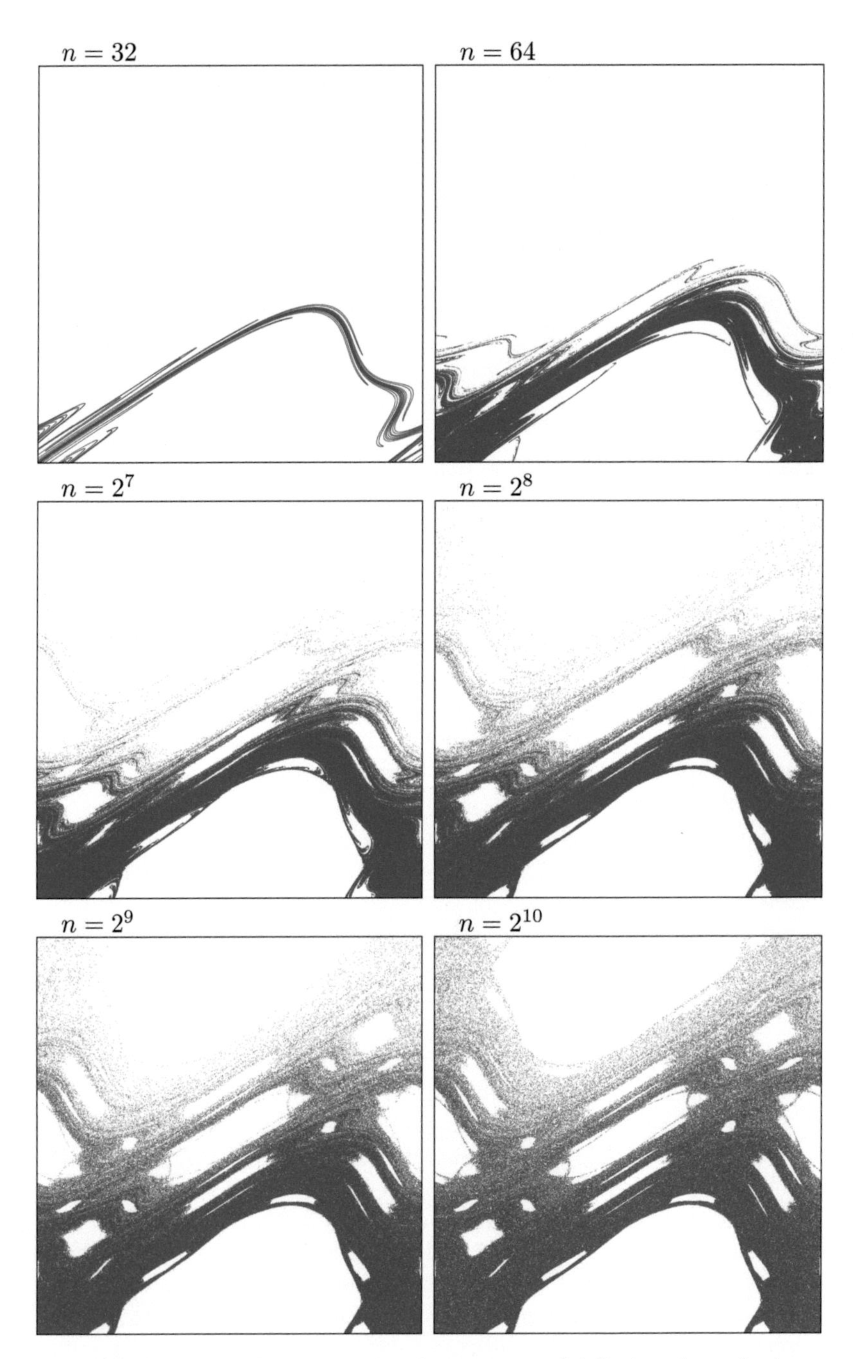

Figure 10.21 Illustrating the phenomenon of diffusion along the heteroclinic tangles: the map ϕ^n applied to a set of 2^{20} points very close to $x = y = 0$. Here, $\varepsilon = 1.25$.

tends to be more uniform, and the uniformity further increases, adding more and more iterations.

The reader may observe that the number of iterations is not very high, which is true. However, it may be expected that if the length of the chain of heteroclinic connections increases, then the number of iterations needed to observe diffusion will significantly increase, too. The latter fact is illustrated in Fig. 10.22, for a value $\varepsilon = 0.98$ which is bigger than, but not far from the critical value $\varepsilon_{\rm c} = 0.9716\ldots$ at which the last torus is broken. The reader will remark that the resonant region of period 2 is reached by a limited number of points only after some 2^{22} iterations, and that reaching the next resonances of period 3 and 1 requires some 2^{26} iterations. The considerable slowdown of diffusion is evident.

For values of ε still higher than the breaking threshold of the golden torus but closer and closer to it, the number of iterations which is necessary to obtain a similar portrait of diffusion will rapidly become extremely high, since it is expected to increase exponentially with the inverse of the difference $\varepsilon - \varepsilon_{\rm c}$, tending to infinity when $\varepsilon \to \varepsilon_{\rm c}$. Thus, observing the same phenomenon will become impractical.

The slowdown phenomena illustrated here are known as *stickiness*. It has been shown that they depend on the intricate structure of ordered islands and hyperbolic periodic orbits, indeed. In particular, the investigation concerning the mechanism of breaking of invariant tori has revealed the existence of structures, similar to Cantor sets, which are in some sense the remnants of tori and have been called *Cantori* (by clear assonance with *KAMtori*). We shall not enter this subject: see, for example, [2], [167] and [62]; a description of the phenomenon with a number of references may be found in [50], Section 2.7.

10.3.3 The Case of Higher Dimension

Before closing the discussion concerning the standard map, let us spend a few words about the case of higher dimension. For example, we may consider an integrable canonical system with three degrees of freedom (three harmonic oscillators, or three perturbed rotators) with small coupling terms.[5] The energy surface is a five-dimensional invariant manifold, but the invariant tori are three-dimensional objects. The set of invariant tori fills most of the energy surface (in the sense of measure), but its complement is open and connected. Therefore, the useful property that invariant tori in a system of two degrees of freedom (or in the standard map) isolate invariant regions is

---

[5] For a study of both numerical and analytical of systems of coupled rotators see, for example, [18][19].

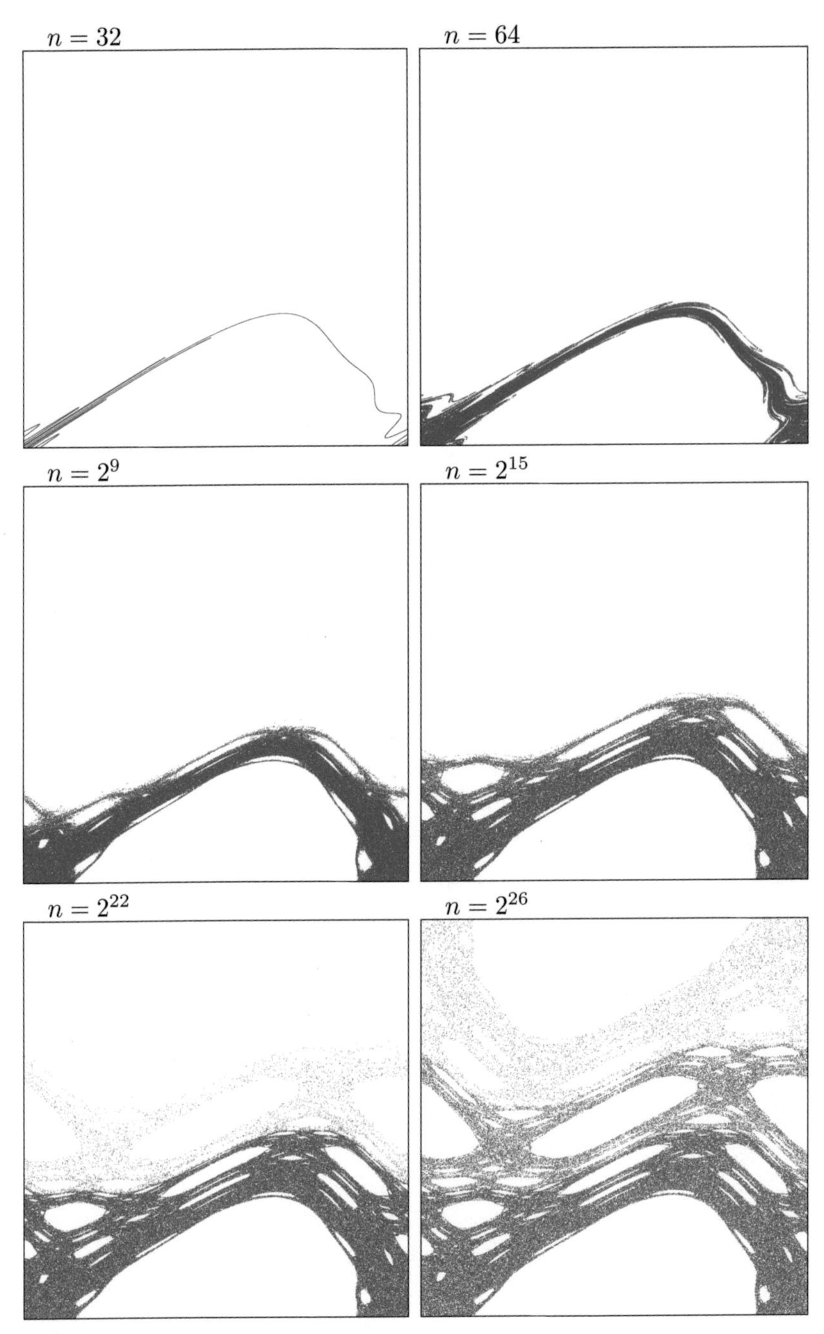

Figure 10.22 Illustrating the phenomenon of diffusion along the heteroclinic tangles: the map ϕ^n applied to a set of 2^{20} points very close to $x = y = 0$. Here, $\varepsilon = 0.98$.

lost: the existence of invariant tori puts no limits on the diffusion of orbits, no matter how small the perturbation.

The complement of the set of invariant tori may contain periodic orbits as well as tori of dimension lower than the number n of degrees of freedom. Both these objects can be either elliptic or hyperbolic; in the latter case they possess stable and unstable manifolds, and the existence of heteroclinic intersections cannot be excluded.

The first example which illustrates the existence of chains of heteroclinic intersections was proposed by Arnold [6]. It is constructed by adding a periodic time-dependent perturbation to a system with two degrees of freedom. In his example the fixed points of the Poincaré map are replaced by two-dimensional invariant tori which are hyperbolic, but are separated objects – as predicted by Kolmogorov's theorem.[6] The need for a higher dimension comes from his aim to prove that diffusion occurs as a general fact provided that the system has at least three degrees of freedom, due to homoclinic intersection between unstable and stable whiskers. After Arnold's paper, the phenomenon of orbits which wander in the complement of the set of invariant tori has been named *Arnold diffusion*. Arnold's paper marked the beginning of a long-standing work aimed at proving that heteroclinic intersections are indeed typical in perturbed systems. We do not enter such a long discussion; we just recall the references [124], [125], [24] already quoted.

There is a question which deserves to be discussed: *how much time is required for an orbit to visit all the points of the chain?* The relevant point is that the number n of homoclinic intersections which are involved grows very large – actually exponentially large in his example – when the perturbation becomes smaller and smaller. On the other hand, looking at the example of Fig. 10.13, a heteroclinic orbit must spend many iterations in order to pass from a neighbourhood of A to the neighbourhood of B. Similarly, a diffusion orbit, such as an orbit starting from a neighbourhood of P_1 in the chain of Fig. 10.14, needs many iterations in order to reach the intersection $W_1^{(u)} \cap W_2^{(s)}$ and then to reach a neighbourhood of P_2. If a chain is composed of an exponentially large number of heteroclinic intersections, then the number of iterations needed for an orbit to traverse all the chain grows exponentially large. This is indeed the claim of Nekhoroshev's theorem.

10.4 Stability in the Large

The property of superexponential stability of Kolmogorov's invariant tori discussed in Section 10.1.3 shows that an invariant torus should not be

[6] Arnold named *whiskers*, the stable and unstable manifolds emanating from the invariant tori. After him, the expression "whiskered tori" became common in the literature.

considered as an isolated object. Its existence entails the existence of a strongly stable neighbourhood. However, everything appears as a local property, which still leaves the global behaviour of the orbits somehow obscure but intriguing. In addition, the discussion of Section 10.2 shows that there are regions dominated by the resonance where the orbits may exhibit a chaotic behaviour.

A complete description of the global behaviour of orbits is off limits. However, we may at least try to shed some light on the connection between the two main theorems which have been the subjects of Chapters 8 and 9, namely the theorems of Kolmogorov and of Nekhoroshev. The aim is to show that *for autonomous systems* exponential stability and superexponential stability are strongly related to the existence of invariant tori of Kolmogorov.[7]

10.4.1 Statement of the Theorem

We go back once again to the general problem of dynamics, namely a canonical system with Hamiltonian

$$H(p,q) = H_0(p) + \varepsilon f(p,q,\varepsilon) \ , \quad (p,q) \in \mathscr{D} = \mathscr{G} \times \mathbb{T}^n \ , \tag{10.16}$$

where $\mathscr{G} \subset \mathbb{R}^n$, which we may assume to be closed. We briefly recall the hypotheses of Nekhoroshev's theorem.

(i) The Hamiltonian is assumed to be a real analytic bounded function on the complex extended domain $\mathscr{D}_{\varrho,2\sigma} = \mathscr{G}_\varrho \times \mathbb{T}^n_{2\sigma}$, and to be bounded on the border of the domain.

(ii) $f(p,q)$ is assumed to depend analytically on the perturbation parameter ε in a neighbourhood of $\varepsilon = 0$.

(iii) For all p in $\mathscr{G}_\varrho$, the unperturbed Hamiltonian $H_0(p)$ is assumed to satisfy

$$\bigl\|C(p)v\bigr\| \le M\|v\| \ , \quad \bigl|\langle C(p)v, v\rangle\bigr| \ge m\|v\|^2 \quad \text{for all } v \in \mathbb{R}^n \tag{10.17}$$

with positive constants $m \le M$.

Let us also introduce the following.

Definition 10.8: *An n-dimensional torus $\mathcal{T}$ is said to be (η, T)-stable in case one has $\mathrm{dist}(\phi^t P, \mathcal{T}) < \eta$ for all $|t| < T$ and for every $P \in \mathcal{T}$.*

Theorem 10.9: *Consider the Hamiltonian (10.16), with the hypotheses (i), (ii) and (iii). Then there exists $\varepsilon^* > 0$ such that for all $\varepsilon < \varepsilon^*$ the following statement holds true: there is a sequence $\bigl\{\mathscr{D}^{(r)}\bigr\}_{r\ge 0}$ of nested*

---

[7] If the system is non-autonomous with a non-periodic dependence on time, then Kolmogorov's theorem does not apply. On the other hand, if the time dependence can be described as a non-periodic perturbation, then exponential stability in Nekhoroshev's sense may still be proved: see, for example, [81].

subsets of $\mathscr{D}$, with $\mathscr{D}^{(0)} = \mathscr{D}$, and two sequences $\{\varepsilon_r\}_{r\geq 0}$ and $\{\varrho_r\}_{r\geq 1}$ of positive numbers satisfying

$$\varepsilon_0 = \varepsilon\ , \quad \varepsilon_r = \mathcal{O}(\exp(-1/\varepsilon_{r-1}^a))\ ,$$
$$\varrho_0 = \varrho\ , \quad \varrho_r = \mathcal{O}(\varepsilon_{r-1}^{1/4})\ ,$$

with a positive constant $a < 1$, such that for every $r \geq 0$, we have:

(i) $\mathscr{D}^{(r+1)} \subset \mathscr{D}^{(r)}$;

(ii) $\mathscr{D}^{(r)}$ *is a set of n-dimensional tori diffeomorphic to* $\mathscr{G}_{\varrho_r}^{(r)} \times \mathbb{T}^n$ *endowed with canonical action-angle coordinates* $(p^{(r)}, q^{(r)})$;

(iii) $\mathrm{Vol}(\mathscr{D}^{(r+1)}) > (1 - \mathcal{O}(\varepsilon_r^a))\,\mathrm{Vol}(\mathscr{D}^{(r)})$;

(iv) for every $p^{(r)} \in \mathscr{G}^{(r)}$ *the torus* $p^{(r)} \times \mathbb{T}^n \subset \mathscr{D}^{(r)}$ *is* $(\varrho_{r+1}, 1/\varepsilon_{r+1})$*-stable;*

(v) $\mathscr{D}^{(\infty)} = \bigcap_r \mathscr{D}^{(r)}$ *is a set of invariant tori for the flow ϕ^t, and, moreover, we have* $\mathrm{Vol}(\mathscr{D}^{(\infty)}) > (1 - \mathcal{O}(\varepsilon_0^a))\,\mathrm{Vol}(\mathscr{D}^{(0)})$.

10.4.2 Sketch of the Proof

The proof is based on iteration of Nekhoroshev's theorem; thus, our first task is to recast the statement as an iterative algorithm. To this end we should extract some extra information on the non-resonant region from the proof in Chapter 9. Here we provide a sketch of the proof; a detailed exposition may be found in [85].

Let us denote by $\mathscr{D}^{(0)} = \mathscr{G}^{(0)} \times \mathbb{T}^n$ the initial domain $\mathscr{D}$ and, similarly, by ε_0, ϱ_0 the initial values of the parameters and by m_0 and M_0 the initial constants in (10.17). The hypotheses made allow us to apply Nekhoroshev's theorem, thus concluding that every torus $p^{(0)} \in \mathscr{G}^{(0)}$ is $(\varrho_0, \mathcal{O}(\exp(1/\varepsilon_0))$ stable. This requires the full geometric construction of Section 9.4.

Let us now focus our attention on the non-resonant region $\mathscr{G}^{(1)}$. According to point (iv) in Section 9.4.1 it is constructed by subtracting from $\mathscr{G}^{(0)}$ everything which belongs to a resonant zone of multiplicity one. Therefore, it is characterized by the absence of resonances of order lower than rK_0 with some $K_0 \sim 1/\sigma$ and with $r = \mathcal{O}(1/\varepsilon_0^a)$. Moreover, $\mathscr{G}^{(1)}$ possesses a complex extension to a complex domain $\mathscr{G}_{\varrho_1}$ with $\varrho_1 = \mathcal{O}(\varepsilon_0^{1/4})$. The analytic part of the proof of Nekhoroshev's theorem introduces canonical coordinates $(p^{(1)}, q^{(1)})$ in the extended domain $\mathscr{D}_{\varrho_1,\sigma_1} = \mathscr{G}_{\varrho_1} \times \mathbb{T}_{\sigma_1}^n$; this is claim (ii) for $r = 1$. In the new coordinates, the Hamiltonian takes a form similar to the original form (10.16); that is, we can separate an unperturbed part $H_0'(p)$ which is in non-resonant Birkhoff normal form, plus a perturbation of size $\varepsilon_1 = O(\exp(-1/\varepsilon^a))$; this is stated in Proposition 9.4, taking into account that we limit our attention to a non-resonant domain. The reader may note that $\mathscr{D}^{(1)}$ may be not connected, but new coordinates and the corresponding Hamiltonian are constructed in every connected component;

the same estimates apply in every component. All this is Nekhoroshev's theorem.

Here begins the iterative part of the proof. We want to estimate the volume of the non-resonant region $\mathscr{G}^{(1)}$, showing that it has a large relative measure with respect to $\mathscr{G}^{(0)}$. Recall that $\mathscr{G}^{(1)}$ is constructed by subtracting a finite number of resonant zones of multiplicity 1 from $\mathscr{G}^{(0)}$. Considering the image $\omega(\mathscr{G}^{(0)})$ through the frequency map ω, we estimate the volume taken out from this image with a straightforward adaptation of the argument in Section A.2.1. Then we estimate the volume subtracted from $\mathscr{G}^{(0)}$ using hypotheses (10.17). This is claim (iii).

Having thus performed the previous step for $r = 0$, we land on a non-resonant domain $\mathscr{D}^{(1)} = \mathscr{G}^{(1)} \times \mathbb{T}^n \subset \mathscr{D}^{(0)}$ which has the same properties of $\mathscr{D}^{(0)}$ with a different value of $\varrho_1 = \mathcal{O}(\varepsilon^{1/4})$. The new Hamiltonian on $\mathscr{D}^{(1)}$ is analytic and bounded and satisfies the hypotheses (10.17) with new constants $m_1 = m_0 - \mathcal{O}(\varepsilon^{1/4})$ and $M_1 = M_O + \mathcal{O}(\varepsilon^{1/4})$ and with a perturbation of size $\varepsilon_1 = \mathcal{O}\big(\exp(-1/\varepsilon^a)\big)$.

The procedure can be iterated, thus allowing us to construct by recurrence a sequence of domains $\mathscr{D}^{(r)}$ with parameters $\varepsilon_r = \mathcal{O}\big(\exp(1/\varepsilon_{r-1})\big)$ and with $\varrho_r = \mathcal{O}(\varepsilon_{r-1}^{1/4})$, as claimed, together with sequences m_r and M_r.

The problem now is to prove that we can consistently iterate the procedure infinitely many times, which means that we should make a suitable choice of the relevant parameters. In particular, we should prove that if ε_0 is small enough, then the sequence ε_r decreases to zero fast enough, so that Nekhoroshev's theorem can be applied at every step. Here there are two points which deserve a further explanation.[8]

---

[8] Here we follow the construction of Arnold [3], consisting precisely in eliminating step by step an open set of domains around resonant manifolds corresponding to Fourier modes $|k| \le N$, with N increasing to infinity. However, Arnold adapts the quadratic scheme of convergence of Kolmogorov, that is, with a perturbation which, ignoring the impact of small divisors, decreases step by step as $\varepsilon_r = \mathcal{O}(\varepsilon_{r-1}^2)$. Actually, part of such a rapid decrease is spent in order to control the small divisors; a suitable choice of some parameters makes the perturbation to decrease as $\varepsilon_r = \mathcal{O}(\varepsilon_{r-1}^a)$ with $1 < a < 2$; for example, $a = 3/2$. Our construction replaces the quadratic scheme with Nekhoroshev's theorem, thus giving $\varepsilon_r = \mathcal{O}\big(\exp(-1/\varepsilon_{r-1})\big)$ as claimed in the statement of the theorem. As in Arnold's scheme, we prove that for $r \to \infty$ there is a surviving set of invariant tori; in a strict sense, we are not authorized to conclude that such a set is the same as Arnold's, though they look similar. What we gain is the statement at claim (iv), namely that the diffusion around the set of invariant tori is superexponentially slow; this is indeed our main claim, which cannot be proved on the basis of the quadratic scheme of Arnold.

The first point concerns the limit $\mathscr{D}^{(\infty)} = \mathscr{G}^{(\infty)} \times \mathbb{T}^n$. At every step $\mathscr{G}^{(r)}$ is a non-empty closed domain which is a subset of $\mathscr{G}^{(r-1)}$; the limit $\bigcap_r \mathscr{G}^{(r)}$ is closed and has an empty interior. However, at every step the volume of $\mathscr{G}^{(r)}$ is estimated as $\mathrm{Vol}(\mathscr{G}^{(r)}) = \bigl(1 - \mathcal{O}(\varepsilon_r^a)\bigr)\,\mathrm{Vol}(\mathscr{G}^{(r-1)})$. We prove that if ε_0 is sufficiently small, then the sequence ε_r decreases fast enough, so that the infinite product $\prod_r \bigl(1 - \mathcal{O}(\varrho_r)\bigr)$ remains positive (and close to one). This assures that $\mathrm{Vol}(\mathscr{G}^{(\infty)})$ remains large, thus proving claim (iii).

The second point concerns claim (v), namely that $\mathscr{D}^{(\infty)}$ is a set of invariant KAM tori. It may appear puzzling, because $\mathscr{G}^{(\infty)}$ is not open. However, we use the argument of Arnold, which may be summarized as follows.

Let $\tilde{\mathscr{D}}^{(r)} = \mathscr{G}^{(r)}_{\varrho_r} \times \mathbb{T}^n$, so that $\mathscr{D}^{(r)} \subset \tilde{\mathscr{D}}^{(r)}$. In view of $\varrho_r \to 0$, we clearly have $\mathscr{D}^{(\infty)} = \bigcap_{r\geq 0} \tilde{\mathscr{D}}^{(r)}$. Let $P \in \mathscr{D}^{(\infty)}$, so that $P \in \mathscr{D}^{(r)}$ for every r. Denote by $(p^{(r)}, q^{(r)})$ the coordinates of P in $\mathscr{G}^{(r)}_{\varrho_r} \times \mathbb{T}^n$. Then we have $\phi^t P \in \tilde{\mathscr{D}}^{(r)}$ for $|t| < T_r$, say, with T_r monotonically increasing to infinity. Since r is arbitrary and $\tilde{\mathscr{D}}^{(s)} \subset \tilde{\mathscr{D}}^{(r)}$ for $s > r$, we also have $\phi^t P \in \tilde{\mathscr{D}}^{(r)}$ for $|t| < T_s$ for all s, that is, $\phi^t P \in \tilde{\mathscr{D}}^{(r)}$ for all t and for all r. Thus, by definition of $\mathscr{D}^{(\infty)}$, we get $\phi^t P \in \mathscr{D}^{(\infty)}$ for all t. This shows that the set $\mathscr{D}^{(\infty)}$ is invariant for the Hamiltonian flow. In view of the form of the Hamiltonian in $\tilde{\mathscr{D}}^{(r)}$ we also have $\mathrm{dist}\bigl(\phi^t P, (p^{(r)}, q^{(r)} + \omega^{(r)}(p^{(r)})t)\bigr) < \mathcal{O}(\varepsilon_r^{1/2})$ for $|t| < \mathcal{O}(\varepsilon_r^{-1/2})$, where $\omega^{(r)}(p^{(r)})$ is the unperturbed frequency at step r. On the other hand, by construction we have $\omega^{(r)} \to \omega^{(\infty)}$ with $\omega^{(\infty)}$ non-resonant. Therefore, we also have $\mathrm{dist}(\phi^t x, (p^{(r)}, q^{(r)} + \omega^{(\infty)}(p^{(r)})t)) < \mathcal{O}(\varepsilon_r^{1/2})$ for $|t| < \mathcal{O}(\varepsilon_r^{-1/2})$. Since r is arbitrary, we conclude that the orbit $\phi^t x$ has non-resonant frequencies $\omega^{(\infty)}$, so that it is dense on a torus, which belongs to $\mathscr{D}^{(\infty)}$.

10.5 Some Final Considerations

Let us attempt to collect the contents of this last chapter in a possible description of the dynamics of an autonomous, near-to-integrable Hamiltonian system, namely, once again, the general problem of dynamics. We should stress that the literature on this subject is so wide that collecting all results in a single book would be an overwhelming task; a fortiori, trying to explain it in a short section is a hopeless attempt. Thus, let us make only a few considerations based on the contents of the present notes, also including a few (partial) considerations concerning recent research.

There are three facts which are usually at the basis of discussions concerning the coexistence of order and chaos in dynamics. The first one is basically concerned with the role of resonance in generating chaos. The existence of stable and unstable manifolds and of homoclinic and heteroclinic orbits lies at the very root of the chaotic behaviour; we owe this discovery to Poincaré.

The second fact is the persistence of quasi-periodic motions on invariant tori, a discovery that we owe to Kolmogorov. It is worthwhile to recall once more that the predominance of invariant tori when the perturbation is very small has been initially interpreted as the ultimate answer to the problem of stability, for example for the planetary system. But we should also point out that the long-time simulation of the dynamics of the planets over times exceeding the estimated age of the solar system have shown that chaos is there, though well concealed [146][147].

The third fact is the enormous amount of numerical experiments which highlight the pervasive existence of chaos, even on space and time scales which are not immediately identified. One of the first observed phenomena is the existence of a sharp threshold which separates a macroscopically ordered behaviour from an almost completely chaotic one. Although most studies are concerned with low-dimensional models, transition phenomena have been observed also in systems with many degrees of freedom, which are of interest in Statistical Mechanics. The dynamics of lattices such as, for example, the FPU model are such examples. The question in this case is how far the theoretical results discussed in the present notes may shed light on the experimental results.

There is a widespread opinion, in particular among physicists, that the existence of invariant tori in systems with many degrees of freedom is not so relevant (see, for instance, [151]). Such an opinion is supported by the known fact that invariant tori do not isolate separated regions in phase space, thus allowing for the so-called Arnold's diffusion. Moreover, Kolmogorov's theorem applies, provided that the perturbation does not exceed a threshold, ε^* say, which goes to zero for $n \to \infty$. From the viewpoint of Physics the natural question is the following: *does the threshold have any meaning for a physical system; for example, for the planetary system?* The problem is that, using the estimates produced by the first proofs of Arnold and Moser (who, to do justice to the authors, did not pretend that their estimates should be directly applicable), one discovers that in the case of the solar System the mass of Jupiter should be of the order of that of a proton, or even less. Only in recent years have we been able to prove realistic results for low-dimensional systems, based on computer-assisted proofs ([33], [34], [36], [37], [158]). For KAM theory for infinite-dimensional systems see, for example, [130], [131].

As we have pointed out, Nekhoroshev's theorem seems to be in a better position from the viewpoint of Physics. The theorem claims nothing about the existence of invariant tori, nor does it exclude the presence of chaos. The fascinating fact is that it puts strong limits on the speed of diffusion in open subsets of the phase space, claiming that a somewhat visible diffusion takes exponentially long times; this is physically relevant, even if infinite time is out of reach. However, in this case the drawback is that the threshold of applicability may be too small to be meaningful for a physical system, and that it decreases to zero in the limit $n \to \infty$. As for Kolmogorov's theorem, only in recent years have we obtained some results which may be considered to be realistic; again, the proof is computer assisted ([33], [84], [92], [197], [98]). The case of large or infinite-dimensional systems may be handled by further relaxing our requests. For example, we further relax our requests by, say, focusing attention on some interesting quantities ([22], [97]) or on some interesting initial states ([21], [12]).

There is a further point which may bother us: it is concerned with the amazingly sharp transition from order to chaos which is commonly observed in numerical simulations; the figures for perturbed oscillators in Chapter 5 and for the standard map in the present chapter are only a few examples of what can be found in hundreds of papers, after the pioneering ones of Contopoulos, Hénon and Heiles, Chirikov. The exponentially long time of Nekhoroshev may appear still not enough to explain the sharp character of such a transition.

The fact which emerges from the discussion of the present chapter is that invariant tori play a role which goes farther than merely being the predominant behaviour for small perturbations. The phenomenon of superexponential stability of Kolmogorov's tori seems to be quite relevant. We may figure out the following picture. An invariant Kolmogorov torus is the head of a structure which dominates the dynamics in its neighbourhood. It appears that such a structure has a typical radius which depends on the size of the perturbation, and decreases to zero as the perturbation increases towards the critical size corresponding to the destruction of the torus. Inside this radius the effective perturbation turns out to be as small as $\exp(-1/\varrho)$, which represents a local improvement with respect to previous estimates (see, for instance, [178]). Therefore, many other tori are allowed to survive, being in some sense the children of the leading torus. The dynamics is strongly affected by the existence of such a structure: the stability time actually increases as $\exp\bigl(\exp(1/\varrho)\bigr)$, considerably longer than the exponential time of Nekhoroshev. Thus, although the KAM tori are not isolating, they form, nevertheless, a kind of impenetrable structure that the orbits can neither escape nor enter for an exceedingly long time. Inside that structure the behaviour of the system can be said to be effectively integrable, though not in the sharp sense of the theorems of Liouville and of Arnold and Jost. The sharp transition to chaos may be interpreted as a consequence of such a strong closeness to integrability of regions containing Kolmogorov's tori.

A similar analysis can be worked out in the case of an elliptic equilibrium or (using a Poincaré section) of a periodic orbit. The analytical method is essentially the same, and the Birkhoff normal form may be constructed up to a finite order, as we have done formally in Chapter 5. Around the equilibrium, or a periodic orbit, there are invariant tori which in the figures appear to gather close to the equilibrium. We can prove indeed that superexponential stability applies also in that case. The structure of invariant tori breaks down when resonances are strong enough to destroy them; this happens indeed when the distance from the equilibrium is large enough. However, the resonances create new families of elliptic periodic orbits which in turn create stable regions around them, on a smaller and smaller scale. This is the phenomenon of 'boxes into boxes', which is hardly described in a global sense.

The real trouble with this world of ours is not that it is an unreasonable world, nor even that it is a reasonable one. The commonest kind of trouble is that it is nearly reasonable, but not quite. Life is not an illogicality; yet it is a trap for logicians. It looks just a little more mathematical and regular than it is; its exactitude is obvious, but its inexactitude is hidden; its wildness lies in wait.

G. K. Chesterton

Appendix A
The Geometry of Resonances

This appendix collects some information used in the notes, together with some further information that may have an independent interest.

The first section is concerned with discrete subgroups and resonance moduli. The aim is to complete the technical details used in the proof of the Arnold–Jost theorem, Section 3.4. It includes the construction of a basis of a discrete subgroup and the relation between different bases of the same subgroup.

The second section is concerned with resonance and non-resonance relations. The aim is to complete the discussion of Chapters 7 and 8. It includes, in particular, a discussion of the Diophantine condition of strong non-resonance and its relations with other conditions, in particular, condition $\boldsymbol{\tau}$ used in the notes and the Bruno condition.

A.1 Discrete Subgroups and Resonance Moduli

A subset $G \subset \mathbb{R}^n$ is said to be a *discrete subgroup* in case it is a group with respect to the vector sum and it contains only isolated points. In particular, it is closed with respect to the algebraic operation of sum, that is, if $\mathbf{a} \in G$ and $\mathbf{b} \in G$, then $\mathbf{a} \pm \mathbf{b} \in G$. Of course, this also means that $k\mathbf{a} \in G$ for every integer k.

Example A.1: *Discrete subgroups.* The set $G = \{0\}$ containing only the origin of $\mathbb{R}^n$ is a trivial subgroup. A definitely more interesting example is the set $G = \{m_1\mathbf{e}_1 + \cdots + m_k\mathbf{e}_k\,,\ (m_1, \ldots, m_k) \in \mathbb{Z}^k\}$ generated by k linearly independent vectors $\mathbf{e}_1, \ldots, \mathbf{e}_k$ in $\mathbb{R}^n$, with $0 < k \leq n$. *E.D.*

The following proposition states that this example is, in fact, the general one.

Proposition A.2: *Let $G \subset \mathbb{R}^n$ be a non-trivial discrete subgroup of $\mathbb{R}^n$. Then there exists a positive $k \leq n$ and k independent vectors $\mathbf{e}_1, \ldots, \mathbf{e}_k$, such that*

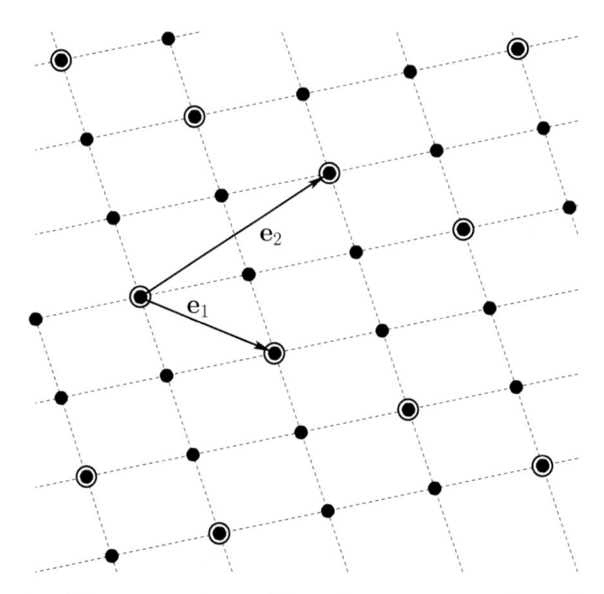

Figure A.1 Illustrating the problem of constructing a basis for a two-dimensional discrete group. The elements of the group are represented as dots. The vectors $\mathbf{e}_1, \mathbf{e}_2$ are independent but are not a basis. For integer combinations of these two vectors generate only the elements of the group represented by circled dots.

$$G = \{m_1\mathbf{e}_1 + \cdots + m_k\mathbf{e}_k \ , \ (m_1, \ldots, m_k) \in \mathbb{Z}^k\} \ . \tag{A.1}$$

We shall say that $\{\mathbf{e}_1, \ldots, \mathbf{e}_k\}$ is a *basis* of G and that G has dimension k. The trivial case $G = \{0\}$ is characterized by $\dim G = 0$.

The statement of the proposition looks quite simple and natural, but the proof requires some careful considerations. The difficulty is mainly due to the fact that only integer combinations of the elements of the basis are allowed.

In its simplest terms, the problem is the following. Consider a one-dimensional discrete group G. It is an easy matter to realize that all elements of G can be constructed as integer multiples of a given vector $\mathbf{e}_1$. Therefore, a basis can be constructed by taking either $\mathbf{e}_1$ or $-\mathbf{e}_1$. Any other choice allows us to construct only a subset G' of G (more precisely, a subgroup of G). For instance, the vector $2\mathbf{e}_1$ generates $-2\mathbf{e}_1$, $4\mathbf{e}_1$, $-4\mathbf{e}_1 \ldots$, but not $3\mathbf{e}_1$. Avoiding such a situation is an elementary matter, of course. But things can be trickier in dimensions higher than one. An example is illustrated in Fig. A.1. If the discrete subgroup contains all dots represented in the plane, the vectors $\mathbf{e}_1, \mathbf{e}_2$ represented in the figure are not a basis of the group: they generate only the points represented as circled dots.

A.1.1 Construction of a Basis of a Discrete Subgroup

Let G be a discrete subgroup of $\mathbb{R}^n$ and let $\Pi = \mathrm{span}(G)$ be the subspace of $\mathbb{R}^n$ generated by G. We say that $k = \dim \Pi$ is the dimension of G. This definition is justified because no more than k independent vectors can be found in G, and, on the other hand, we can always find k independent vectors in G which are a basis of Π. Indeed, let $\mathbf{e}_1, \ldots, \mathbf{e}_k$ be a basis of Π, and let, for example, $\mathbf{e}_1 \notin G$. Then there is $\mathbf{v} \in G$ with $\mathbf{v} \notin \mathrm{span}(\mathbf{e}_2, \ldots, \mathbf{e}_k)$. For if there is no such $\mathbf{v}$, then it must be $G \subset \mathrm{span}(\mathbf{e}_2, \ldots, \mathbf{e}_k)$ and we should conclude $\Pi \subset \mathrm{span}(\mathbf{e}_2, \ldots, \mathbf{e}_k)$, contradicting the hypothesis that $\mathbf{e}_1, \ldots, \mathbf{e}_k$ is a basis of Π. Therefore, $\mathbf{v}, \mathbf{e}_2, \ldots, \mathbf{e}_k$ is still a basis of Π. With at most k such substitutions we construct the desired basis of Π which contains only vectors in G.

To a set $\{\mathbf{e}_1, \ldots, \mathbf{e}_k\} \subset G$ of independent vectors we associate the parallelogram $D(\mathbf{e}_1, \ldots, \mathbf{e}_k)$ defined as

$$D(\mathbf{e}_1, \ldots, \mathbf{e}_k) = \{\mu_1 \mathbf{e}_1 + \cdots + \mu_k \mathbf{e}_k \ : \ 0 \le \mu_j < 1 \,,\ j = 1, \ldots, k\} \ . \tag{A.2}$$

Lemma A.3: *A set $\{\mathbf{e}_1, \ldots, \mathbf{e}_k\} \subset G$ of independent vectors is a basis of G if and only if*

$$D(\mathbf{e}_1, \ldots, \mathbf{e}_k) \cap G = \{0\} \ . \tag{A.3}$$

Proof. Let $G' = \{m_1 \mathbf{e}_1 + \cdots + m_k \mathbf{e}_k \,,\ (m_1, \ldots, m_k) \in \mathbb{Z}^k\}$ (see Fig. A.1). Since G is a group, we have $G' \subset G$, and in particular $G' = G$ in case $\mathbf{e}_1, \ldots, \mathbf{e}_k$ is a basis of G. We prove that $G' = G$ if and only if condition (A.3) is fulfilled. If $G' = G$, then $D(\mathbf{e}_1, \ldots, \mathbf{e}_k) \cap G' = \{0\}$ by definition of $D(\mathbf{e}_1, \ldots, \mathbf{e}_k)$ and of G'. Conversely, let $D(\mathbf{e}_1, \ldots, \mathbf{e}_k) \cap G = \{0\}$ and assume, by contradiction, that $G' \neq G$. Then there exists a non-zero $\mathbf{v} \in G' \setminus G$, so that $\mathbf{v} = \lambda_1 \mathbf{e}_1 + \cdots + \lambda_k \mathbf{e}_k$ where at least one of the λ's is not an integer. By the group property of G we can also take $0 \le \lambda_j < 1$, so that $\mathbf{v} \in D(\mathbf{e}_1, \ldots, \mathbf{e}_k) \cap G$, contradicting the hypothesis. We conclude that $G' = G$. *Q.E.D.*

Proof of Proposition A.2. Let $k = \dim(G)$. If $k = 0$, then $G = \{0\}$ and there is nothing else to prove. So, let us assume $k > 0$.
Let $g\{\mathbf{e}_1, \ldots, \mathbf{e}_k\} \subset G$ be independent vectors. If condition (A.3) is fulfilled, then $\{\mathbf{e}_1, \ldots, \mathbf{e}_k\}$ is a basis. If not, $D(\mathbf{e}_1, \ldots, \mathbf{e}_k) \cap G$ contains a finite number of points. For the closure of $D(\mathbf{e}_1, \ldots, \mathbf{e}_k)$ is a compact subset of $\mathbb{R}^n$, so it contains a finite number of points of G because G has no accumulation points.
We prove that we can find another basis $\mathbf{e}'_1, \ldots, \mathbf{e}'_k$ such that $D(\mathbf{e}'_1, \ldots, \mathbf{e}'_k) \cap G$ contains strictly fewer points than $D(\mathbf{e}_1, \ldots, \mathbf{e}_k) \cap G$. This implies that with a finite number of steps we can construct a basis $\mathbf{e}_1, \ldots, \mathbf{e}_k$ satisfying

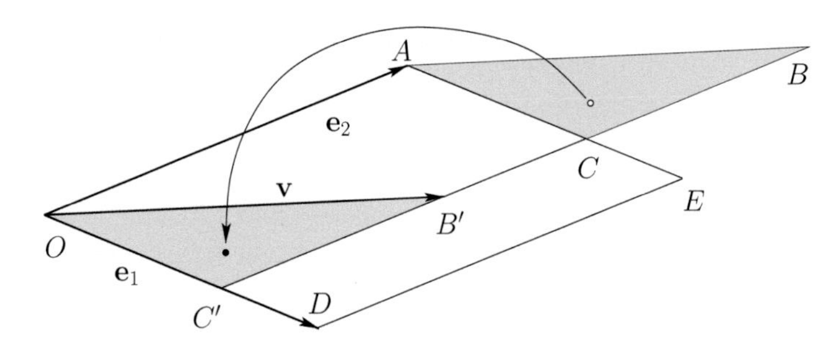

Figure A.2 Illustrating the construction of the basis $\mathbf{e}'_1 = \mathbf{v}$, $\mathbf{e}'_2 = \mathbf{e}_2$. The set $D(\mathbf{e}_1, \mathbf{e}_2)$ is the parallelogram $OAED$, without the segments AE and ED; the set $D(\mathbf{e}'_1, \mathbf{e}'_2)$ is the parallelogram $OABB'$, without the segments AB and BB'; the set $D(\beta_1\mathbf{e}_1, \mathbf{e}_2)$ is the parallelogram $OACC'$, without the segments AC and CC'. The one-to-one mapping between $D(e'_1, e'_2)$ and $D(\beta_1\mathbf{e}_1, \mathbf{e}_2)$ is constructed by translating the triangle ABC over the triangle $OB'C'$; the mapping is one-to-one because the segment AC is mapped on OC', while AB is not included in $D(\mathbf{e}'_1, \mathbf{e}'_2)$.

(A.3), which proves the proposition. So, let us show how to construct the new basis.

Let $0 \neq \mathbf{v} \in D(\mathbf{e}_1, \ldots, \mathbf{e}_k) \cap G$, and suppose that $\mathbf{v} = \beta_1\mathbf{e}_1 + \cdots + \beta_k\mathbf{e}_k$ with $0 \neq \beta_1 \geq \beta_2 \geq \cdots \geq \beta_k$. Define a new basis $\mathbf{e}'_1, \ldots, \mathbf{e}'_k$ by replacing $\mathbf{e}_1$ with $\mathbf{v}$, namely, set $\mathbf{e}'_1 = \mathbf{v}$ and $\mathbf{e}'_2 = \mathbf{e}_2, \ldots, \mathbf{e}'_k = \mathbf{e}_k$. By the way, this is a basis in view of $\beta_1 \neq 0$. We prove that (see Fig. A.2)

$$\sharp\big(D(\mathbf{e}'_1, \ldots, \mathbf{e}'_k) \cap G\big) = \sharp\big(D(\beta_1\mathbf{e}_1, \ldots, \mathbf{e}_k) \cap G\big) < \sharp\big(D(\mathbf{e}_1, \ldots, \mathbf{e}_k) \cap G\big) .$$

The second inequality is trivial, for in view of $0 < \beta_1 < 1$, we clearly have $\big(D(\beta_1\mathbf{e}_1, \ldots, \mathbf{e}_k) \cap G\big) \subset \big(D(\mathbf{e}_1, \ldots, \mathbf{e}_k) \cap G\big)$, so that $\mathbf{v} \notin D(\beta_1\mathbf{e}_1, \ldots, \mathbf{e}_k) \cap G$, while $\mathbf{v} \in D(\mathbf{e}_1, \ldots, \mathbf{e}_k) \cap G$.

The first equality is proven by explicitly constructing a one-to-one mapping between $D(\mathbf{e}'_1, \ldots, \mathbf{e}'_k) \cap G$ and $D(\beta_1\mathbf{e}_1, \ldots, \mathbf{e}_k) \cap G$. To this end we proceed as follows. Recalling $\mathbf{e}'_1 = \mathbf{v} = \beta_1\mathbf{e}_1 + \cdots + \beta_k\mathbf{e}_k$, every $\mathbf{w}' \in D(\mathbf{e}'_1, \ldots, \mathbf{e}'_k)$ can be written as

$$\mathbf{w}' = \mu_1\beta_1\mathbf{e}_1 + (\mu_2 + \mu_1\beta_2)\mathbf{e}_2 + \cdots + (\mu_k + \mu_1\beta_k)\mathbf{e}_k \ , \quad 0 \leq \mu_j < 1 \ .$$

In particular, we have $0 \leq \mu_1\beta_1 < \beta_1$ and $0 \leq \mu_j + \mu_1\beta_j < 1 + \beta_j$ for $1 < j \leq k$. Setting $\nu_1 = \mu_1\beta_1$ and $\nu_j = \mu_j + \mu_1\beta_j \pmod 1$, we associate to $\mathbf{w}'$ a unique $\mathbf{w} \in D(\beta_1\mathbf{e}_1, \ldots, \mathbf{e}_k)$ defined as

$$\mathbf{w} = \nu_1\mathbf{e}_1 + \cdots + \nu_k\mathbf{e}_k \ .$$

This is unique in view of the inequalities. We prove that the inverse map is unique by giving its explicit expression

$$\mathbf{w}' = \mu_1\mathbf{e}_1 + \cdots + \mu_k\mathbf{e}_k \ ,$$

with

$$\mu_1 = \frac{\nu_1}{\beta_1} \ , \quad \mu_j = 1 + \nu_1 - \frac{\nu_1\beta_j}{\beta_1} \ (\mathrm{mod}\ 1) \ , \quad 1 < \mathrm{j} \leq \mathrm{k} \ .$$

For $\mathbf{w}'$ is clearly unique, and it belongs to $D(\mathbf{e}'_1, \ldots, \mathbf{e}'_k)$ in view of the inequalities $\nu_1/\beta_1 < 1$ and $-1 < \nu_1\beta_j/\beta_1 < 1 - \nu_1\beta_j/\beta_1$. *Q.E.D.*

A.1.2 Unimodular Matrices

We now investigate to which extent we can change the basis of a discrete subgroup.

A *unimodular matrix* is a square matrix with integer entries and determinant ± 1. The identity matrix is, of course, unimodular, and so are the product of two unimodular matrices and the inverse of a unimodular matrix. Therefore, unimodular matrices form a group.

Lemma A.4: *Let $\{\mathbf{e}_1, \ldots, \mathbf{e}_k\}$ be a basis of a k-dimensional discrete subgroup G of $\mathbb{R}^n$. Any collection $\{\mathbf{e}'_1, \ldots, \mathbf{e}'_k\} \subset G$ of k independent vectors is a basis of G if and only if there is a unimodular matrix $\mathsf{M} = \{M_{i,j}\}_{i,j=1,\ldots,k}$ such that*

$$\mathbf{e}'_j = \sum_{j=1}^{k} M_{j,k}\mathbf{e}_k \ , \quad i = 1, \ldots, k \ . \tag{A.4}$$

Proof. Let M be a unimodular matrix. Then the vectors $\mathbf{e}'_1, \ldots, \mathbf{e}'_k$ are linearly independent and generate a subgroup G'. We prove that $G' = G$. For $G' \subset G$ because any $\mathbf{v} \in G'$ can be written as $\mathbf{v} = \sum_i v_i\mathbf{e}'_i = \sum_i(\sum_j M_{i,j}v_i)\mathbf{e}_j$ with integers $v_1, \ldots, v_k$, and so it is a combination of $\mathbf{e}_1, \ldots, \mathbf{e}_k$ with integer coefficients; on the other hand, the inverse M^{-1} of M is unimodular, and so the same argument proves also $G \subset G'$. We conclude that $G' = G$.

Conversely, let $\{\mathbf{e}'_1, \ldots, \mathbf{e}'_k\}$ be a basis of G. Then there is a non-singular matrix M transforming the basis $\{\mathbf{e}_1, \ldots, \mathbf{e}_k\}$ to the basis $\{\mathbf{e}'_1, \ldots, \mathbf{e}'_k\}$ according to (A.4). Since $\mathbf{e}'_j \in G$, the matrix M must have integer entries. On the other hand, since $\{\mathbf{e}'_1, \ldots, \mathbf{e}'_k\}$ is a basis of G, the inverse M^{-1} of M must have integer entries, too. This implies that both M and M^{-1} have integer determinants, and by $\det \mathsf{M} \cdot \det \mathsf{M}^{-1} = 1$ we get $\det \mathsf{M} = \pm 1$. *Q.E.D.*

A.1.3 Resonance Moduli

The set $\mathbb{Z}^n$ is a discrete subgroup of $\mathbb{R}^n$, trivially. Therefore, subgroups of $\mathbb{Z}^n$ are, in turn, discrete subgroups of both $\mathbb{Z}^n$ and $\mathbb{R}^n$. We investigate here some algebraic facts related to this situation.

A *resonance module* $\mathscr{M}$ is a discrete subgroup of $\mathbb{Z}^n$ satisfying

$$\mathrm{span}(\mathscr{M}) \cap \mathbb{Z}^n = \mathscr{M} \ . \tag{A.5}$$

This imposes a severe restriction on a discrete subgroup in order to qualify it as a resonance module.

Example A.5: *Resonance modules* Trivial examples of resonance modules are the origin $0 \in \mathbb{Z}^n$ and $\mathbb{Z}^n$ itself. By the way, the only n-dimensional discrete subgroup of $\mathbb{Z}^n$ which is a resonance module is $\mathbb{Z}^n$ itself; for if $G \subset \mathbb{Z}^n$ is an n-dimensional subgroup of $\mathbb{Z}^n$, then $\mathrm{span}(G) = \mathbb{R}^n$, and $\mathbb{R}^n \cap \mathbb{Z}^n = \mathbb{Z}^n$. Less trivial examples are the one-dimensional subgroup of $\mathbb{Z}^2$ generated by the vector $(1,1)$, and the two-dimensional subgroup of $\mathbb{Z}^3$ generated by the basis $\{(1,1,0)\,,\,(0,1,1)\,\}$.
A trivial example of a discrete subgroup of $\mathbb{Z}^2$ which is not a resonance module is generated by the basis $\{(2,1)\,,\,(1,2)\}$. For, the point $(1,1)$ belongs to $\mathbb{Z}^n$, but not to the subgroup. A less trivial example is the two-dimensional subgroup of $\mathbb{Z}^3$ generated by the basis $\{(2,2,1)\,,\,(1,2,2)\}$. *E.D.*

A.1.4 Basis of a Resonance Module

We investigate here the relations between the basis of a resonance module in $\mathbb{Z}^n$ and the basis of $\mathbb{Z}^n$. We shall use the notation $\mathrm{Vol}(\mathbf{u}_1,\ldots,\mathbf{u}_k)$ to denote the k-dimensional volume of the parallelepiped with sides $\mathbf{u}_1,\ldots,\mathbf{u}_k$.

Lemma A.6: *A collection $\{\mathbf{e}_1,\ldots,\mathbf{e}_n\}$ of independent vectors in $\mathbb{Z}^n$ is a basis of $\mathbb{Z}^n$ if and only if $\mathrm{Vol}(\mathbf{e}_1,\ldots,\mathbf{e}_n) = 1$.*

Proof. The canonical basis $\mathbf{e}_1 = (1,0,\ldots,0),\ldots,\mathbf{e}_n = (0,\ldots,0,1)$ obviously satisfies $\mathrm{Vol}(\mathbf{e}_1,\ldots,\mathbf{e}_n) = 1$. By Lemma A.4, any other basis $\{\mathbf{e}'_1,\ldots,\mathbf{e}'_n\}$ has the form (A.4), with a unimodular matrix M. In view of $\det\mathsf{M} = \pm 1$, we conclude $\mathrm{Vol}(\mathbf{e}'_1,\ldots,\mathbf{e}'_n) = 1$. *Q.E.D.*

Corollary A.7: *Let $\{\mathbf{e}_1,\ldots,\mathbf{e}_n\}$ be a basis of $\mathbb{Z}^n$. Then the matrix*

$$\mathsf{M} = \begin{pmatrix} \mathbf{e}_1 \\ \vdots \\ \mathbf{e}_n \end{pmatrix} ,$$

the rows of which are the vectors of the basis, is a unimodular matrix.

This is an obvious statement, since the matrix M transforms the canonical basis of $\mathbb{Z}^n$ into the basis $\{\mathbf{e}_1,\ldots,\mathbf{e}_n\}$.

Lemma A.8: *Let $\{\mathbf{e}_1, \ldots, \mathbf{e}_k\}$ be $k < n$ independent vectors in $\mathbb{Z}^n$. Then the following conditions are equivalent:*

(i) $\{\mathbf{e}_1, \ldots, \mathbf{e}_k\}$ *is a basis of a resonance module;*

(ii) there are $n-k$ independent integer vectors $\mathbf{g}_1, \ldots, \mathbf{g}_{n-k}$ such that the n vectors $\{\mathbf{e}_1, \ldots, \mathbf{e}_k, \mathbf{g}_1, \ldots, \mathbf{g}_{n-k}\}$ form a basis of $\mathbb{Z}^n$.

Proof. We first prove that (i) implies (ii). Let $\{\mathbf{e}_1, \ldots, \mathbf{e}_k, \mathbf{g}_1, \ldots, \mathbf{g}_{n-k}\}$ be a basis of $\mathbb{R}^n$. If $\mathrm{Vol}(\mathbf{e}_1, \ldots, \mathbf{e}_k, \mathbf{g}_1, \ldots, \mathbf{g}_{n-k}) = 1$, then this is the desired basis. If not, by Lemma A.3 there is a non-zero vector $\mathbf{v} \in \mathbb{Z}^n$ such that

$$\mathbf{v} = \mu_1 \mathbf{e}_1 + \cdots + \mu_k \mathbf{e}_k + \mu_{k+1} \mathbf{g}_1 + \cdots + \mu_n \mathbf{g}_{n-k},$$

with $0 \le \mu_j < 1$, $j = 1, \ldots, n$. In particular, $\mathbf{v} \notin \mathrm{span}(\mathbf{e}_1, \ldots, \mathbf{e}_k)$, because $\{\mathbf{e}_1, \ldots, \mathbf{e}_k\}$ is a basis of a resonance module. Therefore, one at least of the coefficients $\mu_{k+1}, \ldots, \mu_n$ must be non-zero. Suppose $\mu_n \neq 0$, and construct a new basis $\{\mathbf{e}'_1, \ldots, \mathbf{e}'_k, \mathbf{g}'_1, \ldots, \mathbf{g}'_{n-k}\}$ by just replacing $\mathbf{g}_{n-k}$ with $\mathbf{v}$. Then

$$\begin{aligned} \mathrm{Vol}(\mathbf{e}'_1, \ldots, \mathbf{e}'_k, \mathbf{g}'_1, \ldots, \mathbf{g}'_{n-k}) &= \mathrm{Vol}(\mathbf{e}_1, \ldots, \mathbf{e}_k, \mathbf{g}_1, \ldots, \mathbf{g}_{n-k-1}, \mu_n \mathbf{g}_{n-k}) \\ &< \mathrm{Vol}(\mathbf{e}_1, \ldots, \mathbf{e}_k, \mathbf{g}_1, \ldots, \mathbf{g}_{n-k}) . \end{aligned}$$

On the other hand, all these volumes are integers, so that the volume is decreased at least by one. Since $\mathrm{Vol}(\mathbf{e}_1, \ldots, \mathbf{e}_k, \mathbf{g}_1, \ldots, \mathbf{g}_{n-k})$ is finite, with a finite number of steps we can satisfy $\mathrm{Vol}(\mathbf{e}'_1, \ldots, \mathbf{e}'_k, \mathbf{g}'_1, \ldots, \mathbf{g}'_{n-k}) = 1$, that is, we can construct a basis of $\mathbb{Z}^n$.

We prove now that (ii) implies (i). Let G be the subgroup generated by $\mathbf{e}_1, \ldots, \mathbf{e}_k$, and let $\mathbf{v} \in \mathrm{span}(G) \cap \mathbb{Z}^n$. Since $\mathbf{v} \in \mathbb{Z}^n$ we have

$$\mathbf{v} = \mu_1 \mathbf{e}_1 + \cdots + \mu_k \mathbf{e}_k + \mu_{k+1} \mathbf{g}_1 + \cdots + \mu_n \mathbf{g}_{n-k}$$

with $\mu_1, \ldots, \mu_n$ integers; on the other hand, since $\mathbf{v} \in \mathrm{span}(G)$, we have $\mu_{k+1} = \cdots = \mu_n = 0$. Therefore, $\mathbf{v} \in G$, and (i) follows. *Q.E.D.*

Corollary A.9: *Assume that $\{\mathbf{e}_1, \ldots, \mathbf{e}_n\}$ are independent vectors in $\mathbb{Z}^n$. Then $\mathrm{Vol}(\mathbf{e}_1, \ldots, \mathbf{e}_n) = 1$ if and only if $D(\mathbf{e}_1, \ldots, \mathbf{e}_n) \cap \mathbb{Z}^n = \{0\}$, where $D(\mathbf{e}_1, \ldots, \mathbf{e}_n)$ is defined by (A.2).*

Proof. By Lemma A.6, both conditions are necessary and sufficient in order $\{\mathbf{e}_1, \ldots, \mathbf{e}_n\}$ be a basis of $\mathbb{Z}^n$. *Q.E.D.*

A.2 Strong Non-resonance

In view of the discussion in Section A.1 we may adopt the following definition. *To a vector $\omega \in \mathbb{R}^n$, we associate the resonance module $\mathscr{M}_\omega$ defined as*

$$\mathscr{M}_\omega = \Pi_\omega^\perp \cap \mathbb{Z}^n \ ,$$

namely the intersection of the $(n-1)$-dimensional plane $\Pi_\omega^\perp \subset \mathbb{R}^n$ orthogonal to ω with $\mathbb{Z}^n$. The vector ω is said to be non-resonant in case $\dim \mathscr{M}_\omega = 0$.

It should be mentioned that the discussion of non-resonance conditions has a strong connection with the theory of Diophantine approximations. However, a satisfactory discussion of the latter arguments would exceed the limits of the present note: there are already too many pages. Hence we shall restrict the discussion to topics that are of direct interest here.

The first theorems concerning Diophantine approximation are due to Johann Peter Gustav Lejeune Dirichlet (1805–1859); see, in particular, [61]. More recent comprehensive books on the subject of continued fractions and Diophantine approximations are, for example, [32], [127] or [198]. For an introductory discussion of the problems of small divisors in different contexts, together with many references, see, for example, [169].

The inequality that we know as the Diophantine condition was essentially known, in qualitative form, to Dirichlet, and was used by Poincaré, as we have seen in Section 7.3.2. The formulation used by Kolmogorov in his original proof, and recalled in the statement of Theorem 8.1, is due to Siegel, who introduced it in quantitative form in his papers [203] and [204].[1] More recently, other strong non-resonance conditions have been introduced, weaker than the Diophantine one. The most known one is the so-called *Bruno condition*, introduced by Alexander Dmitrievich Bruno.[2] The present section is devoted to a short discussion of the relations between the preceding conditions and condition $\boldsymbol{\tau}$ enunciated in Section 8.3.6, formula (8.82).

In the rest of the section we shall consider a non-increasing real sequence of positive numbers $\{\alpha_r\}_{r\geq 0}$. It will be convenient to normalize the sequence, so that we have

$$1 = \alpha_0 \geq \alpha_1 \geq \alpha_2 \geq \ldots \,, \quad \alpha_r > 0 \,. \tag{A.6}$$

a sequence of this type has been used as characterizing the small divisors. See, for example, Section 5.4, formula (5.47); Section 7.4.1, formula (7.46)

---

[1] The problem investigated in [203] can be so stated: prove that an analytic map of the complex plane into itself which has the origin as a fixed point can be analytically conjugated to its linear part in a neighbourhood of the origin. The problem had been proposed in 1871 by Schröder [199], who had also found the formal solution as a power series; it is indeed the simplest occurrence of a series with small divisors in a dynamical problem. The convergence was proved by Siegel assuming a Diophantine-type condition. In [204] Siegel was able to extend his previous result to the case of linearization of an analytic system of differential equations in $\mathbb{C}^n$ in the neighbourhood of an equilibrium. The latter problem had been proposed by Poincaré in his thesis [184]; Poincaré also solved it in a simple but meaningful case which does not involve small divisors. A proof of Siegel's theorems using the method of Lie series may be found in [93][96].

[2] In the literature the name is also spelled as Bryuno or, more frequently, as Brjuno. The spelling Bruno is adopted here because Alexander Bruno himself, in a private discussion with the author, insisted that this is the correct spelling.

and Fig. 7.4; Section 8.3.3, formula (8.71) which includes also a normalization factor. In all these cases the sequence α_r has the additional property $\alpha_r \to_{r\to\infty} 0$. However, most considerations in what follows apply also to the case of a sequence with non-zero limit, though being somewhat trivial in that case.

A.2.1 A Result from Diophantine Theory

The aim here is to estimate the relative measure of the set of frequencies ω which satisfy a condition of *strong non-resonance*, in some sense. The problem is stated in more precise terms as follows. Let us say that a frequency vector $\omega \in \mathbb{R}^n$ is strongly non-resonant in case one can find a positive sequence $\psi(s)$, where s is a positive integer, such that the inequality

$$\bigl|\langle k, \omega\rangle\bigr| \ge \psi(|k|) \quad \text{for } 0 \ne k \in \mathbb{Z}^n \tag{A.7}$$

holds true; here the notation $|k| = \sum_l |k_l|$ has been used. Given an open bounded subset $\mathscr{D} \subset \mathbb{R}^n$, the question is whether one can determine $\psi(s)$ in such a way that the subset of strongly non-resonant frequencies in $\mathscr{D}$, namely

$$\Omega = \bigl\{\omega \in \mathscr{D} : \bigl|\langle k, \omega\rangle\bigr| \ge \psi(|k|)\,,\ |k| > 0\bigr\}\ ,$$

is not empty or, better, has positive measure in $\mathscr{D}$.

A straightforward procedure to determine such a sequence $\psi(s)$ is the following. Pick a non-zero $k \in \mathbb{Z}^n$ and consider

$$\tilde{\Omega}_k = \bigl\{\omega \in \mathscr{D} : \bigl|\langle k, \omega\rangle\bigr| < \psi(|k|)\bigr\}\ ,$$

namely the set of the ω which are close to resonance with k. Such a set is contained in a strip of width smaller than $\sqrt{n}\psi(|k|)/|k|$ around the plane through the origin orthogonal to k, intersected with $\mathscr{D}$ (the factor $\sqrt{n}$ is due to the relation $|k| \le \sqrt{n}\|k\|$, where $\|\cdot\|$ is the Euclidean norm). Thus, its measure $\mathrm{Vol}(\tilde{\Omega}_k)$ is bounded by

$$\mathrm{Vol}(\tilde{\Omega}_k) \le 2\sqrt{n}C\frac{\psi(|k|)}{|k|}\ ,$$

where C depends only on the domain $\mathscr{D}$; for example, $C = (\mathrm{diam}\,\mathscr{D})^{n-1}$. Then the measure of the complement of Ω in $\mathscr{D}$ cannot exceed

$$\mathrm{Vol}\biggl(\bigcup_{k\ne 0}\tilde{\Omega}_k\biggr) \le \sum_{k\ne 0}\mathrm{Vol}(\tilde{\Omega}_k) \le 2\sqrt{n}C\sum_{k\ne 0}\frac{\psi(|k|)}{|k|}\ .$$

Writing now

$$\sum_{k\ne 0}\frac{\psi(|k|)}{|k|} = \sum_{s>0}\sum_{|k|=s}\frac{\psi(s)}{s}\ ,$$

and using the fact that the number of vectors $k \in \mathbb{Z}^n$ satisfying $|k| = s$ does not exceed $2^n s^{n-1}$, one finally gets[3]

$$\mathrm{Vol}\Big(\bigcup_{k\neq 0} \tilde{\Omega}_k\Big) \le 2^{n+1}\sqrt{n}C \sum_{s>0} s^{n-2}\psi(s) \ .$$

Then it is enough to choose $\psi(s) = \gamma s^{-\tau}$ with suitable constants $\gamma > 0$ and $\tau > n-1$ in order to get that the complement of Ω in $\mathscr{D}$ has a measure which is small with γ. Such a result, although obtained with rough estimates, is optimal for what concerns the value of τ. Indeed, in view of an approximation theorem by Dirichlet, for $\tau < n-1$, the set Ω is empty, while for $\tau = n-1$, the set Ω is non-empty, but has zero measure.

We conclude that the *Diophantine condition*

$$\big|\langle k, \omega\rangle\big| \ge \gamma |k|^{-\tau} \quad \text{for } 0 \neq k \in \mathbb{Z}^n \ , \ \gamma > 0 \ , \ \tau > n-1 \tag{A.8}$$

is satisfied by a set of frequencies with large relative measure (in the Lebesgue sense).

A.2.2 The Condition of Bruno

In [30] Alexander Bruno introduced the remarkable condition

$$\text{Б} = -\sum_{r\ge 1} \frac{1}{2^r} \log \alpha_{2^r-1} < +\infty \ . \tag{A.9}$$

Originally, the condition was introduced as an improvement of the results of Siegel in [204]: it is weaker than the Diophantine condition, indeed.[4] Later the same condition, or some form of it, has been applied to other problems, including KAM theory.

---

[3] Let us denote by $J_{n,s}$ the number of vectors $k \in \mathbb{Z}^n$ such that $|k| = s$. Let us show that the recursive formula $J_{n,s} = J_{n-1,s} + 2J_{n-1,s-1} + \cdots + 2J_{n-1,0}$ applies. For any vector k with $|k| = s$ may be written as $k = (k_1, \tilde{k})$ with $\tilde{k} \in \mathbb{Z}^{n-1}$ such that $|\tilde{k}| \le s$ and with $k_1 = \pm(s - |\tilde{k}|)$. For $n = 1$, we have $J_{1,0} = 1$ and $J_{1,s} = 2$ for $s > 0$, and for $n > 1$ we have $J_{n,0} = 1$. For $n = 2$, it is immediately evident that $J_{n,s} \le 2^2 s$. For $n > 2$, by a recursive procedure, assume that $J_{n-1,s} \le 2^{n-1}s^{n-2}$ and calculate $J_{n,s} \le 2^{n-1}s^{n-2} + 2[2^{n-1}(s-1)^{n-2} + \cdots + 2^{n-1} + 1] \le 2^{n-1}s^{n-2} + 2^n(s-1)^{n-1} + 1 \le 2^n s^{n-1}$. This is the desired relation.

[4] It is remarkable that for the problem of Schröder–Siegel (a map of the complex plane with the origin a fixed point) the optimality of the condition of Bruno has been proved by Jean Christophe Yoccoz [213]. For the problem of Kolmogorov the question of optimality is still open.

Let us prove that the Bruno condition is equivalent to our condition $\boldsymbol{\tau}$. To this end we prove that the constant Γ of condition $\boldsymbol{\tau}$ satisfies

$$\Gamma < \text{Б} < 2\left(\Gamma - \frac{\log \alpha_1}{2}\right) . \tag{A.10}$$

In view of the sequence α_r being non-increasing we have

$$\alpha_{2^{r-1}} \geq \alpha_{2^{r-1}+1} \geq \cdots \geq \alpha_{2^r-1} .$$

Therefore, we calculate

$$-\sum_{k=2^{r-1}}^{2^r-1} \frac{\log \alpha_k}{k(k+1)} \leq -\log \alpha_{2^r-1} \sum_{k=2^{r-1}}^{2^r-1} \frac{1}{k(k+1)} = -\frac{\log \alpha_{2^r-1}}{2^r} .$$

This obviously implies the lower bound in (A.10) since

$$\Gamma = -\sum_{k\geq 1} \frac{\log \alpha_k}{k(k+1)} = -\sum_{r\geq 1} \sum_{k=2^{r-1}}^{2^r-1} \frac{\log \alpha_k}{k(k+1)} < -\sum_{r\geq 1} \frac{\log \alpha_{2^r-1}}{2^r} = \text{Б} .$$

For the upper bound we use the chain of inequalities

$$\alpha_{2^r-1} \geq \cdots \geq \alpha_{2^{r+1}-1} ;$$

we calculate

$$-\frac{\log \alpha_{2^r-1}}{2^r} = -2 \log \alpha_{2^r-1} \sum_{k=2^r}^{2^{r+1}-1} \frac{1}{k(k+1)} \leq -2 \sum_{k=2^r}^{2^{r+1}-1} \frac{\log \alpha_k}{k(k+1)} ,$$

which leads to

$$-\sum_{r\geq 1} \frac{1}{2^r} \log \alpha_{2^r-1} \leq -2 \sum_{r\geq 1} \sum_{k=2^r}^{2^{r+1}-1} \frac{\log \alpha_k}{k(k+1)} = -2\left(\Gamma - \frac{\log \alpha_1}{2}\right) .$$

This concludes the proof of the equivalence between condition $\boldsymbol{\tau}$ and the Bruno condition.

A.2.3 Some Examples

Let us attempt to better understand the consequences of condition $\boldsymbol{\tau}$. We do it through some examples which also illustrate the relations with the Diophantine condition. To this end, in view of (A.8), we consider the sequence (ignoring the unessential constant γ)

$$\alpha_r = r^{-\tau} , \quad r \geq 1 ,$$

which represents a lower bound for divisors generated by Diophantine frequencies.

We first prove that *Diophantine frequencies satisfy condition* $\boldsymbol{\tau}$. To this end, we calculate

$$-\sum_{r\geq 1}\frac{\ln\alpha_r}{r(r+1)}=\tau\sum_{r\geq 1}\frac{\ln r}{r(r+1)}<\infty\ ,$$

which shows that the claim is true. By the way, this also shows that if the sequence α_r satisfies condition $\boldsymbol{\tau}$, then so does the sequence $\alpha_r\delta_r^2$ with $\delta_r\sim r^{-2}$ which appears in our estimates for the case of Kolmogorov, for it just adds to the estimate (8.82) a convergent series.

A second claim is: *condition* $\boldsymbol{\tau}$ *is weaker than the Diophantine one.* For example, the sequence $\alpha_r=e^{-r/\ln^2 r}$ is clearly not Diophantine and gives

$$-\sum_{r\geq 1}\frac{\ln\alpha_r}{r(r+1)}=\sum_{r\geq 1}\frac{1}{(r+1)\ln^2 r}<\infty\ ,$$

thus satisfying condition $\boldsymbol{\tau}$. It goes without saying that the set of frequencies satisfying condition $\boldsymbol{\tau}$ also has large measure, since it contains all Diophantine frequencies.

Finally we show that *there are sequences which violate condition* $\boldsymbol{\tau}$. For instance, if $\alpha_r=e^{-r}$, then we have

$$-\sum_{r\geq 1}\frac{\ln\alpha_r}{r(r+1)}=\sum_{r\geq 1}\frac{1}{r+1}=\infty.$$

It may be useful to add a final, perhaps pedantic remark. The sequence α_r defined as in (8.71) depends on an arbitrary integer $K\geq 1$. What we are, in fact, considering there is the subsequence $\vartheta_r=\alpha_{rK}$ (including $\vartheta_0=1$) of a sequence such as (A.6). It is an easy matter to see that *the subsequence* $\vartheta_r=\alpha_{rK}$ *satisfies condition* $\boldsymbol{\tau}$ *if and only if the sequence* α_r *satisfies the same condition.* To this end let us rearrange condition $\boldsymbol{\tau}$ as

$$-\sum_{r\geq 1}\frac{\ln\alpha_r}{r(r+1)}=-\sum_{s\geq 0}\sum_{j=1}^{K}\frac{\ln\alpha_{sK+j}}{(sK+j)(sK+j+1)} \tag{A.11}$$

and remark that in view of the non-increasing character of the sequence α_r we have (using also $\alpha_0=1$)

$$-\ln\alpha_{sK}\leq-\ln\alpha_{sK+j}\leq-\ln\alpha_{(s+1)K}\quad\text{for } s\geq 0 \text{ and } j=1,\ldots,K\ . \tag{A.12}$$

Note also that we have

$$\begin{aligned}\sum_{j=1}^{K}\frac{1}{(sK+j)(sK+j+1)}&=\frac{1}{sK+1}-\frac{1}{(s+1)K+1}\\&=\frac{1}{K\left(s+\frac{1}{K}\right)\left(s+1+\frac{1}{K}\right)}\ .\end{aligned}$$

Assume that the sequence α_r satisfies condition $\boldsymbol{\tau}$. From the left inequality of (A.12) we get

$$-\frac{\ln \alpha_{sK}}{K\left(s+\frac{1}{K}\right)\left(s+1+\frac{1}{K}\right)} \le -\sum_{j=1}^{K} \frac{\ln \alpha_{sK+j}}{(sK+j)(sK+j+1)} .$$

In view of (A.11), we get (use the trivial inequality $sK+1 \le s(K+1)$ and remove the null term $s=0$ from the sum)

$$-\frac{1}{K+1}\sum_{s\ge 1} \frac{\ln \alpha_{sK}}{s(s+1)} \le -\sum_{r\ge 1} \frac{\ln \alpha_r}{r(r+1)} ,$$

so that the sequence α_{sK} satisfies condition $\boldsymbol{\tau}$.

Conversely, assume that the sequence α_{sK} satisfies condition $\boldsymbol{\tau}$. From the right inequality in (A.12) we get

$$-\sum_{j=1}^{K} \frac{\ln \alpha_{sK+j}}{(sK+j)(sK+j+1)} \le -\frac{\ln \alpha_{(s+1)K}}{K\left(s+\frac{1}{K}\right)\left(s+1+\frac{1}{K}\right)} ,$$

and from (A.11) we get (put $r=s+1$ in the sum at the right member)

$$-\sum_{r\ge 1} \frac{\ln \alpha_r}{r(r+1)} \le -\frac{1}{K}\sum_{r\ge 1} \frac{\ln \alpha_{rK}}{r(r+1)} ,$$

so that the sequence α_r satisfies condition $\boldsymbol{\tau}$.

Appendix B
A Quick Introduction to Symplectic Geometry

The present appendix intends to provide the reader with some concepts of symplectic geometry which are used in the notes. An exhaustive exposition of the argument would exceed the limits of the present text. On the other hand, as said in Chapter 1, the notes follow a traditional approach which characterizes many classical treatises. Only on a few occasions has the language of symplectic geometry been used, either with the scope of allowing the reader to make a quick connection with the discussion in some recent books, or in order to simplify the proof of some relevant statements.

B.1 Basic Elements of Symplectic Geometry

Symplectic geometry makes essential use of the bilinear antisymmetric form induced by the matrix J introduced in Section 1.1.3, formula (1.18), namely

$$\mathsf{J} = \begin{pmatrix} 0 & \mathsf{I} \\ -\mathsf{I} & 0 \end{pmatrix} ,$$

where I is the identity matrix.

Symplectic geometry is characterized by the skew symmetric matrix J in the same sense as the identity matrix I characterizes Euclidean geometry. Indeed, Euclidean geometry is characterized by transformations preserving the bilinear symmetric form

$$\langle \mathbf{x}, \mathsf{I}\mathbf{y} \rangle = \sum_j x_j y_j \ . \tag{B.1}$$

Here, the innocuous matrix I is added in order to emphasize the parallel with the symplectic form which will be introduced shortly. This form is often called the *inner product*, and is preserved by the group of orthogonal transformations, namely matrices U satisfying

$$\mathsf{U}^\top \mathsf{U} = \mathsf{I} \ . \tag{B.2}$$

Similarly, symplectic geometry is characterized by transformations preserving the bilinear form

$$[\mathbf{x}, \mathbf{y}] = \langle \mathbf{x}, \mathsf{J}\mathbf{y} \rangle \ , \tag{B.3}$$

where J is the matrix defined by (1.18). This is called the *symplectic product*. It is antisymmetric, that is, $[\mathbf{x}, \mathbf{y}] = -[\mathbf{y}, \mathbf{x}]$, and non-degenerate (i.e., $[\mathbf{x}, \mathbf{y}] = 0$ for all $\mathbf{y} \in \mathbb{R}^{2n}$ implies $\mathbf{x} = 0$).

B.1.1 The Symplectic Group

Let us consider a linear mapping in $\mathbb{R}^{2n}$,

$$x = \mathsf{U}y \ , \tag{B.4}$$

where U is a $2n \times 2n$ non-singular matrix. The matrix U is said to be *symplectic* in case

$$\mathsf{U}^\top \mathsf{J} \mathsf{U} = \mathsf{J} \ . \tag{B.5}$$

This condition is actually equivalent to

$$\mathsf{U} \mathsf{J} \mathsf{U}^\top = \mathsf{J} \ . \tag{B.6}$$

For by (B.5) we have $\mathsf{U}^\top = \mathsf{J}\mathsf{U}^{-1}\mathsf{J}^{-1}$, and so also $\mathsf{U}\mathsf{J}\mathsf{U}^\top = \mathsf{U}\mathsf{J}^2\mathsf{U}^{-1}\mathsf{J}^{-1} = -\mathsf{J}^{-1} = \mathsf{J}$ in view of $\mathsf{J}^2 = -\mathsf{I}$. By the way, this shows that if U is symplectic, then $\mathsf{U}^\top$ is symplectic, too.

The set of symplectic matrices is a group with respect to matrix multiplication. For the identity matrix I is clearly symplectic; if U and V are symplectic, then $(\mathsf{U}\mathsf{V})^\top \mathsf{J}(\mathsf{U}\mathsf{V}) = \mathsf{V}^\top\mathsf{U}^\top\mathsf{J}\mathsf{U}\mathsf{V} = \mathsf{V}^\top\mathsf{J}\mathsf{V} = \mathsf{J}$, hence $\mathsf{U}\mathsf{V}$ is symplectic; if U is symplectic then $\mathsf{J} = \mathsf{I}\mathsf{J}\mathsf{I} = (\mathsf{U}^{-1})^\top\mathsf{U}^\top\mathsf{J}\mathsf{U}\mathsf{U}^{-1} = (\mathsf{U}^{-1})^\top\mathsf{J}\mathsf{U}^{-1}$, hence U^{-1} is symplectic. Finally, if U is symplectic, so is $\mathsf{U}^\top$.

Going back to the canonically conjugated coordinates (q, p), it is immediately seen that the symplectic product of two $2n$ vectors $\mathbf{z} = (\mathbf{q}, \mathbf{p})$ and $\mathbf{z}' = (\mathbf{q}', \mathbf{p}')$ is given by the expression $\sum_{j=1}^{n} (q_j p'_j - q'_j p_j)$. That is, the symplectic product is obtained by projecting the parallelogram with sides $\mathbf{z}, \mathbf{z}'$ onto each of the planes q_j, p_j, and then adding up algebraically the oriented areas of all these projections.

The set of linear transformations which preserve the symplectic product is characterized as the group of symplectic matrices. We have indeed $[\mathsf{U}\mathbf{x}, \mathsf{U}\mathbf{y}] = \langle \mathsf{U}\mathbf{x}, \mathsf{J}\mathsf{U}\mathbf{y} \rangle = \langle \mathbf{x}, \mathsf{U}^\top\mathsf{J}\mathsf{U}\mathbf{y} \rangle$, and this coincides with $[\mathbf{x}, \mathbf{y}]$ if and only if U satisfies (B.5).

B.1.2 Symplectic Spaces and Symplectic Orthogonality

A *symplectic space* is a real linear vector space V equipped with a bilinear antisymmetric non-degenerate form $[\cdot, \cdot]$. We shall consider here only vector spaces of finite dimension. Two vectors $\mathbf{x}, \mathbf{y}$ are said to be symplectic-orthogonal in case $[\mathbf{x}, \mathbf{y}] = 0$. We shall write $\mathbf{x} \angle \mathbf{y}$. To any subspace W of V, we associate the set

$$W^{\angle} = \{\mathbf{x} \in V \mid \mathbf{x} \angle \mathbf{y} \text{ for all } \mathbf{y} \in W\} \ ; \tag{B.7}$$

by the bilinearity of the symplectic form $W^{\angle}$ is a subspace, which is said to be *symplectic-orthogonal*[1] to W. We emphasize that the concept of symplectic orthogonality presents sharp differences with respect to the concept of orthogonality in Euclidean geometry. Here are three main differences.

(i) Every vector is self-symplectic-orthogonal, since the symplectic product is antisymmetric; therefore, any one-dimensional subspace is self-symplectic-orthogonal, too.

(ii) The restriction of the symplectic product to a subspace is still a bilinear form, but in general it fails to be non-degenerate. For instance, the restriction to a one-dimensional subspace is clearly degenerate.

(iii) The subspaces W and $W^{\angle}$ need not be complementary.

Two basic properties which are common to both geometries are given by the following.

Lemma B.1: *Let W be a subspace of a symplectic space V. Then*

$$\dim W + \dim W^{\angle} = \dim V \tag{B.8}$$

and

$$\left(W^{\angle}\right)^{\angle} = W \ . \tag{B.9}$$

Proof. If $\dim W = 0$ or $\dim W = \dim V$, then the statement is trivial. So, let us suppose that $\dim W = m$ with $0 < m < \dim V$. Denoting $n = \dim V$, let $\{\mathbf{u}_1, \ldots, \mathbf{u}_n\}$ be a basis of V, with $\{\mathbf{u}_1, \ldots, \mathbf{u}_m\}$ a basis of W. Writing a generic vector $\mathbf{v} \in V$ as $\mathbf{v} = \sum_{j=1}^{n} v_j \mathbf{u}_j$, the symplectic product takes the form $[\mathbf{v}, \mathbf{w}] = \sum_{j,k} a_{jk} v_j w_k$, where $a_{jk} = [\mathbf{u}_j, \mathbf{u}_k] = -a_{kj}$ is an element of a non-degenerate matrix. If $\mathbf{w} \in W$, then $w_{n-m+1} = \cdots = w_n = 0$, because by hypothesis $\{\mathbf{u}_1, \ldots, \mathbf{u}_m\}$ is a basis of W. If, moreover, $\mathbf{v} \in W^{\angle}$, then the relation of symplectic orthogonality is

$$\sum_{j=1}^{m} \beta_j w_j = 0 \ , \quad \beta_j = \sum_{k=1}^{n} a_{jk} v_k \ .$$

Since $\mathbf{w} \in W$ is arbitrary, the first equation implies $\beta_1 = \cdots = \beta_m = 0$ and $\beta_{m+1}, \ldots, \beta_n$ arbitrary; so, there are $n - m$ independent solutions. Since the matrix $\{a_{jk}\}$ is non-degenerate, the second relation guarantees the existence

[1] Some authors use the name *left orthogonality*. See, for instance, [9].

of exactly $n - m$ independent vectors symplectic-orthogonal to the subspace W, so that $\dim W^{\angle} = n - m$, as claimed.
Coming to (B.9), if $\mathbf{w} \in W$, then $\mathbf{v} \angle \mathbf{w}$ for all $\mathbf{v} \in W^{\angle}$, because by definition every element of $W^{\angle}$ is symplectic-orthogonal to every element of W, and so also to $\mathbf{w}$, so that $W \subset (W^{\angle})^{\angle}$. On the other hand, from (B.8) we know that $\dim(W^{\angle})^{\angle} = \dim W$, so that (B.9) follows. *Q.E.D.*

We now go deeper into the concept of symplectic orthogonality, pointing out some properties which do not appear in Euclidean geometry. To this end, we first need some definitions.

A subspace W of a symplectic space V is said to be:

(i) *isotropic* in case $W \subset W^{\angle}$;

(ii) *coisotropic* in case $W \supset W^{\angle}$, namely, if its symplectic-orthogonal subspace is isotropic;

(iii) *Lagrangian* in case $W = W^{\angle}$, namely, if it is both isotropic and coisotropic;

(iv) *symplectic* in case the symplectic product restricted to W is still non-degenerate.

These definitions are easily illustrated by considering the space $\mathbb{R}^{2n}$. We denote by $\{\mathbf{e}_1, \ldots, \mathbf{e}_n, \mathbf{d}_1, \ldots, \mathbf{d}_n\}$ the canonical basis of $\mathbb{R}^{2n}$ and by $(q_1, \ldots, q_n, p_1, \ldots, p_n)$ the coordinates, so that a vector $\mathbf{x} \in \mathbb{R}^{2n}$ is represented as $q_1\mathbf{e}_1 + \cdots + q_n\mathbf{e}_n + p_1\mathbf{d}_1 + \cdots + p_n\mathbf{d}_n$. The symplectic bilinear form is defined by the relations

$$[\mathbf{e}_j, \mathbf{e}_k] = [\mathbf{d}_j, \mathbf{d}_k] = 0 \ , \quad [\mathbf{e}_j, \mathbf{d}_k] = \delta_{j,k} \ , \quad j, k = 1, \ldots, n \ .$$

We shall refer to the basis $\{\mathbf{e}_1, \ldots, \mathbf{e}_n, \mathbf{d}_1, \ldots, \mathbf{d}_n\}$ satisfying the latter relations as the *canonical symplectic basis.*

Example B.2: *Arithmetic planes.* Let us call the *arithmetic plane* the subspace spanned by any subset of the vectors $\{\mathbf{e}_1, \ldots, \mathbf{e}_n, \mathbf{d}_1, \ldots, \mathbf{d}_n\}$. Formally: given any subsets J, K of $\{1, \ldots, n\}$, we consider the plane spanned by the vectors $\{\mathbf{e}_j\}_{j \in J} \cup \{\mathbf{d}_k\}_{k \in K}$. The following examples are easily understood:

(i) the arithmetic plane spanned by any subset of the vectors $\{\mathbf{e}_1, \ldots, \mathbf{e}_n\}$ is an isotropic subspace;

(ii) the direct sum of any of the arithmetic planes of the point (i) with the arithmetic plane $\mathrm{span}(\mathbf{d}_1, \ldots, \mathbf{d}_n)$ is a coisotropic subspace;

(iii) the arithmetic plane $\mathrm{span}(\mathbf{e}_1, \ldots, \mathbf{e}_n)$ is a Lagrangian subspace;

(iv) the arithmetic two-dimensional plane $\mathrm{span}(\mathbf{e}_j, \mathbf{d}_j)$ is symplectic for every $j \in \{1, \ldots, n\}$; further symplectic subspaces are generated by direct sum of such arithmetic planes.

In the examples (i)–(iii) the role of the vectors $\mathbf{e}_j$ and $\mathbf{d}_j$ can be exchanged, of course. *E.D.*

A more interesting example is the following:

Example B.3: *Lagrangian arithmetic planes.* Consider any partition of the indexes $\{1,\ldots,n\}$ into two disjoint subsets J and K (i.e., $J\cap K=\emptyset$ and $J\cup K=\{1,\ldots,n\}$). Then the arithmetic plane spanned by the n vectors $\{\mathbf{e}_j\}_{j\in J}\cup\{\mathbf{d}_k\}_{k\in K}$ is a Lagrangian plane. There are 2^n different Lagrangian planes which are generated this way.[2] *E.D.*

B.1.3 Canonical Basis of a Symplectic Space

The preceding examples are, in fact, quite general. For any symplectic space can be equipped with a canonical symplectic basis. This is stated by the following.

Proposition B.4: *Let V be a symplectic space. Then $\dim V$ is even, say $2n$, and there exists a canonical basis $\{\mathbf{e}_1,\ldots,\mathbf{e}_n,\mathbf{d}_1,\ldots,\mathbf{d}_n\}$ satisfying*

$$[\mathbf{e}_j,\mathbf{e}_k]=[\mathbf{d}_j,\mathbf{d}_k]=0\ ,\quad [\mathbf{e}_j,\mathbf{d}_k]=\delta_{jk}\ ,\quad j,k=1,\ldots,n\ . \tag{B.10}$$

The proof depends on some properties which are of independent interest and are isolated in the following two lemmas.

Lemma B.5: *A subspace W of V is symplectic if and only if the subspaces W and $W^{\angle}$ are complementary. In this case $W^{\angle}$ is a symplectic subspace.*

Proof. Let W be a symplectic subspace. Then $W\cap W^{\angle}=\{0\}$. For we have $[\mathbf{v},\mathbf{w}]=0$ for all $\mathbf{w}\in W$ in view of $\mathbf{v}\in W^{\angle}$, and this implies $\mathbf{v}=0$ in view of $\mathbf{v}\in W$ and of the non-degeneracy of the symplectic product. On the other hand, by (B.8), we have $\dim(W\oplus W^{\angle})=\dim W+\dim W^{\angle}=\dim V$, so that W and $W^{\angle}$ are complementary.
Conversely, let W and $W^{\angle}$ be complementary, and let $\mathbf{v}\in W$ be such that $[\mathbf{v},\mathbf{w}]=0$ for all $\mathbf{w}\in W$. By definition of $W^{\angle}$ this implies $[\mathbf{v},\mathbf{w}]=0$ for all $\mathbf{w}\in V$, and so also $\mathbf{v}=0$ by non-degeneracy. We conclude that W is symplectic.
It remains to prove that $W^{\angle}$ is symplectic. Since W and $W^{\angle}$ are complementary, if $[\mathbf{v},\mathbf{w}]=0$ for all $\mathbf{w}\in W^{\angle}$, then $\mathbf{v}=0$ by the same argument as presented earlier. *Q.E.D.*

Lemma B.6: *Let V be a symplectic space. Then $\dim V\geq 2$, and there exists a decomposition of V in two complementary symplectic subspaces V_1 and V_2, with $\dim V_1=2$ and $V_2=V_1^{\angle}$. Moreover, in V_1 there exists a symplectic canonical basis $\mathbf{e}_1,\mathbf{d}_1$.*

Proof. Let $\mathbf{e}_1\neq 0$ be an arbitrary vector. Then, by non-degeneracy, there exists a vector $\mathbf{d}_1$ independent of $\mathbf{e}_1$ such that $[\mathbf{e}_1,\mathbf{d}_1]\neq 0$. This proves that

[2] There are 2^n different partitions of n objects into two disjoint subsets.

the dimension must be at least 2. In view of linearity, by a trivial rescaling we can choose $\mathbf{d}_1$ such that $[\mathbf{e}_1, \mathbf{d}_1] = 1$.
Let $V_1 = \text{span}(\mathbf{e}_1, \mathbf{d}_1)$, so that $\dim V_1 = 2$ and $\mathbf{e}_1, \mathbf{d}_1$ is a basis of V_1. We prove that it is a symplectic subspace. To this end, first check that the decomposition of any vector $\mathbf{w} \in V_1$ over the basis $\mathbf{e}_1, \mathbf{d}_1$ is $\mathbf{w} = [\mathbf{w}, \mathbf{d}_1]\mathbf{e}_1 - [\mathbf{w}, \mathbf{e}_1]\mathbf{d}_1$; this is elementary. Suppose now that $[\mathbf{w}, \mathbf{v}] = 0$ for all $\mathbf{v} \in V_1$. Then we have, in particular, $[\mathbf{w}, \mathbf{e}_1] = [\mathbf{w}, \mathbf{d}_1] = 0$, and so also $\mathbf{w} = 0$ in view of the preceding decomposition. This proves that the restriction of the symplectic form to the subspace V_1 is non-degenerate. Therefore, V_1 is a symplectic subspace, and $\mathbf{e}_1, \mathbf{d}_1$ is a canonical basis of it.
In view of Lemma B.5, the subspace $V_2 = V_1^{\angle}$ is symplectic and complementary to V_1. *Q.E.D.*

Proof of Proposition B.4. If $\dim V = 2$, just apply once Lemma B.6. If $\dim V > 2$, then apply recursively Lemma B.6. With an obvious change of notation, first write $V = V_1 \oplus V'$, where V_1 admits a symplectic canonical basis $\mathbf{e}_1, \mathbf{d}_1$, and V' is symplectic, with $\dim V' = \dim V - 2$. Next, apply again the lemma to V', getting $V' = V_2 \oplus V''$ with V_2 admitting a canonical basis $\mathbf{e}_2, \mathbf{d}_2$. Proceeding the same way, end up with a decomposition $V = V_1 \oplus \cdots \oplus V_n$ (recall that we assume that V is a vector space of finite dimension), where V_j is a two-dimensional symplectic subspace equipped with a canonical basis $\mathbf{e}_j, \mathbf{d}_j$ satisfying $[\mathbf{e}_j, \mathbf{e}_j] = [\mathbf{d}_j, \mathbf{d}_j] = 0$ and $[\mathbf{e}_j, \mathbf{d}_j] = 1$ $(j = 1, \ldots, n)$. This implies that $\{\mathbf{e}_1, \ldots, \mathbf{e}_n, \mathbf{d}_1, \ldots, \mathbf{d}_n\}$ is a basis for V, and so also $\dim V = 2n$. By construction, all subspaces $V_1, \ldots, V_n$ are pairwise symplectic-orthogonal, which implies $[\mathbf{e}_j, \mathbf{e}_k] = [\mathbf{d}_j, \mathbf{d}_k] = [\mathbf{e}_j, \mathbf{d}_k] = 0$ for $j \neq k$. This proves (B.10). *Q.E.D.*

B.1.4 Properties of the Subspaces of a Symplectic Space

We establish two properties which will be relevant in the discussion concerning integrable systems.

Lemma B.7: *If W is isotropic, then $\dim W \leq n$; if W is coisotropic, then $\dim W \geq n$; if W is Lagrangian, then $\dim W = n$.*

An immediate consequence of this lemma is the following corollary.

Corollary B.8: *An isotropic subspace W is Lagrangian if and only if $\dim W = n$. The same holds true for a coisotropic subspace.*

Proof of Lemma B.7. Just use Lemma B.1, formula (B.8). If W is isotropic, that is, $W \subset W^{\angle}$, then $\dim W \leq \dim W^{\angle}$, which implies $\dim W \leq n$. If W is coisotropic, just reverse the argument. If W is Lagrangian, it is both isotropic and coisotropic, so that both the inequalities $\dim W \leq n$ and $\dim W \geq n$ apply. *Q.E.D.*

Lemma B.9: *Let V be a symplectic space and $\{\mathbf{e}_1, \ldots, \mathbf{e}_n, \mathbf{d}_1, \ldots, \mathbf{d}_n\}$ be* *a canonical basis. Then any Lagrangian subspace W of V is complementary to at least one of the Lagrangian arithmetic planes of Example B.3.*

Proof. Consider the Lagrangian plane $D = \mathrm{span}\{\mathbf{d}_1, \ldots, \mathbf{d}_n\}$, and let $P = D \cap W$ and $m = \dim P$. Since P is a subspace of D, we have $0 \leq m \leq n$, and there exist $n - m$ vectors in $\{\mathbf{d}_1, \ldots, \mathbf{d}_n\}$ spanning a $(n - m)$-dimensional arithmetic plane complementary to P in D. That is, there is a subset $K \subset \{1, \ldots, n\}$, with $\#K = n - m$, such that the arithmetic plane $D_K = \mathrm{span}\{\mathbf{d}_k\}_{k \in K}$ satisfies $D_K \cap P = \{0\}$ and $D_K \oplus P = D$ (the set K need not be unique). Let now $J = \{1, \ldots, n\} \setminus K$, so that $\{J, K\}$ is a partition of $\{1, \ldots, n\}$, and let $E_J = \mathrm{span}\{\mathbf{e}_j\}_{j \in J}$. We prove that the Lagrangian subspace $L = E_J \oplus D_K$ is complementary to W, namely, it is the Lagrangian arithmetic plane we are looking for. Since $\dim W = \dim L = n$, it is enough to prove that $W \cap L = \{0\}$. This is seen as follows. On the one hand, by $P \subset W$ and $W \angle W$ we have $P \angle W$; on the other hand, by $D_K \subset L$ and $L \angle L$ we have $D_K \angle L$. Using these relations, we get[3] $D = P \oplus D_K \angle W \cap L$, and so $W \cap L \subset D$ in view of D being Lagrangian. Therefore, we get $W \cap L = (W \cap D) \cap (L \cap D) = P \cap D_K = \{0\}$, by construction of D_K. *Q.E.D.*

---

[3] Let $\mathbf{v} \in W \cap L$. Since $\mathbf{v} \in W$, we have $\mathbf{v} \angle P$; since $\mathbf{v} \in L$, we have $\mathbf{v} \angle D_K$. We conclude $\mathbf{v} \angle P \oplus D_K$.

References

[1] M. Abramowitz, I. A. Stegun: *Handbook of Mathematical Functions*, Dover Publishing (1972).

[2] S. Aubry: *The new concept of transitions by breaking of analyticity in a crystallographic model*, in A. R. Bishop, T. Schneider (eds.): *Solitons and Condensed Matter Physics*, Springer Series in Solid-State Sciences, vol. 8, pp. 264–77. Springer (1978).

[3] V. I. Arnold: *Proof of a theorem of A. N. Kolmogorov on the invariance of quasi–periodic motions under small perturbations of the Hamiltonian*, Usp. Mat. Nauk **18**(5), 13 (1963); Russ. Math. Surv. **18**(5), 9 (1963).

[4] V. I. Arnold: *Small denominators and problems of stability of motion in classical and celestial mechanics*, Usp. Math. Nauk **18**(6), 91 (1963); Russ. Math. Surv. **18**(6), 85 (1963).

[5] V. I. Arnold: *A theorem of Liouville concerning integrable problems of dynamics*, Sibirsk. Math. Zh. **4**(2), 471–4 (1963).

[6] V. I. Arnold: *Instability of dynamical systems with several degrees of freedom*, Sov. Math. Dokl. **5**(1), 581–5 (1964).

[7] V. I. Arnold, A. Avez: *Ergodic Problems of Classical Mechanics*, W. A. Benjamin (1968).

[8] V. I. Arnold: *Équation differéntielles ordinaires*, Ed. MIR (1974). English translation: *Ordinary Differential Equations*, Massachusetts Institute of Technology Press (1978). Translations from the Russian original (1974).

[9] V. I. Arnold: *Méthodes mathématiques de la mécanique classique*, Ed. MIR (1976). English translation: *Mathematical Methods of Classical Mechanics*, Graduate Texts in Mathematics **60**, Springer-Verlag (1978). Translations from the Russian original (1974).

[10] V. I. Arnold: *Chapitres supplémentaires de la théorie des équations différentielles ordiaires*, Ed. MIR (1988). English translation: *Geometrical Methods in the Theory of Ordinary Differential Equations*, Springer-Verlag (1988). Translations from the Russian original (1978).

[11] V. I. Arnold, Valery V. Kozlov, Anatoly I. Neishtadt: *Mathematical Aspects of Classical and Celestial Mechanics*, Encyclopedia of Mathematical Sciences, 3rd ed. Springer (2006).

[12] D. Bambusi: *Asymptotic stability of breathers in some Hamiltonian networks of weakly coupled oscillators,* Comm. Math. Phys. **324**(2), 515–47 (2013).

[13] R. Barrar: *A proof of the convergence of the Poincaré-von Zeipel procedure in celestial mechanics*, Am. J. Math. **88**(1), 206–20 (1966).

[14] R. Barrar: *Convergence of the von Zeipel procedure*, Cel. Mech. **2**(4), 494–504 (1970).

[15] J. Barrow–Green: *Poincaré and the Three Body Problem*, American Mathematical Society (1997).

[16] G. Benettin, L. Galgani, A. Giorgilli, J. M. Strelcyn: *A proof of Kolmogorov's theorem on invariant tori using canonical transformations defined by the Lie method*, Il Nuovo Cimento **79** B(2), 201–23 (1984).

[17] G. Benettin, L. Galgani, A. Giorgilli: *A proof of Nekhoroshev's theorem for the stability times in nearly integrable Hamiltonian systems.* Cel. Mech. **37**(1), 1–25 (1985).

[18] G. Benettin, L. Galgani, A. Giorgilli: *Numerical investigations on a chain of weakly coupled rotators in the light of classical perturbation theory*, N. Cim. **89** B(2), 103–19 (1985).

[19] G. Benettin, L. Galgani, A. Giorgilli: *Classical Perturbation Theory for systems of weakly coupled rotators*, N. Cim. **89** B(2), 89–102 (1985).

[20] G. Benettin, G. Gallavotti: *Stability of motions near resonances in quasi-integrable Hamiltonian systems,* Journ. Stat. Phys. **44**(3), 293 (1986).

[21] G. Benettin, J. Fröhlich, A. Giorgilli: *A Nekhoroshev-type theorem for Hamiltonian systems with infinitely many degrees of freedom,* Comm. Math. Phys. **119**(1), 95–108 (1988).

[22] G. Benettin, L. Galgani, A. Giorgilli: *Realization of holonomic constraints and freezing of high frequency degrees of freedom in the light of classical perturbation theory, part II*, Comm. Math. Phys. **121**(4), 557–601 (1989).

[23] G. Benettin, A. Giorgilli: *On the Hamiltonian interpolation of near to the identity symplectic mappings with application to symplectic integration algorithms*, J. Stat. Phys. **74**(5), 1117–44 (1994).

[24] P. Bernard, V. Kaloshin, K. Zhang: *Arnold diffusion in arbitrary degrees of freedom and crumpled 3-dimensional normally hyperbolic invariant cylinders,* Acta Math. **217**(1), 1–79 (2016).

[25] J. Bertrand: *Théorème relatif au mouvement d'un point attiré vers un centre fixe*, Comptes Rendus de l'Academie des Sciences, tome LXXVII, 849–53 (1873).

[26] G. D. Birkhoff: *Proof of Poincaré's Geometric Theorem*, Trans. Am. Math. Soc. **14**(1), 14–22 (1913).

[27] G. D. Birkhoff: *Dynamical Systems*, American Mathematical Society (1927).

[28] M. Born: *The Mechanics of the Atom*, Frederick Ungar Publishing Company (1927).

[29] M. Braun: *On the applicability of the third integral of motion*, J. Differ. Equ. **13**(2), 300–18 (1973).

[30] A. D. Bruno: *Analytical form of differential equations*, Trans. Moscow Math. Soc. **25**, 131–288 (1971); **26**, 199–239 (1972).

[31] H. Bruns: *Bemerkungen zur Theorie der allgemeinen Störungen*, Astronomische Nachrichten **CIX**, 215–22 (1884).

[32] J. W. S. Cassels: *An Introduction to Diophantine Approximation*, Cambridge University Press (1957).

[33] A. Celletti, A. Giorgilli: *On the stability of the Lagrangian points in the spatial restricted problem of three bodies*, Cel. Mech. **50**(1), 31–58 (1990).

[34] A. Celletti, C. Falcolini: *Construction of invariant tori for the spin-orbit problem in the Mercury-Sun system*, Cel. Mech. **53**(2), 113–27 (1992).

[35] A. Celletti, L. Chierchia: *A Constructive Theory of Lagrangian Tori and Computer-Assisted Applications*, Dynamics Reported, vol. 4, Springer-Verlag, 60–129 (1995).

[36] A. Celletti, L. Chierchia: *On the stability of realistic three body problems*, Comm. Math. Physics **186**(2), 413–49 (1997).

[37] A. Celletti, A. Giorgilli, U. Locatelli: *Improved estimates on the existence of invariant tori for Hamiltonian systems,* Nonlinearity **13**(2), 397–412 (2000).

[38] E. Charpentier, A. Lesne, N. K. Nikolski: *L'Héritage de Kolmogorov en mathématiques*, Éditions Belin (2004). English translation: *Kolmogorov's Heritage in Mathematics*, Springer-Verlag (2007).

[39] T. M. Cherry: *On integrals developable about a singular point of a Hamiltonian system of differential equations*, Proc. Camb. Phil. Soc. **22**(4), 325–49 (1924).

[40] T. M. Cherry: *On integrals developable about a singular point of a Hamiltonian system of differential equations, II*, Proc. Camb. Phil. Soc. **22**(4), 510–33 (1924).

[41] T. M. Cherry: *On the solutions of Hamiltonian systems in the neighborhood of a singular point*, Proc. London Math. Soc., Ser. 2, **27**(1), 151–70 (1926).

[42] B. V. Chirikov: *Research concerning the theory of nonlinear resonance and stochasticity*, Preprint N 267, Institute of Nuclear Physics, Novosibirsk (1969). English translation in CERN Trans. 71-40 (1971). Accessed at https://cds.cern.ch/record/325497?ln=en.

[43] B. V. Chirikov: *A universal instability of many dimensional oscillator systems*, Phys. Rep. **52**(5), 263–379 (1979).

[44] P. Collas, *Algebraic solution of the Kepler Problem using the Runge-Lenz vector*, Am. J. Phys. **38**(2), 253–5 (1970).

[45] G. Contopoulos: *A third integral of motion in a Galaxy*, Z. Astrophys. **49**, 273–91 (1960).

[46] G. Contopoulos: *Resonant case and small divisors in a third integral of motion, I*, Astron. J. **68**(10), 763–79 (1963).

[47] G. Contopoulos: *Resonant case and small divisors in a third integral of motion, II*, Astron. J. **70**(10), 817–35 (1965).

[48] G. Contopoulos: *Resonance phenomena and the non-applicability of the third integral,* in F. Nahon, M. Hénon (eds.): *Les nouvelles méthodes de la dynamique stellaire*, Bull. Astron. Ser. 3, **2**, 223–42 (1967).

[49] G. Contopoulos, L. Galgani, A. Giorgilli: *On the number of isolating integrals in Hamiltonian systems*, Phys. Rev. A **18**(3), 1183–9 (1978).

[50] G. Contopoulos: *Order and Chaos in Dynamical Astronomy,* Springer-Verlag (2002).

[51] G. Contopoulos, C. Efthymiopoulos, A. Giorgilli: *Non-convergence of formal integrals of motion*, J. Phys. A: Math. Gen. **36**(32), 8639–60 (2003).

[52] G. Contopoulos, C. Efthymiopoulos, A. Giorgilli: *Non-convergence of formal integrals of motion II: improved estimates for the optimal order of truncation*, J. Phys. A: Math. Gen. **37**(35), 10831–58 (2004).

[53] G. Contopoulos: *Adventures in Order and Chaos*, Astrophysics and Space Science Library, vol. **313**, Kluwer Academic Publishers (2004).

[54] G. Contopoulos: *A review of the 'third' integral,* in M. Sansottera, U. Locatelli (eds.): *Modern Methods in Hamiltonian Perturbation Theory*, Math. Eng. 2(3), 472–511 (2020).

[55] Jean le Rond D'Alembert: *Recherches sur differens points importans du système du monde*, David l'ainé libraire (1754–6).

[56] T. Dauxois: *Fermi, Pasta, Ulam, and a mysterious lady,* Phys. Today **6**, 55–7 (2008).

[57] C. E. Delaunay: *Théorie du mouvement de la lune*, Mem. **28**, French Academy of Sciences (1860).

[58] A. Delshams and P. Gutierrez: *Effective stability and KAM theory*, J. of Diff. Eq. **128**(2), 415–90 (1996).

[59] A. Deprit: *Canonical transformations depending on a small parameter*, Cel. Mech. **1**(1), 12–30 (1969).

[60] E. Diana, L. Galgani, A. Giorgilli, A. Scotti: *On the direct construction of formal integrals of a Hamiltonian system near an equilibrium point*, Boll. Un. Mat. It. **11**, 84–9 (1975).

[61] L. G. P. Dirichlet: *Verallgemeinerung eines Satzes aus der Lehre von den Kettenbrüchen nebst einigen Anwendungen auf die Theorie der Zahlen,* S. B. Preuss. Akad. Wiss. 9395 (1842).

[62] C. Efthymiopoulos, G. Contopoulos, N. Voglis, R. Dvorak: *Stickiness and cantori*, J. Phys. A **30**(23), 8167–86 (1997).

[63] C. Efthymiopoulos, G. Contopoulos, M. Katsanikas: *Analytica invariant manifolds near unstable points and the structure of chaos*, Cel. Mech. **119**(3), 331–56 (2014).

[64] A. Einstein: *Zum Quantensatz von Sommerfeld und Epstein,* Deutsche physikalische Gesellschaft, Verhandlungen **19**(9), 82–92 (1917). English translation: *On the Quantum Theorem of Sommerfeld and Epstein,* The collected papers of Albert Einstein: The Berlin years, writings, 1914–1917, vol. 6, Doc. 45, Princeton University Press (1997).

[65] L. H. Eliasson: *Absolutely convergent series expansion for quasi-periodic motions*, report 2–88, Department of Mathematics, University of Stockholm (1988); later published in MPEJ **2**, 1–33 (1996).

[66] E. Fermi, J. Pasta, S. Ulam: *Studies of nonlinear problems*, Los Alamos document LA-1940 (1955). doi: https://doi.org/10.2172/4376203

[67] J. Figueras, A. Haro, A. Luque: *Rigorous computer-assisted application of KAM theory: a modern approach*, Found. Comput. Math. **17**(5), 1123–1193 (2017).

[68] H. Flaschka: *The Toda lattice. II. Existence of integrals,* Phys. Rev. B **9**(4), 1924–5 (1974).

[69] J. Ford, S. D. Stoddard, J. S. Turner: *On the integrability of the Toda lattice*, Prog. Theor. Exp. Phys. **50**(5), 1547–60 (1973).

[70] A. P. Fordy: *Hamiltonian symmetries of the Henon–Heiles system*, Phys. Lett. A **97**(1), 21–3 (1983).

[71] G. Gallavotti: *Twistless KAM tori*, Comm. Math. Phys. **164**(1), 145–56 (1994).

[72] G. Gallavotti, G. Gentile: *Majorant series convergence for twistless KAM tori*, Ergod. Theory Dyn. Syst. **15**(5), 857–69 (1995).

[73] G. Gallavotti (ed.): *The Fermi–Pasta–Ulam problem: a status report*, Lecture Notes in Physics **728**, Springer (2008). doi: https://doi.org/10.1007/978-3-540-72995-2

[74] F. Gantmacher: *Lectures in Analytical Mechanics*, Mir Publishers (1975).

[75] J. W. Gibbs: *Elementary Principles of Statistical Mechanics*, Charles Scribner's Sons (1902).

[76] A. Giorgilli, L. Galgani: *Formal integrals for an autonomous Hamiltonian system near an equilibrium point*, Cel. Mech. **17**(3), 267–80 (1978).

[77] A. Giorgilli: *A computer program for integrals of motion*, Comp. Phys. Comm. **16**(3), 331–43 (1979).

[78] A. Giorgilli, L. Galgani: *Rigorous estimates for the series expansions of Hamiltonian perturbation theory*, Cel. Mech. **37**(2), 95–112 (1985).

[79] A. Giorgilli: *Rigorous results on the power expansions for the integrals of a Hamiltonian system near an elliptic equilibrium point*, Ann. Inst. H. Poincaré, **48**(4), 423–39 (1988).

[80] A. Giorgilli, A. Delshams, E. Fontich, L. Galgani, C. Simó: *Effective stability for a Hamiltonian system near an elliptic equilibrium point, with an application to the restricted three body problem*, J. Diff. Eq. **77**(1), 167–198 (1989).

[81] A. Giorgilli, E. J. Zehnder: *Exponential stability for time dependent potentials*, ZAMP **43**(5), 827–55 (1992).

[82] A. Giorgilli, U. Locatelli: *Kolmogorov theorem and classical perturbation theory*, ZAMP **48**(2), 220–61 (1997).

[83] A. Giorgilli, U. Locatelli: *On classical series expansions for quasi–periodic motions*, MPEJ, **3** paper 5 (1997).

[84] A. Giorgilli, Ch. Skokos:*On the stability of the Trojan asteroids*, Astron. Astroph. **317**(1), 254–61 (1997).

[85] A. Giorgilli, A. Morbidelli: *Invariant KAM tori and global stability for Hamiltonian systems*, ZAMP **48**(1), 102–34 (1997).

[86] A. Giorgilli: *Small denominators and exponential stability: from Poincaré to the present time,* Rend. Sem. Mat. Fis. Milano, **68**(1), 19–57 (1998).

[87] A. Giorgilli: *Classical constructive methods in KAM theory*, PSS, **46**(10), 1441–51 (1998).

[88] A. Giorgilli, U. Locatelli: *A classical self-contained proof of Kolmogorov's theorem on invariant tori*, in C. Simó (ed.): *Hamiltonian systems with three or more degrees of freedom*, NATO ASI series C: Math. Phys. Sci. **533**, Kluwer Academic Publishers, 72–89 (1999).

[89] A. Giorgilli: *Unstable equilibria of Hamiltonian systems*, Disc. and Cont. Dynamical Systems, **7**(4), 855–71 (2001).

[90] A. Giorgilli: *Notes on exponential stability of Hamiltonian systems*, in *Dynamical Systems, Part I: Hamiltonian Systems and Celestial Mechanics*, Pubblicazioni del Centro di Ricerca Matematica Ennio De Giorgi, Pisa, 87–198 (2003).

[91] A. Giorgilli: *Classical constructive methods in KAM theory*, Planetary and Space Science **46**(1), 1441–51 (1998).

[92] A. Giorgilli, U. Locatelli, M. Sansottera: *Kolmogorov and Nekhoroshev theory for the problem of three bodies*, Cel. Mech. and Dyn. Astr. **104**(2), 159–75 (2009).

[93] A. Giorgilli and S. Marmi: *Convergence radius in the Poincaré-Siegel problem*, DCDS Series S **3**(4), 601–21 (2010).

[94] A. Giorgilli, M. Sansottera: *Methods of algebraic manipulation in perturbation theory*, in *Chaos, Diffusion and Non-integrability in Hamiltonian Systems – Applications to Astronomy*, Proceedings of the 3rd La Plata International School on Astronomy and Geophysics, P. M. Cincotta, C. M. Giordano, C. Efthymiopoulos (eds.), Universidad Nacional de La Plata and Asociación Argentina de Astronomía Publishers, La Plata, Argentina (2012).

[95] A. Giorgilli: *On a theorem of Lyapounov*, Rendiconti dell'Istituto Lombardo Accademia di Scienze e Lettere, Classe di Scienze Matematiche e Naturali, **146**, 133–60 (2012).

[96] A. Giorgilli, U. Locatelli, M. Sansottera: *Improved convergence estimates for the Schröder–Siegel problem*, Annali di Matematica Pura e Applicata **194**(4), 995–1023 (2014).

[97] A. Giorgilli, S. Paleari, T. Penati: *An extensive adiabatic invariant for the Klein–Gordon model in the thermodynamic limit,* Annales Henri Poincaré **16**(4), 897–959 (2015).

[98] A. Giorgilli, U. Locatelli, M. Sansottera: *Secular dynamics of a planar model of the Sun-Jupiter-Saturn-Uranus system; effective stability into the light of Kolmogorov and Nekhoroshev theories*, Regul. Chaotic Dyn. **22**(1), 54–77 (2017).

[99] A. Giorgilli: *Perturbation methods in Celestial Mechanics,* in G. Baù, A. Celletti, C. B. Gales, G. Gronchi (eds.): *Satellite Dynamics and Space Missions*, Springer INDAM series **34**, 51–114. doi: https://doi.org/10.1007/978-3-030-20633-8 (2019).

[100] A. Giorgilli: *D'Alembert fisico e matematico*, Rendiconti dell'Istituto Lombardo Accademia di Scienze e Lettere, Classe di Scienze Matematiche e Naturali **152**, 247–92 (2020).

[101] A. Giorgilli, N. Guicciardini: *La legge gravitazionale dell'inverso del quadrato nei Principia di Newton*, Rendiconti dell'Istituto Lombardo Accademia di Scienze e Lettere, Classe di Scienze Matematiche e Naturali **153** (2021).

[102] H. Goldstein, *Prehistory of the "Runge-Lenz" vector*, Am. J. Phys. **43**(8), 737–8 (1975).

[103] J. M. Greene: *A method for determining a stochastic transition,* J. Math. Phys. **20**(6), 1183 (1979).

[104] W. Gröbner: *Die Lie–Reihen und Ihre Anwendungen*, VEB Deutscher Verlag der Wissenschaften (1967). Italian translation: *Serie di Lie e loro applicazioni*, Ed. Cremonese, 1973.

[105] F. G. Gustavson: *On constructing formal integrals of a Hamiltonian system near an equilibrium point*, Astron. J. **71**(8), 670–86 (1966).

[106] W. R. Hamilton: *On the Application to Dynamics of a General Mathematical Method Previously Applied to Optics,* Report of the Fourth

Meeting of the British Association for the Advancement of Science, 513–18 (1834).

[107] W. R. Hamilton: *On a General Method in Dynamics*, Philosophical Transactions of the Royal Society, part II, 247–308 (1834).

[108] W. R. Hamilton: *Second Essay on a General Method in Dynamics*, Philosophical Transactions of the Royal Society, part I, 95–144 (1835).

[109] P. C. Hemmer: *Dynamic and Stochastic Types of Motion in the Linear Chain*, Det. Fysiske Seminar Trondheim no. 2 (1959).

[110] M. Hénon, C. Heiles: *The applicability of the third integral of motion: some numerical experiments*, Astron. J. **69**(1), 73–9 (1964).

[111] M. Hénon: *Integrals of the Toda lattice,* Physical Review B **9**(4), 1921–2 (1974).

[112] J. Henrard: *The algorithm of the inverse for Lie transform*, in B. D. Tapley, V. Szebehely (eds.): *Recent Advances in Dynamical Astronomy,* pp. 248–57. Reidel (1973).

[113] J. Henrard, J. Roels: *Equivalence for Lie transforms*, Cel. Mech. **10**(4), 497–512 (1974).

[114] A. Henrici, T. Kappeler: *Global action-angle variables for the periodic Toda lattice*, Int. Math. Res. Notices (2008). doi: https://doi.org/10.1093/imrn/rnn031

[115] P. Holmes: *Proof of non-integrability for the Hénon–Heiles Hamiltonian near an exceptional integrable case*, Physica D **10**(2), 335–47 (1982).

[116] G. Hori: *Theory of general perturbations with unspecified canonical variables*, Publ. Astron. Soc. Japan **18**(4), 287–96 (1966).

[117] J. Hubbard, Y. Ilyashenko: A proof of Kolmogorov's theorem, DCDS A **4**(4), 1–20 (2003).

[118] P. Iglesias: *Les origines du calcul symplectique chez Lagrange,* Le journal de maths des élèves **1**(3), 153–161 (1995).

[119] H. Ito: *Non-integrability of Henon–Heiles system and a theorem of Ziglin*, Kodai Math. J. **8**(1), 120–38 (1983).

[120] C. G. Jacobi: *Ueber die Reduction der Integration der partiellen Differentialgleichungen erster Ordnung zwischen irgend einer Zahl Variablen auf die Integration eines einzigen Systemes gewhnlicher Differentialgleichungen*, Crelle's J. **XXVII**, 97–162 (1837).

[121] C. G. Jacobi: *Sur la réduction de l'intégration des Équations differentielles du premier ordre entre un nombre quelconque de variables à l'intégration d'un deul système d'équations differentielles ordinaires,* Liouville's J. **III**, 60–96 (1837).

[122] C. G. Jacobi: *Suite du mémoire sur la réduction de l'intégration des Équations differentielles du premier ordre entre un nombre quelconque*

de variables à l'intégration d'un deul système d'équations differentielles ordinaires, Liouville's J. **III** 161–201 (1837).

[123] R. Jost: *Winkel- und Wirkungsvariable für allgemeine mechanische Systeme*, Helv. Phys. Acta **41**(6), 965–968 (1968).

[124] V. Kaloshin, M. Levi: *Geometry of Arnold Diffusion*, SIAM Rev. **50**(4), 702–20 (2008).

[125] V. Kaloshin, M. Levi: *An example of Arnold diffusion for near-integrable Hamiltonians*, Bull. Am. Math. Soc. **45**(3), 409–27 (2011).

[126] L. V. Kantorovich: *Functional analysis and applied mathematics*, Uspekhi Mat. Nauk **3**(6), 89–185 (1948).

[127] A. Ya. Khinchin: *Continued Fractions,* University of Chicago Press (1964).

[128] A. N. Kolmogorov: *Preservation of conditionally periodic movements with small change in the Hamilton function*, Dokl. Akad. Nauk SSSR **98**, 527 (1954). English translation in: Los Alamos Scientific Laboratory translation LA-TR-71-67; reprinted in: Lecture Notes in Physics **93**.

[129] V. Kozlov: *Integrability and nonintegrability in Hamiltonian mechanics*, Uspekhi Mat. Nauk **38**(1), 3–67 (1983); English translation in Russian Math. Surveys **38**(1), 1–76 (1983).

[130] S. B. Kuksin: *Perturbation theory of conditionally periodic solutions of infinite-dimensional Hamiltonian systems and its applications to the Kortewegde Vries equation,* Mat Sb (N.S.) **136/178**(3/7), 396–412, 431. Translation in Math USSR-Sb **64**(2):397413 (1988).

[131] S. B. Kuksin: *Fifteen years of KAM for PDE. Geometry, topology, and mathematical physics,* Am. Math. Soc. Transl. Ser. **212**(2), 237–58 (2004).

[132] J. L. Lagrange: *Récherches sur la nature et la propagation du son*, Miscellanea Taurinensia, tome I (1759). Reprinted in: *Oeuvres de Lagrange*, Gauthier–Villars (1867), tome I, pp. 39–148.

[133] J. L. Lagrange: *Nouvelles récherches sur la nature et la propagation du son*, Miscellanea Taurinensia, tome I (1760–1761). Reprinted in: *Oeuvres de Lagrange*, tome I, pp. 151–332, Gauthier-Villars (1867)

[134] J. L. Lagrange: *Solution de différents problèmes de calcul intégral*, Miscellanea Taurinensia, tome II (1762–1765). Reprinted in: *Oeuvres de Lagrange*, tome I, pp. 471–678, Gauthier-Villars (1867).

[135] J. L. Lagrange: *Sur l'altération des moyens mouvements des planètes*, Nouveaux Mémoires de l'Académie Royale des Sciences et Belles-Lettres de Berlin (1776). Reprinted in: *Oeuvres de Lagrange*, tome IV, p. 255, Gauthier-Villars (1867).

[136] J. L. Lagrange: *Recherches sur les équations séculaires des mouvements des nœuds et des inclinaisons des planètes*, Mémoires de

l'Académie des Sciences de Paris, année 1774 (presented; published 1778).

[137] J. L. Lagrange: *Théorie des variations séculaires del éléments des planètes. Première partie contenant les principes et les formules générales pour déterminer ces variations*, Nouveaux mémoires de l'Académie des Sciences et Belles-Lettres de Berlin (1781). Reprinted in: *Oeuvres de Lagrange*, tome V, pp. 125–207, Gauthier-Villars (1870).

[138] J. L. Lagrange: *Théorie des variations séculaires del éléments des planètes. Seconde partie contenant la détermination de ces variations pour chacune des planètes pricipales*, Nouveaux mémoires de l'Académie des Sciences et Belles-Lettres de Berlin (1782). Reprinted in: *Oeuvres de Lagrange*, tome V, pp. 211–489, Gauthier-Villars (1870).

[139] J. L. Lagrange: *Méchanique Analytique*, chez La Veuve Desaint (1788).

[140] J. L. Lagrange: *Mémoire sur la théorie des variations des éléments des planètes et en particulier des variations des grands axes de leurs orbites*, Mémoires de la première classe de l'Institut de France (1808). Reprinted in: *Oeuvres de Lagrange*, tome VI, pp. 713–68, Gauthier-Villars (1870).

[141] J. L. Lagrange: *Mémoire sur la théorie générale de la variation des constantes arbitraires dans tous les problèmes de la Mécanique*, Mémoires de la première classe de l'Institut de France (1808). Reprinted in: *Oeuvres de Lagrange*, tome VI, pp. 769–805, Gauthier-Villars (1870).

[142] J. L. Lagrange: *Second mémoire sur la théorie générale de la variation des constantes arbitraires dans tous les problèmes de la Mécanique*, Mémoires de la première classe de l'Institut de France (1809). Reprinted in: *Oeuvres de Lagrange*, tome VI, pp. 807–18, Gauthier-Villars (1870).

[143] J. L. Lagrange: *Traité de Mécanique Analytique*, 3rd ed. Reprinted in: *Oeuvres de Lagrange*, tomes XI–XII, Gauthier-Villars (1870).

[144] C. Lanczos: *The Variational Principles of Mechanics*, University of Toronto Press (1970).

[145] P. S. de Laplace: *Mémoire sur les solutions particulières des équations différentielles et sur les inégalités séculaires des planètes,* Mémoires de l'Académie des Sciences de Paris, année 1772 (presented 1774) (1775). Reprinted in: *Œuvres de Laplace*, tome VIII, p. 325.

[146] J. Laskar: *A numerical experiment on the chaotic behaviour of the solar system*, Nature **338**(6212), 237–8 (1989).

[147] J. Laskar: *The chaotic motion of the solar system: a numerical estimate of the size of the chaotic zones*, Icarus **88**(2), 266 (1990).

[148] J. Laskar: *Lagrange et la Stabilité du Sytème Solaire*, in G. Sacchi Landriani, A. Giorgilli (eds.): *Sfogliando la Méchanique Analitique*, LED edizioni (2008).

[149] P. Lax: *Integrals of nonlinear equations of evolution and solitary waves,* New York University, Courant Institute of Mathematical Sciences, Report NYO-1480–87, Jan. 1968. Reprinted in: Commun. Pure Appl. Math. **21**(5), 467–90 (1968).

[150] T. Levi Civita, U. Amaldi: *Lezioni di Meccanica Razionale,* Zanichelli (1923).

[151] A. J. Lichtenberg, M. A. Lieberman: *Regular and Stochastic Motion*, Springer-Verlag (1983).

[152] J. Liouville: *Sur l'intégration des équations differentielles de la dynamique,* Journal de mathématiques pures et appliquées 1re série, tome XX, 137–8 (1855).

[153] J. E. Littlewood: *On the equilateral configuration in the restricted problem of three bodies*, Proc. London Math. Soc. **3**(9), 343–72 (1959).

[154] J. E. Littlewood: *The Lagrange configuration in celestial mechanics*, Proc. London Math. Soc. **s3-9**(4), 525–43 (1959).

[155] A. Lindstedt: *Beitrag zur integration der differentialgleichungen der störungstheorie*, Mém. Acad. Imp. des Sciences St. Pétersbourg, **31**(4), 1–20 (1883).

[156] R. De la Llave: *A tutorial on KAM theory,* in: *Smooth Ergodic Theory and Its Applications,* Proc. Sympos. Pure Math. **69**, 175–292, American Mathematical Society (2001).

[157] U. Locatelli, E. Meletlidou: *Convergence of Birkhoff normal form for essentially isochronous systems*, Meccanica **33**(2), 195–211 (1998).

[158] U. Locatelli, A. Giorgilli: *Invariant tori in the Sun–Jupiter–Saturn system*, DCDS-B **7**(2), 377–98 (2007).

[159] P. Lochak, C. Meunier: *Multiphase Averaging for Classical Systems with Applications to Adiabatic Theorems*, Springer-Verlag (1988).

[160] P. Lochak: *Canonical perturbation theory via simultaneous approximations*, Usp. Math. Nauk **47**(6), 59–140 (1992). English translation in Russ. Math. Surv. **47**(6), 57–133 (1992).

[161] P. Lochak, A. I. Neishtadt: *Estimates of stability time for nearly integrable systems with a quasiconvex Hamiltonian,* Chaos **2**(4), 495–499 (1992). doi: https://doi.org/10.1063/1.165891

[162] P. Lochak: *Tores invariants à torsion évanescente dans les systèmes hamiltoniens proches de l'intégrable*, Comptes Rendus de l'Académie des Sciences - Series I - Mathématique, **327**(9), 833–6 (1998).

[163] P. Lochak: *Arnold Diffusion; a Compendium of Remarks and Questions,* in C. Simó (ed.): *Hamiltonian Systems with Three or More Degrees of Freedom*, NATO ASI series C: Math. Phys. Sci. **533**, Kluwer Academic Publishers 168–83 (1999).

[164] P. Lochak, J.-P. Marco, D. Sauzin: *On the Splitting of Invariant Manifolds in Multidimensional Near-Integrable Hamiltonian Systems,* Mem. Am. Math. Soc. **163**(775) (2003).

[165] E. N. Lorenz: *Deterministic non-periodic flow*, J. Atmos. Sci. **20**(2), 130–41 (1963).

[166] A. M. Lyapunov: *The General Problem of the Stability of Motion* (in Russian), doctoral dissertation, University of Kharkov (1892). French translation in: *Problème général de la stabilité du mouvement*, Annales de la Faculté des Sciences de Toulouse, deuxième série, tome IX, 203–474 (1907). Reprinted in: Ann. Math. Stud. **17** (1949).

[167] R. S. Mackay, J. D. Meiss, I. C. Percival: *Transport in Hamiltonian systems,* Phys. D **13**(1), 55–81 (1984).

[168] F. Magri: *The seeds of Hamiltonian mechanics in the writings of Lagrange*, in G. S. Landriani, A. Giorgilli (eds.): *Sfogliando la Méchanique Analytique*, pp. 175–91, Istituto Lombardo di Scienze e Lettere, edizioni LED (2008).

[169] S. Marmi: *An Introduction to Small Divisors Problems*, Istituti editoriali e poligrafici inernazionali (2000).

[170] A. Morbidelli, A. Giorgilli: *Superexponential stability of KAM tori*, J. Stat. Phys. **78**(5), 1607–17 (1995).

[171] J. K. Moser: *Nonexistence of integrals for canonical systems of differential equations*, Comm. Pure Appl. Math. **8**(3), 409–36 (1955).

[172] J. K. Moser: *Stabilitätsverhalten kanonisher differentialgleichungssysteme,* Nachr. Akad. Wiss. Göttingen, Math.-Phys. Kl. IIa, 87–120 (1955).

[173] J. K. Moser: *On the generalization of a theorem of A. Liapounoff*, Comm. Pure Appl. Math. **11**(2), 257–71 (1958).

[174] J. K. Moser: *On invariant curves of area-preserving mappings of an annulus*, Nachr. Akad. Wiss. Göttingen, Math.-Phys. Kl. II 1962, 1–20 (1962).

[175] J. K. Moser: *Convergent series expansions for quasi-periodic motions*, Math. Ann. **146**(2), 136–76 (1967).

[176] J. K. Moser: *Stable and Random Motions in Dynamical Systems*, Princeton University Press (1973).

[177] J. K. Moser, E. J. Zehnder: *Notes on Dynamical Systems*, Courant Lecture Notes in Mathematics, vol. **12** (2005).

[178] A. I. Neishtadt: *On the accuracy of conservation of the adiabatic invariant*, J. Appl. Math. Mech. (PMM) **45**(1), 58–63 (1981).

[179] N. N. Nekhoroshev: *Exponential estimates of the stability time of near-integrable Hamiltonian systems,* Russ. Math. Surv. **32**(6), 1 (1977).

[180] N. N. Nekhoroshev: *Exponential estimates of the stability time of near-integrable Hamiltonian systems, 2,* Trudy Sem. Petrovs. **5**, 5 (1979).

[181] I. Newton: *Methodus fluxionum et serierum infinitorum,* unpublished manuscript (1676). English translation by J. Colson: *The Method of Fluxions and Infinite Series*, printed by Henry Woodfall (1736).

[182] I. Newton: *Philosophiæ Naturalis Principia Mathematica*, autore Is. Newton, Trin. Coll. Cantab. Soc. Matheseos Professore Lucasiano, & Societatis Regalis Sodali, Londini, jussu Societatis Regiæ ac typis Josephi Streater, Anno MDCLXXXVII (1687).

[183] A. D. Perry, S. Wiggins: *KAM tori are very sticky: Rigorous lower bounds on the time to move away from an invariant Lagrangian torus with linear flow,* Phys. D **71**(1), 102–21 (1994).

[184] H. Poincaré: *Sur les propriétés des fonctions definies par les équations aux différences partielles*, Thèses présentées à la Faculté des Sciences de Paris (1879). Reprinted in: *Oeuvres de Henri Poincaré*, tome I, pp. IL–CXXII, Gauthier-Villars (1928).

[185] H. Poincaré, *Mémoire sur les courbes définies par une équation différentielle*, J. Math. Pures Appl. **7**, 375–422 (1881) and **8**, 251–96 (1882).

[186] H. Poincaré: *Sur les résidus des intégrales doubles*, Acta Math. **1**(1), 321–80 (1887).

[187] H. Poincaré: *Sur le problème des trois corps et les équations de la dynamique*, Acta Math. **13**(1), 1–270 (1890).

[188] H. Poincaré: *Les méthodes nouvelles de la mécanique céleste*, Gauthier-Villars (1892).

[189] H. Poincaré: *Analysis situs*, J. Ec. Polytech. **1**, 1–121 (1895).

[190] H. Poincaré: *Leçons de Mécanique Céleste* professées a la Sorbonne, tome I, Théorie générale des perturbations planetaires, Gauthier-Villars (1905).

[191] H. Poincaré: Sur un théorème de géométrie, Rendiconti del Circolo Matematico di Palermo, **33**, 375–407 (1912).

[192] L. Pontriaguine: *Équations différentielles ordinaires*, Ed. MIR (1975).

[193] J. Pöschel: *Nekhoroshev's estimates for quasi-convex Hamiltonian systems*, Math. Z. **213**(1), 187 (1993).

[194] J. Roels, M. Hénon: *Recherche des courbes invariantes d'une transformation ponctuelle plane conservant les aire*, Bull. Astron. **32**, 267–85 (1967).
[195] H. Rüssmann: *Non-degeneracy in the perturbation theory of integrable dynamical systems*, in M. M. Dodson, J. A. G. Vickers (eds.): *Number Theory and Dynamical Systems*, pp. 5–18, Cambridge University Press (1989).
[196] H. Rüssmann: *On the frequencies of quasi periodic solutions of analytic nearly integrable Hamiltonian systems*, in S. Kuksin, V. Lazutkin, J. Pschel (eds.): *Seminar on Dynamical Systems*, PNDLE 12, 160–83, Birkaüser Verlag (1994).
[197] M. Sansottera, U. Locatelli, A. Giorgilli: *On the stability of the secular evolution of the planar Sun-Jupiter-Saturn-Uranus system*, Math. Comput. Simul. **88**(1), 1–14 (2013).
[198] W. M. Schmidt: *Diophantine Approximation,* Lect. Notes Math. **785**, Springer (1980).
[199] E. Schröder: *Über iterierte Functionen*, Math. Ann. **3**, 296–322 (1871).
[200] G. Servizi, G. Turchetti, G. Benettin, A. Giorgilli: *Resonances and asymptotic behaviour of Birkhoff series*, Phys. Lett. A **95**(1), 11–14 (1983).
[201] M. B. Sevryuk: *Reversible Systems,* Lect. Notes Math. **1211**, Springer-Verlag (1986).
[202] C. L. Siegel: *On the integrals of canonical systems*, Ann. Math. **42**(3), 806–822 (1941).
[203] C. L. Siegel: *Iteration of analytic functions*, Ann. Math. **43**(4), 607–612 (1942).
[204] C. L. Siegel: *Über die Normalform analytischer Differentialgleichungen in der Nähe einer Gleichgewichtslösung*, Nachr. Akad. Wiss. Göttingen, Math.-Phys. Kl. Math.-Phys.-Chem. Abt., 21–30 (1952).
[205] C. L. Siegel, J. K. Moser: *Lectures on Celestial Mechanics*, Springer-Verlag (1971).
[206] F. Tisserand: *Traité de mécanique célèste*, Gauthier-Villard (4 volumes: 1891–1896).
[207] M. Toda: *Vibration of a chain with nonlinear interaction,* J. Phys. Soc. Japan **22**, 431–436 (1977).
[208] E. T. Whittaker: *On the adelphic integral of the differential equations of dynamics*, Proc. Roy. Soc. Edinburgh, Sect. A **37**, 95–109 (1916).
[209] E. T. Whittaker: *A Treatise on the Analytical Dynamics of Particles and Rigid Bodies*, 4th ed., Cambridge University Press (1937). Previous editions published in 1904, 1917 and 1927.
[210] J. Williamson: *On the algebraic problem concerning the normal forms of linear dynamical systems*, Amer. J. Math. **58**, 141–163 (1936).

[211] A. Wintner: *On the linear conservative dynamical systems*, Annali di Matematica Pura ed Applicata, ser. 4, **13**(1), (1934–1935).
[212] A. Wintner: *The Analytical Foundations of Celestial Mechanics*, Princeton University Press (1941).
[213] J. C. Yoccoz: *Théorème de Siegel, nombres de Bruno et polynômes quadratiques*, Astérisque **231**, 3–88 (1995).
[214] J. C. Yoccoz: *Une erreur féconde du mathématicien Henri Poincaré*, La Gazette des mathematiciens **107**, 19–27 (2006).
[215] E. Zehnder: *Homoclinic points near elliptic fixed points*, Commun. Pure Appl. Math. **26**(2), 131–182 (1973).

Index

action
  construction, 100
  integral, 43, 101
action-angle, 38, 64, 84, 92, 98, 243
  algorithm, 102
  Delaunay, 103
  not unique, 99
adiabatic invariant, 326
Arnold diffusion, 399
Arnold–Jost theorem, 102
autonomous, 14
averaging method, 326

bifurcation diagram, 153
Birkhoff
  complete stability, 192, 326, 365
  normal form, 234
boxes into boxes, 406

canonical equations, 1, 10, 13
canonical transformation, 31
  area-preserving, 38
  completely canonical, 32
  condition, 32
  criterion, 31, 37, 42, 43, 45
  exchange of coordinates, 34, 47
  extended point, 49
  generating function, 31
  Hamiltonian flow, 39
  identity, 48
  Lie series, 205
  near-the-identity, 49, 199, 209
  scaling, 33, 38, 48
  time-depending, 52
  translation, 33

Cantorus, 397
Catalan sequence, 317, 337
Cauchy
  derivatives, 212
  estimate, 227
  generalized, 215, 227
  method of majorants, 218, 322
  Poisson bracket, 215
central forces
  first integrals, 19
  Hamilton–Jacobi method, 60
  Hamiltonian, 6
  Liouville method, 82
  quadrature, 28
centre point, 153
chaos, 403
  (re)-discovery, 169
  chaotic sea, 176
  discovery, 169
  transition to, 169, 170
circle map, 120
commutation
  of flows, 69
  of vector fields, 72, 89
commutator, 11
complexification
  action-angle, 213
  of domain, 213
configuration space, 3
constant of motion, 14
coordinates
  angular, 91
  canonically conjugated, 1
  Delaunay, 103
  induced by flow, 72
  Liouville, 76

deformation, 339, 346
Delaunay variables, 103
differential form, 42
diffusion, 340, 384, 395, 404
  Arnold's example, 399
  channel, 347, 348
Diophantine
  approximation theory, 414
  condition, 193, 254
  frequencies, 268
Dirichlet
  approximation theorem, 361
  pigeonhole principle, 120
discrete subgroup, 407
  basis, 408
  dimension, 408
dynamical variable, 10, 36

energy integral, 14
entrance time, 356
equation
  Hamilton–Jacobi, 31, 57
  variational, 67
equilibrium, 22
  elliptic, 148, 159, 324
  elliptic, nonlinear, 163, 233
  focus, 148, 160
  hyperbolic, 148, 159, 160, 371
  linear approximation, 147
  node, 148
  point, 147
  saddle, 148, 159
exit time, 356
expansion
  formal, 181
  Fourier, 136, 137, 140, 144
  parameter, 135
  polynomial, 163
  truncated, 183
  without parameter, 256
exponential stability, 194

fast convergence, 270
fast drift plane , 339
first integral, 14, 144, 238
  algebraic manipulation, 183
  asymptotic character, 186
  deformation, 190
  direct vs. indirect, 240
  energy, 15
  equations, 135
  for a linear system, 158
  formal, 165
  level curves, 183
  noise, 191
  non-existence, 135
  quadrature, 21
  quantitative estimates, 187
  truncated, 144, 189, 333, 339
first integrals
  Toda lattice, 113
focus point, 153
forced oscillations
  Hamiltonian, 9
formal calculus, 196
  power series, 197
  small parameter, 197
Fourier series
  decay of coefficients, 226
  splitting, 256, 290, 331
FPU problem, 168, 177, 404
free particle
  first integrals , 18
  Hamilton–Jacobi method, 58
  Hamiltonian, 5

galactic model
  Contopoulos, 169
  Hamiltonian, 6, 169
  Hénon–Heiles, 169
general problem of dynamics,
    117, 268, 324, 400, 403
  degenerate, 252
  non-degenerate, 251
generating function, 45
  first form, 46

for Lie series, 199
general form, 50
identity transformation, 48
near-the-identity, 49
scaling transformation, 48
time-depending, 55
usual form, 47
generating sequence, 206, 235
geography of resonances
cylinder, 343, 347
extended block, 343, 348
geometric properties, 352
resonant block, 343, 345
resonant extended plane, 343, 347
resonant manifold, 342, 344
resonant plane, 343, 347
resonant region, 342, 345
resonant zone, 342, 344
geometry
Euclidean, 421
symplectic, 421

Hamilton
equations, 1
flow, 39
function, 4
Hamiltonian
autonomous, 2
function, 2
non-autonomous, 2
quadratic, 16, 108, 153
reversible, 166
harmonic oscillator
action-angle variables, 86, 163
Hamiltonian, 5
Liouville method, 81
phase portrait, 16
scaling transformation, 38
harmonic repulsor
phase portrait, 16
heteroclinic
chain, 384, 393, 399
intersection, 393
orbit, 403
points, 383
homoclinic orbit, 403
existence, 379
infinitely many, 383
homoclinic tangle, 389, 390
representation, 383
homological equation, 135, 235, 237, 259, 262, 270, 279, 300

integrable
by quadratures, 64
formally, 168
system, 63, 118
system, perturbed, 117
integral invariant
absolute, 42
relative, 43
invariant torus, 404
destruction, 389, 405
Kolmogorov, 369
last, 389
on a map, 386
stickiness, 369
involution system, 65
complete, 77, 92, 99, 167

Jacobi identity, 11

KAM theory, 267
Kepler problem
action-angle variables, 105
degenerate, 133
Delaunay coordinates, 103, 106
first integrals, 20, 103
Kolmogorov
classical algorithm, 292
invariant torus, 369
Kolmogorov's theorem, 268, 320, 363
classical method, 271, 289
Newton method, 270

quadratic algorithm, 275
Kronecker
flow, 119, 123, 269
map, 124, 127

Lagrange
bracket, 40
formalism, 1, 3, 32
function, 4
triangular equilibria, 179
Lax pair, 114
Lie derivative, 11, 13, 199
Lie series, 195
algorithm, 199
analyticity, 218
analyticity of composition, 221
canonical transformation, 205
composition, 202, 259, 275
exchange theorem, 200
formal calculus, 198
normal form, 254, 259
operator, 199
properties, 200
triangular diagram, 201
Lie transform, 195, 365
algorithm, 206
analyticity, 222
canonical transformation, 209
exchange theorem, 207, 237
inverse, 208
normal form, 235, 251, 332
operator, 206
properties, 207
triangular diagram, 207
Lindstedt series, 321
cancellations, 322
linear chain, 108, 122
first integrals, 110
frequencies, 110
normal modes, 112
linear system
diagonalization, 149
evolution operator, 150
Hamiltonian, 154
Liouville
coordinates, 76, 77, 93, 98
integration algorithm, 79
theorem, 64, 78
list of indices, 305
partial ordering, 305
selection rule, 308
special, 305

matrix
exponential of, 150
identity, 421
jacobian, 35
skew symmetric, 10
symmetric, 151
symplectic, 156, 421, 422
unimodular, 100, 122, 245, 411

N–body problem
first integrals, 18
Hamiltonian, 7
near-the-identity transformation
Lie series, 205
Lie transform, 209
time-ε flow, 50
Nekhoroshev's theorem, 328, 363, 404
Newton method, 270
node point, 153
degenerate, 153
noise, 340, 346
non-degeneracy, 132
non-resonance, 122, 126, 413
Bruno condition, 416
condition $\boldsymbol{\tau}$, 290, 299, 307, 311, 417
Diophantine condition, 268, 416
domain, 334, 341, 366
nonlinear equilibrium
stability, 178
norm

polynomial, 187
supremum, 212
weighted Fourier, 227, 257
normal form, 147, 332
diagonal, 149, 156
Kolmogorov, 269, 291
linear system, 148
local domain, 330
non-resonant, 244
nonlinear elliptic equilibrium, 237
Poincaré–Birkhoff, 233, 365, 367
quadratic Hamiltonian, 153
remainder, 332
resonant, 245, 265
truncated, 258
numerical exploration, 168, 181, 363, 404

one degree of freedom
action-angle variables, 85
area-preserving, 38
Hamilton–Jacobi method, 59
Hamiltonian, 16
Liouville method, 80
oscillation period, 25
quadrature, 21, 25
orbit, 2
asymptotic, 371
heteroclinic, 364, 383
homoclinic, 364, 379, 383
on a torus, 119
periodic, 118
quasi-periodic, 118
order and chaos
boxes into boxes, 390
coexistence, 172, 177, 363, 403
threshold, 404
transition, 405
ordered island, 176, 389, 390, 392

pendulum
action-angle variables, 87
forced, 9
Hamiltonian, 6
kicked, 386
period, 26
periodically forced, 371
phase portrait, 16
phase average, 129
phase space, 1
extended, 2, 53
flow, 2
Poincaré
classification of equilibria, 153
formal calculus, 197
last geometric theorem, 175, 389
non-existence theorem, 135, 137, 252
normal form, 234
one-period map, 371
problem of dynamics, 117, 268
section, 124, 171
théorie des conséquents, 372
point transformation, 48
extended, 48
Poisson bracket, 11
Cauchy estimate, 215
fundamental, 36
preservation, 36
properties, 11
Poisson theorem, 14

quadratic method, 271
quadrature, 21, 26
quasi-periodic motion, 131

resonance, 125
complete, 121, 125
module, 121, 125, 331, 412, 413
module, basis, 412
multiplicity, 121
overlapping, 342, 348

partial, 121, 126
relations, 121
rotation angles, 121
rotating frame, 7
rotation angle, 120, 124
rotation number, 386, 389, 390
Runge–Lenz vector, 20

saddle point, 153
secular terms, 321
separatrices, 16, 87, 174, 181, 250, 371, 389, 391
small divisor, 136, 270, 279, 289
accumulation, 253, 262, 304
small parameter, 117, 197
stability
asymptotic, 324
complete, 145, 192, 326
exponential, 194, 326, 368
local, 339
long-time, 324
perpetual, 323, 326
superexponential, 326, 364, 369, 405
stable/unstable manifolds, 371
existence, 374
homoclinic intersection, 379
intersection, 378, 383
perturbed, 375
standard map, 386
stationary group, 91
stickiness, 397
strange attractor, 168
superconvergence, 271
symplectic
arithmetic plane, 424
bilinear antisymmetric form, 421
bilinear symmetric form, 421
canonical basis, 424, 425
coisotropic subspace, 424, 426
complementary subspace, 423, 425
geometry, 421
isotropic subspace, 424, 426
Lagrangian subspace, 52, 424, 426
left-orthogonality, 423
matrix, 35, 37, 421, 422
orthogonality, 423
product, 422
space, 422
subspace, 423
transformation, 31
system
anisochronous, 132
autonomous, 2
essentially isochronous, 134, 265
integrable, 63, 118
integrable, perturbed, 117
isochronous, 133
one degree of freedom, 16, 59

theorem
Arnold–Jost, 65, 102
Kolmogorov's, 268, 320
Liouville's, 64, 78
Nekhoroshev's, 328, 401
Poincaré's, 135, 137
Poisson's, 14
superexponential stability, 400
third integral, 170
three-body problem
Hamiltonian, 7
restricted circular, 8, 180
triangular equilibria, 179
time average, 129
Toda
lattice, 113
torus
angles, 92, 98
periods in a neighbourhood, 93
two-dimensional, 119
vector fields on, 89
transformation

canonical, 31
of coordinates, 31
orthogonal, 421
symplectic, 31
trigonometric polynomial, 129, 141, 255, 290

two-body problem
Hamiltonian, 7

uncoupled rotators, 133

vector fields
on a manifold, 88

Printed in the United States
by Baker & Taylor Publisher Services